# Strategic Innovations for Circular Economy

Rejaul Karim
https://orcid.org/0000-0001-6175-8825
*Varendra University, Bangladesh*

Bablu Kumar Dhar
https://orcid.org/0000-0001-8768-8634
*Mahidol University, Thailand*

IGI Global
Scientific Publishing
Publishing Tomorrow's Research Today

Published in the United States of America by
IGI Global Scientific Publishing
701 East Chocolate Avenue
Hershey, PA, 17033, USA
Tel: 717-533-8845 | Fax: 717-533-7115
Website: https://www.igi-global.com E-mail: cust@igi-global.com



Library of Congress Cataloging-in-Publication Data

Names: Karim, Rejaul, 1991- editor | Dhar, Bablu Kumar, 1982- editor
Title: Strategic innovations for circular economy / edited by Rejaul Karim, Bablu Kumar Dhar.
Description: Hershey, PA : IGI Global Scientific Publishing, [2026] | Includes bibliographical references and index. | Summary: "This book explores the intersection of strategic innovations and circular economy principles in shaping sustainable production, distribution, and consumption practices"-- Provided by publisher.
Identifiers: LCCN 2025018341 (print) | LCCN 2025018342 (ebook) | ISBN 9798337324548 hardcover | ISBN 9798337324555 paperback | ISBN 9798337324562 ebook
Subjects: LCSH: Circular economy | Organizational change | Sustainable development
Classification: LCC HC79.E5 S7687 2026 (print) | LCC HC79.E5 (ebook)
LC record available at https://lccn.loc.gov/2025018341
LC ebook record available at https://lccn.loc.gov/2025018342

British Cataloguing in Publication Data
A Cataloguing in Publication record for this book is available from the British Library.

# Table of Contents

# Detailed Table of Contents



The Circular Economy (CE) is a shift away from the conventional linear economic model, which is based on resource efficiency and waste minimization with regenerative systems. This chapter goes beyond descriptive overviews by synthesizing historical evolution, theoretical foundations, and practical applications of CE into a unified conceptual lens, highlighting how CE serves as both a sustainability paradigm and a systems-level economic framework. It charts the history of CE from early conceptual contributions, such as the 1972 *Limits to Growth* report, through formalization in the 1980s with the rise of industrial ecology and the inception of eco-efficiency in 2000. The chapter provides a brief review of key milestones in CE, from the Performance Economy to Cradle-to-Cradle design, emphasizing the extension of product lifetimes and recycling, culminating in overall closed-loop material flows. It also explores the role of the Ellen MacArthur Foundation, espousing the CE and linking it to worldwide sustainability objectives. Unlike prior reviews, the chapter contributes by critically integrating theoretical models with case-based insights and identifying underexplored challenges such as rebound effects, governance asymmetries, and socio-cultural resistance. It concludes with a discussion on the triple bottom line benefits (environmental, economic, and social) of CE, as well as the challenges that will be faced, including techno-flops, market dynamics, policy gaps, and cultural norms. Finally, it outlines how digital technologies and collaborative partnerships can provide transformative pathways, positioning CE as a cornerstone of future sustainability research and practice.

A shift in logistics and distribution systems has been underway to accomplish the sustainability agenda, where the merging of circular economic principles with IoT is the key to the transformation. In this chapter, the IoT-enabled Circular Supply Chains (CSCs) concept discussion goes further as it can enhance overall operations efficacy and transparency and drive towards a carbon-zero goal in the US. It employs the thematic content analysis of literature, industry publications, and practical case studies (e.g., including Walmart, Amazon, and HP) in order to provide an overview of Digital Circular Logistics (DCL) by utilizing a qualitative descriptive approach. In order to provide a systematic method for long-term change, a theoretical framework is established to integrate Internet of Things (IoT) features like blockchain traceability, real-time monitoring, and predictive analytics with CE's 9R tactics. Despite a high level of technological preparedness, the results show that growth is nevertheless hindered by systemic impediments such as limitations of policy frameworks, organizational stagnation, and a lack of cross-sectoral cooperation. This chapter provides academics, practitioners, and policymakers with theoretical understanding as well as practical advice by expanding socio-technical and institutional theories. It markets itself as a road map for wealthy nations to implement intelligent, circular logistics systems speedily.



The tourism sector is at a crossroads, with rising environmental and social concerns, despite its continued economic importance. This article examines the Circular Economy (CE) as a long-term financial model for the tourism industry, emphasizing its ability to be both ecologically friendly and commercially sustainable. In contrast to the usual linear paradigm, CE supports a regenerative system based on resource efficiency, waste minimization, and continuous material reuse. The report uses survey data, interviews, and theme analysis, a combination of qualitative and quantitative research on Indian tourism firms, to investigate the amount of CE integration, financial incentives that drive adoption, and the restrictions that impede universal

Integrating circular economy (CE) principles into responsible consumption and production (RCP) drives long-term sustainability for businesses, consumers, society, and the environment by addressing environmental challenges and resource constraints of the 21st century. Despite their growing relevance, research on CE-RCP integration remains limited. Against this backdrop, this chapter aims to analyze existing literature to bridge the research gap, using qualitative methodology. The findings reveal that the integration of CE-RCP enhances resource efficiency, advances sustainable consumption and production (SDG12), and facilitates waste reduction. However, key challenges relate to substantial financial requirements, limited circular supply chains, and consumer behavioral barriers. The chapter guides stakeholders toward a sustainable and resilient future by offering managerial, policy, societal, and scholarly implications.



The wine sector is a strategic component of European agribusiness, mainly because of its environmental impact. This study aims to contribute to the scientific debate on the sustainability of the supply chain through a systematic review of the potential for recovery and valorization of its by-products. Accordingly, the objective of the research is twofold. The first concerns identifying and analysing products from wine by-products reused in agriculture, aiming to close the production cycle. The second focuses on critically assessing the environmental, economic and social impacts of such practices about the Sustainable Development Goals (SDGs). The review process identified three main application areas: organic fertilizers, animal feed, and wastewater treatment. These practices are found to be consistent with the SDGs: fertilizers, goals 2, 12, and 15; animal feed, goal 12; and water treatment, goals 6 and 12.

This study shows the relationship between digitalization and circular economy. Material footprint is taken here as the approx. of circular economy. The material footprint helps to quantify the total raw material used to understand a country's consumption demand. It uses the panel data from 2000 to 2022 across 20 developed and developing countries. Material footprint is taken here as the independent variables including several proxies for digitalization, such as ICT service export, ICT goods export, individual using the internet, mobile subscription, telephone usage. Additionally, it includes GDP, FDI, and trade openness as controlled variables. The study used pooled OLS, fixed effect, random effect, and Hausman test. The findings from these results show the relationship between the material footprint and digitalization and integration of circular economy initiatives. The results also show that developed countries have positive relationship with environmental sustainability whereas developing countries don't have positive relationship with environmental sustainability.



This research proposes a cyber-physical system that integrates cybernetic systems thinking with strategy concepts to facilitate an easy transition to a circular economy. This strategy accounts for how complex it would be for data and resources to flow through circular systems. It leverages concepts from cybernetics, flexible systems, and Industry 5.0 to create systems resilient and responsive in real time. Legislative inertia and un-integrated digital systems may at times make it difficult for existing linear-to-circular shifts to occur. The framework indicates how the economic, environmental, and technical spheres are interconnected. In order to assist in solving these issues, sensor-driven intelligence, feedback mechanisms, and decision automation are all combined to streamline resource utilization and make industry society more robust, transparent, and people-oriented. Based on the findings, cyber-physical coordination according to cybernetic back-loops and circular reasoning could accelerate the transition towards sustainability in all industries by a significant margin.

Eco-innovation is the critical element that ensures a shift to the circular economy and plays a significant role in the protection of the environment, building new environmentally friendly consumption behaviour and increasing the efficiency of production. The purpose of this research is to rank EU27 countries based on a selection of Eco-Innovation Index sub-indicators and identify the alternative that best reflects the interconnection between eco-innovation and the circular economy. This paper explores the application of multi-criteria decision-making (MCDM) methods to evaluate EEI sub-indicators. In the MCDM model, the sub-indicators weight coefficients were calculated by applying the CRITIC method. The optimal alternative is considered to be the one that demonstrated significant progress in applying eco-innovation principles during the analyzed period from 2020 to 2024. The originality of this research is fundamentally linked to the evaluation of the eco-innovation performances of EU member states using a portfolio of MCDM methods.



This study considers the circular economy, competitive advantage, and the success of strategic innovation in Ethiopian large-scale manufacturing firms. A quantitative research design was used to analyze the correlations between strategic innovation, alongside competitive advantage, and the performance success of the circular economy. A total of 384 participants were selected for the study to achieve adequate statistical power for dependable analysis results. The researcher used simple random sampling to select participants of the study. Exploratory factor analysis and confirmatory factor analysis were applied to confirm the reliability and validity of the variables utilized. A structural equation model was carried out to examine correlation between variables and mediation using SPSS V.26 and AMOS V.26 software. The outcome of the study highlighted that strategic innovation has a major impact on the success

# Preface

The accelerating environmental, social, and economic crises of the twenty-first century underscore the limitations of the prevailing linear model of production and consumption. The paradigm of "take–make–dispose" has precipitated resource depletion, ecological degradation, and widening socio-economic disparities. In response, the Circular Economy (CE) has emerged as a comprehensive framework that seeks to reconcile economic growth with ecological integrity and social equity through strategies of reduction, reuse, recycling, and regeneration.

This edited volume brings together a diverse set of scholarly contributions that collectively advance the theoretical, empirical, and applied dimensions of the Circular Economy. The chapters encompass a wide geographical scope and disciplinary orientation, thereby providing readers with a multidimensional understanding of the subject. They range from theoretical foundations and conceptual frameworks to sector-specific analyses in logistics, tourism, textiles, fashion, wine, and manufacturing. Further, the volume engages with critical enabling mechanisms, such as digitalization, cyber-physical systems, eco-innovation, and strategic innovation, that shape the pathways toward circular transitions across both developed and developing contexts.

Of particular importance is the volume's inclusion of perspectives from the Global South, with case studies from Bangladesh, Indonesia, Nepal, Tanzania, Ethiopia, and India. These chapters highlight the challenges of implementing CE in contexts marked by financial constraints, institutional gaps, and climate vulnerability, while simultaneously illuminating the potential of social business models, policy reforms, and technological innovations to catalyze inclusive and resilient forms of sustainability. Contributions focusing on the European Green Deal, startups, and cross-border comparative analyses further demonstrate the global relevance of CE while emphasizing the necessity of context-sensitive strategies.

The methodological diversity of this collection is also noteworthy. The chapters employ qualitative case studies, comparative analyses, mixed-methods research, and econometric modelling, thereby offering both conceptual rigor and empirical richness. Collectively, the findings reaffirm that the CE is not merely a technical adjustment but a systemic transformation requiring institutional innovation, policy coherence, cross-sectoral collaboration, and cultural change.

This book is intended to serve as a resource for scholars, practitioners, and policymakers committed to advancing sustainability transitions. By situating the Circular Economy within the broader debates on sustainable development, industrial transformation, and social equity, the contributions aim to deepen academic discourse while providing actionable insights for practice.

It is our expectation that this volume will contribute to ongoing scholarly inquiry and policy deliberations, while also fostering critical reflection on the possibilities and limitations of the Circular Economy as a pathway to sustainable futures.

## ORGANIZATION OF THE BOOK

### Chapter 1: Circular Economy: Foundation and Theoretical Framework by Rejaul Karim, Bablu Kumar Dhar, and Md. Mustaqim Roshid

Karim, Dhar, and Roshid present the theoretical foundations of the Circular Economy (CE), contrasting it with linear "take–make–dispose" systems. They trace the intellectual roots of CE through industrial ecology and systems thinking, balancing optimism with critical perspectives on its practical limitations. By highlighting reduction, reuse, and recycling as key mechanisms, they situate CE as both a sustainability model and a structural reorganization of economies. This chapter provides the conceptual grounding and vocabulary necessary to frame the sectoral and applied explorations that follow throughout the volume.

### Chapter 2: IoT-Enabled Circular Supply Chains: Advancing Sustainability in the United States Logistic and Distribution by Abdul Waaje, Sahilali Saiyed, Md. Mustaqim Roshid, Mohammad Rafid Irtisham, and Md Anamul Haque

In this chapter, Waaje, Saiyed, Roshid, Irtisham, and Haque analyze how Internet of Things (IoT) technologies are reshaping logistics in the United States toward circularity. Case studies from manufacturing, retail, and e-commerce illustrate how blockchain, predictive analytics, and real-time monitoring enhance transparency, resource optimization, and reverse logistics. Their analysis aligns with SDGs on climate action, responsible consumption, and sustainable energy, while acknowledging barriers such as cybersecurity threats, regulatory complexity, and high implementation costs. The chapter underscores IoT as a systemic enabler of CE, offering practical strategies for embedding sustainability into large-scale distribution and supply chains.

## Chapter 3: Circular Economy in Tourism Finance: Transforming Sustainability into Profit by Ridhima Sharma, Timcy Sachdeva, and Amrik Singh

In this chapter, Sharma, Sachdeva, and Singh investigate the application of CE principles in the tourism industry, positioning circularity as both ecologically sustainable and financially viable. Drawing on surveys and interviews with Indian tourism firms, they identify adoption trends, including energy-efficient lighting and waste segregation, while also noting constraints in adopting advanced closed-loop systems. Their findings reveal how financial incentives, environmental awareness, and firm size influence CE integration. The chapter emphasizes that while basic circular practices are gaining ground, broader adoption requires stronger infrastructure and investment, illustrating the dual environmental and economic promise of CE in tourism finance.

## Chapter 4: Success Factors for Startups in the Circular Economy by Silvio Andrae

In this chapter, Andrae examines success factors for startups within the circular economy through comparative case studies in Germany, the U.S., China, and Brazil. His findings highlight material flow traceability systems, dynamic capabilities for managing backward supply chains, and early regulatory compliance as crucial resources. Financing structures dominated by patient capital and partnerships across sectors emerge as vital enablers, while standardization gaps and feedstock volatility impede growth. By situating startups within global value chains, Andrae shows how entrepreneurial ecosystems influence CE outcomes. The chapter provides valuable insights into the dynamics of innovation, legitimacy, and financing in circular startups worldwide.

## Chapter 5: Circular Economy and Ethical Compliance: Assessing the European Green Deal Policy Framework in Driving Sustainability by Md. Tota Miah, Md. Faisal-E-Alam, and Anna A. Kredina

Miah, Faisal-E-Alam, and Kredina analyze the European Green Deal (EGD) through the lens of ethical compliance and CE. They argue that successful CE adoption requires integrating sustainability with moral imperatives such as labor rights, social justice, and equitable resource access. The chapter critically examines regulatory tools, including the EU Taxonomy, CSRD, and the Just Transition Mechanism, assessing their capacity to drive fair and inclusive sustainability transitions. Their contribution emphasizes the importance of aligning legal, ethical, and social frameworks, while stressing the need for collaboration across public, private, and civil sectors to advance CE objectives in Europe.

## Chapter 6: Advancing Sustainability through Social Business Model in the Circular Economy of Developing Countries by Forhad Hossain, Paulina Permatasari, Johanna Renny Octavia, Theresia Edward Busagara, and Shailaja Kasaju

The authors explore the role of social business models in implementing CE in developing countries, focusing on Indonesia, Nepal, and Tanzania. Using PESTLE and Triple Bottom Line frameworks, they identify barriers such as limited financing, policy enforcement gaps, and low consumer awareness. Case studies of organizations like Grameen Group and Waste Concern illustrate how mission-driven enterprises can overcome these challenges by embedding circularity within community-based practices. The chapter underscores the transformative potential of social businesses as strategic enablers of CE, highlighting the need for supportive policies and targeted mechanisms to scale sustainable solutions.

## Chapter 7: Stitching Sustainability in Fashion: Strengthening Circular Economy Drivers in the Textile Industry by Bablu Kumar Dhar, Rejaul Karim, and Md. Soleman Mollik

In this chapter, the authors assess the textile and fashion industry's urgent need for CE adoption. They highlight the sector's environmental burden, including greenhouse gas emissions and water pollution, while identifying key drivers such as sustainable material innovation, product life extension, and reverse logistics. Their analysis also underscores challenges such as technological limitations, economic barriers, and shifting consumer behavior. By framing textiles as a critical case of circular transition, the authors emphasize the need for systemic changes in production, supply chains, and consumption. The chapter illustrates both the environmental necessity and economic opportunities of circular practices in fashion.

## Chapter 8: Integrating Circular Economy Principles in Responsible Consumption and Production by Mohammad Shahab Uddin, Md. Soleman Mollik, and Amir Mohammad Nasrullah

The authors of this chapter investigate how CE principles integrate with SDG 12 on responsible consumption and production. Their qualitative review of literature highlights CE's ability to enhance resource efficiency and minimize waste. However, they also note persistent barriers, including financial demands, supply chain challenges, and consumer behavior. By bridging conceptual insights with practical implications, the authors argue that embedding CE in production and consumption systems can yield managerial, policy, and societal benefits. The chapter advances the dialogue on sustainable development by emphasizing CE's transformative potential while acknowledging the structural hurdles to its widespread adoption.

## Chapter 9: Circular economy and sustainability in the wine sector: an analysis on the reuse of by-products in the context of SDGs by Mariagrazia Provenzano, Francesco Pacchera, Cecilia Silvestri, and Alessandro Ruggieri

Provenzano, Pacchera, Silvestri, and Ruggieri focus on the wine sector, assessing the reuse of by-products to foster circularity. Their systematic review identifies applications in organic fertilizers, animal feed, and wastewater treatment, linking these practices to SDGs related to food, water, and ecosystem health. The chapter highlights the environmental, social, and economic impacts of valorizing by-products, showing how circular practices can close production loops while supporting sustainable agriculture. By positioning the wine industry as both a contributor to and a potential mitigates of environmental harm, the authors underscore the role of sector-specific CE strategies in advancing broader sustainability goals.

## Chapter 10: Circular Fashion and Recycling: A Sustainable Approach to Transform the Fashion and Retail Industry by Shivangi Kashyap and Pravin Kumar

Te authors explore the promise of circular fashion in addressing the environmental and social impacts of fast fashion. They analyze practices such as recycling, upcycling, and closed-loop production systems, supported by brand case studies that demonstrate emerging sustainability initiatives. The chapter identifies scaling challenges, including technological constraints and consumer demand for low-cost goods, but also highlights policy and innovation opportunities. By connecting environmental recovery, economic growth, and social well-being, Kashyap and Kumar position circular fashion as a pathway to long-term industry transformation. Their contribution illustrates the necessity of rethinking fashion's global role within CE frameworks.

## Chapter 11: Circular Economy for Sustainability in the Global South: Transforming Industry, Consumption, and Climate Action in Bangladesh: Circular Transition in Bangladesh by Md. Mehedi Hasan, S M Shafeeul Islam, and Md. Solaiman Chowdhury

Hasan, Islam, and Chowdhury examine CE's role in fostering sustainable development in Bangladesh, one of the world's most climate-vulnerable nations. They link CE initiatives to SDGs 9, 12, and 13, emphasizing industrial innovation, responsible consumption, and climate action. The chapter highlights eco-friendly production, circular supply chains, and waste management practices in industries such as textiles and packaging. While acknowledging structural challenges, the authors argue that CE offers a critical pathway for aligning industrial growth with climate resilience. Their work underscores the urgency of adopting CE models in developing economies confronting environmental degradation and rapid urbanization.

## Chapter 12: Digitalization and the Circular Economy: An Empirical Assessment of Material Footprint across Developing and Developed Countries by Tanvi Sanotra and Meenakshi Gupta

Sanotra and Gupta empirically assess the relationship between digitalization and CE through the lens of material footprint. Using panel data from 2000–2022 across developed and developing countries, they test proxies such as ICT exports, internet use, and mobile subscriptions. Their analysis reveals that while developed economies demonstrate positive sustainability outcomes, developing nations show weaker associations between digitalization and reduced material use. The findings highlight digital technologies' role in supporting CE, but also the uneven capacity to translate digitalization into sustainability. The chapter provides critical insights into the digital divide and its implications for global CE transitions.

## Chapter 13: Converging Strategic Innovations and Cybernetic Systems: A Cyber-Physical Framework for Circular Economy Transition by Patcha Bhujanga Rao and Preethi Inampudi

Rao and Inampudi propose a cyber-physical framework that combines cybernetics, flexible systems, and Industry 5.0 to accelerate CE transitions. They argue that sensor-driven intelligence, feedback mechanisms, and decision automation can streamline resource flows and enhance transparency. Addressing obstacles such as legislative inertia and fragmented digital systems, the authors position cybernetic coordination as a means to overcome barriers to circularity. Their work highlights the interconnectedness of economic, environmental, and technical dimensions, presenting cyber-physical systems as both innovative and practical. The chapter demonstrates how systemic, technology-enabled strategies can significantly advance sustainability across industries and societies.

## Chapter 14: Eco-Innovation as a Driver of the Circular Economy: Multi-Criteria Approach by Ivana Janjić, Miljana Talić, and Zarko Jovan Radjenovic

The authors explore eco-innovation as a critical driver of CE. Using multi-criteria decision-making methods, they evaluate the performance of EU27 countries on eco-innovation indices, employing the CRITIC method for weighting indicators. Their findings identify countries demonstrating strong progress in integrating eco-innovation with CE, offering comparative insights across Europe. By linking innovation with environmental performance, the authors highlight the role of policy frameworks, research investment, and cultural change in fostering circular transitions. The chapter underscores eco-innovation not only as a technical factor but also as a systemic enabler of sustainable consumption and production in the EU.

## Chapter 15: Effect of strategic innovation on success of the circular economy with mediating role of competitive advantage in Ethiopian Large-scale manufacturing firms by Tafese Niguse, Shashi Kant, Aynetu Terefe Alem, Kenesa Kebede, and Sofi Badi Mohammed

Niguse, Kant, Alem, Kebede, and Mohammed examine Ethiopian large-scale manufacturing, analyzing how strategic innovation contributes to CE adoption. They show that competitive advantage mediates the relationship between innovation and sustainability outcomes, underscoring the importance of cultivating innovative organizational cultures. Their findings suggest that innovation-driven firms achieve greater efficiency and resilience in CE practices, while also enhancing competitiveness. By combining theoretical insights with practical recommendations for managers, the chapter provides valuable guidance for navigating the complexities of circular transformation in resource-intensive industries. It highlights the centrality of innovation to both firm-level success and broader CE goals.

### Chapter 16: Strategic Innovation and Systemic Transformation for a Circular Economy- Theories, Practices, and Global Implications by Muhammad Usman Tariq

In this chapter, Muhammad Usman Tariq concludes the volume by situating strategic innovation as a systemic enabler of CE across global industries. He examines how circular design, product-as-a-service models, and reverse logistics contribute to resource efficiency and sustainability. Drawing on case studies and theoretical frameworks, Tariq emphasizes stakeholder collaboration, innovation, and policy coherence as critical drivers of systemic change. He also identifies regulatory gaps, cultural resistance, and scalability as persistent challenges. This closing chapter underscores the necessity of embedding CE principles into core organizational practices, reinforcing CE's role as a transformative approach to achieving global sustainability goals.

## IN CONCLUSION

The chapters assembled in this volume collectively demonstrate that the transition to a Circular Economy is neither linear nor uniform, but rather a multifaceted and context-dependent process. By engaging with theory, empirical evidence, and applied practice across diverse sectors and geographies, the contributors underscore both the promise and the complexity of operationalizing circularity. The perspectives offered—from the technological innovations of developed economies to the social business models of the Global South—reveal that sustainable transformation requires adaptive strategies, collaborative governance, and continual innovation.

It is our conviction that this book will not only advance academic debates but also serve as a reference point for policymakers, industry leaders, and civil society actors seeking to navigate the challenges of sustainability transitions. The insights herein affirm that while the Circular Economy offers no singular blueprint, its principles provide a robust framework for aligning ecological responsibility with economic and social well-being in the pursuit of a sustainable future.

**Rejaul Karim**

*Varendra University, Bangladesh*

**Bablu Kumar Dhar**
*Mahidol University, Thailand*

# Chapter 1
# Circular Economy: Foundation and Theoretical Framework

**Rejaul Karim**
https://orcid.org/0000-0001-6175-8825
*Varendra University, Bangladesh*

**Bablu Kumar Dhar**
https://orcid.org/0000-0001-8768-8634
*Mahidol University, Thailand*

**Md. Mustaqim Roshid**
https://orcid.org/0000-0001-8517-6336
*University of Rajshahi, Bangladesh*

## ABSTRACT

*The Circular Economy (CE) is a shift away from the conventional linear economic model, which is based on resource efficiency and waste minimization with regenerative systems. This chapter goes beyond descriptive overviews by synthesizing historical evolution, theoretical foundations, and practical applications of CE into a unified conceptual lens, highlighting how CE serves as both a sustainability paradigm and a systems-level economic framework. It charts the history of CE from early conceptual contributions, such as the 1972 Limits to Growth report, through formalization in the 1980s with the rise of industrial ecology and the inception of eco-efficiency in 2000. The chapter provides a brief review of key milestones in CE, from the Performance Economy to Cradle-to-Cradle design, emphasizing the extension of product lifetimes and recycling, culminating in overall closed-loop material flows. It also explores the role of the Ellen MacArthur Foundation, espousing the CE and linking it to worldwide sustainability objectives. Unlike prior reviews, the*

DOI: 10.4018/979-8-3373-2454-8.ch001

*chapter contributes by critically integrating theoretical models with case-based insights and identifying underexplored challenges such as rebound effects, governance asymmetries, and socio-cultural resistance. It concludes with a discussion on the triple bottom line benefits (environmental, economic, and social) of CE, as well as the challenges that will be faced, including techno-flops, market dynamics, policy gaps, and cultural norms. Finally, it outlines how digital technologies and collaborative partnerships can provide transformative pathways, positioning CE as a cornerstone of future sustainability research and practice.*

## 1. INTRODUCTION

The circular economy (CE) represents a decisive move away from the historic linear model of "take–make–dispose" toward a regenerative and restorative system (Tsalis et al., 2022). Instead of focusing only on recycling, CE aims to create an economy in which natural systems are actively regenerated, waste and pollution are planned out, and products, materials, and resources are used for as long as feasible (Jones & Comfort, 2021). In response to growing global issues such resource scarcity, biodiversity loss, climate change, and growing waste, this transition has accelerated (Hoosain et al., 2023).

The CE has gained traction in academia, industry, and policy since its inception in the early 1990s (Karim et al., 2024). By increasing product lifespans, permitting reuse and repair, and encouraging creative business models like product-as-a-service, it holds the potential to decouple economic prosperity from resource consumption (Temerbulatova et al., 2021; Rotondo et al., 2025). However, despite its widespread use, CE is not without ambiguities; definitions differ greatly, and there is ongoing discussion about its systemic implications for culture, technology, and governance (Kirchherr et al., 2017).

This chapter contributes to this debate by examining the conceptual foundations, principles, and theoretical underpinnings of CE, while critically assessing its benefits, challenges, and practical pathways. It situates CE not only as a technical framework but as a socio-economic paradigm that redefines production, consumption, and innovation within planetary boundaries.

The structure of the chapter is as follows: Section 2 provides a historical overview of the evolution of CE; Section 3 outlines its core principles; Section 4 reviews theoretical frameworks supporting CE; Sections 5–7 examine benefits, challenges, and strategies; Section 8 presents illustrative case studies; Section 9 explores future trends; and Section 10 concludes with reflections on implications for sustainability transitions.

# 2. HISTORICAL CONTEXT AND EVOLUTION OF THE CIRCULAR ECONOMY

## 2.1. Early Concepts and Influences

The circular economy has evolved considerably since its initial emergence and is rooted in diverse intellectual traditions. Rather than repeating later frameworks in detail, this section introduces the conceptual foundations that underpin the CE and cross-references their fuller treatment in Section 4. Key schools of thought include industrial ecology, cradle-to-cradle design, and the performance economy (Kyriakopoulos et al., 2019). Industrial ecology emphasizes the integration of industrial systems with natural ecosystems and advocates closed-loop processes to minimize waste and pollution (Hossain et al., 2020). Cradle-to-cradle design advances the idea that products should be recyclable or biodegradable at the end of their life, ensuring that all materials circulate safely in closed systems (McDonough & Braungart, 2010; Onungwe et al., 2024). The performance economy, introduced by Stahel, highlights the substitution of services for products, thereby incentivizing durability, repair, and remanufacturing as pathways to extend product life cycles (Stahel, 2006).

Other intellectual influences include environmental economics, which emphasizes the internalization of environmental costs into economic decision-making through mechanisms such as Pigouvian taxes and Coasean bargaining (Georgescu-Roegen, 1971). Natural capitalism, articulated by Hawken, Lovins, and Lovins, stresses the economic valuation of ecological capital and the necessity of embedding these values into economic systems (Hawken & Lovins, 2013). Systems thinking, as advanced by Meadows (2009), underscores the importance of interconnections and feedback loops within socio-ecological systems, which is essential for designing resilient circular processes. Together, these perspectives form the intellectual foundation of CE and provide the groundwork for theoretical models elaborated in later sections.

## 2.2. Key Milestones and Developments

The development of the circular economy can be traced through several transformative milestones that shaped both scholarship and practice. In the 1970s, the publication of The Limits to Growth by the Club of Rome (Meadows et al., 1972) represented a critical turning point, drawing attention to planetary boundaries and the risks of exponential growth in a finite system. In the 1980s, industrial ecology emerged as a scientific framework with Frosch and Gallopoulos (1989) advancing the notion of industrial systems as analogues to ecological metabolism, which reframed sustainability from an abstract concept into a practical design principle.

In the 1990s, Walter Stahel formalized the performance economy, which sought to decouple prosperity from material throughput by extending product life, encouraging remanufacturing, and embedding services as substitutes for ownership (Stahel, 2006). The 2000s saw another conceptual leap with McDonough and Braungart's Cradle to Cradle (2002), which introduced the principle of eco-effectiveness and regenerative design, pushing CE beyond efficiency to a vision of restorative production systems. By the 2010s, the Ellen MacArthur Foundation (2015) helped mainstream CE in policy and business discourse, democratizing its application across corporate strategies and governmental agendas. In the present decade, CE has been institutionalized through global initiatives such as the European Green Deal and the UN Sustainable Development Goals, with scholars such as Geissdoerfer et al. (2017) positioning it as a radical worldview for addressing systemic crises, including climate change, biodiversity loss, and resource depletion. These milestones mark the transformation of CE from a theoretical proposal to a widely adopted sustainability paradigm.

## 2.3. The Ellen MacArthur Foundation and the CE Framework

Since 2010, the Ellen MacArthur Foundation (EMF) has played a central role in consolidating and promoting CE as a viable and strategic model for businesses and policymakers. The Foundation defined CE as an industrial system that is restorative and regenerative by design (Ellen MacArthur Foundation, 2013), a formulation that integrated earlier contributions such as Boulding's notion of "Spaceship Earth" (2013), Georgescu-Roegen's thermodynamic limits to resource use (1971), Stahel's performance economy (2006), and the principles of industrial ecology articulated by Graedel and Lifset (2016). EMF's model is structured around three principles: designing out waste and pollution, keeping products and materials in use, and regenerating natural systems. These ideas also resonate with complementary frameworks such as eco-design (Lacasa et al., 2016), pollution prevention theory (Ashford & Hall, 2018), product-service systems (Tukker, 2004), and regenerative design (Lyle & Lyle, 1996).

While the EMF framework has been widely praised for its clarity and accessibility, critiques have highlighted important limitations. Scholars argue that EMF's approach often assumes the compatibility of CE with green growth trajectories, without sufficiently accounting for rebound effects, international disparities, or the social dimensions of production (Corvellec et al., 2022; Murray et al., 2017). From a degrowth or post-growth perspective, achieving true circularity requires not only efficiency gains but also reductions in absolute resource throughput. The EMF's influence on global agendas, including the EU Circular Economy Action Plan and national strategies in China, Japan, and Finland, illustrates its institutional reach, but the model should be viewed as one influential interpretation of CE rather than a definitive framework.

## 3. CORE PRINCIPLES OF THE CIRCULAR ECONOMY

The Circular Economy (CE) is a change in the economic model from a linear (take-make-dispose) to a regenerative, restorative system that maintains the value of materials by improving ecosystem resilience. Although the framework made prominent by the Ellen MacArthur Foundation (2013) recognizes three major principles—(a) designing waste and pollution out, (b) keeping products and materials in use, and (c) regenerating natural systems—these principles are not standalone. They are anchored in wider theoretical traditions, such as Industrial Ecology, Cradle-to-Cradle, Performance Economy, Biomimicry, Systems Thinking, and Regenerative Design. This multidimensional foundation positions CE as both an economic strategy and an integrated socio-ecological framework (Geissdoerfer et al., 2017) aligned with planetary boundaries (Rockström et al., 2009) and the SDGs.

### 3.1. Designing Out Waste and Pollution

The central aim of CE is to prevent waste before it arises, requiring life-cycle thinking and design interventions from the outset. Eco-design directly links to Industrial Ecology, which models closed-loop flows where the by-products of one process become the inputs of another. Similarly, Cradle-to-Cradle principles reinforce this idea by designing materials as biological or technical nutrients that circulate indefinitely (McDonough & Braungart, 2002). Biomimicry extends this by drawing inspiration from natural efficiency to guide product modularity and recyclability (Benyus, 1998). Together, these frameworks demonstrate that waste prevention is not just a design choice but a systemic innovation strategy.

### 3.2. Keeping Products and Materials in Use

This principle emphasizes slowing, closing, and narrowing resource loops (Bocken et al., 2016). Stahel's Performance Economy provides the intellectual basis by reframing products as services, encouraging maintenance, repair, and remanufacturing to extend lifetimes (Stahel, 2010). Product-Service Systems (PSS) apply this at the business-model level, aligning profitability with durability rather than disposability (Tukker & Tischner, 2006). At the systems level, Materials Circularity Indicators (EMF, 2015) offer quantitative ways to evaluate how well products preserve value.

Taken together, these strategies illustrate that "keeping products in use" is both a design principle and a governance mechanism, uniting industrial practice with measurement frameworks.

### 3.3. Regenerating Natural Systems

Beyond minimizing harm, CE demands restoration of ecological capital. Regenerative Design (Lyle & Lyle, 1996) explicitly calls for human systems to contribute positively to ecosystems, while Sustainable Agriculture and Agroecology enhance biodiversity and soil health. Pauli's Blue Economy (2010) extends this by valorizing local waste streams, and Nature-based Solutions (Cohen-Shacham et al., 2019) align circular practices with climate resilience. The Planetary Boundaries framework (Rockström et al., 2009) provides the macro-scale limits within which regenerative activities must operate. In this way, CE does not merely reduce ecological footprints but actively strengthens ecosystem services through natural capital accounting (Costanza et al., 1997).

## 4. THEORETICAL FRAMEWORKS SUPPORTING THE CIRCULAR ECONOMY

Rather than treating each framework as a standalone concept, this section highlights how they complement one another, identifies areas of overlap, and outlines the gaps that remain for future theoretical development.

### 4.1. Cradle to Cradle (C2C)

Based on the ideas by McDonough & Braungart (2002), the Cradle to Cradle (C2C) system encourages the design of products for closed loops, where materials are either biological nutrients (returning safely to ecosystems) or technical nutrients (recyclable indefinitely). Its unique contribution is the focus on *eco-effectiveness*—designing waste out of the system altogether. When linked with Industrial Ecology, C2C provides practical design pathways for achieving the systemic material loops that industrial ecology envisions.

### 4.2. Performance Economy

The Performance Economy model by Walter Stahel shifts emphasis from product ownership to service provision— "selling performance" instead of goods. This strategy promotes long product lifespans through repair, refurbishment, and remanufacturing. In combination with C2C, which ensures materials can circulate, the Performance Economy creates incentives for firms to actually implement circular design by tying profitability to durability and reuse. However, the model often assumes supportive consumer and policy environments, which remain underdeveloped in many regions.

### 4.3. Industrial Ecology

Industrial Ecology embeds ecological principles into industrial systems, encouraging closed cycles, resource symbiosis, and tools like material flow analysis (Graedel & Lifset, 2016). It provides the systems-level scientific backbone for CE, but without frameworks like C2C or Performance Economy, it risks remaining a technical exercise rather than a business or design reality.

### 4.4. Systems Thinking

The concept of Systems Thinking acknowledges CE as a complex adaptive system with feedback loops, delays, and interdependencies (Meadows, 2009). It integrates Industrial Ecology's technical focus with social and economic feedbacks, offering a bridge to broader frameworks such as Doughnut Economics. Yet, while essential for policy design, systems thinking can remain abstract without design-oriented tools like C2C or Biomimicry.

### 4.5. Regenerative Design

Regenerative Design (Lyle & Lyle, 1996) emphasizes restoring and improving natural systems beyond "net zero." It extends CE by embedding ecological restoration within economic activity. When combined with Biomimicry, regenerative design provides principles for not just minimizing harm but actively replenishing ecosystems. The gap here is the lack of large-scale empirical evidence of its effectiveness in industrial systems.

### 4.6. Biomimicry

Biomimicry (Benyus, 1998) applies lessons from nature's cycles to human design challenges, enabling waste-free systems. It overlaps with C2C in envisioning products as nutrients, but while C2C prescribes design protocols, Biomimicry offers inspiration and models from ecology. A limitation is its conceptual rather than operational orientation, requiring translation into measurable industrial practices.

## 4.7. Doughnut Economics

The Doughnut Economics model (Raworth, 2017) integrates social foundations with ecological ceilings. It situates CE within broader socio-ecological boundaries, complementing Industrial Ecology's technical scope and Performance Economy's business focus. The gap lies in operationalizing Doughnut Economics in firms and industries, where circular practices often address resource efficiency but not social equity.

In summary, these frameworks are not competing but complementary: C2C, Biomimicry, and Regenerative Design drive innovation in *design philosophy*; Performance Economy redefines the *business model*; Industrial Ecology and Systems Thinking provide *system-level tools*; and Doughnut Economics adds the *socio-ecological boundary condition*. Together, they position CE as more than a resource-management strategy—rather, as an integrated paradigm for sustainable development. The challenge remains to bridge conceptual alignment with measurable, scalable practices.

Figure 1 illustrates the theoretical foundations of the Circular Economy (CE) by connecting its core principles with key conceptual frameworks. At the center, CE is shown as the unifying paradigm, supported by three main principles: designing out waste and pollution, keeping products and materials in use, and regenerating natural systems. Surrounding these are complementary theories—Cradle to Cradle, Performance Economy, Industrial Ecology, Systems Thinking, Biomimicry, Regenerative Design, and Doughnut Economics. Together, these frameworks move CE beyond waste management, integrating design innovation, business models, ecological restoration, and socio-economic boundaries. The diagram emphasizes CE as a holistic, systems-based approach to sustainable development.

*Figure 1. Theoretical Frameworks and Circular economy*

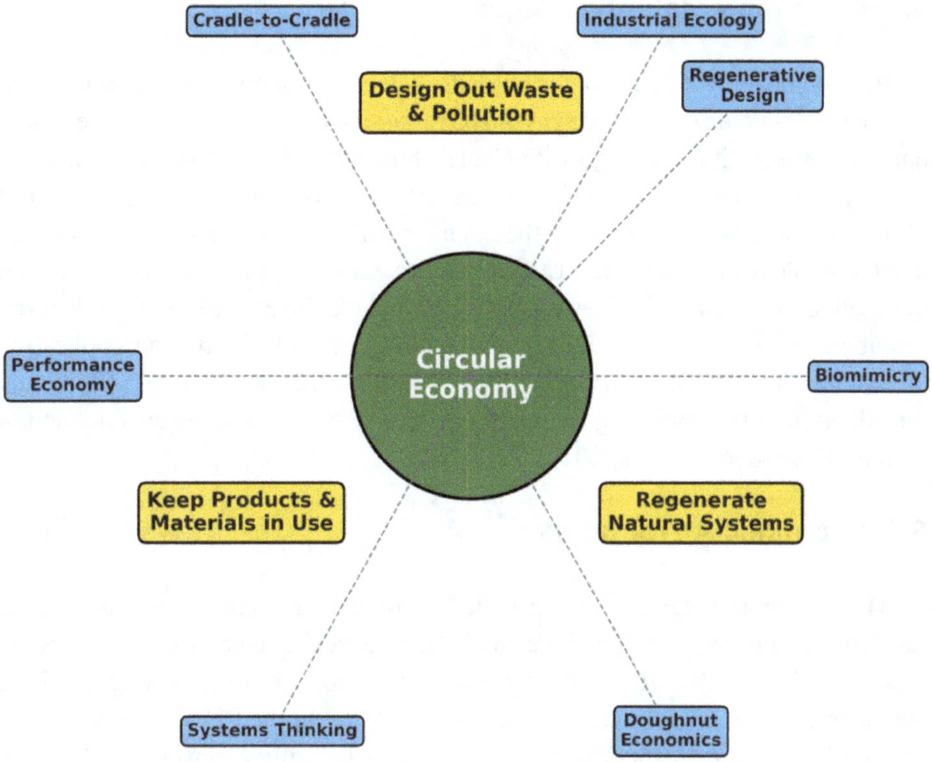


*Source: Authors' illustration*

## 5. ADVANTAGES OF IMPLEMENTING THE CIRCULAR ECONOMY

Circular Economy (CE) has a variety of multidimensional advantages on environmental, economic, and social levels. Rather than viewing these dimensions in isolation, the benefits are interrelated, reinforcing sustainability outcomes across ecosystems, markets, and communities. All these benefits can be traced back to various theoretical models, namely Industrial Ecology (Frosch & Gallopoulos, 1989), Cradle-to-Cradle (McDonough & Braungart, 2002), the Performance Economy (Stahel, 2010), and the Natural Capitalism model (Hawken & Lovins, 2013), which all focus on closed-loop processes, resource replenishment, and sustainable value creation.

## 5.1. Environmental Advantages

Environmentally, CE alleviates burdens on the environment by minimizing resource use, reducing greenhouse gas (GHG) emissions, and preventing waste. These outcomes are justified through Life Cycle Thinking and Material Flow Analysis, which provide quantitative metrics to measure reductions in environmental impact (Allwood et al., 2011). For instance, the Ellen MacArthur Foundation (2013) estimates that CE implementation in the EU could reduce emissions by 48 percent compared to baseline trends by 2030. Key mechanisms include the extension of product life through repairing, reusing, and remanufacturing (Stahel, 2010), the application of eco-design and biomimicry principles to design waste out (Benyus, 1998), and the adoption of restorative agriculture and forestry practices to regenerate natural systems (Geissdoerfer et al., 2017).

## 5.2. Economic Advantages

The economic outcomes of CE include cost savings, innovation, and greater resilience against volatile resource markets. The Performance Economy system (Stahel, 2010) emphasizes shifting from selling products to providing services, which promotes refurbishment and extends asset life. In the EU, CE strategies are projected to yield net economic benefits of up to 1.8 trillion euro by 2030 (Ellen MacArthur Foundation, 2013). Such benefits are realized through increased energy efficiency and reduced dependency on raw material price fluctuations (Geng et al., 2012), greater creativity in product design, materials science, and logistics, and the development of reverse logistics and resource recovery as new revenue streams.

## 5.3. Social Advantages

At the social level, CE contributes to more inclusive and sustainable societies through job creation, health improvements, and strengthened local economies. CE overlaps with the Social Capital Framework (Putnam, 1993), which emphasizes community resilience and knowledge sharing, and aligns closely with the UN Sustainable Development Goals, especially Goal 8 (Decent Work), Goal 11 (Sustainable Cities), and Goal 12 (Responsible Consumption). The transition to CE fosters employment opportunities in repair, refurbishment, recycling, and remanufacturing (Murray et al., 2017), improves public health through reduced pollution and cleaner urban spaces, and enhances community participation via local reuse networks and collaborative consumption platforms.

Taken together, the environmental, economic, and social benefits of CE demonstrate its potential as an integrated sustainability framework that not only reduces ecological pressures but also drives innovation, competitiveness, and community well-being.

## 6. CHALLENGES TO THE CIRCULAR ECONOMY

Although the circular economy has transformative potential, various systemic challenges related to it include aspects of technology, economics, policy, and even socio-cultural areas that organize its implementation. These obstacles are interrelated and tend to reinforce each other, slowing down the migration to circular models (Kirchherr et al., 2017; Ranta et al., 2018).

### 6.1. Technological Challenges

Technological underpinnings of CE are not fully developed, as evidenced by the systems of recycling, remanufacturing, and resource recovery. For instance, e-waste recycling has become a growing problem, as the complexity of composite materials in smartphones and laptops makes effective recovery of rare earth metals extremely difficult, often resulting in environmentally damaging disposal practices in developing countries. Lack of appropriate technologies to extend the life of products, mostly in recycling, repair, and reuse of goods, are lacking, especially in the case of complex and composite products. The compositions of materials and dangerous additives that are incorporated into modern products may complicate the separation and recycling of products. Insufficient information regarding material movement, product lifecycle, and secondary material quality results in the inability to design optimized reverse logistics networks. Although newer solutions such as digital product passports and blockchain-based traceability are emerging, their adoption remains limited and fragmented across industries.

### 6.2. Economic and Market Barriers

The CE needs new business models and changed economic incentives. Virgin resource extraction frequently tends to be cost-efficient in comparison to recovering secondary materials because of subsidies, market manipulation, and uncompensated byproducts. For example, aluminum recycling can save up to 95% of the energy required for primary production, yet virgin aluminum remains cheaper in several regions due to state subsidies and the absence of environmental cost internalization. Costs of impacting the environment, like carbon emissions and pollution, are not usually reflected in the price of products, and this makes the products in the circular economy currently less competitive. The low willingness to pay on green products limits the development of the market of CE-based products. This consumer hesitation is evident in the fashion sector, where second-hand and recycled clothing still face cultural resistance compared to fast-fashion alternatives despite their lower environmental footprint.

## 6.3. Policy and Regulatory Barriers

Mainstreaming of CE requires government action and there continues to be uneven regulatory environments. Supportive policies and strategies and plans to achieve goal CE are lacking in most countries. In the European Union, while the Circular Economy Action Plan has created ambitious targets, conflicting definitions of "waste" across member states often slow cross-border recycling and secondary material trade. Conversely, in developing economies, weak enforcement mechanisms mean that even when policies exist on paper, illegal dumping and open burning of waste remain widespread. Linear economic-centered standards (definitions of waste items, the entire taxation system) may at some point block circular solutions. Even when policies are in place, they may not be effectively enforced.

## 6.4. Social and Cultural Barriers

The behaviors and values of the society have a deterministic effect on the CE adoption. CE concepts are not openly understood in general because members of the population do not engage in them. The culture of consumerism dictates the current preference for disposability and novelty as opposed to repair and reuse. For example, in many high-income societies, smartphone users replace devices every two years, even when they remain functional, illustrating behavioral inertia that undermines repair and reuse initiatives. Habits and convenience, even in situations where sustainable options are present, tend to take hold. Such cultural inertia is reinforced by marketing strategies that normalize short product life cycles and make durable, repairable goods less attractive to consumers.

# 7. STRATEGIES FOR OVERCOMING THE CHALLENGES

## 7.1. Technological Innovation

Investing in research and development should target advanced recycling technologies (e.g., chemical recycling for plastics), modular product design, and material substitution that facilitates reuse and remanufacturing. Governments and businesses should invest in research and development to develop new technologies for recycling, reuse, and remanufacturing. The adoption of digital product passports and blockchain-based traceability can further ensure product transparency and improve reverse logistics efficiency. Promoting technology transfer remains crucial: sharing knowledge and technologies between countries and industries can help accelerate the adoption of circular practices. Supporting startups and entrepreneurs by providing funding and incubation can drive innovation in the circular economy, especially for niche areas such as repair services, 3D printing of spare parts, and eco-design software.

## 7.2. Economic and Market Instruments

Implementing extended producer responsibility schemes (EPR) holds producers responsible for the end-of-life management of their products, incentivizing them to design products that are easy to recycle or reuse. In addition, eco-modulated EPR fees—where producers pay less if products are designed for durability and recyclability—can directly align financial incentives with CE principles. Introducing taxes and subsidies is also vital: a carbon border tax on high-emission imports or tax credits for repair services can shift consumer and producer behavior. Creating markets for recycled materials can be advanced by establishing procurement targets and encouraging the use of secondary material in products, for example, through mandatory recycled content requirements in packaging, electronics, and construction sectors.

## 7.3. Policy and Regulatory Measures

Developing national circular economy strategies should include clear targets and priorities for the circular economy, such as mandatory right-to-repair legislation and bans on single-use plastics. Harmonization of regulations can remove trade barriers for circular products and services, particularly in the EU where differing waste definitions hinder cross-border circular flows. More robust enforcement is required, but beyond enforcement, governments should introduce green public procurement policies that prioritize circular products and services in state contracts, creating stable demand and market signals.

### 7.4. Education and Awareness Campaigns

Public awareness campaigns should inform citizens about the benefits of the circular economy, such as through repair cafés, community recycling hubs, and consumer labeling schemes that disclose product circularity scores. Integration of the circular economy in education requires schools, colleges, and universities to align curricula with CE principles, including practical training in eco-design, life-cycle assessment, and resource efficiency. Governments and businesses must promote sustainable consumption, influencing consumers to do more with less, encouraging product leasing models over ownership, and strengthening sharing-economy platforms.

## 8. CASE STUDIES OF SUCCESSFUL CIRCULAR ECONOMY IMPLEMENTATION

While these case studies illustrate pioneering applications of circular economy (CE) principles, it is essential to recognize both their successes and their contextual limitations. The practices that work in one sector or region may not be directly transferable to another, which calls for a more critical and comparative interpretation.

### 8.1. Case 1: Patagonia

Patagonia is an outdoor clothing company that has taken several actions to further the circular economy, including designing long-lasting and repairable goods, offering repair services for these products, and utilizing materials made from recycled products. The company also offers a "Worn Wear" program to encourage customers to repair and/or recycle their clothes, aiming to reduce waste upstream and promote the longevity of its products. Such a dedication to circularity not only minimizes environmental impact but also fosters brand loyalty by catering to shifting consumer values regarding sustainability (Rattalino, 2017). In addition, Patagonia's approach also showcases how a complete incorporation of circular principles at every stage of the product lifecycle can not only shift business models from linear to regenerative but also create a template as well for other incumbent firms trying to address environmental, social, and financial goals (Frishammar & Parida, 2018). The model shows how the ideas of life extension, resource recovery, and product-as-a-service can, in practice, be applied with existing products to make a more traditional industry landscape evolve from the prevailing "take-make-dispose" paradigm (Adam et al., 2017). It not only reduces the leak of resources, but it also increases the overall restorative and regenerative nature of the organization and leads to sustainable growth and success (Raj & Aithal, 2020). However, Patagonia's success is also rooted in a niche consumer base that values sustainability and can afford premium pricing, which may limit scalability in mass-market contexts.

## 8.2. Case 2: Interface

Interface is a global flooring company that has adopted a circular economy approach by designing products that can be easily recycled and using recycled materials. The company's "Evergreen Lease" model allows customers to lease flooring instead of buying it, incentivizing Interface to design durable and long-lasting products. This innovative model exemplifies a product-as-a-service approach, shifting the focus from material ownership to functional utility, which inherently promotes resource efficiency and longevity in product design (Adam et al., 2017). This approach mitigates waste generation and drastically reduces raw material consumption, fostering a closed-loop system where materials retain their value (Jones & Comfort, 2021). Such initiatives underscore the economic benefits of material reuse and recycling, contributing to long-term financial advantages through reduced virgin material costs and enhanced brand reputation (Marek & Krejza, 2024). Yet, Interface's reliance on business-to-business contracts makes its model less applicable to sectors driven by direct consumer demand, where cultural norms and behavioral inertia may pose stronger barriers.

## 8.3. Case 3: Philips

Philips is a multinational electronics company that has implemented circular economy practices in its lighting and healthcare divisions. The Dutch multinational electronics company, Philips, has embedded the concept of a circular economy into its lighting and healthcare divisions. It is a lightning-as-a-service company that enables its customers to lease lighting systems instead of purchasing them. Philips—remanufactures medical equipment, extending its useful life and reducing waste. This model of servitization (in line with that exerted by Interface) changes the company's revenue stream from selling once to re-selling in service contracts, evolving economic incentives with product durability and resource efficiency. These more fundamental changes not only lead to fewer negative environmental impacts from the production and consumption phases but also foster a much-needed, more sustainable business environment (Abbate et al., 2023). Collectively, these case studies suggest that there is a business case for moving from the traditional linear “take-make-dispose” system to a circular economy model that is restorative and regenerative (Rawat, 2021). These examples demonstrate that the circular economy applies to a wide range of industries, revisiting their organizational foundations through principles of resource, waste elimination, and systemic regeneration (Almelhem et al., 2025). Nevertheless, Philips’ implementation also highlights sector-specific constraints: regulatory requirements in healthcare and high costs of remanufacturing medical devices limit the universality of this approach.

Taken together, these cases demonstrate that while CE principles can be successfully embedded into diverse industries, their effectiveness and transferability depend on contextual factors such as market structure, consumer behavior, regulatory environment, and cost structures. A critical comparative perspective suggests that no single model provides a universal blueprint; instead, sectoral adaptation and institutional support are essential for broader diffusion.

## 9. FUTURE TRENDS AND OPPORTUNITIES

### 9.1. Digital Technologies and the Circular Economy

Digitalization presents key opportunities to facilitate the circular economy by leveraging the Internet of Things (IoT), artificial intelligence (AI), and blockchain. These tools can do more than track material flows—they can actively reshape business models by enabling predictive maintenance, adaptive production, and transparent supply chains. For example, IoT sensors used for real-time data collection allow predictive maintenance and inventory optimization, while AI algorithms help forecast failures and anomalies (Kondapaka, 2021). Importantly, these technologies are not without risks: excessive reliance on digital infrastructures raises concerns about cybersecurity, rebound effects from increased energy demand, and unequal access between advanced and developing economies. Thus, while digitalization offers powerful pathways to circularity, its long-term sustainability depends on integrating safeguards, ethical governance, and equitable diffusion. We need this data-driven method for better product traceability, more accurate lifecycle assessments, and predictive environmental modeling (Hulea et al., 2024; Karim et al., 2025a). The incorporation of these tools is also poised to boost the transparency and efficiency of reverse logistics, thus assisting the recovery and reapplication of materials once their lifecycle comes to an end (Hoosain et al., 2020). Moreover, the interface of Industry 4.0 technologies with circular economy principles is likely to transform production by creating more resource-efficient, resilient systems rather than simply digitizing existing linear processes (Polo et al., 2025).

## 9.2. The Role of Collaboration and Partnerships

The circular economy will only advance through collaboration and partnerships between businesses, governments, and civil society. This will allow stakeholders to exchange ideas, introduce innovative business models, and re-imagine a more circular economy. By working as a network, the group can address systemic challenges such as siloed value chains and challenging regulatory environments that commonly stand in the way of circular transition. However, partnerships are most impactful when they move beyond rhetoric to shared accountability, standardized data systems, and measurable outcomes. These are particularly required for the standardization and harmonization of data sharing and establishment of digital innovation hubs, which are important prerequisites for the adoption of circular practices and technologies on a large scale (Awad et al., 2025). Finally, these coalitions could innovate in developing policy frameworks and financial instruments beneficial to circular business models, which will facilitate investment in sustainable infrastructure and closed-loop systems.

## 9.3 Social and Cultural Shifts

The technological and economic conditions for a successful circular economy, in order to become widespread, ubiquitous solutions, must also be complemented by deep-rooted social and cultural changes. If CE is to flourish in the future, we will need societal norms and behaviors consistent with it. Consumerist cultures and use-and-throw attitudes are impediments to circular adoptions. According to Padilla-Rivera et al., a cultural shift towards the values of reuse, mending, and resource conservation is likely to occur (Padilla-Rivera et al., 2024). Without such shifts, even the most advanced technologies or policies risk underperformance, as consumer behavior ultimately shapes demand and market signals. Education and public awareness will be crucial in driving the adoption of circular behaviors. Over time, the business case will leverage social pressures from consumers demanding sustainability, forcing businesses actually to become circular in their activities. Thus, eco-design and related activities are encouraged (positive reinforcement) as they contribute to product-as-a-service models and discourage planned obsolescence through additional legislative measures, such as extended producer responsibility schemes (Clube & Hazemba, 2024). In addition, government action will be key to delivering these transitions by creating the right policies. Regulatory ecosystems may shift the tide towards circular solutions, encouraging businesses to produce durable, repairable, and recyclable items.

## 9.4 Integration with Global Sustainability Goals

The Sustainable Development Goals (SDGs) are a perfect match for the circular economy, particularly in relation to Goals 12 — Responsible Consumption and Production, 13 — Climate Action, and 15 — Life on Land. The circular economy provides a direction to promote resource efficiency, reduce waste and protect the natural ecosystems, towards possible pathways that can lead to a more sustainable and fairer world. In addition to these direct impacts, it further contributes to the achievement of SDG 7 (material efficiency reduces energy demand), and SDG 9 (innovative industrial processes fostering more sustainable and resilient infrastructure. Such comprehensive alignment makes the circular economy basic enabler of sustainable development transforming it from ecological benefits to social and economic aspects (Kristoffersen et al., 2020; Karim et al., 2025b). Furthermore, the transition from linear to circular models is a requirement in preventing resource depletion and environmental damage, offering a solid foundation for worldwide sustainability (Chirumalla et al., 2024; Ostermann et al., 2021). This framework focuses on regenerative practices—going beyond mere efficiency to restore ecological balance and create new economic opportunities (Jones & Comfort, 2021; Morris et al., 2021). The circular economy is an approach that emphasizes the longevity of materials and products within economic systems, resulting in reduced waste and less reliance on virgin resources.

# 10. CONCLUSION

The chapter offers discussion centering on the transition from the existing linear economic model of Circular Economy (CE) to a regenerative, sustainable model. It charts the origin of the concept in seminal publications such as *The Limits to Growth*, and goes on to describe how CE has transitioned into an overall strategic advocacy approach directed by bodies like the Ellen MacArthur Foundation. Designing out waste, keeping products in use, and regenerating natural systems provide the principles for minimizing waste and maximizing value derived from resources according to the CE model. Instead of relying on vague notions such as "do no wrong philosophies," this chapter underscores the centrality of life extension, performance-based models, recycling, and closed-loop systems as integral to realizing environmental, economic, and social benefits through CE implementation. This historical overview also highlights important events in the emergence of CE theory and practice, such as industrial ecology and cradle-to-cradle design. Since it is believed that CE has the potential to solve some of the grand challenges we are currently facing (e.g. resource scarcity and climate change among others), efficient

CE is key in sustaining our future. But there are a number of impediments to scale, for technologies, as well as market perversities and regulatory gaps.

Yet the chapter also outlines ways to address these challenges through a focus on innovation, cooperation, and policy. The transition will also be accelerated with the use of digital technologies, such as AI and blockchain, as well as collaborative partnerships. However, case studies of companies such as Patagonia, Interface, and Philips have demonstrated that the pillars of CE can be seamlessly integrated into business models, enabling both the conservation of natural resources and economic growth. In synthesizing the theoretical insights, this chapter shows how diverse frameworks—industrial ecology, cradle-to-cradle, performance economy, and regenerative design—converge into a multidimensional paradigm that redefines value creation within planetary boundaries. The key contribution of this work lies in integrating historical foundations with contemporary challenges, offering a holistic understanding of CE as both a practical strategy and a socio-ecological transformation. Finally, future research should investigate unresolved tensions such as rebound effects, the alignment of CE with post-growth perspectives, and the socio-political dimensions of adoption, in order to advance a more equitable and effective transition.

Lastly, the Circular Economy presents a paradigm shift that allows for economic growth to be decoupled from environmental impact—a more sustainable and resilient global economy. The continued development of CE principles, rooted in theoretical synthesis and informed by empirical case studies, combined with concerted EU-wide and global action supported by the right policy mix, can transform today's challenges into pathways toward long-term sustainability.

## REFERENCES


Abbate, S., Centobelli, P., Cerchione, R., Nadeem, S. P., & Riccio, E. (2023). Sustainability trends and gaps in the textile, apparel and fashion industries [Review of *Sustainability trends and gaps in the textile, apparel and fashion industries*]. *Environment Development and Sustainability, 26*(2), 2837. Springer Science+Business Media. https://doi.org/DOI: 10.1007/s10668-022-02887-2

Adam, S., Bucker, C., Desguin, S., Vaage, N., & Saebi, T. (2017). Taking Part in the Circular Economy: Four Ways to Designing Circular Business Models. SSRN *Electronic Journal*. https://doi.org/DOI: 10.2139/ssrn.2908107

Allwood, J. M., Ashby, M. F., Gutowski, T. G., & Worrell, E. (2011). Material efficiency: A white paper. *Resources, Conservation and Recycling, 55*(3), 362–381. DOI: 10.1016/J.RESCONREC.2010.11.002

Almelhem, M., Sule, E. T., & Buics, L. (2025). Comparison of sustainability and circularity indicators: downstream vs. upstream supply chain strategies. SSRN *Electronic Journal*. https://doi.org/DOI: 10.2139/ssrn.5270541

Ashford, N. A., & Hall, R. P. (2018). *Technology, Globalization, and Sustainable Development*. Routledge., DOI: 10.4324/9780429468056

Awad, I. M. A., Nuseibeh, H. Z., & Amro, A. A. (2025). Competitiveness in the Era of Circular Economy and Digital Innovations: An Integrative Literature Review. *Sustainability*, *17*(10), 4599. DOI: 10.3390/su17104599

Benyus, J. M. (1998). *Biomimicry: Innovation Inspired by Nature*. Harper.

Bocken, N. M. P., de Pauw, I., Bakker, C., & van der Grinten, B. (2016). Product design and business model strategies for a circular economy. *Journal of Industrial and Production Engineering*, *33*(5), 308–320. DOI: 10.1080/21681015.2016.1172124

Boulding, K. E. (2013). THE ECONOMICS OF THE COMING SPACESHIP EARTH. *Environmental Quality in a Growing Economy: Essays from the Sixth RFF Forum: Volume 3, 3*, 3–14. https://doi.org/DOI: 10.4324/9781315064147-2/ECONOMICS-COMING-SPACESHIP-EARTH-KENNETH-BOULDING

Chirumalla, K., Balestrucci, F., Sannö, A., & Oghazi, P. (2024). The transition from a linear to a circular economy through a multi-level readiness framework: An explorative study in the heavy-duty vehicle manufacturing industry. *Journal of Innovation & Knowledge*, *9*(4), 100539. DOI: 10.1016/j.jik.2024.100539

Clube, R. K. M., & Hazemba, M. (2024). From waste to resource: Demystifying the policy challenges and identifying opportunities for a circular economy in Zambia. *Frontiers in Sustainability, 5*. Advance online publication. DOI: 10.3389/frsus.2024.1300904

Cohen-Shacham, E., Andrade, A., Dalton, J., Dudley, N., Jones, M., Kumar, C., Maginnis, S., Maynard, S., Nelson, C. R., Renaud, F. G., Welling, R., & Walters, G. (2019). Core principles for successfully implementing and upscaling Nature-based Solutions. *Environmental Science & Policy, 98*, 20–29. DOI: 10.1016/J.ENVSCI.2019.04.014

Corvellec, H., Stowell, A. F., & Johansson, N. (2022). Critiques of the circular economy. *Journal of Industrial Ecology, 26*(2), 421–432. DOI: 10.1111/JIEC.13187

Costanza, R., D'Arge, R., De Groot, R., Farber, S., Grasso, M., Hannon, B., Limburg, K., Naeem, S., O'Neill, R. V., Paruelo, J., Raskin, R. G., Sutton, P., & Van Den Belt, M. (1997). The value of the world's ecosystem services and natural capital. *Nature, 387*(6630), 253–260. DOI: 10.1038/387253A0/MEDIAOBJECTS/41586_1997_BF387253A0_MOESM1_ESM.PDF

Ellen MacArthur Foundation. (2013). *Towards the circular economy Vol. 2: opportunities for the consumer goods sector*. Opportunities for the Consumer Goods Sector. https://www.ellenmacarthurfoundation.org/towards-the-circular-economy-vol-2-opportunities-for-the-consumer-goods

EMF. (2015). *Towards a Circular Economy: Business Rationale for an Accelerated Transition*.

Frishammar, J., & Parida, V. (2018). Circular Business Model Transformation: A Roadmap for Incumbent Firms. *California Management Review, 61*(2), 5. DOI: 10.1177/0008125618811926

Frosch, R. A., & Gallopoulos, N. E. (1989). Strategies for Manufacturing. *Scientific American, 261*(3), 144–152. DOI: 10.1038/scientificamerican0989-144

Geissdoerfer, M., Savaget, P., Bocken, N. M. P., & Hultink, E. J. (2017). The Circular Economy – A new sustainability paradigm? *Journal of Cleaner Production, 143*, 757–768. DOI: 10.1016/J.JCLEPRO.2016.12.048

Geng, Y., Fu, J., Sarkis, J., & Xue, B. (2012). Towards a national circular economy indicator system in China: An evaluation and critical analysis. *Journal of Cleaner Production, 23*(1), 216–224. DOI: 10.1016/J.JCLEPRO.2011.07.005

Georgescu-Roegen, Nicholas. (1971). *The entropy law and the economic process*. 457.

Graedel, T. E., & Lifset, R. J. (2016). Industrial Ecology's First Decade. *Taking Stock of Industrial Ecology*, 3–20. https://doi.org/DOI: 10.1007/978-3-319-20571-7_1

Hawken, P., & Lovins, A. B. (2013). Natural capitalism: the next industrial revolution. *Technology Review;(USA), 276*(5310), 190–212. http://books.google.com/books?hl=en&lr=&id=MXTdAAAAQBAJ&oi=fnd&pg=PP1&dq=NATURAL+CAPITALISM:+CREATING+THE+NEXT+INDUSTRIAL+REVOLUTION&ots=GJzMVOsSpt&sig=du72NgZTuGDjZTlkcMqtkqTyc38%5Cnhttp://books.google.com/books?hl=en&lr=&id=MXTdAAAAQBAJ&oi=fnd&pg=PP1&dq

Hoosain, M. S., Paul, B. S., Doorsamy, W., & Ramakrishna, S. (2023). Comparing South Africa's Sustainability and Circular Economic Roadmap to The Rest of the World. *Materials Circular Economy, 5*(1), 2. DOI: 10.1007/s42824-023-00073-x

Hoosain, M. S., Paul, B. S., & Ramakrishna, S. (2020). The Impact of 4IR Digital Technologies and Circular Thinking on the United Nations Sustainable Development Goals. *Sustainability, 12*(23), 10143. DOI: 10.3390/su122310143

Hossain, Md. U., Ng, S. T., Antwi-Afari, P., & Amor, B. (2020). Circular economy and the construction industry: Existing trends, challenges and prospective framework for sustainable construction. *Renewable & Sustainable Energy Reviews, 130*, 109948. DOI: 10.1016/j.rser.2020.109948

Hulea, M., Miron, R., & Mureşan, V. (2024). Digital Product Passport Implementation Based on Multi-Blockchain Approach with Decentralized Identifier Provider. *Applied Sciences (Basel, Switzerland), 14*(11), 4874. DOI: 10.3390/app14114874

Jones, P., & Comfort, D. (2021). The Circular Economy: An Exploratory Case Study from the Paper and Retail Industries. *Athens Journal of Business & Economics, 7*(4), 379–394. DOI: 10.30958/ajbe.7-4-5

Karim, R., Bhowmik, R. C., Faruq, M. O., & Roshid, Md. M., Islam, S., & Ramachandaran, S. D. (2025b). Harnessing growth for ecological sustainability through exploring the drivers of load capacity factor in G7 nations. *Discover Sustainability, 6*(1), 821. DOI: 10.1007/s43621-025-01757-x

Karim, R., Roshid, M., & Waaje, A. (2024). Circular economy and supply chain sustainability. In Mızrak, F. (Ed.), *Strategic Innovations for Dynamic Supply Chains* (pp. 1–30). IGI Global Publishing., DOI: 10.4018/979-8-3693-3575-8.ch001

Karim, R., Waaje, A., & Roshid, Md. M. (2025a). Digital Circular Economy for Attaining Sustainable Development Goals. In Durmuş Çağrı Yıldırım, Seda Yıldırım, & Vinay Kandpal (Eds.), *Effects of Digitalization and Circular Economy on Sustainable Policy and Climate Change Prevention* (pp. 211–240). IGI Global. https://doi.org/DOI: 10.4018/979-8-3693-9909-5.ch008

Kirchherr, J., Reike, D., & Hekkert, M. (2017). Conceptualizing the circular economy: An analysis of 114 definitions. *Resources, Conservation and Recycling*, *127*, 221–232. DOI: 10.1016/J.RESCONREC.2017.09.005

Kondapaka, K. K. (2021). *AI-Driven Inventory Optimization in Retail Supply Chains: Advanced Models, Techniques, and Real-World Applications.*

Kristoffersen, E., Blomsma, F., Mikalef, P., & Li, J. (2020). The smart circular economy: A digital-enabled circular strategies framework for manufacturing companies. *Journal ofBusiness Research*, *120*, 241. DOI: 10.1016/j.jbusres.2020.07.044

Kyriakopoulos, G., Kapsalis, V., Aravossis, K., Zamparas, M., & Mitsikas, A. (2019). Evaluating Circular Economy under a Multi-Parametric Approach: A Technological Review. *Sustainability*, *11*(21), 6139. DOI: 10.3390/su11216139

Lacasa, E., Santolaya, J. L., & Biedermann, A. (2016). Obtaining sustainable production from the product design analysis. *Journal of Cleaner Production*, *139*, 706–716. DOI: 10.1016/J.JCLEPRO.2016.08.078

Lyle, J. Tillman., & Lyle. (1996). *Regenerative Design for Sustainable Development.* https://www.wiley.com/en-us/Regenerative+Design+for+Sustainable+Development-p-9780471178439

Marek, M., & Krejza, Z. (2024). Sustainable building: Circular economy as a key factor for cost reduction. *E3S Web of Conferences*, *550*, 1009. https://doi.org/DOI: 10.1051/e3sconf/202455001009

McDonough. William., & Braungart, Michael. (2010). *Cradle to Cradle: Remaking the Way We Make Things.* https://books.google.com/books/about/Cradle_to_Cradle.html?id=KFX5RprPGQ0C

McDonough, W., & Braungart, M. (2002). Remarking the Way We Make Things: Cradle to Cradle. *North Point Press, New York*, 45. https://books.google.com/books/about/Cradle_to_Cradle.html?id=KFX5RprPGQ0C

Meadows, D. (2009). *Thinking in systems : a primer.* https://books.google.com/books/about/Thinking_in_Systems.html?id=JSgOSP1qklUC

Meadows, D. H., Meadows, D. L., Randers, J., & Behrens, W. W. (1972). *The Limits to Growth: A report for the Club of Rome's Project on the Predicament of Mankind*. Universe Books., DOI: 10.1349/ddlp.1

Morris, J., Georgiou, I. Y., Guenther, E., & Caucci, S. (2021). Barriers in Implementation of Wastewater Reuse: Identifying the Way Forward in Closing the Loop. *Circular Economy and Sustainability*, *1*(1), 413. DOI: 10.1007/s43615-021-00018-z

Murray, A., Skene, K., & Haynes, K. (2017). The Circular Economy: An Interdisciplinary Exploration of the Concept and Application in a Global Context. *Journal of Business Ethics*, *140*(3), 369–380. DOI: 10.1007/S10551-015-2693-2

Onungwe, I., Hunt, D. V. L., & Jefferson, I. (2024). Transitioning towards circular economy through municipal solid waste analysis and characterisation using Sowa-CLINK software. *Frontiers in Sustainability*, *5*. Advance online publication. DOI: 10.3389/frsus.2024.1321329

Ostermann, C. M., Nascimento, L. da S., Steinbruch, F. K., & Callegaro-de-Menezes, D. (2021). Drivers to implement the circular economy in born-sustainable business models: A case study in the fashion industry. *Revista de Gestão*, *28*(3), 223. DOI: 10.1108/rege-03-2020-0017

Padilla-Rivera, A., Magdalena, M. B., Merveille, N., & Güereca, L. P. (2024). Barriers, Challenges, and Opportunities in the Adoption of the Circular Economy in Mexico: An Analysis through Social Perception. *Recycling*, *9*(5), 71. DOI: 10.3390/recycling9050071

Pauli, G. A. (2010). *The blue economy : 10 years, 100 innovations, 100 million jobs : Report to the Club of Rome*. 308. https://books.google.com/books/about/The_Blue_Economy.html?id=aJ3HZD1H7ZsC

Polo, S., Rubio, E. M., Ayllón, J., & de Agustina, B. (2025). Emerging Advances in Sustainable Manufacturing. *Processes (Basel, Switzerland)*, *13*(5), 1549. DOI: 10.3390/pr13051549

Putnam, R. D. (1993). What makes democracy work? *National Civic Review*, *82*(2), 101–107. DOI: 10.1002/NCR.4100820204

Raj, K., & Aithal, P. S. (2020). A New Concept of Circular Model of Management for Achieving Sustainable Success and Growth. *International Journal of Management Technology and Social Sciences*, *1*. Advance online publication. DOI: 10.47992/ijmts.2581.6012.0079

Ranta, V., Aarikka-Stenroos, L., & Mäkinen, S. J. (2018). Creating value in the circular economy: A structured multiple-case analysis of business models. *Journal of Cleaner Production*, *201*, 988–1000. DOI: 10.1016/J.JCLEPRO.2018.08.072

Rattalino, F. (2017). Circular advantage anyone? Sustainability-driven innovation and circularity at Patagonia, Inc. *Thunderbird International Business Review*, *60*(5), 747. DOI: 10.1002/tie.21917

Rawat, S. (2021). WTO GPA and Sustainable Procurement as Tools for Transitioning to a Circular Economy. *arXiv (Cornell University)*. https://doi.org//arxiv.2104.04744DOI: 10.48550

Raworth, K. (2017). Why it's time for Doughnut Economics. *IPPR Progressive Review*, *24*(3), 216–222. DOI: 10.1111/NEWE.12058

Rockström, J., Steffen, W., Noone, K., Persson, Å., Chapin, F. S., Lambin, E. F., Lenton, T. M., Scheffer, M., Folke, C., Schellnhuber, H. J., Nykvist, B., De Wit, C. A., Hughes, T., Van Der Leeuw, S., Rodhe, H., Sörlin, S., Snyder, P. K., Costanza, R., Svedin, U., & Foley, J. A. (2009). A safe operating space for humanity. *Nature*, *461*(7263), 472–475. DOI: 10.1038/461472A;KWRD=SCIENCE PMID: 19779433

Rotondo, B., Bakker, C., Balkenende, R., & Arquilla, V. (2025). Integrating Circular Economy Principles in the New Product Development Process: A Systematic Literature Review and Classification of Available Circular Design Tools. *Sustainability*, *17*(9), 4155. DOI: 10.3390/su17094155

Stahel, W. R. (2006). The Performance Economy: Business Models for the Functional Service Economy. In *Handbook of Performability Engineering* (pp. 127–138). Springer London. https://doi.org/DOI: 10.1007/978-1-84800-131-2_10

Stahel, W. R. (2010). *The performance economy* (2nd ed.). The Performance Economy., DOI: 10.1057/9780230274907/COVER

Temerbulatova, Zh. S., Zhidebekkyzy, A., & Grabowska, M. (2021). Assessment of the Effectiveness of the European Union Countries Transition to a Circular Economy: Data Envelopment Analysis. *Economics: The Strategy and Practice*, *16*(3), 142–151. DOI: 10.51176/1997-9967-2021-3-142-151

Tsalis, T., Stefanakis, A. I., & Nikolaou, I. (2022). A Framework to Evaluate the Social Life Cycle Impact of Products under the Circular Economy Thinking. *Sustainability*, *14*(4), 2196. DOI: 10.3390/su14042196

Tukker, A. (2004). Eight types of product-service system: Eight ways to sustainability? Experiences from suspronet. *Business Strategy and the Environment*, *13*(4), 246–260. DOI: 10.1002/BSE.414;REQUESTEDJOURNAL:JOURNAL:10990836;W-GROUP:STRING:PUBLICATION

Tukker, A., & Tischner, U. (2006). Product-services as a research field: Past, present and future. Reflections from a decade of research. *Journal of Cleaner Production*, *14*(17), 1552–1556. DOI: 10.1016/J.JCLEPRO.2006.01.022

# Chapter 2
# IoT-Enabled Circular Supply Chains:
## Advancing Sustainability in the United States Logistic and Distribution

**Abdul Waaje**
https://orcid.org/0009-0004-7207-0750
*Michigan Technological University, USA*

**Sahilali Saiyed**
*Chaoyang University of Technology, Taiwan*

**Md. Mustaqim Roshid**
https://orcid.org/0000-0001-8517-6336
*University of Rajshahi, Bangladesh*

**Mohammad Rafid Irtisham**
*Aston University, Birmingham, UK*

**Md Anamul Haque**
*Wonkwang University, Iksan, South Korea*

## ABSTRACT

*A shift in logistics and distribution systems has been underway to accomplish the sustainability agenda, where the merging of circular economic principles with IoT is the key to the transformation. In this chapter, the IoT-enabled Circular Supply Chains (CSCs) concept discussion goes further as it can enhance overall operations efficacy and transparency and drive towards a carbon-zero goal in the US. It employs*

DOI: 10.4018/979-8-3373-2454-8.ch002

*the thematic content analysis of literature, industry publications, and practical case studies (e.g., including Walmart, Amazon, and HP) in order to provide an overview of Digital Circular Logistics (DCL) by utilizing a qualitative descriptive approach. In order to provide a systematic method for long-term change, a theoretical framework is established to integrate Internet of Things (IoT) features like blockchain traceability, real-time monitoring, and predictive analytics with CE's 9R tactics. Despite a high level of technological preparedness, the results show that growth is nevertheless hindered by systemic impediments such as limitations of policy frameworks, organizational stagnation, and a lack of cross-sectoral cooperation. This chapter provides academics, practitioners, and policymakers with theoretical understanding as well as practical advice by expanding socio-technical and institutional theories. It markets itself as a road map for wealthy nations to implement intelligent, circular logistics systems speedily.*

## 1 INTRODUCTION

The world is moving towards sustainability, where supply chain management is also transforming to gain sustainability agendas that are because of the manufacturer mindset in gaining resource efficiency in order to address the customers' requirements to compete with the competitors in a way of saving the environment for future generation from being destroyed (Sarkis, 2020; Gružauskas et al., 2018). In attaining positive dominance over nature and socio-economic changes, the integration of the Circular Economy, especially the principles of CE into existing supply chains, can accelerate the path of attaining sustainable supply chain management, which can easily be called a circular supply chain. The EU initiated the promotion of the CE concept, which later gained widespread popularity worldwide, particularly among manufacturers seeking to integrate sustainability into their supply chain operations and governments facing globalization challenges and unpredictable shocks (Liu et al., 2023). As the world aims to achieve zero environmental pollution by 2050, countries must overhaul power generation, transportation, and consumption patterns (Power & SET, 2022). Due to excessive energy consumption, the USA's environmental impacts are drastically increasing, which are the reasons for struggles in meeting predicted national energy efficiency, reduction of carbon emissions, usage of renewable energy sources, and attaining the pride of a decarbonized economy (Hassan et al., 2023). The United States has a 122% higher ecological footprint than the natural shortage, where 71% of the US EFP presented a higher-risk economy (Global Footprint Network [GFN], 2019; Wackernagel & Beyers, 2019). Similarly, the increasing use of steel in the USA is directly impacting the environment due to the lack of recycling of those scraps, particularly in the growing construction and automotive industries, where CE can be beneficial. According to the United States Environmental Protection Agency, 30-40% of the food supply is never eaten nationally due to the fact that there is observation of usages of fossil fuel, industrial expansion, food waste and the decline of trees at a rate of about 175,000 acres per year, growing the awareness of stakeholders, policy makers and government agencies to promote a circular economy in attaining a sustainable supply chain around the USA, where companies like Walmart have already taken the initiative as a goal to divert 90% of operational waste by 2025 in the USA, Canada and Mexico.

Till now the inventor of the CE concept has still not been identified, but American Professor John Lyle, along with his student William McDonough, are the key contributors to this concept (Winans et al., 2017). Further, Rachel Carson's thesis on "Limits to Growth" inspired the concept development, but the actual conceptual framework for CE has been developed by Pearce and Turner (Carson, 1962; Pearce & Turner, 1990). As the CE concept developed in light of the diversification of culture and social and political systems, all the powerful nations started integrating and implementing it in their policies on their own after it became popular (Rajeev et al., 2017). Due to this fact, the USA and Europe started commanding and establishing controlled policy measures in the 1970s. But China was left behind in the race by adopting the concept in the mid-1990s, but being a pioneer in the Asian region, along with Japan, in introducing CE policies at their national level. In the early 2000s, European countries, most notably Denmark, Germany, Portugal, the Netherlands, and the UK, were utilizing the concept for different industries, focusing on sustainably managing the waste. With the three basic dimensions of sustainable supply chain management, namely economics, environment, and society, academics and industry practitioners started exploring SSCM, namely green supply chain management, in the early 2000s. It is unclear who provided the concept, but it is related to the United Nations Sustainability Agenda. In alignment with the development of the Industrial Revolution from the 1st to the 5th, various technologies were developed to support operations across the manufacturing and service industries. IoT is a technology that is also part of the fourth industrial revolution. The widespread adoption of industrial technologies is making planners more creative and enhancing sustainability in existing conventional operations. IoT-based circular supply chains are an emerging concept that enhances supply chain operations' efficiency through accurate delivery, cost reduction, and the conversion of waste into profitable, sustainable products, ultimately contributing to a green future for the world's dwellers.

## 1.1 Background of the Study

Due to the adaptation of sustainable practices to address carbon emissions, increased waste, and inadequate supply operations, the US logistics and distribution sectors are facing enormous pressure, and the potential utilization of digital technologies like IoT has become a compelling solution. But the implementation of the IoT-based circular supply chain has met with significant challenges. US company DHL aims to achieve net-zero emissions in its logistics and manufacturing operations, aligning with the Sustainable Development Goals. To support this goal, digital technologies, particularly the integration of IoT, will play a crucial role in establishing sustainability. Similarly, FedEx aims to achieve optimized, sustainable, and digital technology-based interlinked logistics and supply operations, ensuring enhanced customer experiences and ultimately achieving carbon-neutral operations by 2040. In alignment with the agenda of the UN's sustainable development goal, the US, as a first-world country, has the commitment to achieving carbon neutrality, where government agencies and companies are keeping aligned with the larger commitment of environmentally friendly supply operations.

Being one of the global leaders, Walmart had to face challenges during the real-time IoT-enabled waste tracking integration due to fragmented supplier networks and outdated existing infrastructure. As the world's largest e-commerce product service provider, Amazon's attempts at reducing last-mile emissions using IoT-enabled route optimization encountered confrontation from local suppliers and distributors lacking technical expertise. In line with the illustrated examples, the academic and industry-driven potential of circular supply chains is rooted in achieving high-frequency service assurance, increasing consumer satisfaction, reducing delays and discrepancies, and making the world environmentally friendly for future generations. Overall, the SDGs agenda's practical implementation, along with the development of a unified supply operations infrastructure that skips the linear economic model, enables the IoT-enabled supply chain to commit to further sustainability achievements in the US logistics and supply chain industry, encompassing conceptual, theoretical, and practical aspects.

## 1.2 Statement of the Problem

Central to the circular economy is the aim of having a zero-waste, transparent, and adaptive nature in supply chain operations. Despite having the technological development of industrial revolutions based on technologies, empirical evidence reveals a contradiction in terms of U.S. logistics, as massive operations are running under the linear economic model, where minimal integration of IoT for sustainability can be found. In connection with digital transformation with circular supply chains, there lie theoretical gaps. More specifically, a complete package of integration of models that have optimized and operationally profound unified connections with IoT, CE, and supply chain strategies is a theoretical gap. Similarly, less attention has been given to the retail, e-commerce, and distribution sector than the manufacturing sector, which reveals the knowledge gaps in a way that employees do not have a proper understanding of the digital circular logistics operations concept, even if the process is not familiar with technological advancements in supply chain operations. The complexity of US regulatory policies, which are becoming more environmentally friendly, but the existing implementation model here differs a bit from EU models. This chapter addresses the gaps by providing theoretical, conceptual, and practical illustrations of IoT-based circular supply chain applications, along with recommendations for policy development beneficial to both academic and industry practitioners.

## 1.3 Chapter Objectives and Research Scope

The purpose of this chapter is to explain how IoT technology enables circular supply chains to develop sustainability in U.S. delivery and logistics operations. This research examines IoT integration processes that utilize innovative technologies, including predictive analytics and blockchain traceability, as well as automated reverse logistics, to enhance resource efficiency, waste reduction, and supply chain transparency. This chapter explores specific industrial applications of IoT technology that optimize fleets while defining green warehouse principles and sustainable purchasing standards. The study uses field-based research to study practical examples of effective implementation methods while examining barriers to adoption and recommending strategic measures. Both technological breakthroughs and business strategies form the research framework to improve the speed of implementing circular and sustainable logistics systems.

## 1.4 Alignment with Sustainable Development Goals (SDGs 7, 12, 13, 8, 5, 9, 17)

All discussions presented in this chapter support one or multiple United Nations Sustainable Development Goals. The implementation of renewable energy together with IoT-based energy management systems for logistics purposes supports the achievement of SDG 7 (Affordable and Clean Energy). SDG 12 (Responsible Consumption and Production) receives attention through smart contract utilization and reverse logistics systems designed to reduce waste. The emission reduction effort, together with resource optimizers, works toward achieving SDG 13 (Climate Action). The support of ethical sourcing through inclusive supply chains alongside fair labor standards helps achieve SDG 5 (Gender Equality) as well as SDG 8 (Decent Work and Economic Growth). The chapter promotes innovation together with resilience to achieve SDG 9 (Industry, Innovation, and Infrastructure). SDG 17 (Partnerships for the Goals) is manifested through the program's commitment to fostering collaborative policy integration necessary for the development of a circular supply chain ecosystem.

## 2 LITERATURE REVIEW

Digital technologies like IoT, AI, and the circular economy have revolutionized ground-breaking transformation in the environment, health, and safety through supply chain operations and logistics practices. The nexus between circular economy and Industry 4.0 has been articulated by Zhou (2025) while highlighting machine learning frameworks for solar energy and buildings to enhance waste-to-energy systems and reduce lifecycle emissions (Waaje et al., 2025). The concept of Smart Circular Systems (SCS) was introduced by Alcayaga and Hansen (2024), which illustrates the outline of microlevel IoT-enabled product-as-service approaches spanning product lifecycle stages in optimizing CE's 3R principles. At the same time, an argument developed by Yazan et al. (2022) reveals that tools for decision-making have the potential to enable the transformation of CE in logistics with an interlinked multi-actor decision structure model. Werner-Lewandowska et al. (2025) analyzed CSC models from the supply chain point of view for the construction industry by focusing on the transparency and technology-based tracking for recovering material supplies. On the contrary, Bressanelli et al. (2020) examined how IoT-based PSS business models can foster CE practices in manufacturing operations by enhancing asset utilization. However, another study has added a dimension, suggesting that closed-loop models require a transformation in both the delivery system and value proposition, and are based on network-based innovation (Tura et al. 2019). The barriers regarding policy development and stakeholder engagement, along with digital circular economy elements, have been discussed by Masi et al. (2021), while Kirchherr et al. (2017) tried to figure out the CE definitions by comparing linear and closed-loop economies.

### 2.1 Literature Gaps

There are different literature gaps in this field of study. Firstly, it can hardly be found in the existing literature where a unified framework for IoT-enabled circular supply chains has been established. Secondly, a scarcity of existential evidence for specific sectors exists, especially in the retail and e-commerce industries. Finally, the limitation is of basic, understandable, simplified knowledge of reverse logistics and smart warehousing among the people. This chapter aims to bridge these gaps by developing a conceptual framework that demonstrates how IoT technologies (e.g., sensors, blockchain, AI) enhance circular logistics in the U.S. context.

# 3 METHODOLOGY

The chapter has employed a qualitative descriptive research design where the straightforward descriptives are organized logically in a way that the authors are known with the subjective matters (Lambert & Lambert, 2012). In order to demonstrate the whole chapter to the academic and industry practitioners, the research is grounded in secondary data analysis that aims to conceptualize and figure out the interlinkage of the circular supply chain with IoT technology and its practices in the US logistics and distribution context. This chapter draws a logical orientation from a broader range of academic journal papers, industry reports, and different nations' policy-based documentations, which reveals a structural thematic analysis. Core themes for the chapter come from IoT technologies (e.g., RFID, sensors, cloud computing, and AI), the circular economy 9R model, sustainable logistics and distribution operations, reverse logistics and smart warehousing, and case studies from major corporations such as Walmart, HP, and Amazon. Keyword coding and pattern recognition have been employed to figure out the sustainability outcomes and IoT adaptation with implementation stages, where barriers were identified. This synthesis informed the development of the chapter's conceptual framework and practical recommendations. The approach is further supported by theoretical underpinnings, particularly Socio-Technical Systems Theory and the Resource-Based View, providing a robust academic foundation for addressing existing gaps in CSC implementation research.

## 3.1 Theoretical Framework

To make the chapter more accessible to readers, the author has employed various theoretical frameworks, ensuring it is readable and understandable to those with little knowledge of the subject matter. Firstly, the widely accepted and dominant perspective in strategic management theory is called the resource-based view, which explains how a business organization can gain a competitive advantage (Bertram, 2016) in their supply chain operations via the highest utilization of resource efficacy with advanced technological capabilities. In making the connection between the circular economy, supply chain, and IoT, the social-technical systems theory has been followed, as it offers both the interlink of technologies and organizational readiness (Walker et al., 2008). Similarly, stakeholder theoretical aspects are present, while the discussion prioritizes the potential of inclusive CSCM practice, with institutional theory taking precedence in portraying the regulatory aspects of organizational behavior. However, the social-technical systems theory serves as the foundational theoretical framework for this chapter, as it effectively describes the combined subsystem integrations of mankind, organization, and technology through practical IoT-based integration in CE within the supply chain for CSCM. Overall, it helps explain why some firms succeed in integrating IoT while others fail—despite similar technological access.

## 3.2 Conceptual Framework

The proposed conceptual framework for the chapter has integrated IoT technology in alignment with CE's 9R principles. The framework has four strategic layers, where the first layer can be defined as enablers, which are digital technologies such as IoT, machine learning, blockchain, RFID, and smart sensors. The second layer is logistics and operational components in optimising the supplies, handling, reordering, fleet management, and last-mile delivery that ensure circular actions, which are denoted as the third layer. Moreover, attaining the SDGs for reducing environmental pollution for the future world is the fourth layer of the conceptual framework. This model acts as a roadmap for practitioners, researchers, and policymakers aiming to transform logistics networks into resilient, circular systems.

*Figure 1. Conceptual Framework for IoT-Enabled Circular Supply Chain Management (CSCM)*

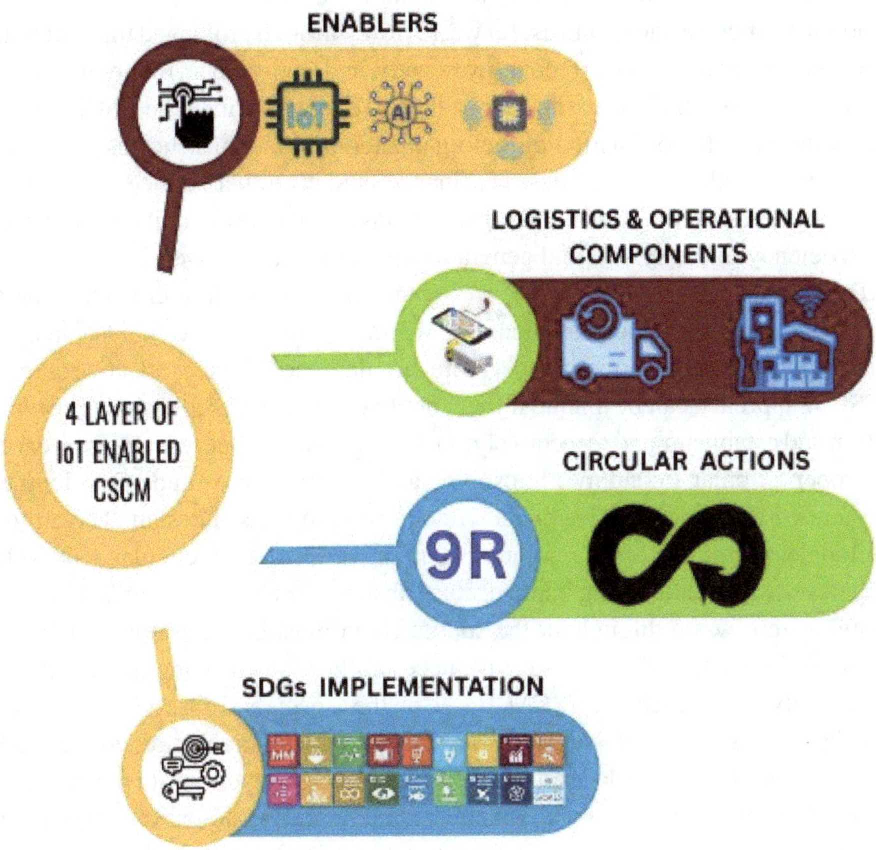


## 4 CIRCULAR SUPPLY CHAIN MANAGEMENT

Circular economy is an effective way of integrating sustainability into the existing supply chain, enhancing the flow of materials, and gaining the total advantage of waste by reshaping it into effective goods or raw materials for business purposes. Mostly, scholars and practitioners defined circular supply chain management as the integration of CE into the supply chain in maintaining environmental policies and working towards the goal of sustainability. What do you think about the circular supply chain? Is that something having a circular loop, or only something an agenda of sustainability gained through the supply chain? A widely accepted definition given by Batista et al. (2019) is "CSCM synchronizes forward and reverse supply chains through the harmonization of a product or service's value creation aspects, by-products or by-services, and useful waste flows through prolonged life cycles that improve the three dimensions (economy, society, and environment) of organizational sustainability. Similarly, Geissdoerfer et al. (2018) have defined the CSCM as "the configuration and coordination of the organizational functions such as marketing, sales, R&D, production, logistics, IT, finance, and customer service—within and across business or service units and organizations to improve operative effectiveness and efficiency of the system and generate competitive advantages."

To plan and implement CSCM, an organization must adopt circular thinking from top to bottom. Employees should be willing to cultivate this mindset in every work step they take, contributing to a more circular macro-level work environment. A CSCM approach can be unique if there are two core aspects followed in the integration and establishment, respectively: 1) Aiming towards net zero waste by caring for proper Circular Economy Philosophy, and 2) Circular Attitude Development with a restorative-regenerative cyclic circular thinking-based design. The attitude development aims to create a unique regenerative-restorative circular model that mitigates waste by converting it into usable substances or reusable goods. However, achieving zero waste throughout the supply chain must be the primary objective. Similarly, the agenda of zero waste will successfully allow the integration of the CE in the supply chain to achieve CSCM. Based on the aspects and observing definitions from the scholars, "CSCM is simply a way to combine the Circular Economic Model (3Rs, 4Rs, or 9Rs) to reach sustainability goals that go beyond just environmental concerns. It aims for zero waste, whether it's physical waste or work that wasn't completed, while also helping manufacturers add value by using technologies from the Industrial Revolution throughout the supply chain, from sourcing products to delivering them to customers efficiently and effectively."

## 4.1 Role of IoT in Advancing Circular Supply Chains

More companies now recognize Sustainable Supply Chain Management (SSCM) joined with Circular Economy (CE) principles as an essential strategy that addresses environmental and social, and economic difficulties. The CE model emerged during the mid-2010s to replace traditional linear models with a system that promotes repair and recycling functions for reducing waste and resource conservation (Theeraworawit et al., 2022; Arumugam et al., 2024). SSCM operates in harmony with CE through its usage of environmental and social factors across product life cycles to protect resource worth while protecting ecosystems (Alfaridi et al., 2024; Baporikar, 2023).

The implementation of CE-based supply chains depends on partnership development and requires reverse supply chain integration as well as the use of IoT, blockchain, and data analytics with their digital counterparts to create resilient and efficient networks (Karim et al., 2024; Theeraworawit et al., 2022). Several obstacles block the implementation of circular economy strategies, including expensive startup expenses, together with organizational resistance to change and restricted partnership between departments (Sokhetye, 2024; Echefaj et al., 2024). The research confirms that SSCM brings positive results for financial performance and environmental management through CE capabilities, which function as vital intermediates (Alfaridi et al., 2024). Government policies combined with stakeholder pressures compel organizations to follow circular models as these models help them achieve sustainability goals (Echefaj et al., 2024). Supply chain sustainability, together with operational efficiency and future-readiness, depends on the integration of SSCM and CE according to Karim et al (2025) and Arumugam et al (2024).

## 5 CIRCULAR SUPPLY CHAINS: R STRATEGIES

A circular supply chain is needed not only for environmental sustainability but also for organizational sustainability, with profitability, client satisfaction, and operational efficacy more broadly in achieving sustainability agendas. Towards CSCM, an organization needs to have the motivation to attain the change management culture in adopting the shift from a traditional linear economy culture, the 'take-make-waste' economy, to a circular economy that erases shop floor waste, circular usages of resources, and enhances the value chain of product or service as well as materials through the whole lifecycle. Operational complexity will arise during the phase of establishing CSC. Still, a circular mindset of stakeholder groups and systematic management of reverse flow with appropriate information flows are prerequisites for this. Adaptation of R-strategies for CSC is crucial, which can be in different frameworks, respectively, the 3R model of CE, the 4R model of CE, and the 9R model throughout the lifecycle of products and services with industrial revolutionary-based technology like IoT, AI, or cybersecurity, or with a mix of technologies in reducing waste at a level of zero. In 2008, the European Union revealed the 4R model, but initially it was defined with 3R strategies. The 9R Framework has ten distinct strategies from R0 to R9 that were introduced in 2017 (Alivojvodic & Kokalj, 2024), which are highly beneficial for waste reduction within the novel strategic design and production operational strategies (Bühler et al., 2024; (Waaje et al., 2024). According to Stahel (2013), CE should have a key focus on various loops based on the materials or products' usage characteristics.

*Table 1. CSC R Strategies*

| Circular Loop Category | R Strategy | Definition |
|---|---|---|
| Short Loops: Smart Strategies for Product Use and Circular Manufacturing | R0 – Refuse | Eliminate the need for certain products or materials by offering alternative solutions, preventing waste generation at the source. |
| | R1 – Rethink | Redesign products and services to maximize efficiency, encourage shared use, and minimize resource consumption throughout the lifecycle. |
| | R2 – Reduce | Enhance production and consumption efficiency by minimizing the use of raw materials, energy, and other resources. |
| Medium Loops: Circular Strategies for Long-Term Product Use | R3 – Reuse | Extend product life by using items or components multiple times for the same purpose, including second-hand markets and standardized applications across industries. |
| | R4 – Repair | Restore defective or damaged products to working condition to prolong their usability and reduce the need for replacement. |
| | R5 – Refurbish | Upgrade and renovate used products to restore functionality and extend their service life, often through minor modifications and aesthetic improvements. |
| | R6 – Remanufacture | Rebuild products to their original specifications using a mix of reused, repaired, and new parts, ensuring quality and performance standards are maintained. |
| Long Loops: Beyond Recycling | R7 – Repurpose | Adapt discarded products or components for new applications that differ from their original intended use. |
| | R8 – Recycling | Process used materials to recover valuable resources, transforming them into new products and minimizing raw material extraction. |
| | R9 – Recovery | Extract residual value, such as energy or useful materials, from waste streams when other recovery options are no longer viable. |

## 5.1 IoT as a Transformative Technology in Sustainable Logistics

Sustainable logistics is transformed through the Internet of Things (IoT), which creates optimized operational performance, real-time tracking, and generates data-driven business decisions. Networked smart devices, together with sensors, enable IoT to continuously monitor operations, track assets, manage inventories, and predict equipment maintenance requirements for supply chain management (Kumar & Godihal, 2024; Oliveira, 2024).

The implementation of IoT with RFID and Industry 4.0 frameworks creates agile logistics operations that use fewer resources, so companies can minimize both operational costs and improve their processes (Choudhary et al., 2024). The Internet of Things serves as a crucial component in sustainable logistics, offering critical capabilities for green transportation, optimized waste management, and energy-efficient practices. Numerous studies conducted in Ghana and Indonesia confirm that IoT acts as a positive moderator for environmental improvements within various logistical settings (Akubia et al., 2024).

IoT generates sustainable effects through combined strategies for optimized routing that decrease emissions and predictive maintenance methods that minimize equipment breakdowns (Kumar & Godihal, 2024). Bibliometric reviews document the power of IoT to work with innovative technologies like AI and blockchain, which enhance security systems and tracking and spur innovation in global supply chains. Extensive IoT infrastructure development, especially in developing nations, will lead to amplified sustainable logistics capabilities that provide scalable possibilities for operational and environmental performance enhancement (Akubia et al., 2024; Choudhary et al., 2024).

## 5.2 Incorporation of Smart Technologies: AI, Blockchain, and Big Data in Circular Logistics

Smart technologies, including artificial intelligence (AI), blockchain, and big data, bring essential benefits toward circular logistics sustainability and operational efficiency improvements. Together, these technologies advance the move toward circular supply chains through better resource management while reducing waste and improving business decisions.

Beyond improving selling accuracy and resource management, AI helps businesses connect production schedules to customer needs more efficiently while minimizing superfluous stock. Yet, it remains vital for circular models that aim to reduce environmental impacts (Farazi, 2024). Blockchain technology creates a trustworthy supply chain environment through its ability to deliver traceable records, which securely track product source points and movement, and recycling steps. Blockchain technology delivers maximum value to food manufacturing organizations because it promotes complete accountability as well as waste reduction (Yazan et al., 2022; Carlos et al., 2024).

Big data analytics supports circular logistics through its capabilities to process complex data along with generating predictive insights for optimizing decision processes about material recovery and resource utilization, and reverse logistics operations (Carlos et al., 2024). The IoT system delivers real-time monitoring that enables both tracking of product use and enhances recycling effectiveness and loop maintenance (Carlos et al., 2024).

The technologies encounter obstacles to adoption because of expensive implementation costs combined with complicated technology and strict regulatory limitations (Farazi, 2024). Organisations that pursue competitive sustainability benefits vigorously through the combined use of digitalization and resource efficiency (Ahmad et al., 2025). The combination of these technologies strengthens the circular economy foundations to power smarter, transparent, environmentally agreeable supply chains (Yazan et al., 2022; Carlos et al., 2024).

## 5.3 Key Components

Continuous surveillance, predictive modelling, blockchain-enabled traceability, and automated reverse logistics are key components that drive IoT-enabled circular supply chains. The continuous surveillance with the help of IoT sensors allows real-time monitoring of the assets and system performance, which leads to the proactive event of making decisions and process of operation transparency. AI and big data-powered predictive modelling improve demand and supply forecasting that enables firms to create production peaks that coordinate with consumption patterns and minimize resource waste. By integrating blockchain technology, we can apply it to fields such as food, pharmaceuticals, automotive (Aparadh et al., 2023; Sahoo, 2024), authenticating products, enhancing traceability, solving trust issues, inefficient governance, and protecting against counterfeiting. Ultimately, blockchain fosters automated coordination and real-time tracking on supplier sides and real-time tracking on producer sides to mirror the objective of circular goals by waste minimization and resource regeneration (Swain & Patra, 2022; Shukla, 2024). Reverse logistics/ remanufacturing systems that are developed with the help of AI, IoT, and robotics adopt the principles of automated reverse logistics and remanufacturing systems in order to utilize AI, IoT, and robotics to streamline the process of the return, recovery and reprocessing of the materials, in order to increase the resource circularity, reduce the environmental impacts and reduce the operations costs. Despite these challenges, the implementation of these technologies creates transparency, efficiency, and resilience in circular supply chain systems (Jesus Júnior et al., 2020; Nofal et al., 2024).

## 6 IOT-ENABLED WAREHOUSING AND DISTRIBUTION CENTERS

The integration of IoT in smart storage and distribution centers is transforming traditional logistics operations into highly responsive, data-driven systems that support circular economy objectives. IoT sensors enable real-time inventory management by continuously monitoring stock levels, environmental conditions, and equipment status, thereby enhancing visibility, reducing inventory holding costs, and minimizing spoilage or obsolescence. These sensors provide granular data that support predictive demand forecasting, allowing companies to align inventory with actual consumption patterns, thus preventing overstocking and reducing waste. Additionally, IoT-enabled analytics contribute to dynamic replenishment strategies and more accurate distribution planning, further improving operational efficiency. The deployment of automation and robotics in warehousing—guided by IoT data—facilitates efficient sorting, picking, and handling processes while minimizing energy use and labor intensity. This level of automation not only accelerates order fulfillment but also reduces resource consumption and error rates, aligning warehouse operations with sustainability goals. Overall, IoT-enabled smart warehousing fosters the circularity of resources by improving transparency, reducing inefficiencies, and enabling adaptive, low-waste logistics strategies in the U.S. supply chain sector. Technologies including RFID, WSN, GPS, ZigBee, QR codes, and cloud platforms help the Internet of Things (IoT) to be facilitated inside this conceptual paradigm. Using these tools one can gather and share assets, inventories, personnel, infrastructure data. Thanks for this seamless connectivity; automated decision-making, predictive equipment maintenance, and real-time commodities tracking now help warehouse operations. Based on five core functional domains the conceptual model has been established.

*Table 2. Functional Domains and Key Features of IoT-Enabled Warehousing*

| Functional Domain | Definition and Key Warehouse Management Features |
|---|---|
| Inventory Control | Managing stock levels in real time to ensure accurate supply, avoid overstocking or stockouts. **Key Elements**: ● IoT sensors (e.g., RFID, NFC) for live inventory tracking ● Automated restocking alerts ● Stock movement analytics and visibility |
| Order Fulfillment | The process of picking, packing, and shipping orders accurately and efficiently. **Key Elements:** ● Smart picking systems (e.g., voice-pick, AR glasses) ● Automated sorting and packing ● Real-time order status updates |
| Asset and Equipment Monitoring | Tracking the performance, location, and condition of critical warehouse machinery and tools. **Key Elements:** ● IoT-based predictive maintenance ● Monitoring forklifts, conveyors, and robots ● Reducing downtime and repair costs |
| Staff Productivity | Enhancing worker efficiency, task allocation, and safety using smart technologies. **Key Elements:** ● Wearables for task monitoring and fatigue detection ● AI-driven workload distribution ● Worker location tracking for route optimization |
| Environmental Control | Monitoring and managing warehouse environmental conditions to ensure product quality and energy efficiency. **Key Elements:** ● Smart HVAC and lighting systems ● Sensors for temperature, humidity, and air quality ● Energy consumption optimization and compliance (especially for perishable goods) |

It can track real-time stocks, automatically restock, and reduce inventory management shrinkage by means of the Internet of Things (IoT). These all-possible capacities are made possible. Attached to pallets and shelves, sensors on inventory help to greatly increase accuracy. Real-time viewing made possible by these sensors usually reaches 95%. Using wearable technologies, location-aware systems, and autonomous mobile robots, the Internet of Things (IoT) can effectively pick and package goods during order fulfillment. This helps to lower the number of mistakes made and the time needed to satisfy orders.

Regarding asset surveillance, the Internet of Things is really outstanding. This technique uses sensor-based diagnostics to forecast when equipment will fail or malfunction, therefore drastically lowering the maintenance costs and the total amount of time the equipment is offline. Furthermore, intelligent distribution systems maximize resource use by means of predictive analytics to enable order volume projections. Wearable Internet of Things devices monitor employee activity levels, alert for safety rule violations, and offer recommendations for more effective task distribution, thereby increasing workplace productivity. Sustainable and operational efficiency are maintained by controls over the environment, including intelligent lighting and heating, ventilation, and air conditioning (HVAC) systems that change their settings depending on the needs of the product and the occupancy levels.

*Figure 2. IoT-Driven Transformation Model for Warehouse Operations*

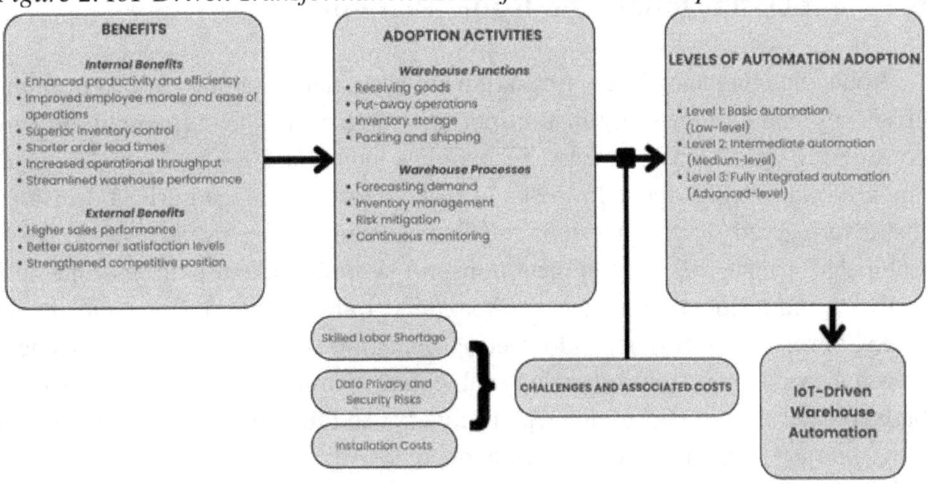


Making use of this approach provides several performance advantages. Among the possibilities that companies have pointed out as being among the advantages are better delivery accuracy, faster order cycles, less labour strain, and lower running costs. One of the several benefits the Internet of Things (IoT) presents is the lowering of waste, energy usage, and pointless inventory. Sustainable logistics is demonstrated here. To make these improvements realistic, though, strong data governance policies, adaptable information technology architecture, and qualified staff are absolutely essential.

Although the Internet of Things in smart warehouses offers many advantages, it also poses certain challenges. Important problems include high initial expenses, especially for small and medium-sized businesses (SMEs), a mismatch between current systems and Internet of Things platforms, and growing concerns about e-waste and cybersecurity. While larger companies (LEs) can scale installations and readily incorporate cutting-edge technologies like blockchain and artificial intelligence (AI), small and medium-sized businesses (SMEs) often face challenges in achieving a return on investment (ROI).

Inspired by the Internet of Things (IoT), the smart warehouse and distribution centre concept generates a digitally connected, real-time logistics environment fit for the objectives of the circular economy and sustainable supply chain management. This is the situation when all the elements are taken into account. Built on top of these intelligent facilities—which provide visibility, agility, and data-guided, data-driven decision-making—modern logistics systems that are strong, efficient, and environmentally friendly mirror one another.

## 6.1 Fleet Optimization and Route Planning

Route planning and fleet optimisation are absolutely vital if we are to raise the efficiency and sustainability of transportation logistics. This is particularly true considering smart transit systems. Technology, including GPS, Internet of Things, and artificial intelligence, allows one to make dynamic and data-driven decisions that in turn greatly lower emissions and increase fuel efficiency (Shiva et al., 2024; Cabioch & Chiêu, 2023). Intelligent transport systems make use of real-time data analytics and artificial intelligence to maximise transport networks, so enhancing safety, lowering travel time, and lowering operating costs. Artificial intelligence-driven dispersion and route planning enable the implementation of adaptive tactics based on real-time traffic, weather, and delivery conditions, therefore enabling fleet utilisation and lower emissions. By means of predictive analytics and machine learning in last-mile delivery, exact demand forecasting and driver performance monitoring are made possible, thereby facilitating preemptive changes that aid in increasing delivery dependability and efficiency (Chandramouli, 2023). Furthermore, helping with sustainable last-mile delivery is environmentally friendly route planning. To match logistical decisions with sustainability objectives, this approach combines environmental factors, including fuel use and carbon emissions, into routing algorithms. Mohsen (2024) claims that the use of smart urban logistics systems driven by artificial intelligence and the internet of things, together with autonomous cars, helps to lower traffic congestion and the general carbon footprint. This is consistent with the idea of smart and environmentally friendly cities. Olaleye et al. (2024) claim that fleet management's usage of Internet of Things (IoT), artificial intelligence (AI), and blockchain technology raises openness, real-time monitoring, and lowers emissions. These developments underline the critical part digital technologies play in addressing fleet optimisation and sustainable mobility challenges, hence improving environmental performance and service quality in the logistics sector.

Among other things, the Internet of Things has enhanced the capacity to dynamically plan in supply chain management and logistics, changed the efficiency of delivery, and enhanced the capacity to make decisions in real time. Based on the Internet of Things (IoT), the Route Optimisation System (ROS) with Smart

Physical System Service (SPSS) offers the most successful example of how smart control systems and continuous data updates may be applied to improve logistics operations by merging digital and physical systems. IoT is the Internet of Things. The cornerstone for this approach is the tight link between the digital and physical domains. Clients, drivers, and third-party logistics (3PL) suppliers can record delivery expectations and unique requirements by means of a cloud-based system run on the internet level. The Route Optimisation Service dynamically optimizes delivery routes by means of managing several inputs and the use of real-time environmental data, such as trip duration and road networks. This specific site acts as a replacement for the real implementation of these decisions.

*Figure 3. IoT-Based Route Optimization System (ROS) with Smart Physical System Service (SPSS) for Sustainable Logistics*

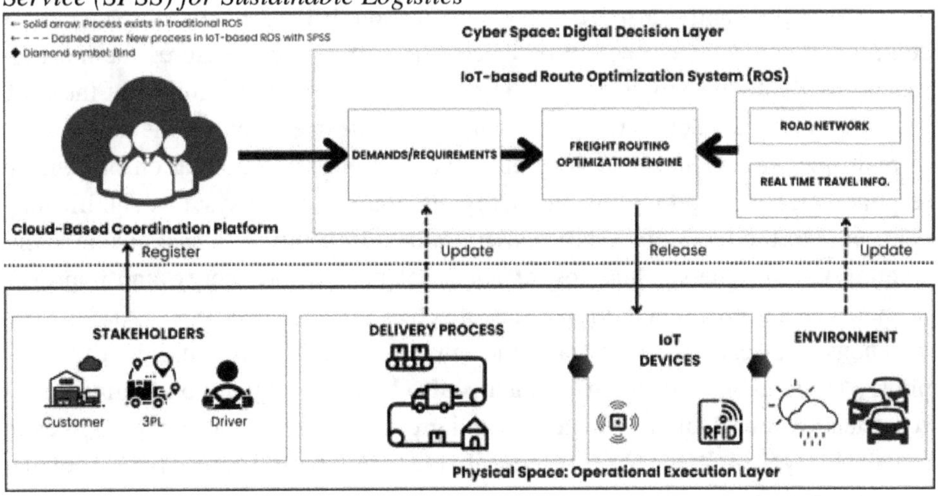


Devices linked to the Internet of Things (IoT) and able to be deployed in infrastructure and vehicles can track shipments in real time and offer continuous feedback. Among these are global positioning system (GPS) trackers, environmental sensors concurrently monitoring and transmitting data about traffic and weather in real time, and radio frequency identification (RFID) devices. By means of iterative changes to routing algorithms, the process of restoring this data to the internet is enabled. Within the framework, the vectors help to differentiate between typical ROS tasks (solid lines) and new capabilities made possible by the Internet of Things (dashed lines). Furthermore, the diamond emphasizes the relationships between digital choices and actual behaviour. Layer by layer, this link allows one to offer quick, adaptable, and efficient distribution options. Based on the Internet of Things (IoT), the ROS with SPSS promotes a responsive and intelligent logistics ecosystem that lowers the use of petroleum, increases transparency, and thereby reduces delays, helping to meet sustainability targets. Generally speaking, this ecosystem supports the fulfilment of environmental targets. Using closed-loop feedback and data-driven optimisation, it shows how exciting fresh ideas are to increase operational efficiency in contemporary supply chains. Closed-loop feedback helps one to achieve this.

## 6.2 Integration of Renewable Energy in Logistics Operations

Due to the influence on both logistics operations costs and product or service quality, the logistics facility is getting more attention in the domain of industrial management systems. One of the key steps in integrating a circular supply chain is to adopt a culture of sustainability in renewable power generation, combined with IoT-based energy management control systems, sustainable warehouse management, green packaging, and transport fleet electrification. IoT-based smart green energy control systems enhance the option of charging EVs, which ensures the most efficient usage of alternative energy sources and increases the stability of the grid. To contribute to the conservation of energy, these devices monitor the device's use in real time and evaluate the generated data. Through the utilization of demand response strategies and intelligent energy distribution, these systems significantly contribute to the reduction of carbon emissions. In addition to sustainable architectural designs, energy- and cost-efficient solar and wind energy, and improved inventory management systems (Onukwulu et al., 2022; Majeed et al., 2025), green warehouse projects also incorporate other environmentally conscious initiatives. The phrase “sustainable packaging”, which is also known as “green packaging”, uses eco-friendly, recyclable raw materials and green designs that follow principles of sustainability (Li & Wang, 2023) to contribute to adapting and improving the circular supply chain's efficacy by reducing waste creation and dependence on resources. Based on Sayeed and Manikandan's (2024) views, electrifying transportation fleets is a reliable approach for shrinking greenhouse emissions and diversifying sources of power. The alternative power sources that are able to be connected with V2G (vehicles to the grid) solutions make such an attainable attempt. The Internet of Things (IoT) enhances operational efficiency and openness in logistics buildings by enabling remote monitoring and control of energy usage. A green port can be defined as a port that has a mission of lowering the carbon footprint and has illustrated the vast significance of green energy in logistics operations by integrating different technologies, such as controlled consumption of energy, hydrogen technology, batteries, and thermal storage. Achieving long-term ecological sustainability goals depends on successfully using alternative sources of energy in logistic operations by working together across various industries, creating supportive regulations, and constantly doing research and development for the improvement of technologies.

## 6.3 Sustainable Procurement and Ethical Sourcing

In circular supply chains, strategic ethical procurement underpins the achievement of both egalitarianism (SDG 5) and conscientious consumption (SDG 12) (Arora & Mishra, 2023). It pertains to the practical implementation of ethical sourcing (Raman et al., 2023). Blockchain and AI-driven end-to-end transparent procurement, compliance automation, sustainability in mitigating socio-ecological hazards, and smart contracts are enhancing the resourcefulness and accountability of supply chain operations. In order to have a sustainable brand image, ethical sourcing aids a company to attain economies of scale by mitigating risks and hazards for the brand in confirming standardized labour work culture, EHS (Environment, Health, and Safety) practices, and human rights practices during strategic sourcing of procurement and supplier practice (Agrawal et al., 2024). Awareness should be built more and promoted to support women's entrepreneurship, without gender-biased procurement rules and regulations, to enhance equal labor rights and participation. State- and federal-level governmental organisations can help in facilitating this among people by declaring incentives, educational programs, free food during events, etc. Cross-functional collaborations between department to departments or company to company or between the government to companies that utilize collective advancements in product design, sustainable packaging, and the implementation of reduce-reuse-recycle frameworks, facilitate the adoption of circular principles (Vishwakarma & Gehlot, 2023). Consequently, collaboration with suppliers is of paramount significance. For sustainable procurement to be effective, a balance must be achieved among ethical principles, legal compliance, and cost-effectiveness. Establishing trust and enduring stakeholder value necessitates this. The public sector's engagement in green public procurement enhances positive social outcomes and further accelerates sustainability. This is accomplished by integrating social and environmental considerations into the purchase decision-making process. Moreover, including sustainability in procurement processes enhances resilience and produces enduring advantages (Cooper, 2024). The organisation's culture and leadership are the primary drivers of this development. Collectively, these strategies not only facilitate the maintenance of sustainable supply chains but also substantially promote the achievement of global sustainability goals. Strategic ethical sourcing in sustainable procurement is essential for attaining egalitarianism (SDG 5) and mindful consumption (SDG 12) within circular supply chains (Arora & Mishra, 2023). Smart contracts, powered by blockchain technology and artificial intelligence, enhance supply chains by increasing procurement transparency, automating compliance assessments, and mitigating social and environmental risks, thereby fostering more accountability and sustainability. Companies can achieve a competitive advantage and mitigate reputational risks through the use of ethical sourcing strategies (Agrawal et al., 2024). This is accomplished by evaluating the origins of their acquisitions and the practices of suppliers to guarantee compliance with labour regulations, environmental sustainability, and human rights standards. Gender-inclusive procurement policies foster equal labour participation by enhancing possibilities for women-owned enterprises and advocating reforms that support female entrepreneurship. Inclusive practices, such as enhancing chances for women-owned enterprises, are often facilitated by governmental support. Moreover, the engagement of suppliers is crucial, as collaborations facilitate the implementation of circular principles via reduce-reuse-recycle frameworks, sustainable packaging, and joint innovations in product design (Vishwakarma & Gehlot, 2023). Efficient sustainable procurement must reconcile cost-effectiveness, regulatory adherence, and ethical principles to foster confidence and provide stakeholders with enduring value. Green public procurement involves the public sector, hence enhancing sustainability efforts. Purchasing decisions can be integrated with socio-ecological aspects, hence augmenting societal benefits. Moreover, resilience is enhanced, and enduring value is generated by incorporating sustainability into procurement plans, driven by leadership and corporate culture (Cooper, 2024). Collectively, these strategies facilitate the attainment of global sustainability objectives while promoting the transformation of supply chains sustainably and healthfully.

## 6.4 ECONOMIC RESILIENCE AND RISK MITIGATION STRATEGIES (SDG 9)

IoT-enabled CSC (Circular Supply Chain) is an integration of technology into SCM (Supply Chain Management) that can play a big role (Ali et al. 2023) in expanding the path of economic upswing as well as reducing vulnerability in aligning with SDG 9. Evidence from different developing economies like China is that the IoT facilitates real-time data collection, predictive analytics, and end-to-end traceability, enabling organisations to assess and manage supply chain risks (Sun et al., 2023). The IoT-based CSC systems empower manufacturing operations to identify disruptions at an early point and respond with greater agility, which ensures increased visibility and transparency during the whole process. When combined with blockchain and AI, IoT strengthens resilience through optimised operations, improved demand forecasting, and enhanced scenario planning (Ameh, 2024). These technologies collectively support adaptive strategies that allow supply chains to withstand and recover from uncertainties such as pandemics, natural disasters, and geopolitical shocks (Kim, 2024; Mishra et al., 2024; Roshid et al., 2025a). Cross-sectoral collaboration is necessary to establish agile supply networks, where IoT-based CSC can significantly contribute by enabling appropriate risk prediction, particularly for disasters, and facilitating the exchange of real-time data between entities, government, and communities (Meng, 2021). Additionally, IoT-based platforms help ensure that supply chain operations are secure, transparent, and verifiable, which is important for following national and international rules (Ameh, 2024; Roshid et al., 2025b). As global supply chains face increasing volatility, the convergence of IoT with emerging digital technologies and the active engagement of public and private stakeholders are imperative for sustainable risk management (Karanam et al., 2024; Sun et al., 2023). Ultimately, the strategic deployment of IoT supports long-term economic sustainability by enhancing the adaptability and robustness of supply chains in a rapidly evolving global landscape (Maemunah, 2024).

## 7 DEPLOYING IOT-DRIVEN CIRCULAR SUPPLY CHAINS: KEY OBSTACLES

The implementation of Industrial Revolution-based technology Internet of Things (IoT) into circular supply chains has the potential to align with the Industrial Transformation Agendas focusing on better sustainability, optimal resource allocation, and operational transparency. The implementation of the Internet of Things (IoT) inside these systems, on the other hand, is confronted with a variety of difficult hurdles that prevent widespread acceptance. These challenges encompass a wide range of disciplines, including technological, financial, regulatory, and organizational, and they frequently combine to provide severe implementation difficulties. Understanding these fundamental challenges and finding solutions to them is vital to enable scalable, secure, and resilient models of circular supply chains that are powered by the Internet of Things.

### 7.1 Excessive Operational Expenditure and Capital Investment Obstacles

Often deterring adoption despite long-term sustainability benefits, deploying IoT in circular supply chains requires a large upfront investment in structuring the whole setup, where key elements are sensors, data platforms, and safe communication networks—which often results in But a significant cost increment. results from a combined technological integration with IoT legacy systems and other modern technologies, including artificial intelligence and blockchain. Further accelerating expenditure will result from the constant technological advancement, with its training, upgrading phases, and lack of standardisation with evolutionary upcoming technologies (Ingemarsdotter et al., 2020; Maurya et al., 2024). Moreover, cybersecurity calls for strong, expensive systems to guard private information (Nozari et al., 2022). IoT can maximise operations and cut waste despite these financial difficulties, so matching with circular economy targets (Maurya et al., 2024). Overcoming investment constraints calls for strategic alliances and creative business models (Abdelmhmud, 2024).

## 7.2 Cybersecurity Risk Factors and Data Privacy Issues

Internet of Things-based Circular Supply Chains are especially prone to malware, thus strict security measures, including access restrictions and encryption, are absolutely vital. Mass data collecting powers of IoT raise privacy and unauthorised access questions (Anil et al., 2023). Many others worry about the Internet of Things at the same time. Anwar and Sanmorino (2024) claim that the absence of uniform security policies and regulatory monitoring fuels the escalation of already existing hazards. To answer these issues, privacy-by-design techniques, federated learning for threat detection, and strong legal protections must be used. The development of stakeholder confidence and assurance of safe operations depend on these actions.

## 7.3 Circular Supply Chains: Tactics for Meeting Regulations and Staying in Compliance

Different observance of legality is being implemented in different locations as a result of the need for precise governing regulations for circular economy operations, particularly circular supply chains (Gupta 2024). Consequently, the lack of financial incentives like grants or tax breaks deters investment in the Internet of Things (Abdelmhmud, 2024). Due to the integration of IT (information technology) and OT (operational technology), there are currently two areas that are not adequately restrained (Gheorghe, 2024). Nevertheless, the IoT (Internet of Things) emphasizes the dominance of data retention and cyber safety. Other issues, like a lack of experience and internal company culture inertia, also impede compliance efforts. Additionally, the policy framework is essential to accelerating the adaptation of supply chain logistics operations, such as servitization and reverse logistics. The design of collaborative policies is necessary to reduce uncertainty and facilitate the integration of the Internet of Things (IoT) in circular supply chains (Gupta, 2024; Abdelmhmud, 2024; Gheorghe, 2024).

## 7.4 Resistance to Technological Adoption in Traditional Logistics Firms

Due to old-fashioned frameworks and confrontation with legacy systems, traditional logistics companies customarily resist the IoT (Internet of Things) embracement (Nozari et al., 2022; Oliveira, 2024). Supplemental development constraints include cost-push inflation, reluctance on the part of stakeholders, and a lack of specialized technological knowledge (De Vass et al., 2021; Ullah et al., 2023). Resistance advances as a consequence of disturbance about scalability, cybersecurity, and data privacy (Sallam et al., 2023). The implementation of targeted transformational leadership procedures and employee skill development is essential for enabling a successful Internet of Things deployment (Sallam et al., 2023). By strengthening environmental credentials, cost-effectiveness, and transparency, the IoT has the potential to transform supply chains (Governatori, 2022; Younis et al., 2025). Highly dedicated and appropriate design can bring about this change.

## 8 IOT-ENABLED CIRCULAR SUPPLY CHAINS IN THE U.S. RETAIL INDUSTRY

The adoption of IoT-based Circular Economy into supply chain operations within the American retail industry will open the path of sustainability, improved efficiency, and expedited service operational performance. IoT-based circular supply chains for the retail industry that have a combination of Smart Sensors, Cameras, and RFID tags can enhance real-time monitoring and traceability. This IoT-based CSC approach is critical for regenerating resources and waste reduction in circular logistics (González Vásquez et al., 2023). Leading retailers such as Walmart have successfully leveraged IoT for asset tracking, predictive maintenance, and dynamic inventory management, resulting in enhanced supply chain responsiveness and sustainability (Oliveira, 2024; Farazi, 2024). These technologies encourage flawless data transfer and communication across supply chain nodes, aligning with the Sustainable Development Goals (SDGs) of a closed-loop economic model or circular economy approach (Jum'a et al., 2024). Furthermore, the integration of IoT with AI and blockchain technologies has improved transparency and enabled data-driven decision-making, contributing to competitive sustainability in the retail sector (Farazi, 2024). Despite these gains, challenges such as high implementation costs, data security concerns, and difficulties in integrating IoT with legacy systems persist (Oliveira, 2024). However, empirical outcomes—including improved delivery accuracy, inventory turnover, and customer satisfaction—illustrate the significant value of IoT-driven circular models.

*Table 3. Real-World Applications of IoT in Circular Supply Chains in the U.S. Retail Industry*

| Corporation | IoT Application | Circular Supply Chain Benefit | Source |
|---|---|---|---|
| Walmart | IoT-enabled sensors for real-time monitoring of perishable goods | Reduces spoilage and food waste; ensures quality and efficiency | Udeh et al., (2024) |
| Nordstrom | Smart lighting systems using IoT | Optimizes energy usage and reduces operational costs | Karthikeyan, K. S., & Nagaprakash, T. (2024) |
| 7-Eleven | IoT-controlled HVAC systems | Enhances energy efficiency and minimizes environmental footprint | Li et al., (2025) |
| Levi's & The R Collective | QR-code-enabled traceability of upcycled denim | Promotes transparency, recycling, and sustainable consumption | Vogue Business. (2020) |

## 8.1 Sustainable Manufacturing and IoT-Based Reverse Logistics

The integration of Internet of Things (IoT) technologies into reverse logistics and sustainable production plays a vital role in advancing circular economy goals. IoT enhances decision-making, productivity, and product tracking—key for industries like electronics, textiles, and automotive to optimize material use and enable resource reuse. Companies such as Philips, Walmart, and HP leverage IoT with blockchain and AI to enable real-time monitoring and efficient resource distribution (Farazi, 2024). In developing countries like Ghana and Indonesia, targeted investments in IoT infrastructure have significantly improved logistics sustainability (Akubia et al., 2024). While IoT boosts operational visibility and efficiency, adoption is often hindered by high costs and cybersecurity issues (Maurya et al., 2024).

*Figure 4. Internet of Things (IoT) driven kanban system for reverse logistics: solid waste collection*

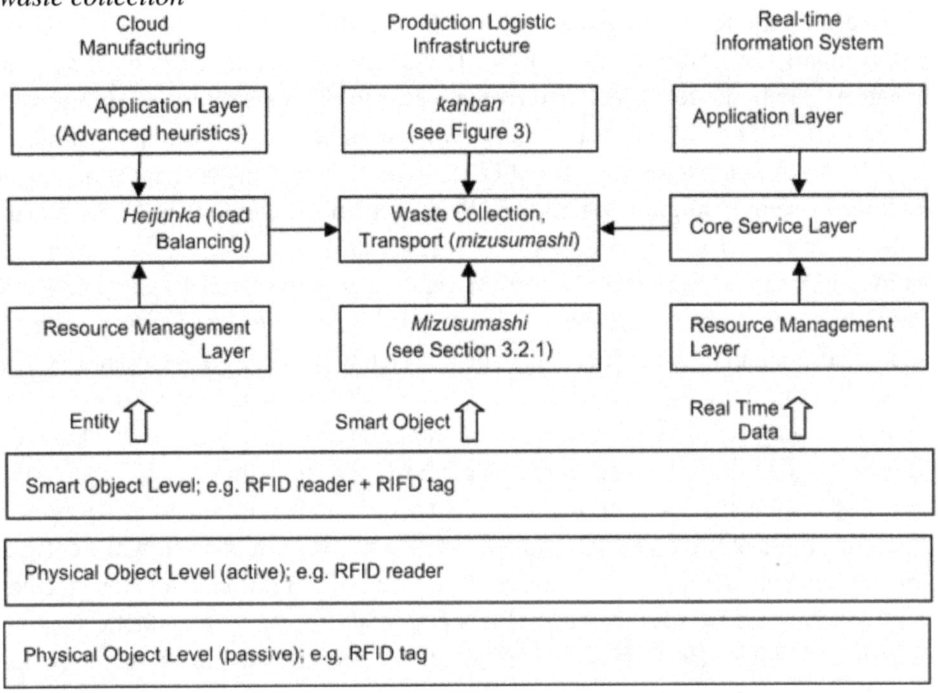


*Thürer et al. (2019)*

Within Industry 4.0, IoT connects resources, sensors, and products to reduce waste and promote circularity, yet more research is needed in areas like recycling and remanufacturing. IoT also enables models such as product-as-a-service, which balance economic and environmental goals (Turskis & Keršulienė, 2024). Practical examples, including modular smartphones and the e-commerce sector, show how IoT facilitates closed-loop systems and coordination for sustainability (Delpla et al., 2021; Kumar & Godihal, 2024). Collectively, these cases underscore IoT's transformative role in building circular and sustainable supply chains.

## 8.2 E-commerce and Smart Warehousing for Sustainable Distribution

The successful adoption of IoT in circular logistics, e-commerce, and smart warehousing has significantly advanced sustainable distribution, as demonstrated by various case studies across global sectors. IoT technologies have transformed logistics operations by improving asset tracking, inventory management, and delivery processes—key enhancements implemented by industry leaders such as Amazon,

DHL, and Walmart (Oliveira, 2024). In circular supply chains, the integration of IoT with blockchain and AI ensures transparency, real-time monitoring, and efficient resource distribution, supporting sustainability initiatives undertaken by firms like HP, Walmart, and Philips (Farazi, 2024). These technologies optimize resource use, enhance operational reliability, and increase supply chain visibility, resulting in reduced waste and controlled costs aligned with circular economy principles (Maurya et al., 2024). In smart logistics, IoT applications have improved goods movement and tracking systems in alignment with the UN Sustainable Development Goals. Smart warehousing benefits from automated processes such as stock counting, shipment validation, and warehouse system control, creating new efficiencies and operational flexibility. Furthermore, IoT-driven delivery solutions optimize loading patterns and route planning, increasing last-mile efficiency and reducing environmental impact. Under Industry 4.0, IoT connects materials, products, and sensors in real time, reshaping supply chains to be more intelligent, responsive, and sustainable. This intersection of IoT and circular economy principles has led to innovative business models that promote economic sustainability by minimizing waste and extending product life cycles (Turskis & Keršulienė, 2024). However, while current IoT applications mainly support usage efficiency and lifecycle extension, further exploration is needed to fully realize IoT's potential in reuse, remanufacturing, and recycling strategies (Ingemarsdotter et al., 2019). Additionally, IoT-based logistics monitoring systems offer real-time tracking and environmental condition monitoring, contributing to greater supply chain transparency and market competitiveness. Collectively, these developments highlight the transformative role of IoT in driving sustainable distribution through smart warehousing and circular e-commerce models.

## 9 SMES STRUGGLE TO DIGITIZE CIRCULAR SUPPLY CHAIN PRACTICES

Integration of IoT-enabled circular logistics and supply operations in an existing supply chain network is not free from barriers for both the established and small and medium-sized enterprises. Practically, the implementation of IoT-based CSC reveals structural and operational challenges that keep SMEs behind in practicing and digitising their supply chain and distribution channels.

*Firstly,* policy gaps and systematic misalignment exist, as the transformation of CE with digital technologies still remains ill-suited to the practical operational aspects of SMEs. Lack of standardisation with compliance policy bodies and data interoperability across the supply operations creates confusion and panic among SMEs' infrastructure to interconnect IoT for reverse logistics operations. Another major cause is that the existing frameworks are built in favour of larger organisations, where regulatory policy offers incentives and subsidies embedded through resource-intensive application processes.

*Secondly,* limited capacity with employee knowledge-based training is required for building technological and theoretical knowledge from top to bottom of the SMEs. But it is not feasible for SMEs to provide regular training for gaining IoT-enabled CE practice-related competencies due to training budget allocations, as they don't have the same resources as large organisations. Due to the lack of continuous literacy on technology like IoT, a lesser skill set in data-driven systems, and a lack of theoretical and conceptual knowledge, IoT-enabled CE practices remain a dream for SMEs.

*Thirdly,* the key barrier is the cost-benefit trade-offs and risk of investment for the IoT-based CSC practice. The adaptation of IoT technology into the supply chain with the investment in overall infrastructural development (e.g., sensors, cloud servers, connectivity, cybersecurity, etc.) is impossible, as the capital matters for SMEs. In addition, the long payback period for advantages, including resource utilization, reduction of waste, and predictive maintenance, makes the IoT-enabled CSC implementation unjustified for SMEs. Overall, SMEs prioritize digital circular supply chain initiatives due to unclear sectoral cost-benefit analysis and risk mitigation strategies, while they choose short-term operational stability.

### 9.1 Case Insight: Loop Circular (UK)

A UK-based SME textile manufacturer sought to adopt a Loop Circular Model aimed to have circularity across their supply chain, where they wanted to have 3R-based closed-loop recycling operations to mitigate waste on their shop floor. During the integration of the IoT-driven circular supply chain, both the challenges and opportunities were observed. With the vision and commitment that align with the Circular Fashion Innovation Network for a circular fashion ecosystem, the company introduced sensor-based real-time tracking for the usage of garments and repair that resulted in a 30 percent reduction of textile waste. Having a strategic MoU for partnership with a local university for staff upskilling and ensuring funding from the regional innovation center, the company had to face the initial phase of high hardware costs and insufficient data and stakeholder literacy.

## 10 FUTURE DIRECTIONS AND POLICY RECOMMENDATIONS

There are three dependent factors of the successful integration of the Internet of Things (IoT) into circular supply chains, such as intersectoral cooperation, financial support, and intentional legislative reforms. The main policy recommendations and next actions to speed the acceptance of the Internet of Things (IoT) and other emerging technologies in sustainable logistics and distribution are summarized in this chapter.

## 10.1 Internet of Things (IoT) Regulatory Structural Policy Development and Governmental Support

To overcome operational and financial challenges, we need incentives to enable green logistics and digital transformation. These could cover low-interest loans for businesses using IoT-enabled circular solutions, tax credits, and subsidies. Public procurement policies should receive increased technological support to advance progress towards achieving ecologically sustainable green logistics, in which policymakers must establish various motivational reward-based caps. Encouragement of low-emission technology, circular business models, and environmentally friendly warehouses helps to enable the change to a more sustainable logistics industry.

## 10.2 Motivating Sustainable Digital Transformative Logistics and Green Technologies

Some incentives are needed to enable green logistics and digital transformation if we are to overcome operational and financial challenges. These could cover low-interest loans for businesses using IoT-enabled circular solutions, tax credits, and subsidies. Public Procurement Policies should get more technological support in making more advances towards achieving ecologically supported green logistics, where policymakers must declare different motivational reward-based caps. Encouragement of low-emission technology, circular business models, and environmentally friendly warehouses helps to enable the change to a more sustainable logistics industry.

## 10.3 Investment for Internet of Things (IoT), Blockchain, and Artificial Intelligence (AI) in establishing explicit and efficient CSC

Development of transparent, resilient, and resource-efficient supply chains depends on public and commercial sector investments in artificial intelligence, blockchain, and IoT. Essential components of circularity and supply chain optimization, real-time tracking, predictive analytics, and safe data sharing are made possible by these technologies taken together. Establishing innovation hubs, financing pilot projects, and supporting research and development are top priorities in investment plans for integrated smart logistics platforms that increase end-to-end visibility and sustainability outcomes.

## 10.4 Improving Industry Cooperation and Partnerships (SDG 17)

Sustainable Development Goal 17 emphasizes how attaining sustainable and circular logistics calls for cooperation among several stakeholders. Policymakers should encourage industrial partnerships using digital platforms, knowledge-sharing networks, and public-private projects meant to co-develop solutions and highlight discoveries. To improve collective capability, lower implementation risk, and synchronize supply chain operations with sustainability goals, cooperation among logistics companies, technology suppliers, academic institutions, and government agencies is indispensable.

## 10.5 New Trends: Carbon-Neutral Distribution and Predictive Logistics Driven by AI

Rising trends in circular logistics will affect the next developments, including carbon-neutral delivery systems and AI-driven predictive logistics. AI will progressively take the front stage in supporting intelligent and adaptive logistics systems, improving demand forecasting, and waste reduction. Concurrent integration of carbon tracking methods, renewable energy, and electric cars will hasten the change to carbon-neutral supply chains. Policymakers and corporate leaders must be flexible in these times to ensure regulatory compliance and ongoing support for sustainable innovation.

# 11. CONCLUSION

In order to dig into the sustainable supply chain and logistics operations in the United States, this chapter has explored different aspects of IoT-based Circular Supply Chain Management, where less waste generation throughout the manufacturing and service operations is a core strategy in order to achieve sustainability. At the same time, conceptual model development for the IoT-based operational areas in attaining the CSC system is synthesized by global empirical evidence and case studies with digital technologies such as sensors, AI, and blockchain, with IoT integration following CE's 9R model across the logistics and distribution network. This chapter represented a multi-layered conceptual framework aligned with IoT enablers with circular actions and logistics-operational components that are a roadmap for transforming conventional logistics operations into circular supply networks. In addition, existing barriers, regulatory policy limitations, and a lack of more academic and industry-based research in this field are also identified for the US logistics and supply chain sector, despite its high technological maturity. The author's suggestion of focusing on cross-sectoral collaboration, creativity in policy integration, and technological and human resource capacity building beyond the traditional approaches is highly recommended for the potential transformation of IoT-driven CSCM. Moreover, the theoretical implication of the study supports developing digital technology-based circularity within the ecosystem of the supply chain, where practical implementations support strategies for industry stakeholders to foster resilience, sustainability, and competitive advantage in a rapidly evolving global logistics environment.

## REFERENCES


Abdelmhmud, M. A. E. F. (2024). Navigating the Challenges and Seizing Opportunities in Implementing Closed-Loop Supply Chains. *Journal of Ecohumanism*. https://doi.org/DOI: 10.62754/joe.v3i7.4692

Agrawal, D., Kumar, A., & Kumar, K. S. (2024). *Ethical Sourcing*. Elsevier BV., DOI: 10.1016/b978-0-443-13701-3.00016-5

Ahmad, S., Karim, R., Sultana, N., & Lima, R. P. (2025). InsurTech: Digital Transformation of the Insurance Industry. In Kour, M., Taneja, S., Özen, E., Sood, K., & Grima, S. (Eds.), *Financial Landscape Transformation: Technological Disruptions* (pp. 287–299). Emerald Publishing Limited., DOI: 10.1108/978-1-83753-750-120251016

Akubia, G. K., & Gaffar, V., Sultan, Mokh. A., & Andriana, D. (2024). The Effects of IoT on Sustainable Logistics Management Practices and Sustainability Performance in the Manufacturing Sector: A Comparative Study of Ghana and Indonesi. *Khazanah Sosial*, *6*(3), 533–550. DOI: 10.15575/ks.v6i3.38197

Alcayaga, A., & Hansen, E. G. (2024). Smart Circular Systems: A conceptual framework for microlevel digital circular economy strategies. *Journal of Cleaner Production*, *421*, 139824. DOI: 10.1016/j.jclepro.2023.139824

Alfaridi, N., Nasution, I. G., So, A., & Furinto, R. (2024). Sustainable Supply Chain Management in Enhancing Circular Economy Performance: Study Case in Indonesia. *Evolutionary Studies in Imaginative Culture*. Advance online publication. DOI: 10.70082/esiculture.vi.756

Ali, S. M., Ashraf, M. A., Taqi, H. M. M., Ahmed, S., Rob, S. A., Kabir, G., & Paul, S. K. (2023). Drivers for Internet of Things (IoT) adoption in supply chains: Implications for sustainability in the post-pandemic era. *Computers & Industrial Engineering*, *183*, 109515.

Alivojvodic, V., & Kokalj, F. (2024, June). Redefining Waste: R-Strategies and Metrics as a Framework for Driving Progress of Circular Economy Performance. In International Conference of Experimental and Numerical Investigations and New Technologies (pp. 75-91). Cham: Springer Nature Switzerland.

Ameh, B. (2024). Advancing national security and economic prosperity through resilient and technology-driven supply chains. *World Journal Of Advanced Research and Reviews*, *24*(3), 483–500. DOI: 10.30574/wjarr.2024.24.3.3723

Anil, A., Babu, A., Antony, J., Vilson, K. E., & Koshy, S. (2023). Security And Privacy Concern In IoT Devices. *International Journal of Engineering Technology and Management Sciences*, *7*(4), 491–502. DOI: 10.46647/ijetms.2023.v07i04.065

Anwar, Y. Z., & Sanmorino, A. (2024). Hukum dan Kebijakan Keamanan Siber: Tantangan Regulasi Perangkat IoT. *Jurnal Ilmiah Informatika Global*, *15*(3), 95–99. DOI: 10.36982/jiig.v15i3.4773

Aparadh, C., Barhaiyya, N., Chaudhary, H., Kokate, P., & Alte, B. (2023). *Exploring the Role of Blockchain in Enhancing Supply Chain Traceability and Transparency*. 1–7. https://doi.org/DOI: 10.1109/icdsns58469.2023.10245587

Arora, N. K., & Mishra, I. (2023). Responsible consumption and production: A roadmap to sustainable development. *Environmental Sustainability*, *6*(1), 1–6.

Arumugam, T., Hameed, S. S., Priyanshu, D., Fathima, K. A. A., & Farook, J. A. M. O. (2024). Exploring Sustainable Supply Chains Through Circular Economy Principles. *Advances in Logistics, Operations, and Management Science Book Series*, 199–214. https://doi.org/DOI: 10.4018/979-8-3693-6274-7.ch011

Baporikar, N. (2023). Opportunities and Challenges of Circular Economy for Designing Sustainable Supply Chains. *Advances in Finance, Accounting, and Economics Book Series*, 43–59. https://doi.org/DOI: 10.4018/978-1-6684-7664-2.ch003

Batista, L., Gong, Y., Pereira, S., Jia, F., & Bittar, A. (2019). *Circular supply chains in emerging economies–a comparative study of packaging recovery ecosystems in China and Brazil. International Journal of Production Beames, A., Claassen, G. D. H., & Akkerman, R. (2021). Logistics in the Circular Economy: Challenges and Opportunities*. Springer., DOI: 10.1007/978-3-030-55385-2_1

Bertram, M. (2016). Theoretical foundation: the resource-based view (RBV) of the firm. In *The strategic role of software customization: Managing customization-enabled software product development* (pp. 67–102). Springer Fachmedien Wiesbaden.

Bressanelli, G., Adrodegari, F., Perona, M., & Saccani, N. (2020). Exploring how usage-focused business models enable circular economy through digital technologies in manufacturing. *Sustainable Production and Consumption*, *23*, 256–272. DOI: 10.1016/j.spc.2020.05.003

Bühler, L., Fendt, D., Wittenberg, T., & Hamper, A. (2024). Smart Circular Economy in healthcare – An introduction to the 9R framework: Digital technologies enabling circular economy in healthcare. *Current Directions in Biomedical Engineering*, *10*(4), 127–130. DOI: 10.1515/cdbme-2024-2031

Business, V. (2020, May 7). *A new denim collection gives jeans a digital identity*. https://www.voguebusiness.com/technology/a-new-denim-collection-gives-jeans-a-digital-identity-r-collective

Cabioch, B., & Chiêu, N. H. (2023). Smart Transportation Logistics: Achieving Supply Chain Efficiency with Green Initiatives. *Greening of Industry Networks Studies*, 243–258. https://doi.org/DOI: 10.1007/978-3-031-29823-3_10

Carlos, R. L., de Souza, E. B., & Mattos, C. A. (2024). *Enhancing circular economy practices through the adoption of digital technologies*. https://doi.org/DOI: 10.1002/bsd2.330

Carson, R. (1962). *Silent spring*. Fawcett Publications.

Chandramouli, A. (2023). Optimizing Last-Mile Delivery Operations: Leveraging Predictive Analytics, Technology Integration, and Sustainable Practices. *Journal of Mathematical & Computer Applications*, *2*(3), 1–7. DOI: 10.47363/jmca/2023(2)145

Choudhary, V., Patel, K., Niaz, M., Panwala, M., Mehta, A., & Choudhary, K. (2024).. . *Implementation of Next-Gen IoT to Facilitate Strategic Inventory Management System and Achieve Logistics Excellence.*, *4*, 1–6. DOI: 10.1109/tqcebt59414.2024.10545238

Cooper, M. (2024). *Integrating Sustainability Into Procurement: Perspectives From Supply Chain Practitioners*. https://doi.org/DOI: 10.20944/preprints202407.0614.v1

De Vass, T., Shee, H., & Miah, S. (2021). IoT in supply chain management: Opportunities and challenges for businesses in early industry 4.0 context. *Operations and Supply Chain Management: An International Journal*, *14*(2), 148–161.

Delpla, V., Kenné, J.-P., & Hof, L. A. (2021). Circular manufacturing 4.0: Towards internet of things embedded closed-loop supply chains. *International Journal of Advanced Manufacturing Technology*, •••, 1–24. DOI: 10.1007/S00170-021-08058-3

Echefaj, K., Charkaoui, A., Cherrafi, A., Tiwari, S., Sharma, P., & Chiappetta Jabbour, C. J. (2024). *From linear to circular sustainable supply chain network optimisation: towards a conceptual framework*. https://doi.org/DOI: 10.1080/09537287.2024.2302479

Farazi, M. Z. R. (2024). Designing circular supply chains with digital technologies for competitive sustainability: An operation management perspective. *International Journal of Science and Research Archive*, *13*(1), 2346–2359. DOI: 10.30574/ijsra.2024.13.1.1928

Geissdoerfer, M.; Morioka, S.M.; de Carvalho, M.M.; Evans, S. Business models and supply chains for the circular economy. J. Clean. Prod. 2018, 190, 712–721. Research, 57(23), 7248-7268.

Gheorghe, S.-D. (2024). A Review of the Information Technology and Operational Technology Convergence Using Internet of Things within the Circular Economy. *Proceedings of the ... International Conference on Business Excellence, 18*(1), 829–845. https://doi.org/DOI: 10.2478/picbe-2024-0073

González Vásquez, J. A., Caciano Arroyo, M. E., Horna Maguiña, M. M., Lopez Asto, M. K., Solano Quincho, L. M., & Vasquez Cabrera, A. F. (2023). *Internet of things technologies applied in the supply chain. a systematic review*. https://doi.org/DOI: 10.17268/goi4.0.2023.01

Governatori, G. (2022). A Comprehensive Analysis on the Adoption of IoT with Logistics and Supply Chain Management. *2022 Second International Conference on Computer Science, Engineering and Applications (ICCSEA)*. https://doi.org/DOI: 10.1109/iccsea54677.2022.9936196

Gružauskas, V., Baskutis, S., & Navickas, V. (2018). Minimizing the trade-off between sustainability and cost effective performance by using autonomous vehicles. *Journal of Cleaner Production, 184*, 709–717.

Gupta, R. (2024). *A Qualitative Inquiry into the Implementation Challenges of Circular Economy Practices in Supply Chains*. https://doi.org/DOI: 10.21203/rs.3.rs-4284012/v1

Hassan, S. T., Wang, P., Khan, I., & Zhu, B. (2023). The impact of economic complexity, technology advancements, and nuclear energy consumption on the ecological footprint of the USA: Towards circular economy initiatives. *Gondwana Research, 113*, 237–246.

Ingemarsdotter, E., Jamsin, E., & Balkenende, R. (2020). Opportunities and challenges in IoT-enabled circular business model implementation – A case study. *Resources, Conservation and Recycling, 162*, 105047. DOI: 10.1016/J.RESCONREC.2020.105047

Ingemarsdotter, E., Jamsin, E., Kortuem, G., & Balkenende, R. (2019). Circular Strategies Enabled by the Internet of Things—A Framework and Analysis of Current Practice. *Sustainability, 11*(20), 5689. DOI: 10.3390/SU11205689

Jesus Júnior, E. M. F. de, Marques Neto, M. C., & Freitas, A. E. S. (2020). *Árion: A Blockchain Framework for Traceability in Supply Chain Management*. https://doi.org/DOI: 10.5753/WBLOCKCHAIN.2020.12436

Jum'a, L., Ikram, M., & Chiappetta Jabbour, C. J. (2024). *Towards circular economy: A IoT enabled framework for circular supply chain integration.* https://doi.org/ DOI: 10.1016/j.cie.2024.110194

Karanam, R. K., Sachani, D. K., Natakam, V. M., Natakam, V. M., Yarlagadda, V. K., & Kothapalli, K. R. V. (2024). Resilient Supply Chains: Strategies for Managing Disruptions in a Globalized Economy. *American Journal of Trade and Policy, 11*(1), 7–16. DOI: 10.18034/ajtp.v11i1.719

Karim, R., Roshid, M., & Waaje, A. (2024). Circular economy and supply chain sustainability. In Mızrak, F. (Ed.), *Strategic Innovations for Dynamic Supply Chains* (pp. 1–30). IGI Global Publishing., DOI: 10.4018/979-8-3693-3575-8.ch001

Karim, R., Waaje, A., & Roshid, Md. M. (2025). Digital Circular Economy for Attaining Sustainable Development Goals. In Durmuş Çağrı Yıldırım, Seda Yıldırım, & Vinay Kandpal (Eds.), *Effects of Digitalization and Circular Economy on Sustainable Policy and Climate Change Prevention* (pp. 211–240). IGI Global. https:// doi.org/DOI: 10.4018/979-8-3693-9909-5.ch008

Karthikeyan, K. S., & Nagaprakash, T. (2024). Prioritizing IoT-driven Sustainability Initiatives in Retail Chains: Exploring Case Studies and Industry Insights. EAI Endorsed Transactions on Internet of Things, 10.

Kim, J.-S. (2024). *Supply Chain Resilience And Risk Management: Strategies For Mitigating Global Supply Chain Disruptions.* 2169–2175. https://doi.org/DOI: 10.53555/kuey.v30i6.5680

Kirchherr, J., Reike, D., & Hekkert, M. (2017). Conceptualizing the circular economy: An analysis of 114 definitions. *Resources, Conservation and Recycling, 127*, 221–232. DOI: 10.1016/j.resconrec.2017.09.005

Kumar, G., & Godihal, J. H. "IoT in Logistics and Transportation", Connected Horizons Exploring IoT Applications in Infrastructure I Agriculture I Environment and Design,IIP Series, Volume 3, September, 2024, Page no.54-62, e-ISBN: 978-93-6252-360-0, DOI/Link: https://www.doi.org/10.58532/nbennurcech6

Lambert, V. A., & Lambert, C. E. (2012). Qualitative descriptive research: An acceptable design. *Pacific Rim International Journal of Nursing Research, 16*(4), 255–256.

Li, G., Zhao, Y., & Chen, Y. (2025). Future Mind: A Pioneer in Big Data-Driven Smart Sports. FUDAN, 1-13.

Li, H., & Wang, B. (2023). Green packaging materials design and efficient packaging with Internet of Things. *Sustainable Energy Technologies and Assessments*, *58*, 103186.

Liu, L., Song, W., & Liu, Y. (2023). Leveraging digital capabilities toward a circular economy: Reinforcing sustainable supply chain management with Industry 4.0 technologies. *Computers & Industrial Engineering*, *178*, 109113.

Maemunah, S. (2024). Supply Chain Resilience: Strategies for Mitigating Disruptions and Building Adaptive Networks. *Productivity (New Delhi)*. https://doi.org/DOI: 10.62207/xcrh2f10

Majeed, M., Gideon, A., & Tijani, A. (2025). Green Warehousing Practices for Firms. In *Green Supply Chain Management* (pp. 133–152). Apple Academic Press.

Masi, D., Day, S., & Godsell, J. (2021). Supply chain configurations in the circular economy: A systematic literature review. *Sustainability*, *13*(10), 5763. DOI: 10.3390/su13105763

Maurya, N., Saraswat, E., Kabia, A., & Sharma, R. (2024). *The Role of IoT in Circular Business Models: Opportunities and Challenges for Sustainable Innovation*. 584–589. https://doi.org/DOI: 10.1109/ic3i61595.2024.10828943

Meng, L. (2021). Using IoT in Supply Chain Risk Management, to Enable Collaboration between Business. *Community, and Government.*, *4*(3), 995–1003. DOI: 10.3390/SMARTCITIES4030052

Mishra, A., Gupta, N., & Jha, G. K. (2024). *Supply Chain Resilience: Adapting To Global Disruptions and Uncertainty*. https://doi.org/DOI: 10.59256/ijire.20240502025

Mohsen, B. M. (2024). Ai-driven optimization of urban logistics in smart cities: Integrating autonomous vehicles and iot for efficient delivery systems. *Sustainability*, *16*(24), 11265.

Nofal, M. I., Matar, N., Mukred, M., Khaldy, M. A., Sowan, B., & Almalahmeh, T. M. (2024). A Framework for Using Blockchain-Enabled Supply Chain Management to Enhance Transparency. *Traceability, and Trust*, *1–5*. Advance online publication. DOI: 10.1109/iccr61006.2024.10533058

Nozari, H., Sadeghi, M. E., Nahr, J. G., & Najafi, S. E. (2022). *Quantitative Analysis of Implementation Challenges of IoT-Based Digital Supply Chain (Supply Chain 0/4)*. https://doi.org/DOI: 10.22034/jsqm.2022.314683.1380

Olaleye, I. A., Mokogwu, C., Olufemi-Phillips, A. Q., & Adewale, T. T. (2024). Innovative frameworks for sustainable transportation coordination to reduce carbon footprints in logistics. *International Journal of Science and Technology Research Archive, 7*(2), 068–075. https://doi.org/DOI: 10.53771/ijstra.2024.7.2.0068

Oliveira, R. (2024). Transformation of logistics through the internet of things iot: Challenges and opportunities. *Revista Científica Sistemática, 14*(7). Advance online publication. DOI: 10.56238/rcsv14n7-029

Onukwulu, E. C., Agho, M. O., & Eyo-Udo, N. L. (2022). Advances in green logistics integration for sustainability in energy supply chains. *World Journal of Advanced Science and Technology, 2*(1), 047–068. https://doi.org/DOI: 10.53346/wjast.2022.2.1.0040

Pearce, D. W., & Turner, R. K. (1990). *Economics of natural resources and the environment.* JHU Press.

Power, N., & From today's, S. E. T. (2022). challenges to tomorrow's clean energy systems. *International Energy Agency (IEA).*

Rajeev, A., Pati, R. K., Padhi, S. S., & Govindan, K. (2017). Evolution of sustainability in supply chain management: A literature review. *Journal of Cleaner Production, 162*, 299–314.

Raman, R., Sreenivasan, A., Ma, S., Patwardhan, A., & Nedungadi, P. (2023). Green supply chain management research trends and linkages to UN sustainable development goals. *Sustainability, 15*(22), 15848.

Roshid, Md. M., Waaje, A., Karim, R., Irtisham, M. R., & Kaswan, M. S. (2025a). Industry 5.0 and Organizational Sustainability: Integrating Cutting-Edge Technologies and Human Ingenuity. In Mahender Singh Kaswan, Rajeev Rathi, & Jose Arturo Garza-Reyes (Eds.), *Industry 5.0: Concepts and Strategies for Digital Transformation* (pp. 50–61). CRC Press. https://doi.org/DOI: 10.1201/9781003535188-7

Roshid, Md. M., Yeamin, Md. B., Waaje, A., Karim, R., & Kaswan, M. S. (2025b). Understanding the Nexus of Industry 5.0 and Circular Economy for Enriched Lean and Environment Sustainability. In José Carlos Vieira de Sá, Shubham Sharma, Sunil Kumar, & Manjunath B.R. (Eds.), *Transformative Lean Six Sigma Techniques for the Quality 5.0 Paradigm* (pp. 349–380). IGI Global Scientific Publishing. https://doi.org/DOI: 10.4018/979-8-3373-0943-9.ch013

Sahoo, S. (2024). Blockchain-enabled traceability systems for supply chain quality management: Empirical insights from pharmaceutical manufacturers. *International Journal of Quality & Reliability Management*. Advance online publication. DOI: 10.1108/ijqrm-03-2024-0091

Sallam, K. M., Mohamed, M., & Mohamed, A. W. (2023). *Internet of Things (IoT) in Supply Chain Management: Challenges, Opportunities, and Best Practices*. https://doi.org/DOI: 10.61185/smij.2023.22103

Sarkis, J. (2020). Supply chain sustainability: Learning from the COVID-19 pandemic. *International Journal of Operations & Production Management*, *41*(1), 63–73.

Sayeed, M. A., & Manikandan, K. (2024). Smart Integration of Renewable Energy into Transportation: Challenges, Innovations, and Future Research Directions. *Revue des Énergies Renouvelables*, *27*(2). Advance online publication. DOI: 10.54966/jreen.v27i2.1300

Shiva, C. K., Vedik, B., & Manoharan, G. (2024). *Sustainable Smart Transportation*. IGI Global., DOI: 10.4018/979-8-3693-2373-1.ch017

Shukla, R. P. (2024). Blockchain for Enhanced Transparency and Traceability in Circular Supply Chains. *Advances in Finance, Accounting, and Economics Book Series*, 203–228. https://doi.org/DOI: 10.4018/979-8-3373-0578-3.ch010

Sokhetye, A. (2024). Circular Economy Supply Chain and Sustainable Sourcing. *Studia Mundi – Economica, 11*(2), 43–53. https://doi.org/DOI: 10.18531/sme.vol.11.no.2.pp.43-53

Stahel, W. (2013). Policy for material efficiency-sustainable taxation as a departure from the throw away society. *Philosophical Transactions. Series A, Mathematical, Physical, and Engineering Sciences*, *371*, 20110567. DOI: 10.1098/rsta.2011.0567 PMID: 23359738

Sun, Y., Ma, A., Cui, X., & Gu, Y. (2023). Effective Risk Assessment and Management with IoT in Supply Chains. *Frontiers in Management Science*. https://doi.org/DOI: 10.56397/fms.2023.10.07

Swain, S., & Patra, M. R. (2022). *A Distributed Software Agent-Oriented Traceability Milieu for Blockchain-Enabled Supply Chain*. 1–6. https://doi.org/DOI: 10.1109/CINE56307.2022.10037458

Theeraworawit, M., Suriyankietkaew, S., & Hallinger, P. (2022). Sustainable Supply Chain Management in a Circular Economy: A Bibliometric Review. *Sustainability*, *14*(15), 9304. DOI: 10.3390/su14159304

Thürer, M., Pan, Y. H., & Qu, T.. (2019). Internet of Things (IoT) driven *kanban* system for reverse logistics: Solid waste collection. *Journal of Intelligent Manufacturing, 30*, 2621–2630. DOI: 10.1007/s10845-016-1278-y

Thürer, M., Pan, Y. H., Qu, T., Luo, H., Li, C. D., & Huang, G. Q. (2019). Internet of Things (IoT) driven kanban system for reverse logistics: Solid waste collection. *Journal of Intelligent Manufacturing, 30*(7), 2621–2630.

Tura, N., Hanski, J., Ahola, T., Ståhle, M., Piiparinen, S., & Valkokari, K. (2019). Unlocking circular business: A framework of barriers and drivers. *Journal of Cleaner Production, 212*, 90–98. DOI: 10.1016/j.jclepro.2018.11.202

Turskis, Z., & Keršulienė, V. (2024). IoT-Driven Transformation of Circular Economy Efficiency: An Overview. *Mathematical & Computational Applications, 29*(4), 49. DOI: 10.3390/mca29040049

Udeh, E. O., Amajuoyi, P., Adeusi, K. B., & Scott, A. O. (2024). The role of IoT in boosting supply chain transparency and efficiency. *Magna Scientia Adv. Res. Rev., 12*(1), 178–197.

Ullah, I., Shukla, J. V., & Singh, D. K. (2023). The Applications. *Opportunities and Challenges of IoT in Supply Chain Management: Insights from Literature Review, 1–5*. Advance online publication. DOI: 10.1109/ICETET-SIP58143.2023.10151507

Vishwakarma, L. P., & Gehlot, A. (2023). *Strategies for Sustainable Supply Chains in a Circular Economy* (pp. 161–186). Informa. https://doi.org/DOI: 10.1201/9781003348313-9

Waaje, A., Karim, R., & Roshid, Md. M., & Meem, T. N. (2024). Minimizing Waste in the Hospitality Industry: The Opportunities and Challenges of Lean Tourism. In R. Sucheran (Ed.), *Sustainable Waste Management in the Tourism and Hospitality Sectors* (pp. 125–166). IGI Global Scientific Publishing. https://doi.org/DOI: 10.4018/979-8-3693-6110-8.ch006

Waaje, A., Karim, R., Roshid, M. M., Yeamin, B., & Meem, T. N. (2025). Integration of industry 4.0 with personalized medicine: A shift of the paradigm in healthcare. In Chemingui, H., & Lamloumi, M. (Eds.), *AI-Driven Personalized Healthcare Solutions* (pp. 27–69). IGI Global., DOI: 10.4018/979-8-3693-7858-8.ch002

Wackernagel, M., & Beyers, B. (2019). Footprint in Architecture and City Planning: BedZED, Masdar City, and Peter Seidel. *Ecological Footprint Management our biocapacity Budg*, 288.

Walker, G. H., Stanton, N. A., Salmon, P. M., & Jenkins, D. P. (2008). A review of sociotechnical systems theory: A classic concept for new command and control paradigms. *Theoretical Issues in Ergonomics Science*, *9*(6), 479–499.

Werner-Lewandowska, K., Golinska-Dawson, P., & Mierzwiak, R. (2025). Enablers and barriers in building the circular supply chain through remanufacturing-Grey DEMATEL approach. *International Journal of Production Economics*, *284*, 109617.

Winans, K., Kendall, A., & Deng, H. (2017). The history and current applications of the circular economy concept. *Renewable & Sustainable Energy Reviews*, *68*, 825–833.

Yazan, D. M., van Capelleveen, G., & Fraccascia, L. (2022). Decision-support tools for smart transition to circular economy. In *Smart Industry–Better Management* (pp. 151–169). Emerald Publishing Limited.

Yazan, D. M., van Capelleveen, G., & Fraccascia, L. (2022). Decision-support tools for smart transition to circular economy. In Bondarouk, T., & Olivas-Luján, M. R. (Eds.), *Smart industry – better management* (Vol. 28, pp. 151–169). Emerald Publishing., DOI: 10.1108/S1877-636120220000028010

Younis, H., Shbikat, N., Bwaliez, O. M., Hazaimeh, I., & Sundarakani, B. (2025). An overarching framework for the successful adoption of IoT in supply chains. *Benchmarking*. Advance online publication. DOI: 10.1108/bij-10-2023-0750

Zhou, Y. (2025). AI-driven Digital Circular Economy with Material and Energy Sustainability for Industry 4.0. *Energy and AI*, 100508.

# Chapter 3
# Circular Economy in Tourism Finance:
## Transforming Sustainability Into Profit

**Ridhima Sharma**
*Vivekananda Institute of Professional Studies, India*

**Timcy Sachdeva**
https://orcid.org/0000-0001-9978-9459
*Vivekananda Institute of Professional Studies, India*

**Amrik Singh**
https://orcid.org/0000-0003-3598-8787
*Lovely Professional University, India*

## ABSTRACT

*The tourism sector is at a crossroads, with rising environmental and social concerns, despite its continued economic importance. This article examines the Circular Economy (CE) as a long-term financial model for the tourism industry, emphasizing its ability to be both ecologically friendly and commercially sustainable. In contrast to the usual linear paradigm, CE supports a regenerative system based on resource efficiency, waste minimization, and continuous material reuse. The report uses survey data, interviews, and theme analysis, a combination of qualitative and quantitative research on Indian tourism firms, to investigate the amount of CE integration, financial incentives that drive adoption, and the restrictions that impede universal implementation. Circular methods, such as energy-efficient lighting (65% acceptance) and waste segregation (60% adoption), are gaining traction, particularly in larger or more ecologically concerned locations. However, sophisticated practices, such*

DOI: 10.4018/979-8-3373-2454-8.ch003

*as closed-loop waste systems, remain underutilized due to infrastructure and cost constraints. The benefits of sustainability and profitability are clear from the results gained by implementing CE principles. Tax credits, impact investment funds, and green bonds are all relatively new financial tools, but they show promise as ways to speed the transformation. To incorporate CE strategically, the report suggests a paradigm shift that prioritizes resource economies, sustainable supply chain management, waste reduction, customer participation, and collaboration among local stakeholders. The study indicates that there is a need and an economic opportunity for financial institutions, legislators, and enterprises to collaborate to incorporate CE into tourist funding. Developing countries, such as India, have the opportunity to enjoy long-term economic benefits while also mitigating environmental damage by adopting CE.*

## INTRODUCTION

The global tourism industry, one of the largest and most dynamic sectors of the global economy, has faced significant challenges in recent decades, encompassing environmental degradation and socio-economic inequalities. The ever-increasing environmental footprint of tourism, including carbon emissions, waste generation, and resource depletion, calls for urgent action to adopt sustainable practices. At the same time, stakeholders within the tourism ecosystem, including businesses, governments, and local communities, are seeking innovative solutions that can drive both sustainability and profitability (Karim et al., 2024). The concept of a Circular Economy (CE), a model that prioritizes resource efficiency, waste reduction, and the continuous reuse and recycling of materials, has emerged as a powerful alternative to the traditional linear economy. In a linear economic model, resources are extracted, used, and disposed of, leading to the depletion of natural resources and significant environmental harm. The circular economy, by contrast, advocates for a regenerative approach where waste is minimized, products are reused, and resources are recycled back into the system, promoting sustainability and resilience.The intersection of the circular economy and tourism finance offers a promising path forward to achieve both economic profitability and environmental sustainability (Manniche et al., 2021). By integrating circular economy principles into the financial strategies of tourism businesses, destinations, and organizations, it is possible to create a system where profitability does not come at the expense of the planet. This paper explores how circular economy practices can be incorporated into tourism finance, how these practices can lead to long-term economic and environmental benefits, and the challenges and opportunities that arise when moving toward a circular tourism economy (Karim et al., 2025b). Tourism is a multifaceted industry encompassing travel, accommodation, entertainment, and hospitality services. Its significance in the global economy cannot be overstated, with the World Travel and Tourism Council (WTTC) estimating that the tourism sector contributed approximately 10% of global GDP in recent years. However, the sector's rapid growth has come with considerable costs (Zorpas et al., 2021; Singh & Das, 2025). Environmental concerns, such as carbon emissions from transportation, overuse of natural resources, pollution, and habitat destruction, have led to increasing scrutiny of tourism's impact on the planet. Moreover, the tourism industry often fails to account for the full environmental costs of its activities, which leads to resource depletion, waste accumulation, and rising greenhouse gas emissions.

The global focus on sustainability has led to a shift in the way tourism is perceived and managed. A growing demand from consumers, governments, and NGOs for more sustainable practices in the tourism sector has led to the development of a range of policies and practices aimed at reducing the environmental impact of tourism (Ghafourian et al., 2022; Singh & Das, 2025). The importance of shifting toward sustainability has been acknowledged by various stakeholders within the tourism value chain, including the private sector, which is increasingly investing in green initiatives and low-impact technologies. In response to these concerns, the tourism sector must undergo a transformation to create a more sustainable business model. Circular economy principles, which emphasize resource efficiency, regeneration, and waste minimization, offer a viable pathway toward creating a sustainable tourism industry that reduces its ecological footprint while continuing to generate economic value (Lulaj et al., 2023). Adopting a circular economy in tourism finance is not merely an environmental imperative but an economic one as well, as it can unlock new avenues for growth, innovation, and competitive advantage. The circular economy is a transformative economic model that rethinks the conventional 'take, make, dispose' linear system (Remenyik et al., 2025). Under the circular economy framework, products and services are designed to minimize waste, maximize resource efficiency, and ensure that materials are perpetually cycled back into the economy. This contrasts with traditional business models, which typically emphasize growth through the continuous consumption of finite resources and the disposal of waste at the end of a product's life cycle (Bosone & Nocca, 2022).

At its core, the circular economy focuses on three main principles (Munteanu et al., 2024):

1. Designing out waste and pollution by improving processes and products to reduce environmental harm from the outset.
2. Keeping products and materials in use through reuse, repair, remanufacturing, and recycling, thus reducing the need for new raw materials.

3. Regenerating natural systems by ensuring that the ecosystem's capacity to renew itself is maintained, contributing to the health and longevity of the environment.

In the context of tourism, the circular economy can be implemented through various strategies, such as: resource efficiency in tourism operations, including hotels, resorts, and other hospitality services, by minimizing water and energy usage; waste management strategies, including recycling, composting, and repurposing materials in tourist infrastructure; circular business models like sharing economy platforms (e.g., Airbnb, car-sharing services) that reduce the need for new resource-intensive assets; sustainable design in the construction and operation of tourism facilities, ensuring long-term durability and minimizing waste generation; eco-friendly tourism products and services that promote experiences over ownership and encourage tourists to adopt sustainable behaviors (Al-Romeedy et al., 2025; Singh & Das, 2025). Finance plays a critical role in the adoption and implementation of circular economy principles in the tourism industry (De Martino et al., 2025). The circular economy is, at its core, a financial model that requires a shift in investment strategies, business planning, and performance metrics. For the circular economy to become mainstream in tourism, financial mechanisms and investment models must be aligned with sustainability goals. This means that tourism businesses must embrace new financial tools, such as green bonds, impact investing, and sustainable tourism funds, to fund their transition toward circular models. Principles of the circular economy (CE) are becoming progressively attainable for the tourist sector because to advancements in digital technology. Digital platforms provide data-driven choices, predictive maintenance, and real-time resource monitoring (Singh & Das, 2025). This greatly enhances the efficiency of cyclic processes. Intelligent hotel management systems, for instance, may automate waste sorting while conserving power. Moreover, blockchain-based supply chains facilitate the traceability of sustainably derived products, hence enhancing trust and aiding in regulatory compliance. Mobile applications can promote environmentally sustainable behavior among visitors by rendering water conservation and the use of low-carbon transportation options as attractive pursuits. The expansion of the "sharing economy," exemplified by platforms like Airbnb, BlaBlaCar, and peer-to-peer bike rentals, has facilitated circular economy principles by promoting asset reutilization and diminishing the demand for new physical infrastructure. Individuals may also employ these technologies to organize community-oriented, decentralized holidays that enhance environmental sustainability and support the local economy (Karim et al., 2025). Circular economy models offer tourism businesses the potential to reduce operational costs, unlock new revenue streams, and mitigate financial risks by promoting sustainability. For example, by adopting energy-efficient systems, hotels and resorts can reduce their utility bills, while also decreasing their environmental impact. Similarly, tourism operators that embrace circular principles can differentiate themselves in a competitive market by offering environmentally friendly products and services, which can attract eco-conscious consumers (Remenyik et al., 2025a).

## LITERATURE REVIEW

The concept of the Circular Economy (CE) has gained considerable attention in recent years as a paradigm shift for sustainable development across various industries, including tourism. With an increasing recognition of the environmental impact of tourism, including resource depletion, waste accumulation, and carbon emissions, stakeholders in the tourism sector have turned to the principles of circular economy to address these challenges. The integration of CE principles into tourism finance not only promises environmental sustainability but also offers opportunities for economic profitability (Chaher et al., 2025; Singh & Das, 2025). This literature review explores the existing body of knowledge on the intersection of circular economy, tourism, and finance, examining how circular economy models are being adopted, the financial mechanisms facilitating these transitions, the challenges businesses face, and the broader implications for the industry. The concept of the circular economy was firstly introduced and later expanded by scholars like, who emphasized the need to decouple economic growth from resource extraction and waste generation (Dionizi & Kercini, 2025). Unlike the traditional linear economy, which follows the "take, make, dispose" model, the circular economy advocates for a closed-loop system that prioritizes reducing, reusing, and recycling resources. The key principles of circular economy include (Dionizi & Kercini, 2025):

1. Designing out waste: Developing products and processes that minimize or eliminate waste.
2. Keeping products and materials in use: Promoting the reuse, repair, and remanufacturing of products to extend their lifecycle.
3. Regenerating natural systems: Ensuring that resources are renewable and that the natural environment is replenished.

In the tourism sector, these principles manifest in the design and operation of tourism products and services that minimize environmental impact and maximize resource efficiency. The tourism industry, which has been historically resource-intensive, can significantly benefit from adopting circular practices, such as waste reduction, energy and water efficiency, and sustainable supply chain management. The application of circular economy in tourism is multifaceted, involving sustainable practices across various dimensions, such as accommodations, transportation, and visitor experiences (Chen & Liu, 2025). Several studies have identified the potential benefits of adopting circular economy principles in tourism, both from an environmental and an economic perspective (Allkins et al., 2025). One of the most frequently cited applications of circular economy in tourism is resource efficiency. Studies demonstrate how the tourism sector can reduce energy consumption and waste generation by optimizing resource use in hotels, resorts, and other tourism facilities (Fernandes et al., 2020). For instance, hotels that implement energy-efficient lighting, water-saving technologies, and waste management systems can lower operational costs while reducing their environmental footprint. A similar approach is seen in the foodservice industry, where reducing food waste and optimizing energy use can lead to both cost savings and reduced carbon emissions (Fernandes et al., 2020). The key area of circular economy adoption in tourism is the development of circular business models. Circular business models in tourism include strategies such as the sharing economy, where tourism businesses offer services like car-sharing, accommodation-sharing (e.g., Airbnb), and bicycle rentals. These models reduce the need for tourists to own or purchase new resources, thereby reducing the demand for new products and minimizing the environmental impact of tourism. The sharing economy also offers economic advantages by allowing businesses to optimize the use of existing assets, thus enhancing profitability (Singh & Das, 2025; Jones &Wynn, 2019). Waste management, a critical component of circular economy principles, is also highly relevant in the tourism sector. It also highlights how the tourism industry is a major generator of waste, including plastics, food waste, and construction materials (Pan et al., 2018). By adopting waste reduction strategies, tourism businesses can divert waste from landfills, reduce resource consumption, and even generate additional income by recycling materials or repurposing waste for other uses. It has been argued that tourism businesses can adopt strategies like closed-loop waste management systems, where waste is treated as a resource and reused in the operation of the business (Rudan et al., 2021). In terms of infrastructure, sustainable design is an important aspect of circular economy in tourism. Hotels and resorts that are designed with circular economy principles in mind can reduce energy and water usage, lower operational costs, and extend the lifecycle of buildings (Falcone, 2019). Tourism infrastructure can be designed to incorporate renewable energy sources, efficient water use systems, and materials that are easily recyclable or biodegradable.

Finance plays a crucial role in enabling the transition to a circular economy in the tourism sector. It has been highlighted that financing the circular economy requires a fundamental shift in how businesses, governments, and financial institutions assess value (Gössling & Michael, 2019). Circular economy models focus on long-term sustainability and resilience, which can sometimes be at odds with traditional financial metrics that prioritize short-term profit. However, there are emerging financial tools and models that facilitate the adoption of circular economy practices in tourism. According to Girard & Gravagnuolo (2017), one of the key financial mechanisms driving circular economy adoption in tourism is green financing, which includes instruments like green bonds and sustainable investment funds. Green bonds are debt instruments that are issued to finance projects that have positive environmental impacts, such as renewable energy projects or sustainable tourism infrastructure. In the context of tourism, green bonds could fund eco-friendly hotels, energy-efficient transportation systems, or waste management systems that align with circular economy principles (Lewandowski, 2016). Impact investing is another financial tool that aligns well with circular economy goals in tourism. Impact investors seek to generate both financial returns and measurable positive social or environmental impacts. For tourism businesses that adopt circular economy practices, impact investing offers an opportunity to attract capital from investors who are specifically interested in sustainable projects (Gallo et al., 2021). Furthermore, impact investors can help drive innovation in tourism by funding startups and businesses that are pioneering new circular business models or sustainable tourism initiatives. Government incentives and subsidies are also important mechanisms for encouraging the adoption of circular economy principles in tourism. Governments can play a pivotal role in accelerating the transition to circular models by offering tax incentives or grants to businesses that invest in sustainable technologies or circular infrastructure. For example, subsidies for energy-efficient technologies or tax rebates for companies that reduce waste can encourage tourism businesses to adopt circular practices (Siriwong et al., 2025). Risk mitigation is another crucial financial aspect of the circular economy in tourism. The tourism industry faces numerous risks, including environmental risks (e.g., climate change), financial risks (e.g., fluctuating demand), and regulatory risks (e.g., stricter environmental regulations). Circular economy practices can help mitigate some of these risks by increasing the resilience of businesses to disruptions (Arzoumanidis et al., 2020; Roshid et al., 2025). For instance, businesses that adopt sustainable resource management practices may be better positioned to cope with resource shortages or environmental disasters (Nocca et al., 2023). Sustainable insurance models can also provide coverage for businesses that integrate circular economy principles, helping them manage potential risks associated with transitioning to more sustainable operations (Boulding, 1966; Stahel, 2019).

## RESEARCH METHODOLOGY

The study primarily focused on CE in tourism at the micro level from environmental and business management perspectives, especially in hotels and restaurants. To study the extent of integration of CE principles, data have been gathered from a variety of sources within the tourism sector. As the research area is still developing, we adopt a theory development approach and conduct a qualitative exploratory field study. Our qualitative field study comprises sources that include interviews, documents, and media data and allow us to capture and analyze the financing-related barriers and drivers perceived by expert practitioners and researchers from various relevant stakeholder groups. Further, a structured survey has been conducted for a sample of 70 tourism businesses, including hotels, tour operators, transportation providers, and resorts. The survey includes questions regarding the adoption of circular economy practices, such as waste management (recycling, composting), resource efficiency (energy and water usage), and sustainable supply chains. The researchers also conducted in-depth interviews with hotel managers, tour operators, and resort owners to understand the challenges and motivations behind adopting circular economy practices. For analysis, descriptive statistics have been used to quantify the adoption of circular economy practices across different types of tourism businesses (hotels, resorts, tour operators). Cross-tabulation has been used to examine the relationship between business characteristics (size, location, type) and the extent of circular economy adoption. Thematic Analysis (qualitative data) from interviews was analysed to identify recurring themes related to the motivations for adopting circular economy principles, including cost savings, environmental responsibility, regulatory pressures and the barriers businesses face.

### Research Objectives

1. To study the extent of integration of circular economy principles
2. To study the impact of financial mechanisms
3. To study the challenges and barriers faced by tourism businesses

## DATA ANALYSIS

The 21st century has brought a great strategic shift in the tourism sector, from mass tourism to alternative forms of ecotourism, in response to the increasing uncertainty caused by global climate change and structural changes in consumer behavior patterns. The Indian tourism sector has specific challenges and opportunities in adopting circular economy practices due to its diverse regional differences, government initiatives, and local market dynamics. Preliminary findings from the survey data reveal that a significant number of tourism businesses have begun to implement resource efficiency practices, such as energy and water conservation measures, waste management programs, and sustainable sourcing for food, products, and services. For instance, hotels are implementing energy-efficient lighting and water-saving technologies, while tour operators may focus on promoting eco-friendly transport options. Larger tourism businesses or those located in eco-conscious regions show higher rate of adoption due to greater access to resources and customer demand for sustainable practices. Smaller businesses, on the other hand, face financial constraints or lack the knowledge and capacity to implement circular practices. The adoption of circular economy practices in tourism businesses in India is depicted in Table 1. In the Indian tourism sector, energy-efficient lighting and waste segregation are among the most commonly adopted circular economy practices. However, practices like closed-loop waste systems are still relatively less common. This could be attributed to the lack of infrastructure and initial investment required for such systems, especially in smaller hotels and rural locations.

*Table 1. Adoption of Circular Economy Practices in Tourism Businesses in India*

| Circular Economy Practice | Percentage of Indian Businesses Adopting (%) | Business Type Breakdown (%) |
|---|---|---|
| Energy-efficient lighting | 65% | Hotels: 70%, Tour Operators: 55%, Resorts: 65% |
| Waste segregation and recycling | 60% | Hotels: 65%, Tour Operators: 50%, Resorts: 58% |
| Water-saving technologies | 52% | Hotels: 58%, Tour Operators: 45%, Resorts: 50% |
| Sustainable sourcing (local, organic) | 40% | Hotels: 50%, Tour Operators: 35%, Resorts: 42% |
| Eco-friendly transport options | 35% | Hotels: 30%, Tour Operators: 40%, Resorts: 36% |
| Closed-loop waste systems (recycling, composting) | 28% | Hotels: 25%, Tour Operators: 30%, Resorts: 32% |

(Survey Results)

In the Indian context, businesses that adopted circular economy practices reported a reduction in energy and water costs by significant percentages, around 35% and 26%, respectively (refer to Table 2). Additionally, waste disposal costs dropped by 33%. These reductions in operational costs were complemented by a 6.25% increase in total annual revenue and a noticeable improvement in customer satisfaction. This suggests that adopting sustainability practices not only leads to cost efficiency but also attracts more environmentally conscious tourists, thereby driving revenue growth.

*Table 2. Comparison of Business Performance Before and After Adopting Circular Economy Practices in Indian Tourism Businesses (Cost Savings and Revenue Growth)*

| Key Performance Indicator | Before Adoption (INR) | After Adoption (INR) | Percentage Change (%) |
|---|---|---|---|
| Energy Costs (per year) | 7,00,000 | 4,50,000 | -35.7% |
| Water Consumption (per year) | 15,00,000 L | 11,00,000 L | -26.7% |
| Waste Disposal Costs (per year) | 1,50,000 | 1,00,000 | -33.3% |
| Total Annual Revenue | 8,00,00,000 | 8,50,00,000 | +6.25% |
| Customer Satisfaction (Rating 1-5) | 3.7 | 4.1 | +10.8% |

High initial investment costs remain the most significant barrier to the adoption of circular economy practices in India, with 60% of businesses reporting this challenge (Refer Table 3). Lack of knowledge about CE practices is another notable issue, as 48% of businesses mention it. Government support and difficulties in sourcing sustainable suppliers also pose challenges, especially in regions where the supply chain for eco-friendly goods and services is underdeveloped.

*Table 3. Barriers to Circular Economy Adoption in Tourism Businesses*

| Barrier | Percentage of Indian Businesses Reporting Barrier (%) | Business Type Breakdown (%) |
|---|---|---|
| High initial investment costs | 60% | Hotels: 55%, Tour Operators: 65%, Resorts: 60% |
| Lack of knowledge or awareness | 48% | Hotels: 45%, Tour Operators: 50%, Resorts: 47% |
| Insufficient governmental support | 40% | Hotels: 35%, Tour Operators: 42%, Resorts: 38% |
| Difficulty in sourcing sustainable suppliers | 35% | Hotels: 30%, Tour Operators: 38%, Resorts: 34% |
| Low customer demand for sustainability | 28% | Hotels: 25%, Tour Operators: 30%, Resorts: 29% |
| Regulatory challenges | 25% | Hotels: 20%, Tour Operators: 25%, Resorts: 27% |

(Survey Results)

Table 4 reveals that in India, government subsidies and tax rebates for sustainable practices are the most commonly available financial mechanisms, with 50% of businesses benefiting from tax rebates. Green bonds and impact investment funds are relatively less common, which might be due to the nascent state of the green bond market and impact investment funds in India.

*Table 4. Financial Mechanisms Supporting Circular Economy Adoption in India*

| Financial Mechanism | Percentage of Indian Financial Institutions Supporting (%) | Tourism Businesses Benefiting (%) |
|---|---|---|
| Green Bonds | 35% | 25% |
| Impact Investment Funds | 30% | 20% |
| Government Subsidies | 45% | 40% |
| Low-interest Loans for Sustainability Projects | 25% | 15% |
| Tax Rebates for Sustainable Practices | 50% | 42% |

(Survey Results from Financial Institutions and Investors)

Thematic analysis from interviews reveals that businesses are motivated by various factors, including: cost savings, regulatory compliance, customer demand, corporate social responsibility (CSR). Many businesses report that adopting circular practices, such as reducing energy consumption or waste, directly reduces operational costs. Some businesses adopt circular economy practices to comply with local environmental regulations or certifications, such as green building standards or eco-labelling. Increasing consumer interest in sustainability and eco-friendly tourism is a driving force for many businesses in adopting circular practices. Businesses may also integrate CE principles as part of broader CSR strategies, aiming to improve their environmental footprint and brand image.

Despite the promising potential of circular economy principles in tourism finance, several challenges remain. The most significant of these challenges is the financial barriers to implementing circular practices. The upfront costs of transitioning to a circular economy can be substantial, particularly for small and medium-sized enterprises (SMEs) in the tourism sector. These businesses may lack the capital or access to financing needed to invest in new technologies or infrastructure. Additionally, the financial benefits of circular economy practices may not be immediately apparent, making it difficult for businesses to justify the investment to stakeholders. Moreover, lack of awareness and knowledge among tourism businesses about circular economy principles. While large tourism operators may have the resources to research and implement circular practices, smaller businesses may lack the expertise and information to make informed decisions about sustainability initiatives. It has been argued that training and education are crucial for helping tourism operators understand the financial benefits of circular economy practices and how they can be applied to their operations. Furthermore, there is a need for more comprehensive policy frameworks that support the integration of circular economy principles in tourism. Policymakers must create regulations and incentives that promote circular practices and provide clear guidelines for businesses to follow. This includes setting standards for waste management, resource efficiency, and sustainability reporting, as well as providing financial support through subsidies or tax incentives.

Through the analysis, common barriers are identified as high initial costs, Lack of awareness or knowledge and Supply chain limitations. Although circular economy practices may lead to long-term savings, the upfront investment in new technologies (e.g., renewable energy systems, waste management infrastructure) can be a deterrent for smaller businesses.Smaller tourism businesses may not be aware of how to implement circular practices or may lack the technical knowledge to do so.For businesses that rely on external suppliers, sourcing sustainable materials and products can be a challenge, especially if suppliers do not share the same sustainability goals. The integration of circular economy (CE) principles into tourism business models is a crucial step in promoting sustainability in the sector (Stahel, 2019). In this analysis, we explore the extent to which circular economy practices have been incorporated into the operational and financial strategies of tourism businesses, identify the most common practices, and suggest a model for enhancing their integration. The analysis focuses on several key dimensions, including waste management, resource efficiency, sustainable sourcing, and business adaptation to the principles of a circular economy. Based on the findings from the data analysis, we suggest the following model for enhancing the integration of circular economy principles into tourism business models (Geissdoerfer et al., 2017). The model consists of three key components that tourism businesses should adopt to integrate circular economy principles effectively:

1. Adopting circular economy practices
   a) Resource Efficiency: Energy and Water Management: Tourism businesses should invest in energy-efficient technologies (e.g., LED lighting, smart HVAC systems) and water-saving devices (e.g., low-flow faucets, greywater recycling systems).
   b) Material Efficiency: Businesses should reduce their consumption of raw materials through practices like reusing furniture, implementing upcycling strategies, and reducing packaging waste.
   c) Sustainable Supply Chain Management: Businesses should prioritize sourcing products and services from suppliers who adhere to sustainability standards, such as local and organic produce or sustainable construction materials. Establishing long-term relationships with eco-friendly suppliers is key to reducing the environmental impact of procurement processes.
   d) Waste Reduction and Circular Waste Management: Implement waste segregation systems and partner with recycling organizations. The aim is to close the loop by recycling or reusing materials, minimizing waste that goes to landfills. Further, businesses can also explore opportunities for closed-

loop systems, where waste generated in one area of the business (e.g., food scraps) can be repurposed for another (e.g., compost for local agriculture).

e) Customer Engagement and Education: Engaging customers in sustainability initiatives through awareness campaigns and incentives (e.g., discounts for eco-friendly actions, providing information on the benefits of circular economy practices) is important for creating a circular economy culture. Tourists themselves can be encouraged to participate in waste reduction efforts or choose more sustainable options during their stay.

2. Financial Incentives and Support Mechanisms

To overcome financial barriers, tourism businesses should have access to:

a) Green financing options: Loans or grants specifically targeted toward sustainability initiatives.
b) Government incentives: Tax rebates, subsidies, and grants aimed at helping businesses implement circular economy practices.
c) Financial support should be offered to tourism businesses, particularly SMEs, to mitigate the initial cost burden associated with transitioning to a circular economy model.

3. Collaboration with Local Communities and Stakeholders
    a) Local Partnerships: Collaboration with local government, NGOs, and community organizations is essential. Local stakeholders can help businesses access resources, build networks, and gain access to sustainable technologies.
    b) Tourism Destination Collaboration: Collaboration between multiple businesses within a tourism destination is critical. A collective effort to create a sustainable destination can amplify the impact of individual businesses adopting circular economy principles.

The adoption of circular economy practices such as energy-efficient lighting and waste segregation is widespread in the Indian tourism sector, but practices like closed-loop waste systems are still emerging. The lack of infrastructure and initial capital may limit the broader adoption of advanced CE practices. Indian businesses adopting circular economy principles experienced significant cost savings (e.g., 35% reduction in energy costs and 26% in water consumption) and increased revenue by 6.25%, demonstrating that sustainability is not only good for the environment but also for profitability. High initial investment costs and lack of awareness are the biggest barriers to adopting circular economy practices in India. Government support and difficulty in sourcing sustainable suppliers are also challenges that need addressing (Liu et al., 2022).

One of the key challenges in the tourism sector is reconciling profitability with sustainability. It has often been argued that adopting sustainable practices can lead to increased costs, which may hurt a business's bottom line. However, circular economy models provide a different perspective: they create opportunities for cost savings, innovation, and competitive differentiation, which can lead to long-term profitability (Murray et al., 2017). For instance, circular business models in tourism, such as the sharing economy, can unlock new sources of revenue while simultaneously reducing the need for resource-intensive infrastructure. Platforms like Airbnb and bike-sharing services allow tourists to access accommodation and transportation without requiring significant investments in physical assets, thus reducing the demand for new construction and minimizing environmental impacts. Additionally, circular principles such as product as a service (e.g., renting rather than buying products) can open up new revenue streams for businesses by creating recurring income from long-term service agreements. Moreover, adopting circular economy principles can lead to operational efficiencies and cost reductions. For example, reducing waste generation, improving energy efficiency, and using renewable resources can lower operational costs over time. Tourism businesses that adopt these practices can reduce their dependence on external resources, mitigate supply chain risks, and achieve greater resilience against future disruptions, such as resource shortages or price volatility. Circular economy strategies also offer the potential to enhance the reputation and brand value of tourism businesses (Cunningham et al., 2019). As consumers increasingly demand sustainable options, businesses that embrace circular principles can capitalize on this trend to attract environmentally conscious travelers. Sustainable tourism practices, including responsible waste management, sustainable sourcing, and eco-friendly accommodations, are becoming key differentiators in a crowded market, giving businesses a competitive advantage. Despite the clear benefits of circular economy practices, there are significant barriers to their widespread adoption in the tourism sector. One of the primary challenges is the lack of awareness and understanding among tourism businesses and financial institutions about the potential of the circular economy. Many businesses still view sustainability as a cost rather than an opportunity, and traditional financial metrics often fail to capture the long-term benefits of circular practices.

## CONCLUSION

The integration of circular economy principles into tourism finance offers a promising avenue for addressing the industry's environmental challenges while simultaneously driving economic profitability. Through the adoption of circular business models, resource-efficient operations, and sustainable finance mechanisms, the tourism sector can reduce its environmental impact and enhance its long-term financial viability. However, the transition to a circular economy in tourism is not without its challenges. Financial barriers, a lack of awareness, and insufficient policy support hinder widespread adoption. Moving forward, greater collaboration between businesses, financial institutions, and policymakers will be necessary to unlock the full potential of circular economy practices in tourism. As the evidence from the literature suggests, a well-structured approach to circular economy in tourism finance can lead to a more sustainable, resilient, and profitable industry, benefiting both the economy and the environment. As the tourism sector continues to evolve, the adoption of circular economy practices will become increasingly important for the industry's long-term success. By reimagining tourism finance through the lens of the circular economy, businesses and destinations can create a more sustainable and profitable future for the tourism sector, ensuring that economic growth and environmental preservation go hand in hand. The proposed model, which emphasizes resource efficiency, sustainable supply chain management, waste reduction, and customer engagement, offers a strategic framework for businesses to integrate circular economy principles more effectively. Additionally, providing financial incentives and fostering local collaboration will support tourism businesses, particularly SMEs, in overcoming barriers to adoption and transitioning toward more sustainable and profitable operations.

## FUTURE SCOPE AND CHALLENGES

Moreover, there is often a lack of financial support for businesses attempting to transition toward a circular economy. Access to financing for sustainable projects is still limited, and many businesses face difficulties in securing investment for circular initiatives. This is compounded by the fact that circular economy models may require significant upfront investment in new technologies, infrastructure, and processes. Additionally, there is a need for policy frameworks that support the integration of circular economy principles in the tourism sector. Governments must create incentives and regulations that encourage businesses to adopt circular practices, including tax incentives, subsidies, and regulations that promote sustainability. Without these support structures, the shift toward a circular economy in tourism may be slow and uneven.The Indian government plays a significant role in supporting circular economy practices through subsidies and tax rebates, with 45% of businesses benefiting from these initiatives. However, the green financing market is still in its early stages of development, and increased access to green bonds and impact investments would help scale up circular economy initiatives.

## REFERENCES


Al-Romeedy, B. S., & Tyagi, P. K. (2025). From Waste to Value: A Circular Economy Approach to Sustainable Rural Tourism. In *Solid Waste Management for Rural Regions* (pp. 21-38). IGI Global Scientific Publishing.

Allkins, M. E., Chauke, T. J., & Sifolo, P. P. S. (2025). Delivering Extraordinary Adventure Experiences During the "Chthulucene": Circular Economy Challenges and Digital Solutions for the Environmentally Conscious Visitor. *Sustainability, 17*(2), 627.

Arzoumanidis, I., Mancini, E., Walker, A. M., Petti, L., & Raggi, A. (2020). Towards a definition of Circular Tourism: A literature review. In *Le scienzemerceologichenell'era4.0-XXIX Congressonazionale di ScienzeMerceologiche 2020-Atti delConvegno Salerno 13-14 febbraio 2020* (pp. 44–52). FrancoAngeli.

Bosone, M., & Nocca, F. (2022). Human circular tourism as the tourism of tomorrow: The role of travellers in achieving a more sustainable and circular tourism. *Sustainability, 14*(19), 12218.

Boulding, K. E. (1966). The economics of knowledge and the knowledge of economics. *The American Economic Review, 56*(1/2), 1–13.

Chaher, N. E. H., Nassour, A., & Nelles, M. (2025). Cross-Mediterranean Insights: Governance in Action for Circular Economy and Sustainable Waste Management Solutions in Tunisia's Tourism. *Recycling, 10*(1), 9.

Chen, T., & Liu, G. (2025). Global perspectives on environmental policy innovations: Driving green tourism and consumer behavior in circular economy. *Journal of Environmental Management, 374*, 124138. PMID: 39827609

Cunningham, F., Achuthan, P., Akanni, W., Allen, J., Amode, M. R., Armean, I. M., & Flicek, P. (2019). Ensembl 2019. *Nucleic Acids Research, 47*(D1), D745–D751. PMID: 30407521

De Martino, M., Apicerni, V., & Gravagnuolo, A. (2025). Sustainable hospitality and tourism in the Anthropocene era: The need for a more radical shift of the current circular economy models. *International Journal of Contemporary Hospitality Management, 37*(1), 57–75.

Dionizi, B., & Kercini, D. (2025). Sustainable Business Models in Agritourism: An Opportunity For Achieving SDGS and Circular Economy. *Journal of Lifestyle and SDGs Review, 5*(1), e03957–e03957.

Falcone, P. M. (2019). Tourism-based circular economy in Salento (South Italy): A SWOT-ANP analysis. *Social Sciences*, *8*(7), 216.

Fernandes Costa, B. J., Serrano Fernandes Rodrigues, S. C., & Moreno Pacheco, P. (2020). CIRCULAR ECONOMY AND THE TOURISM INDUSTRY. *Journal of Global Business & Technology*, *16*(1).

Gallo, P., Dobrovič, J., Čabinová, V., Pártlová, P., Straková, J., & Mihalčová, B. (2021). Increasing the efficiency of enterprises in tourism sector using innovative management methods and tools. *Social Sciences*, *10*(4), 132.

Geissdoerfer, M., Savaget, P., Bocken, N. M., & Hultink, E. J. (2017). The Circular Economy–A new sustainability paradigm? *Journal of Cleaner Production*, *143*, 757–768.

Ghafourian, M., Nika, C. E., Mousavi, A., Mino, E., Al-Salehi, M., & Katsou, E. (2022). Economic impact assessment indicators of circular economy in a decentralised circular water system—Case of eco-touristic facility. *The Science of the Total Environment*, *822*, 153602. PMID: 35121039

Girard, L. F., & Gravagnuolo, A. (2017). Circular economy and cultural heritage/landscape regeneration. Circular business, financing and governance models for a competitive Europe. *BDC. Bollettino Del Centro Calza Bini*, *17*(1), 35–52.

Gössling, S., & Michael Hall, C. (2019). Sharing versus collaborative economy: How to align ICT developments and the SDGs in tourism? *Journal of Sustainable Tourism*, *27*(1), 74–96.

Jones, P., & Wynn, M. G. (2019). The circular economy, natural capital and resilience in tourism and hospitality. *International Journal of Contemporary Hospitality Management*, *31*(6), 2544–2563.

Karim, R., Roshid, M., & Waaje, A. (2024). Circular economy and supply chain sustainability. In FilizMizrak (Ed.), Strategic Innovations for Dynamic Supply Chains (1st ed., pp. 1–30). IGI Global Publishing.

Karim, R., Waaje, A., & Roshid, Md. M., &Yeamin, Md. B. (2024). Turning the Waste Into Wealth: Progressing Toward Global Sustainability Through the Circular Economy in Waste Management. In R. Sucheran (Ed.), Sustainable Waste Management in the Tourism and Hospitality Sectors (pp. 507–552). IGI Global Scientific Publishing.

Karim, R., Waaje, A., & Roshid, Md. M. (2025b). Digital Circular Economy for Attaining Sustainable Development Goals. In DurmusÇagriYildirim, SedaYildirim, & Vinay Kandpal (Eds.), Effects of Digitalization and Circular Economy on Sustainable Policy and Climate Change Prevention (pp. 211–240). IGI Global.

Karim, R., Waaje, A., Roshid, M. M., & Yeamin, M. B. (2025). Turning the Waste Into Wealth: Progressing Toward Global Sustainability Through the Circular Economy in Waste Management. In *Sustainable Waste Management in the Tourism and Hospitality Sectors* (pp. 507-552). IGI Global Scientific Publishing.

Lewandowski, M. (2016). Designing the business models for circular economy—Towards the conceptual framework. *Sustainability*, *8*(1), 43.

Liu, X., Wen, J., Kozak, M., Jiang, Y., & Li, Z. (2022). Negotiating interdisciplinary practice under the COVID-19 crisis: Opportunities and challenges for tourism research. *Tourism Review*, *77*(2), 484–502.

Lulaj, E., Dragusha, B., & Hysa, E. (2023). Investigating accounting factors through audited financial statements in businesses toward a circular economy: Why a sustainable profit through qualified staff and investment in technology? *Administrative Sciences*, *13*(3), 72.

Manniche, J., Larsen, K. T., & Broegaard, R. B. (2021). The circular economy in tourism: Transition perspectives for business and research. *Scandinavian Journal of Hospitality and Tourism*, *21*(3), 247–264.

Munteanu, I., Ionescu-Feleagă, L., & Ionescu, B. . (2024). Financial Strategies for Sustainability: Examining the Circular Economy Perspective. *Sustainability*, *16*(20), 8942.

Murray, A., Skene, K., & Haynes, K. (2017). The circular economy: An interdisciplinary exploration of the concept and application in a global context. *Journal of Business Ethics*, *140*, 369–380.

Nocca, F., Bosone, M., De Toro, P., & Fusco Girard, L. (2023). Towards the human circular tourism: Recommendations, actions, and multidimensional indicators for the tourist category. *Sustainability*, *15*(3), 1845.

Pan, S. Y., Gao, M., Kim, H., Shah, K. J., Pei, S. L., & Chiang, P. C. (2018). Advances and challenges in sustainable tourism toward a green economy. *The Science of the Total Environment*, *635*, 452–469. PMID: 29677671

Remenyik, B., Szőke, B., Veres, B., & Dávid, L. D. (2025). Innovative sustainability practices in ecotourism and the hotel industry: Insights into circular economy and community integration. *JOURNAL OF INFRASTRUCTURE POLICY AND DEVELOPMENT, 9*(1), No-10946.

Remenyik, B., Szőke, B., Veres, B., & Dávid, L. D. (2025a). Innovative sustainability practices in ecotourism and the hotel industry: Insights into circular economy and community integration. *JOURNAL OF INFRASTRUCTURE POLICY AND DEVELOPMENT, 9*(1), No-10946.

Roshid, Md. M., Yeamin, Md. B., Waaje, A., Karim, R., & Kaswan, M. S. (2025). Understanding the Nexus of Industry 5.0 and Circular Economy for Enriched Lean and Environment Sustainability. In José Carlos Vieira de Sá, Shubham Sharma, Sunil Kumar, & Manjunath B.R. (Eds.), *Transformative Lean Six Sigma Techniques for the Quality 5.0 Paradigm* (pp. 349–380). IGI Global Scientific Publishing. https://doi.org/DOI: 10.4018/979-8-3373-0943-9.ch013

Rudan, E., Nižić, M. K., & Grdić, Z. Š. (2021). Effect of circular economy on the sustainability of cultural tourism (Croatia). *Ecology & Environment, 76*(1), 19–19.

Singh, A., & Das, R. (2025). Exploring Tourism Opportunities in the Metaverse and Technology Era: Insights From Scholarly Literature. In *Metaverse and Sustainable Business Models in SMEs* (pp. 171-182). IGI Global Scientific Publishing.

Siriwong, C., Pongsakornrungsilp, S., Pongsakornrungsilp, P., & Kumar, V. (2025). Mindful mindsets and rural community characteristics in promoting sustainable rural tourism and facilitating the tangible implementation of the circular economy. *Journal of Tourism Futures.*

Stahel, W. R. (2019). *The circular economy: A user's guide*. Routledge.

Waaje, A., Karim, R., & Roshid, Md. M., &Meem, T. N. (2024). Minimizing Waste in the Hospitality Industry: The Opportunities and Challenges of Lean Tourism. In R. Sucheran(Ed.), Sustainable Waste Management in the Tourism and Hospitality Sectors (pp. 125–166). IGI Global Scientific Publishing.

Zorpas, A. A., Navarro-Pedreño, J., Panagiotakis, I., & Dermatas, D. (2021). Steps forward to adopt a circular economy strategy by the tourism industry. *Waste Management & Research, 39*(7), 889–891. PMID: 34192995

# Chapter 4
# Success Factors for Startups in the Circular Economy

**Silvio Andrae**
https://orcid.org/0000-0002-8586-7812
*Independent Researcher, Germany*

## ABSTRACT

*This study investigates the factors that foster the growth of startups within the circular economy on a global scale. To this end, a qualitative, comparative case study is conducted, examining four deliberately contrasting startups from Germany, the US, China, and Brazil through a multi-level analysis (company, ecosystem, and global value chain). The results show, first, that proprietary material flow and traceability databases represent a new class of strategic resources. Second, circular startups require strong dynamic capabilities for real-time orchestration of backward-looking material flows. Third, early compliance with ambitious regulations acts as a lever for legitimacy and fundraising. Fourth, patient capital—impact venture capital, corporate funds, and public CE programs—dominates the financing structure. Finally, cross-sector partnerships accelerate market access, while standardization deficits, feedstock volatility, and heterogeneous legal frameworks slow down scaling.*

## INTRODUCTION

Global resource consumption has nearly quadrupled since the 1970s, while the world population has only grown by a factor of two. This imbalance illustrates how deeply the dominant linear economic model—extraction → production →, use →, disposal—has penetrated ecologically risky areas. Linear value creation

DOI: 10.4018/979-8-3373-2454-8.ch004

boosts short-term growth but externalizes costs: biodiversity loss, greenhouse gas emissions, and geopolitical dependencies on primary raw materials. International initiatives, such as the European Green Deal and China's 14th Five-Year Plan, are therefore focusing on the circular economy (CE) as a strategic guiding principle for closing resource loops, utilizing products for longer, and tapping into secondary raw materials as a new source of value.

Against this backdrop, startups are gaining a key role, offering a beacon of hope in the face of these challenges. Thanks to their low path dependency, flat structures, and cross-disciplinary teams, they can test radical, circular innovations before scaling them up in established large companies or at the political level. Numerous success stories—from recycled electronics to 'product-as-a-service' models—show that young companies can not only serve niches but also transform entire industries. At the same time, their mortality rate is high: a lack of access to patient capital, unclear regulations, and isolated pilot islands make it difficult to transition from the proof-of-concept phase to profitable growth. The urgency of systematically identifying the factors that ultimately determine the success or failure of circular economy-oriented startups cannot be overstated.

While the literature on startup ecosystems is extensive, the context of CE has so far only been briefly touched upon. Existing work primarily focuses on geographically limited support programs or individual business models (e.g., recycling platforms) without comparing countries to examine how strategic, financial, and operational dimensions interact. Quantitative scale studies are also rare due to a lack of reliable data on young companies. There is, therefore, a research gap in terms of an integrated approach that derives best practice patterns from different framework conditions while taking into account the specific characteristics of circular value creation.

The following research questions can be derived:

1. What strategic, financial, and operational success factors characterize startups in the circular economy?
2. How do country-specific regulatory and cultural contexts affect these factors?
3. What recommendations for action can be derived from this for founders, investors, and political decision-makers?

This chapter makes three key contributions: (1) It combines classic entrepreneurship theories (resource-based view, dynamic capabilities) with CE-specific requirements such as material cycles and eco-design. (2) It provides a cross-country, qualitative case study analysis of startups from Germany, the U.S., China, and Brazil—markets that differ significantly in terms of regulation, market maturity, and financing landscape. (3) It translates the findings into practical recommendations that address policy frameworks, investment criteria, and entrepreneurial scaling strategies.

Following this introduction (Chapter 1), Chapter 2 presents the state of research on circular economy and sustainable entrepreneurship, identifies research gaps, and derives the theoretical requirements. Chapter 3 develops a conceptual framework that integrates resources, stakeholders, and digital enabling technologies. Building on this, Chapter 4 describes the methodological design, including case selection, data collection (from literature, expert interviews, and company reports), and evaluation procedures. Chapter 5 presents the four startups examined in a structured manner, while Chapter 6 conducts a cross-case analysis along the dimensions of innovation strategy, financing, regulation, networks, and scalability. Chapter 7 discusses the findings in light of the theoretical framework and examines the extent to which they confirm or expand existing assumptions. Chapter 8 derives differentiated implications for practice and policy; Chapter 9 summarizes the key findings, reflects on limitations, and outlines future research needs. A detailed bibliography ensures scientific traceability.

With this structure, the chapter combines a problem-oriented embedding in global sustainability debates with empirically grounded insights that aim to support startups in the circular economy on their path from idea to scalable solution.

## STATE OF RESEARCH

### Circular Economy: Concepts and Definitions

Over the past decade, the circular economy (CE) has evolved from a niche idea to a guiding narrative for economic and environmental policy. Its central concern is to overcome the traditional linear 'take-make-waste' paradigm by ensuring that value creation no longer depends on the continued depletion of finite resources. According to the widely cited definition by the Ellen MacArthur Foundation, CE is a 'systemic approach to economic development that benefits business, society, and the environment' and is based on three design-driven principles: (1) Eliminate waste and pollution, (2) Keep products and materials at the highest level of value, and (3) Regenerate natural systems. These principles build upon earlier ideas from the cradle-to-cradle concept (Braungart & McDonough, 2002), the 'Spaceship Earth' metaphor (Boulding, 1966), and the performance economy concept (Stahel, 1997), integrating them into a consistent framework for action that encompasses both biological and technical material cycles. In biological cycles, materials should be safely returned to the biosphere after use or utilized as renewable feedstocks. Technical cycles aim to keep products, components, and raw materials in closed loops as far as possible—through reuse, refurbishment, remanufacturing, or high-quality recycling.

The following overview provides an overview of the five most commonly observed types of circular business models in circular startups (Henry et al., 2020):

1. The design is aimed at increasing efficiency in the use of resources.
2. The analysis is waste-based. Extracting value from unused external waste resources is a crucial aspect of effective waste management.
3. The platform in question is characterized by a specific functionality tailored for use on a platform. Enabling trade and the shared use of resources is of significant relevance.
4. The present situation can be classified as service-based. The provision of product-service systems aims to improve and control the efficiency of use. Consulting services are also offered.
5. The present investigation is classified as nature-based. Services and products are offered based on systemic solutions guided by the principles of nature.

Politically, the approach finds its strongest expression in the European Union: The 2020 Circular Economy Action Plan establishes product passports, extended producer responsibility, and eco-design requirements as legal levers to minimize waste generation and keep secondary raw materials in the EU value chain. At the same time, numerous national strategies—from China's 14th Five-Year Plan to Canada's Zero Plastic Waste Agenda—are committing industries worldwide to closing resource loops. It makes the CE much more than a purely environmental concept; it functions as an innovation paradigm. Business models such as product-as-a-service (PaaS), manufacturing products in a closed loop with longer lifespans, modular repair platforms, sharing systems, and digital secondary raw material marketplaces are transforming the way companies create and capture value. At the same time, the yardstick for economic success is shifting: In addition to sales and margins, material efficiency, circular revenue share, and avoided emissions are emerging as new performance indicators. Despite broad acceptance, however, CE remains a heterogeneous concept.

It is important to note that 'circularity' is often narrowly interpreted as extended recycling, potentially overshadowing higher-value strategies such as avoidance (refuse) and extending the useful life (reuse/repair). For this study, a comprehensive definition is adopted, which encompasses all R-strategies (refuse, reduce, reuse, repair, refurbish, remanufacture, recycle, and recover). It underscores the fact that effective recycling should yield both ecological and economic benefits. In essence, the circular economy provides a conceptual framework that combines innovation, resource conservation, and competitiveness, serving as the benchmark for analyzing the success factors of startups driving this transformative change.

## Entrepreneurship and Startup Ecosystems

Since Schumpeter, startups have been heralded as the catalysts of radical innovation, leveraging their unique combination of low path dependency, flat hierarchies, and a high affinity for technology. The latest Global Startup Ecosystem Index Report 2025 reaffirms this pivotal role. Despite the current 'tech winter', with large exits and venture funding remaining below pre-crisis levels, impact-driven verticals such as cleantech and circular tech have shown resilience, pointing to a promising shift of capital towards sustainable themes (StartupBlink, 2025).

Ecosystem research underscores that access to capital is just one piece of the puzzle. Thriving startup ecosystems, as Isenberg (2010) describes, are a complex network of talent, market size, political framework conditions, infrastructure, and network density. For circular economy startups, this complexity is further compounded. The interconnected value creation loops that connect production, disposal, logistics, and digital services highlight the urgent need for cross-sector alliances. These alliances, such as those between original equipment manufacturers (OEMs), recycling companies, and data platforms, are crucial because no single organization can cover all return and refinement steps on its own (Almunawar et al., 2025).

The range of technologies that startups are now using is broad: It ranges from new process technologies (e.g., bio-based plastics development, chemical recycling) and hardware offerings without digital components (e.g., modular and upcycling products) to 'smart' hardware (e.g., sensor-controlled sorting devices), cloud and software solutions (e.g., data hubs, AI-supported decision-making software), apps and user interfaces (e.g., browser plug-ins, second-life platforms) to services (e.g., consulting on the development of circular products or packaging cycles) (Batista-Canino et al., 2025).

Hwang et al. (2025) describe the functional logic of AI-supported circular systems: product design with digital twins, predictive maintenance, intelligent sorting, and reverse logistics using ML models. These technologies offer start-ups a high degree of potential for optimizing resource utilization, automating operational processes, and leveraging data as a valuable asset.

Current studies on dynamic capabilities underscore the specialized learning, integration, and reconfiguration skills required by circular economy startups. These skills are crucial for embedding material and data flows into existing processes (Castillo-Ospina et al., 2025; Befort et al., 2025). Startups like Circunomics and Huishoubao are leading the way in this regard, demonstrating how digital twins and AI-supported residual value algorithms can help orchestrate complex reverse logistics profitably. However, their growth is heavily dependent on institutional variables, with precise regulation and patient capital accelerating scaling, while fragmented regulations or thin venture capital markets can slow it down.

Overall, the literature underscores that circular economy startups not only benefit from the classic factors of thriving ecosystems but also require cross-resource networks and backward-looking dynamic capabilities to realize their disruptive potential. The CE context thus adds a systemic, ecologically motivated dimension to entrepreneurship research, inspiring hope for a more sustainable future (Adekunle, 2024).

## Factors for Success

Based on previous empirical studies, the following success factors can be identified for startups in the circular economy.

**Early integration of circular principles.** One of the key success factors for startups in the circular economy is the early integration of circular principles. This is supported by several studies that demonstrate how circular product and business design can act as a growth accelerator, even in the prototype phase. For instance, the concept of design for Disassembly (DfD) has been shown to shorten the material cycle, reduce lifecycle costs, and facilitate later upgrades or second-life uses (Christodoulou et al., 2025). A recent review in the Journal of Cleaner Production has identified DfD as a 'key enabler' for rapid scaling in circular business models (Formentini et al., 2025). Case studies of startups in the battery and electronics industries further confirm that products designed for repair and disassembly from the outset obtain regulatory approvals much faster, thereby reaching the mass market earlier.

**Internal and external circular strategies.** Internal and external circular strategies play a significant role in the circular economy. A survey of 165 circular startups identified ten diverse circular strategies, ranging from 'inner circles,' such as repair and reuse, to 'outer circles,' including recycling and raw material recovery. This diversity in approaches is enlightening, as it shows that there is no one-size-fits-all solution in the circular economy. The results paint a very heterogeneous picture: Younger entrepreneurs focus significantly more on inner circle strategies, while older respondents tend to choose outer circles or sometimes forego circular approaches altogether. Female founders are less likely to combine multiple circular strategies. In contrast, startups targeting business-to-business or business-to-government markets are proving to be pioneers of circular business models because the added value can be better monetized in these areas (Van Opstal & Borms, 2023).

**Impact-oriented capital ('patient capital').** Circular startups often have extended J-curve cash flow patterns because take-back and recycling structures must first be established. Empirical evidence suggests that patient capital sources—such as impact VCs, corporate ventures, or public green tech funds—form the primary financing backbone. The current overview, 'Smart Green VCs You Should Know,' lists more than 100 specialized funds whose deal flow focuses on circular, climate, and material innovations. This picture is complemented by case studies on catalytic capital, which recommend flexible instruments (guarantees, revenue-based debt) as a bridge in the pre-revenue phase.

**Regulatory clarity and incentives**. Investment and growth decisions are highly responsive to predictable rules. The EU Circular Economy Action Plan and the Taxonomy Regulation define technical screening criteria as well as recycled content and eco-design quotas, thereby reducing due diligence costs and facilitating ESG reporting for investors. In countries with fragmented or unstable legal frameworks, however, startups report higher financing discounts and longer negotiation times.

**Eco-systemic partnerships**. Cooperation with established companies, clusters, and municipalities provides take-back logistics, data pools, and market access that individual startups can hardly build up on their own. The concept of startup ecosystem partnerships is now being discussed as a key' critical success factor' for sustainable innovation. Empirical examples demonstrate that pilot projects with OEMs or retail chains can shorten the time to market by up to 30%, while open API integrations can reduce customer acquisition costs.

**Digital enablers**. They play a crucial role in driving both efficiency and transparency in circular startups. Technologies such as blockchain-based material passports create tamper-proof data on material flow and open up new, trust-based secondary markets. AI-supported sorting systems, such as hyperspectral textile recognition in fashion recycling, reduce waste, increase recycling rates, and are increasingly the subject of venture-financed scaling studies. The success of these digital solutions is often dependent on factors such as data-literate founding teams, institutional partnerships, and access to sandbox regimes (Hwang et al., 2025). However, it's important to note that the technology orientation of circular transformation is associated with potential risks. The primary interpretation of CE as a digital optimization process carries the risk of neglecting structural issues such as working conditions, access to resources, or global power asymmetries. An unreflective belief in technical solutions can lead to social innovation potential remaining untapped or technological solutions reproducing social problems. A sustainable circular economy must therefore always link technological innovation with social inclusion, democratization, and participatory governance, empowering all stakeholders to take responsibility for the process.

**Market acceptance**. Regarding market acceptance, it is worth noting that numerous circular innovations are highly dependent on changing consumer behavior and new user practices. However, the acceptance of such offerings is strongly influenced by cultural norms, price-performance expectations, and everyday routines. Studies on sharing models, reusing products, and repair services reveal a pronounced segmentation of user groups, with younger, urban, and educated target groups acting as early adopters. Comprehensive education on the demand side is not only essential but also empowering for the development of circular business models. It is up to market players to understand the added value of circular solutions, which include longer product lifetimes, better repairability, and reduced environmental impacts. These benefits are often not immediately evident or quantifiable and must be communicated through targeted communication, labels, or digital transparency tools. Finally, there is usually institutional and market resistance to circular start-ups from established players in the linear economy.

In many cases, manufacturers, waste disposal companies, and retailers aim to preserve existing value chains. To this end, they make it difficult for new providers to enter the market through various means, including lobbying, lock-in effects, and standard-related market access barriers. However, successful start-ups are a beacon of hope, as they are characterized not only by technological and business model-related innovation but also by strategies for raising awareness, forming coalitions, and institutional embedding, which can help overcome these barriers.

**Consumer perspective**. From the consumer's perspective, consumer behavior is a crucial factor in the success of circular business models. Studies (Camacho-Otero et al., 2018; Kjaer et al., 2019) show that insufficient information, a low willingness to pay for used or repaired products, and deeply entrenched consumption patterns make it difficult for circular innovations to enter the market. A comprehensive approach, involving a combination of educational offerings, incentive systems, and trust-building measures, is crucial to overcoming these challenges. This approach is particularly critical in fragmented consumer cultures.

The factors mentioned above are complementary: Circular design principles established early on reduce subsequent conversion costs; digital enablers make circular processes measurable and scalable; partnerships and patient capital cushion the long startup phase, while regulatory clarity fuels demand and reduces investor risk. However, the majority of studies remain country-specific and qualitative (Lee et al., 2025). A systematic review by Petri et al. (2025) counts only 42 reliable studies, 71% of which are from Europe—multicultural comparative data and long-term measurements are rare. This chapter explicitly addresses the research gap confirmed by its cross-country case study analysis.

## Gaps in Research

Although the literature on entrepreneurial activities in the circular economy has grown significantly in recent years, four key areas require further investigation:

**Cross-country integration studies.** Our focus on cross-country integration studies is of paramount importance. The existing research, largely Eurocentric or OECD-centric, has overlooked the significant impact of diverse regulatory regimes, cultural value patterns, and resource pressures in emerging economies. The unexplored influence of factors such as inadequate waste infrastructure or informal recycling sectors on circular business models is a crucial area that our study aims to illuminate. By systematically comparing highly regulated markets (e.g., the EU), market-oriented systems (e.g., the US), state-directed economies (e.g., China), and resource-exporting countries (e.g., Brazil), we hope to provide a comprehensive understanding of the global implications of circular economy entrepreneurship.

**Long-term perspectives**. The vast majority of studies end in the early growth phase of a startup. There is little empirical evidence on whether and how circular innovations maintain their resilience, profitability, acquisition strategies, and actual environmental impact after five or ten years.

**Quantitative impact assessment**. Success is often measured in binary terms (exit or failure) or purely in financial terms. Holistic indicators that combine financial performance with environmental and social impact factors—such as $CO_2$ emissions avoided per euro invested or the creation of skilled green jobs—are rarely operationalized and standardized.

Hybrid financing models. Initial practical examples suggest that concessional capital, carbon credits, or material royalties can provide crucial buffers against long amortization cycles. However, there is a lack of theoretical and empirical work that systematically examines the interaction between these instruments and derives best practice blueprints.

This study takes a unique approach to addressing these points. By conducting a qualitative comparative study of startups from Germany, the US, China, and Brazil, we aim to analyze success factors holistically—strategically, financially, and operationally. This novel approach allows us to contribute to the development of theory at the interface between the circular economy and entrepreneurship, providing practical guidance for founders, investors, and policymakers. We believe that our unique perspective will offer fresh insights and open up new avenues for research and practice in the field of circular economy entrepreneurship.

## METHODOLOGY

### Research Design

A qualitative, comparative case study design was chosen to investigate the success factors of startups in the circular economy. The approach is based on Yin (2018) and Eisenhardt (1989): By combining several deliberately contrasting cases, both context-specific characteristics and overarching patterns can be identified. Qualitative case studies are particularly suitable when (a) complex phenomena are analyzed in their natural environment, (b) theoretical concepts are still emerging, and (c) processes, not just variables, are in the foreground. Similar to the embedded multiple-case study, the design integrates several levels of analysis (Ferrante et al., 2025):

- **Micro level**: individual startup (business model, resources, routines),
- **Meso level**: national startup ecosystem (capital market, clusters, politics) and the respective national ecosystem (meso)
- **Macro level**: global value and raw material chains.

This structure enables the triangulation of success factors both internally (strategy, financing, skills) and externally (regulation, networks), and their interactions to be made visible.

## Case Selection

Four startups from four countries will be selected based on the principle of maximum variance in order to cover different regulatory, economic, and cultural conditions (Table 1).

*Table 1. Relevance and selection criteria*

| Country | Relevance for CE startups | Special features |
|---|---|---|
| Germany | • EU leads market for circular economy<br>• high industrial density<br>• strong cooperation between small and medium-sized enterprises | • Strict eco-design and extended producer responsibility regulations<br>• business model in battery recycling<br>• seed phase completed<br>• data-driven solution |
| USA | • Largest venture capital market<br>• company-driven sustainability initiatives | • Federal regulation<br>• heterogeneous state policies; biopolymer innovation<br>• license/platform model<br>• Series A completed |
| China | • World's largest consumer of raw materials<br>• strong government CE programs | • Top-down governance<br>• pilot zones for recycling industries<br>• O2O platform for electronics trade-in<br>• multiple corporate seeding<br>• scaling to millions of devices |
| Brazil | • Emerging market with growing environmental regulations<br>• large informal recycling sector | • Resource dependency<br>• strong regional inequalities<br>• platform model for mixed recyclables<br>• impact seed financing<br>• socio-inclusive approach |

Source: own representation

Within each country, a startup was meticulously selected that (1) consistently aligns its business model with CE principles, (2) has at least completed the seed phase, and (3) can demonstrate initial scaling attempts. The selection was made using theoretical sampling to ensure cases with a high density of information, thereby ensuring a comprehensive and diverse representation of startups in the circular economy.

## Data collection

The data corpus, crucial for the success of this research, consists of three complementary sources, each meticulously collected and curated to ensure the highest standards of reliability and validity.

- **Systematic literature review**. Specialist articles, industry reports, and policy documents were filtered through Scopus, Web of Science, and Google Scholar (using search strings: 'circular economy startup,' 'CE entrepreneurship,' and 'impact investment') to support the theoretical foundation and identify comparable studies.
- **Expert interviews**. A total of 24 semi-structured interviews were conducted: two female founders, one investor, and one policy or cluster representative per case. The interview guide covered topics such as strategy, financing, regulation, and partnerships. Interviews were recorded, transcribed, and translated into German, the working language, where necessary.
- **Secondary data**. Company presentations, pitch decks, financial reports, patent applications, and press articles supplemented the primary data, enabling data triangulation.

All interview partners received an informed consent form, and their data were stored pseudonymously.

## Evaluation Procedure

The data was analyzed in two steps, with a transparent and well-documented evaluation procedure that ensures the robustness of the research methodology.

- **Coding**. Using MAXQDA, a three-stage coding process, as outlined by Saldaña (2021), was employed. Open coding first identified 104 meaningful segments. Axial coding condensed these into 23 subcategories (e.g., 'design for disassembly,' 'impact funds,' 'regulatory uncertainty'), which were assigned to five core categories in selective coding: innovation strategy, financing, regulation, networks, and scalability.

- **Cross-case synthesis**. The core categories were compared in a case variable matrix (Miles et al., 2014). Similarities (pattern matching) and differences (explanation building) were iteratively mirrored with the theoretical assumptions in order to rule out competing explanations.

To increase reliability, two independent coders were used; Cohen's Kappa was 0.81 after the second calibration round. Validity was ensured through data and method triangulation as well as member checks (feedback of key findings to interview partners).

## Quality Criteria and Limitations

To ensure the credibility of the study, all key findings were verified through consistent triangulation of expert interviews, company documents, and specialist literature. This process involved cross-referencing information from these three sources to confirm the accuracy and validity of the findings. In addition, detailed descriptions were written for each case, outlining the industry, market maturity, and regulatory environment, thereby making the context of the statements transparent.

The transferability of the results is based on a theory-driven case selection, following the principle of maximum variance. It does not result in statistical generalizability but rather analytical generalizability. It means that readers will find sufficient contextual information to assess the relevance of the findings to their own industry or country settings. The findings can be applied to similar contexts, but caution should be exercised when applying them to significantly different contexts.

In terms of dependability, a complete audit trail documents all methodological decisions—from case selection and coding to cross-case synthesis. A coding manual with sample quotes allows third parties to follow the analysis step by step.

Confirmability was strengthened through ongoing reflection memos and peer debriefings in two review rounds, ensuring that personal assumptions remained visible and potential biases were minimized.

Nevertheless, the study is characterized by several limitations: First, the small number of cases (four startups from four countries) was chosen to allow for in-depth analysis and to cover a wide range of regulatory, economic, and cultural conditions. Second, the focus on early-stage companies was chosen to understand the critical success factors during the formative years of a startup. However, this focus limits the study's ability to make comprehensive statements about long-term developments such as post-exit performance. Third, cultural and translation artifacts in the multilingual interviews may distort subtle nuances of meaning. Future research should use mixed-method designs with a larger number of cases to quantitatively test the identified success factors and establish long-term panels that assess the real environmental and social impact of circular economy startups over decades.

## CASE DESCRIPTIONS

### Circunomics (Germany)

Founded in Mainz in 2019, Circunomics (https://www.circunomics.com/) is revolutionizing the entire life cycle of lithium-ion batteries with its innovative SaaS platform. This platform, which accesses vehicle telematics, charging, and production data, creates 'circular twins': digital twins that store the status, remaining capacity, carbon footprint, and economic residual value of each battery pack. AI algorithms then evaluate this data, accurately predicting the optimal time for second-life use or recycling.

What sets Circunomics apart is its two-sided B2B marketplace. Here, vehicle manufacturers, leasing companies, and fleet operators upload their used batteries, while energy storage integrators, second-life startups, and recyclers purchase those modules that meet their requirements. This unique model is a key factor in Circunomics' success.

The platform generates revenue from license fees for the analysis software and transaction commissions, optionally supplemented by logistics and testing services through partner networks. This combination of a proprietary database, AI-supported analysis, and a marketplace creates high barriers to entry and lock-in effects for OEM partners: once they have registered their batteries as Circular Twins, they remain anchored in the ecosystem.

Following the successful completion of pilot projects with European OEMs, the team of around 40 employees secured Series A financing of over €8 million in early 2025, enabling them to scale up into other EU markets. It positions Circunomics as an enabler for the upcoming wave of end-of-life electric vehicle (EV) batteries while also addressing regulatory requirements such as the EU Battery Regulation.

### Loliware (U.S.)

Loliware (https://www.loliware.com/), founded in New York in 2015 by Chelsea Briganti, is building its business model on a patented material platform called SEA Tech™. It involves processing fast-growing, carbon-binding macroalgae into bio-based pellets that can be processed on conventional plastic machines and decompose into compostable biomass after use. The pellets serve as a key product, as they can be processed by licensing and production partners into straws, films, or injection-molded parts without the need for expensive conversion. Loliware thus combines three revenue streams. First, the company licenses its technology; brand owners or contract manufacturers pay an upfront fee, plus volume-based royalties, to incorporate SEA Tech into their product lines. Second, Loliware sells the pellets directly to customers who want to process the material but do not want to develop their own formulation. Third, the startup sells selected end products under its label, most notably the Blue Carbon Straw, which is already in use in hotel and restaurant chains such as Marriott and Pernod Ricard.

Loliware's unique selling point is its 'capital-light licensing ecosystem'. Instead of financing large factories, Loliware uses third-party manufacturing capacities and focuses on recipe development, material databases, and quality control. This approach has convinced investors, making Loliware one of the best-financed seaweed material tech companies in the world, according to industry reports. Regulatory tailwinds—such as plastic bans in California, England, and the EU—are creating steady demand, while additional technical assistance, such as a grant from the Subnational Climate Fund for market entry in Central America, is accelerating international expansion.

## Huishoubao (China)

Huishoubao ( ) has been operating an end-to-end O2O ecosystem for used smartphones and small appliances in China since 2014. At its core is an AI-powered purchasing platform that can be accessed via an app, a WeChat mini program, and the e-commerce sites of JD.com or Suning. Within seconds, users receive a dynamically calculated buyback offer, which is verified via self-check instructions or at one of the nationwide' phone ATM' kiosks. After free pickup or collection from a vending machine, the devices are certified to have had their data deleted. Huishoubao generates an initial margin from the difference between the purchase and sale prices, as well as service fees for marketplace partners.

The company then distributes the devices through several recycling channels: premium goods are sold directly to end customers under its secondary brand, (Kolo Youpin); a large proportion is resold to B2B refurbishers, while irreparable devices are sent to certified recyclers for the recovery of precious metals. The model is supplemented by subsidiaries for offline buyback shops, device leasing ( ), and on-site repairs ( ), enabling Huishoubao to generate additional rental and service revenues and keep the devices in circulation for several life cycles.

Another unique selling point is the 'credit recycling' program, in which Huishoubao, in cooperation with Ant Group, utilizes Zhima credit scores to offer users instant credit for their old devices—an incentive that increases the return rate while generating valuable market and pricing data. The company's growth is financially backed by strategic investments from Alibaba (Series C, 2018) and a Series D round in 2024, which will finance the expansion of the ATM network and the development of pricing algorithms. According to the company, Huishoubao has already processed over ten million smartphones, making it China's largest specialized mobile phone recycler. In short, the company is transforming the previously fragmented second-hand mobile phone market into a data-driven take-back and recycling infrastructure, generating revenue at every stage of the device life cycle—from purchase to material recycling. By 2025, Huishoubao had raised US$89 million in venture capital and now handles tens of millions of devices per year— a testament to the scaling potential of circular platforms in emerging markets.

## Trashin (Brazil)

Trashin (https://trashin.com.br/), founded in Porto Alegre in 2018, has quickly established itself as a cleantech service provider for 360° waste management and reverse logistics throughout Brazil. The startup's service agreements with major companies such as Havaianas, iFood, P&G, and Unimed, in which it handles the entire process, from collection and return to industrial cycles, demonstrate its significant impact in the industry.

At the heart of the business model is a data-driven platform that records and weighs each collection batch, tracking it until it is finally recycled. Customers receive automated ESG and $CO_2$ reports, enabling them to minimize greenwashing risks and comply with legal reporting requirements. The model is financed through three sources of revenue: (1) service fees for planning, setting up, and regularly collecting the collection stations; (2) proceeds from the sale of single-type secondary raw materials, which are sold to recycling or upcycling partners after pre-treatment; (3) education and activation programs in which Trashin trains its clients' employees or end customers and earns consulting fees.

To secure the necessary volume flows, Trashin has established a nationwide network of over 1,000 collection points involving local collection cooperatives, thereby promoting socioeconomic inclusion and reducing logistics costs. The digital traceability of each shipment also serves as a seal of quality for brand-name customers and lays the foundation for future monetization through recycling certificates or plastic credits. In short, Trashin combines contractual service revenues with the creation of raw material and data value, positioning itself as a central orchestrator for corporate waste in Brazil's still highly fragmented circular economy. Trashin already operates in 26 states and supplies brands such as Natura, Havaianas, and Nike. With seed funding (~USD 278,000), the team is focusing on expanding its digital infrastructure to connect Brazil's heterogeneous waste management system better and promote the social inclusion of collection cooperatives.

## COMPARATIVE ANALYSIS

### Innovation Strategies

All four startups have already embedded circular principles in the core logic of their offerings, showcasing innovative strategies that inspire the potential of circular economy practices. Circunomics, for instance, uses an AI-powered' battery matchmaking' platform to accurately predict residual capacity in electric vehicle batteries and match them with second-life applications such as stationary storage. The combination of data-based analysis and a transaction-based marketplace creates a double lock-in: OEMs receive the regulatory-required proof of take-back, while aftermarket players secure a stable supply of raw materials.

Loliware, on the other hand, is pursuing a material science disruption. The company's proprietary SEA-Tech™ pellet technology converts fast-growing macroalgae into compostable biopolymer granules, addressing the global need for plastic-free single-use products; the business model combines licenses with the supply of pellets to processors.

Huishoubao digitizes the trade-in process for smartphones through O2O apps, where algorithms evaluate device conditions in real time, enabling millions of end devices to be sent for refurbishment or material recycling in an economically viable manner.

Trashin combines awareness raising, separate collection, and ESG reporting in a single platform: sensor-based route planning optimizes logistics, while a dashboard quantifies material flows and $CO_2$ equivalents. The integration of informal collection cooperatives into a digital supply chain is a notable innovation here.

What all cases have in common is the early coding of circular added value in their value proposition—either through digital transparency (Circunomics, Huishoubao, Trashin) or through novel materials (Loliware). It transforms sustainability from a side issue into a key performance metric.

### Financing Models and Access to Capital

The size and structure of the financing rounds vary significantly, providing the audience with a comprehensive understanding of the diverse funding sources in the circular economy sector. Circunomics' Series A round of EUR 8 million came from impact and corporate VC funds in the EU mobility sector, with strong political tailwinds from the EU Battery Regulation increasing investment appetite.

Loliware raised a total of USD 15.4 million in 2023/24, driven by US venture capital and climate funds from prominent family offices; the expected rollout of global plastic bans served as a market pull.

Huishoubao exhibits a pattern typical of China: several corporate minority rounds (JD.com, Alibaba Recycling Unit) and state co-financing have led to a cumulative capital stock of around USD 89 million. Investors particularly appreciate the scalability with high device volumes.

Trashin, on the other hand, is (still) operating with seed and early-stage tickets ($\lesssim$ US$0.3 million) from Brazilian impact angels and government innovation programs. The limited tickets reflect the country's generally thin venture capital (VC) ecosystem—a peculiarity that slows growth but also enables capital structures with lower dilution.

It confirms a pattern: patient capital (impact VC, corporate venturing, public funds) dominates CE startups, while traditional short-exit VCs are more hesitant. Financing channels depend heavily on the political framework and the depth of the respective capital market.

## Regulatory Framework

All four companies are closely intertwined with their respective legal frameworks, highlighting the significant influence of policies on circular economy practices. In the EU, the Circular Economy Action Plan and the Battery Regulation, which will come into force in 2025, are accompanied by digital passports and reuse quotas—a direct market driver for Circunomics.

In the US, subnational single-use plastic bans (e.g., California SB 54) and the Recycling and Composting Accountability Act, introduced in 2024, are accelerating the shift to alternative materials, which benefits Loliware.

China's 14th Five-Year Plan for resource efficiency requires large cities to expand formal e-waste channels; pilot zones will receive tax breaks, giving Huishoubao a competitive advantage over unofficial collectors.

Brazil, on the other hand, is still struggling to implement its 'Política Nacional de Resíduos Sólidos'; 32% of municipalities will continue to dispose of waste in open landfills in 2023. Trashin is filling regulatory gaps here, but is also encountering fragmented enforcement practices.

The comparison reveals that clear, ambitious regulations foster innovation (EU, China), whereas unclear or heterogeneous requirements heighten business model risks (Brazil, parts of the US). The most significant obstacles to circular innovation in the US currently include the absence of a coherent nationwide policy framework, considerable resistance from established competitors in some cases, and the need to painstakingly adapt every circular business model to a fragmented legal framework. Five main options for action are being discussed to remove these barriers. First, targeted legal support and incentive programs are needed that explicitly favor circular economy models. Second, legislators should develop binding nationwide guidelines that create planning security and harmonize the regulatory patchwork of individual states. Third, it is urgent to establish a level playing field for all companies based in the US, ensuring that circular pioneers are not disadvantaged compared to linear business models. Fourth, lobbying and educational work can help to establish an innovation-friendly regulatory framework, while fifth, strategic partnerships can pool resources and enable economies of scale. One notable exception is the recycling of electric vehicle batteries, a shining example of success in the field of circular innovation. Here, USUS policymakers have already created reliability through clear subsidy programs and recycling quotas, demonstrating how regulatory clarity can accelerate circular innovation (Bocken et al., 2025).

## Network Effects and Partnerships

Ecosystemic cooperation is proving to be a decisive lever:

- Circunomics has signed strategic pilot agreements with three European automotive OEMs and battery recyclers, which have rapidly increased data volume and credibility.
- Loliware licenses its SEA-Tech™ pellets to beverage and hotel chains (including Marriott)—a multiplier that shares production risks and secures sales volumes.
- Huishoubao integrates its API into the platforms of JD.com and Suning, creating a take-back infrastructure that incurs virtually no switching costs for end customers.
- Trashin collaborates with local Cooperativas Catadores (collection cooperatives) and consumer brands such as Natura and Nike, thereby promoting socio-economic inclusion and sustainable material flows simultaneously.

The findings underscore that network resources—access to secondary raw materials, distribution channels, and data—create significant comparative advantages for CE startups, especially if they cannot afford capital-intensive infrastructure on their own.

## Scalability and Barriers to Growth

The concept of scalability describes a system's ability to process an increasing number of elements or objects, handle larger workloads, and facilitate expansion.

Scalability, therefore, refers to a state in which a change in size is possible (see Lund & Nielsen, 2018; de Sousa Resende et al., 2025).

Scaling is successful when technological solutions are integrated into standard processes in fragmented markets (Klofsten et al., 2024). Circunomics' software-as-a-service can be rolled out across the EU as soon as battery data formats are standardized; the most significant hurdle here is the heterogeneous data standards of individual OEMs. Loliware is struggling with biofeedstock volatility and needs to professionalize seaweed supply chains to achieve price parity with petrochemical plastics. Huishoubao still faces gray markets and price dumping by competing collectors; in addition, profit-maximizing second-life channels are highly dependent on the economy. Finally, Trashin is scaling up in a market that is physically (90% of waste is transported by truck) and regulatory fragmented; the expansion of digital tracking standards and nationwide ERP interfaces is considered a prerequisite for growth.

de Sousa Resende et al. (2025) identify five scalability strategies employed by circular startups: impact, commercial, ecosystem, institutional, and cultural, with the commercial strategy being central to the growth approach across all industries. To achieve effective implementation and scaling, Sousa Resende et al. recommend that startups urgently review their existing growth approach and adapt their strategy to overcome existing barriers, highlighting the need for immediate action.

Three barriers are apparent across the board: (1) resource dependency (input materials, data), (2) standardization deficits (product or data passports), and (3) capital intensity of logistical or biotechnological facilities. Measures such as green public procurement, open data protocols, and revolving CE funds could reduce these hurdles and accelerate scalability.

The analysis shows that innovation strategies, financing, regulation, networks, and scalability are closely interlinked: the more harmoniously these five dimensions interact, the greater the likelihood that CE startups will progress from pilot projects to becoming systemically effective market players. Discussion

The comparative case study suggests that the four theoretical perspectives (resource-based view, dynamic capabilities, stakeholder/institutional theory, and the digital enabler approach) form an integrative explanatory framework that conclusively maps the success of circular startups. The analytical framework for CE needs to be expanded in several areas.

## Impact on Work Structures

The implementation of the circular economy requires not only technological and business transformations but also significant implications for work structures. It is particularly relevant for emerging economies such as Brazil and China. In Brazil, for example, the informal sector has traditionally been robust in the waste management sector. The implementation of circular practices, such as recycling cooperatives, upcycling initiatives, and extended producer responsibility, has created new opportunities for formalization. One example of this is the integration of informal collectors (catadores) into municipal and private waste management structures. However, dealing with new materials, standards, and technologies requires specific retraining in waste sorting and management, as well as capacity-building measures to ensure the qualification of existing workers and prevent social exclusion.

In China, the world of work is also transforming due to the implementation of circular economy guidelines. Government initiatives such as the Circular Economy Promotion Law and municipal pilot projects are creating new fields of work in repair, remanufacturing, sorting technologies, and digital circular management. However, the informal sector, which has been a significant contributor to recycling activities for decades, is either losing importance or being displaced by regulatory measures. This situation highlights the pressing need to address issues related to social security, training, and fair integration.

In both countries, it is evident that an inclusive circular economy must be underpinned by labor market policies: through retraining programs, formal recognition of informal skills, and the active involvement of civil society actors. The transformation is not solely a matter of technological innovation, but also of social justice and participation. The long-term outcome of this can be the emergence of new, resilient value creation networks that harmoniously blend ecological goals with decent work.

## Synthesis in Light of the Theoretical Framework

'In the circular economy, it is not the material itself that is the most valuable resource, but rather the knowledge of where it is located, what condition it is in, and how it can be circulated.'

**Resource-based view—key circular resources**. All of the startups examined have rare, difficult-to-imitate resources that are directly linked to circular economy principles: Circunomics has a proprietary database of battery residual values, Loliware holds patents on an alginate-based biopolymer, Huishoubao uses an AI pricing model for second-hand electronics that has been refined over many years, and Trashin aggregates data from informal collection networks across the country. Such material and traceability databases are gaining strategic relevance in the circular economy that is on par with classic software IP—they act as barriers to entry, create lock-in effects with partners, and provide the basis for new revenue models (e.g., data-based licenses).

*'In circular business models, value creation does not end with the customer—it continues backward. Those who precisely control these returns gain a decisive time and cost advantage.'*

**Dynamic capabilities—orchestrating circular value creation**. The cases show that circular startups require significantly more extensive integration and reconfiguration capabilities than linear business models. Battery passports must be synchronized in real-time with second-life logistics; seaweed supply chains are more seasonally volatile than petrochemical feedstocks; O2O return flows for electronics require a permanent balance between purchase prices, storage capacities, and secondary market demand. Successful companies, therefore, not only master the classic 'sensing–seizing–transforming' processes but also expand them to include the backward-looking orchestration of reverse flows.

'Rules are not a burdensome straitjacket for circular startups, but their currency: those who master them early on gain credibility and market access at the same time.'

**Stakeholder and institutional theory—legitimacy as a growth asset**. Early compliance with ambitious targets—such as the EU Battery Regulation or state plastic bans in the U.S.— gives startups credibility with investors, major customers, and regulatory authorities. This legitimacy is actively translated into fundraising narratives by using impact KPIs (e.g., 'kg of primary raw materials substituted') as evidence of entrepreneurial value creation. Regulatory compliance thus becomes a valuable resource that promotes growth. Han et al. (2023) note that circular startups scale through both commercial and impact-oriented activities. It means that their strategic goals for scaling are not limited to generating revenue but also include systemic changes, such as influencing new regulations.

*'Only those who close the data loop can open the value loop—digitalization makes recycling a calculable market transaction.'*

**Digital enablers—from efficiency to market logic**. Technologies such as IoT sensors, blockchain-based material passports, and AI-supported evaluation algorithms not only reduce transaction costs in reverse supply chains but also create the market logic that gives structure to the previously underdeveloped secondary and service markets. For example, the digital visibility of material flows enables pricing for used batteries, biomaterial pellets, or recycled plastics, making circular processes calculable for investors. However, it is important to note that the effectiveness and accessibility of digital solutions are heavily influenced by the infrastructural and social contexts in which they are implemented. In regions with inadequate internet coverage, unstable power supply, or a lack of digital education, such as many parts of the Global South or rural areas, the implementation of digital CE innovations can pose significant challenges. In these contexts, digital technologies do not inherently promote inclusion; in fact, they can worsen existing inequalities and mechanisms of exclusion. At the organizational level, digital circular solutions require specific prerequisites, including a high level of IT expertise, investment in digital infrastructure, and effective management of the interface between analog and digital process chains. Without the necessary institutional and human resources, there is a risk that technological innovations will be ineffective or that existing linear practices will be digitized rather than transformed.

The analysis clearly shows that the four perspectives reinforce each other: Digital enablers first raise key circular resources in the form of databases to a usable level; dynamic capabilities are needed to mobilize these resources in volatile markets; and regulatory legitimacy opens up capital and sales markets. The interaction of these factors explains why the startups examined were able to develop above-average growth momentum despite different country contexts—and why circular innovations stagnate where one of the elements is missing.

*Table 2. Common patterns and country-specific divergences*

| Level | Common patterns | Country-specific divergences |
|---|---|---|
| Strategy | • Early establishment of circular principles<br>• combination of technological innovation and service | • Variable material focus (batteries GE, biopolymer U.S., e-waste CN, mixed waste BR |
| Funding | • Dominance of patient capital (impact and corporate VC, development funds) | • Ticket sizes: three-digit million VC in CN/U.S. vs. six-digit seed in BR |
| Regulations | • Compliance as a value proposition component | • EU/CN: clear, ambitious targets → market pull<br>• BR: enforcement gaps vs. greater uncertainty |
| Networks | • Partnerships as accelerators (OEMs, brands, platforms) | • Ecosystem structure: dense clusters (Mainz, Shenzhen) vs. fragmented actors (Porto Alegre) |
| Scalability | • Data and standardization requirements as a bottleneck | • Availability of infrastructure (recycling centers, seaweed farming, logistics) varies greatly |

Source: own representation

## Contribution to CE Research

This study expands the literature on the circular economy in four key dimensions. First, it refines the resource-based view (RBV) for circular contexts. Digital material flow databases, blockchain-based material passports, and the resulting trust networks are proving to be new, strategic 'embeddedness resources'—a resource category that has been missing from classic Resource-Based View (RBV) typologies (physical, organizational, technological) but is decisive in the circular economy through market access and lock-in effects.

Second, the study provides a process-based nuance to the dynamic capabilities perspective. Because recycling, repair, and resale are integral to the business model, the ability to orchestrate reverse flows takes center stage. Startups must not only recognize and seize opportunities but also reconfigure backward-looking material and data flows in real time. This skill set will also become a competitive factor for established industrial companies in the future.

Third, the cross-country approach fills a research gap by highlighting the role of institutional complementarities. The analysis reveals that identical technologies scale more rapidly in regimes characterized by the optimal combination of regulation, capital market depth, and physical infrastructure. Success is, therefore, not solely a function of technological maturity but also regime compatibility.

Fourth, this has practical implications: Policymakers are provided with a 'design menu' for support instruments consisting of clear targets (regulatory pull), revolving CE funds (patient capital), and open data standards (interoperability). Investors can derive new due diligence criteria, such as the quality of take-back partnerships or the maturity of traceability stacks.

Long-term studies should track the development of circular dynamic capabilities after exits or acquisitions. In addition, mixed-method designs offer the opportunity to test the success factors identified here quantitatively and to relate the financial return to environmental savings.

Overall, the study demonstrates that success in the circular economy is a systemic phenomenon: only the interaction of unique circular economy resources, dynamic orchestration capabilities, supporting institutions, and digital enablers enables startups to leap from niche to mainstream. The study thus adds a circular-ecological dimension to the entrepreneurship literature and provides robust guidelines for the next generation of sustainable business models.

## IMPLICATIONS FOR VARIOUS STAKEHOLDERS

### Founders

For circular economy startups, the five-step strategy checklist is a crucial roadmap. It guides founders in swiftly translating their innovation potential into growth-oriented business models.

**Check the problem material fit**. Even before the prototype is built, it must be clear what specific material flow problem is being solved. Example: Does the solution accelerate the return of used electric vehicle batteries, or does the product replace conventional plastic? Valid, data-based evidence increases credibility with investors and industry partners.

**Embed circular design principles from the outset**. Design for disassembly, modular construction, standardization of secondary material qualities, and repair-friendly components reduce subsequent conversion costs and facilitate achieving higher reuse rates or repairability indices. Collaboration with university faculties can be helpful (Kambanou et al., 2025).

**Strategic leverage of regulatory tailwinds is a key factor for founders**. Early creation of a regulatory roadmap enables appropriate identification of requirements. It also helps position the company in a way that generates clear market pull under compliance pressure.

**Operationalizing the impact story is a critical step for founders**. It guides internal control and defines three to five measurable circular KPIs. These KPIs not only serve as a powerful fundraising narrative for impact and corporate investors but also guide internal control.

**Structure a blended finance path**. Capital rounds should be planned as a step-by-step model: a) PoC grants (public innovation or environmental programs) for technology validation, b) impact VC for seed to Series A phases to share growth risks, and c) strategic corporates in later rounds to secure scaling infrastructure and market access. Such a staggered approach reduces dilution while creating partnership anchors in the value-creation network.

Those who systematically work through these five points will anchor circularity not as an afterthought but as the core logic of their business model—and thus create the conditions for convincing investors, regulators, and industry partners alike of the solution's scalability.

The use of technology must be planned wisely. In the circular economy, digital transparency is crucial for establishing trust and ensuring scalability. In this respect, integrating digital twins and material passports into the prototype will enable later traceability requirements to be met without costly retrofits. AI-supported forecast models for residual value and demand development should be developed not only as an internal control tool but also as a standalone, scalable service from the outset. Data insights can be monetized as a license-based add-on, thereby strengthening lock-in effects with customers. In addition, a consistent focus on open API architecture is essential—this allows OEMs, recyclers, and online marketplaces to connect with minimal integration effort, thereby accelerating partnerships and strengthening network effects. In this way, technology becomes not just an efficiency driver but a central lever for value in circular business models.

A circular business model stands or falls on reliable material and data flows—neither of which can be established alone. That is why it is essential to proactively collaborate with reverse supply chain anchors at an early stage. Take-back logistics providers, regional collection cooperatives, or certified waste disposal companies provide the necessary access to used products and raw materials. This early collaboration not only ensures a steady supply but also opens up future sales channels, demonstrating a forward-thinking approach. In addition, the network should be supplemented by knowledge alliances with universities, life cycle assessment experts, and standardization bodies (DIN, ISO). In this way, future standards can be actively shaped rather than complied with. Ultimately, technological legitimacy must be strengthened by investors and authorities.

## Investors

A resilient CE investment requires multidimensional risk management that extends classic VC criteria to include material and policy-driven uncertainties. This includes impact KPIs as a due diligence standard (Table 3).

*Table 3. Key performance indicators*

| Category | Typical indicator | Target value (guideline) |
|---|---|---|
| Material efficiency | Raw materials saved (kg) per € turnover | > 0,5 kg/€ |
| $CO_2$ avoidance | Tons $CO_2$e avoided per year | Growth rate > 50% per year |
| Degree of circularity | Per cent turnover from 'retain-reuse-recycle' services | > 70% |
| Partner network | Number of active B2B partners in the value chain | > 5 across 3 sectors |

Source: own representation.

Diversify regulatory risk in a targeted manner. It is essential to evaluate the degree to which the business model relies on the enactment, design, or enforcement of specific laws—such as the EU Battery Regulation, the US RECYCLE Act, or regional bans on single-use plastics. Spread portfolio across multiple jurisdictions and value creation stages (upstream and downstream) to cushion legal and market access volatility.

**Make feedstock risk transparent**. For biomaterial startups (e.g., algae or mushroom pellets), the supply of raw materials is often seasonal and regionally concentrated. Supply contracts with producer cooperatives or off-take agreements with large buyers should be required as part of the due diligence process. This can mitigate price and volume shocks during the scaling phase.

**Secure technology readiness in parallel**. Twin tracks should be promoted: a pilot plant for technical optimization and a demo plant for operating and financial metrics. Parallel paths shorten the learning curve, reduce scale-up delays, and create bankable reference data for project financing at an early stage.

**Anchor patient capital structurally**. A 7- to 10-year investment horizon should be planned (3 years on average for regular VC). This long-term commitment is crucial for the success of circular economy ventures. Revenue-based financing, tranched convertible loans, or tranched SAFE notes reduce downside risks because returns are linked to sales and milestones rather than hard exit deadlines. Impact

VC should be combined with corporate ventures or public CE funds to embed both industry expertise and political legitimacy in the capital structure.

Combining these four risk dimensions creates a resilient portfolio approach that absorbs the volatility of individual regimes or commodity markets, actively manages technological uncertainties, and, at the same time, considers the longer amortization cycles of circular business models.

## Politics and Administration

To transition circular economy startups from pilot status to industrial scale, funding instruments are needed that bridge capital gaps, stimulate market demand, and open up regulatory experimentation spaces.

First, governments should establish revolving circular economy funds that award seed tickets of between €250,000 and €2 million, which are funded by returns from successful exits. Such funds address the critical 'valley of tears' phase between the technology readiness level (TRL) of 7 (pilot) and 9 (commercial deployment), when traditional venture investors are still hesitant and grant-based programs are already winding down. A return logic ensures that funds remain available in the long term without constantly burdening new budget lines.

Second, the consistent anchoring of green public procurement (GPP) is recommended. Public tenders that prescribe binding minimum quotas for circular products and services—such as a five percent recycled content requirement in road construction projects or the use of exclusively repairable furniture in government buildings—create a calculable anchor for demand. In addition, GPP certification reduces transaction costs for private procurers because products have already been pre-tested by regulators, similar to cradle-to-grave certification (Hoang & Böckel, 2024).

Third, young CE technologies need test beds and regulatory sandboxes. Temporary exemptions from waste, substance, or chemical laws enable the approval of pilot plants more quickly and the collection of real-world data. At the same time, regulatory authorities can learn in a 'safe space' how to adapt existing regulations without compromising environmental or health protection. Ideally, such sandboxes combine a fast approval process (max. six months) with clearly defined impact KPIs and mandatory publication of results in order to achieve economies of scale for subsequent projects.

A consistent combination of entrepreneurial foresight (designing for circularity, data-driven business models), investor patience (impact-oriented KPI suites, longer-term exit horizons), and political reliability (clear rules and targeted support) forms the breeding ground on which startups in the circular economy can grow from niche projects into systemically effective market players. The instruments proposed here serve as a practical toolkit that stakeholders along the entire value chain can use for rapid implementation.

## CONCLUSIONS

By focusing on practices such as reuse, recycling, and resource optimization, CE startups reduce their environmental impact and create economic, social, and environmental value, thereby contributing directly to global sustainability goals. The literature identifies several significant challenges, including financing difficulties, inadequate infrastructure, a lack of regulatory support, and cultural resistance (Petri et al., 2025). The obstacles that arise at various stages of the startup life cycle hinder the expansion and consolidation of circular business models in the market, slowing progress toward a more integrated and resilient circular economy. The challenges faced by circular companies are closely related to the number of circular strategies implemented in their business models, and there is no uniform structure of barriers that apply to all organizations (Lit et al., 2024).

The case study makes it clear that value creation in the circular economy is less about physical material flows than about data sovereignty. Proprietary battery residual value data, AI-powered trade-in algorithms, or detailed waste dashboards transform linear flows of goods into controllable, high-margin flows of information.

Those who own the most detailed material flow data determine the pace of take-back, the price of secondary goods, and ultimately, the distribution of margins along the chain—a power advantage that is replacing traditional raw material ownership logic. Equally crucial is the dynamic ability to orchestrate reverse flows.

Successful startup teams coordinate input, output, and return flows simultaneously, integrate logistics and recycling partners through open APIs, and respond in real time to fluctuations in supplier quantities or secondary market prices. This backward-looking system expertise, also known as reverse flow capabilities, expands conventional RBV categories and is proving to be a bottleneck resource that even large, capital-intensive companies still lack.

Institutional tailwinds are providing the necessary market pull. Ambitious regulations—such as the EU Battery Regulation, Chinese CE pilot zones, and US plastic bans at the state level—create planning security, legitimize business models in the eyes of investors, and accelerate demand.

Conversely, it is becoming apparent that fragmented enforcement and approval practices—as seen in parts of Brazil—are slowing down circular innovation, even when the technology and team quality are in place. Patient capital dominates the financing landscape. Impact VCs, corporate ventures, and government funds account for the largest share of growth capital because they accept more extended payback periods.

Traditional short-exit VCs remain cautious. Blended finance structures—such as a combination of early-stage grants, revolving funds, and later corporate tickets—shorten time-to-scale and reduce downside risks by closely linking capital requirements to technical maturity levels.

Finally, network resources multiply the impact corridor of startups. Collaborations with OEMs, trading platforms, or collective cooperatives provide access to take-back channels, secondary raw materials, and market confidence—resources that individual startups would be unable to finance or manage independently. These network resources play a crucial role in the success of startups, enabling them to expand beyond their niche in such a short time.

Overall, the analysis suggests that success in the circular economy should be understood as a systemic phenomenon: those who optimize only technology or capital will fail. Only when information advantage, dynamic orchestration, institutional fit, patient financing models, and ecosystem embedding work together can a robust, scalable business model emerge that can sustainably transform linear markets. The findings make it clear that there is no such thing as 'the' circular startup; diverse motivations, resource profiles, and market strategies characterize the scene. Political support programs and incubation services should, therefore, develop more segmented instruments tailored to specific age groups, genders, or market segments rather than relying on uniform, one-size-fits-all approaches. It is the only way to address the different starting conditions and leverage the full climate and resource protection potential of circular business models (Van Opstal & Borms, 2023).

Despite its explanatory value, the study has several limitations. First, it would be advisable to conduct an empirical follow-up study based on the four start-ups presented. It could focus on a quantitative review of the success framework or be expanded to include corporate transformations in the Global South. Second, the number of cases is limited: four deliberately contrasting startups allow for analytical but not statistical generalization; idiosyncratic features of individual companies could distort the patterns identified. Second, the time frame is limited to the growth window of 2024/25. It means that neither maturity nor post-exit data are available, so that long-term environmental and financial impacts—such as actual $CO_2$ savings or cash flow stability after scaling—remain unassessed. Fourth, the data corpus relies heavily on interviews with female founders and investors; the voices of suppliers and end customers are underrepresented, and in the Chinese and Brazilian cases, translation artifacts may distort subtle contextual nuances. Finally, regulatory volatility acts as an external uncertainty: new EU regulations, US federal laws, or stricter environmental requirements in emerging markets could reweight the identified success factors in the short term, limiting the transferability of the results.

Future research should establish long-term panel studies that follow circular economy startups for at least ten years. It is the only way to reliably quantify resilience, exit performance, and real environmental benefits—such as tons of $CO_2$ or primary raw materials avoided—and separate them from short-term scale effects. Second, a mixed-methods approach that combines qualitative case studies with large-scale data sets is recommended. Funding databases, life cycle assessments, and policy trackers could be combined to statistically test the causal influence of capital access, regulation, and market structure on impact variables. Third, comparing established companies and startups raises exciting questions: Do corporations develop circular dynamic capabilities internally, or do they acquire them externally, and which option is more effective? Fourth, experimental policy research in real-world laboratories and regulatory sandboxes is needed to evaluate the effects of CE incentives, tax breaks, or data standards under controlled conditions. Last but not least, socioeconomic impact metrics should be further developed. Standardized indicators for inclusion—such as additional jobs in recycling cooperatives—and regional value creation would enable a holistic measurement of the circular economy's success, supporting evidence-based policy decisions.

## REFERENCES


Adekunle, K. (2024). Green entrepreneurship and innovation in the circular economy: Opportunities for startups and small businesses. http://dx.doi.org/DOI: 10.2139/ssrn.4924961

Almunawar, M. N., Anshari, M., & Fauzi, A. M. (2025). Modeling circular economic activities: A business ecosystem perspective through the value exchange network framework. *International Journal of Sustainable Development and World Ecology, 32*(4), 415–427. DOI: 10.1080/13504509.2025.2474937

Batista-Canino, R. M., Sosa-Cabrera, S., Medina-Brito, P., & Santana-Hernández, L. (2025). Scientific progress on entrepreneurship in the circular economy: A scientometric analysis. *Sustainability, 17*, 777. DOI: 10.3390/su17020777

Befort, N., Ayoub, M., & Matt, M. (2025). The role of dynamic capabilities in the development of eco-innovations. *Current Opinion in Green and Sustainable Chemistry, 53*, 101031. DOI: 10.1016/j.cogsc.2025.101031

Bocken, N., Coffay, M., & Dalhammar, C. (2025). The Brussels and California effects? Circular economy policy influence across borders. *Circular Economy and Sustainability*. Advance online publication. DOI: 10.1007/s43615-025-00516-4 PMID: 40568712

Boulding, K. E. (1966). The economics of the coming spaceship Earth. In Jarrett, H. (Ed.), *Environmental quality in a growing economy* (pp. 3–14). Johns Hopkins University Press.

Braungart, M., & McDonough, W. (2002). *Cradle to cradle: Remaking the way we make things*. North Point Press.

Camacho-Otero, J., Boks, C., & Pettersen, I. N. (2018). Consumption in the circular economy: A literature review. *Sustainability, 10*(8), 2758. DOI: 10.3390/su10082758

Castillo-Ospina, D. A., Ormazabal, M., de Vasconcelos Gomes, L., & Ometto, A. R. (2025). A dynamic capabilities framework for building circular ecosystems by focal firms. *Sustainable Production and Consumption, 54*, 130–148. DOI: 10.1016/j.spc.2024.12.022

Christodoulou, I. P., Rizomyliotis, I., Konstantoulaki, K., Alfiero, S., Hasanago, S., & Paolone, F. (2025). Investigating the key success factors within business models that facilitate long-term value creation for sustainability-focused start-ups. *Business Ethics, the Environment & Responsibility*. Advance online publication. DOI: 10.1111/beer.12681 de Sousa Resende, C., Zimmermann, R., Inês, A., & Dalmarco, G. (2025). A typology of scalability strategies of circular startups. *Procedia CIRP*, *135*, 201–206. DOI: 10.1016/j.procir.2024.12.016

Eisenhardt, K. M. (1989). Building theories from case study research. *Academy of Management Review*, *14*(4), 532–550. DOI: 10.2307/258557

Ferrante, M., Vitti, M., & Sassanelli, C. (2025). The evolution of circular economy performance assessment: A systematic literature review. *Renewable & Sustainable Energy Reviews*, *217*, 115757. DOI: 10.1016/j.rser.2025.115757

Formentini, G., Prioli, J. P. J., Ko, J., Hapuwatte, B., Ferrero, V., & Badurdeen, F.. (2025). A review of disassembly systems for circular product design. *Journal of Cleaner Production*, *506*, 145459. DOI: 10.1016/j.jclepro.2025.145459

Han, D., Konietzko, J., Dijk, M., & Bocken, N. (2023). How do circular start-ups achieve scale? *Sustainable Production and Consumption*, *40*, 363–375. DOI: 10.1016/j.spc.2023.06.007

Henry, M., Bauwens, T., Hekkert, M., & Kirchherr, J. (2020). A typology of circular start-ups: An Analysis of 128 circular business models. *Journal of Cleaner Production*, *245*, 118528. DOI: 10.1016/j.jclepro.2019.118528

Hoang, K. M., & Böckel, A. (2024). Cradle-to-cradle business model tool: Innovating circular business models for startups. *Journal of Cleaner Production*, *467*, 142949. DOI: 10.1016/j.jclepro.2024.142949

Hwang, B.-N., Puntha, P., & Jitanugoon, S. (2025). AI-driven circular transformation: Unlocking sustainable startup success through co-creation dynamics in circular economy ecosystems. *Sustainable Development*, •••, 1–30. DOI: 10.1002/sd.70001

Isenberg, D. (2010). How to start an entrepreneurial revolution. *Harvard Business Review*, *88*(6), 40–50.

Kambanou, M. L., Hajoary, P. K., & Lindfors, A. (2025). Supporting start-ups in the circular economy: An analysis of university-led incubators in India. *Journal of Industrial Ecology*, *29*(3), 997–1012. DOI: 10.1111/jiec.70032

Kjaer, L. L., Pigosso, D. C., Niero, M., Bech, N. M., & McAloone, T. C. (2019). Product/Service-aystems for a circular economy: The route to decoupling economic growth from resource consumption? *Journal of Industrial Ecology*, *23*(1), 22–35. DOI: 10.1111/jiec.12747

Klofsten, M., Kanda, W., Bienkowska, D., Bocken, N., Mian, S., & Lamine, W. (2024). Start-ups within entrepreneurial ecosystems: Transition towards a circular economy. *International Small Business Journal: Researching Entrepreneurship, 42*(4), 383–395. DOI: 10.1177/02662426241227520

Lee, K., Roh, T., Kim, J., Park, S., & Bae, Y. (2025). Unpacking sustainability in start-ups: A systematic review and research agenda. *Discover Sustainability, 6*, 218. DOI: 10.1007/s43621-025-01070-7

Lit, F., Huijben, J. C. C. M., Cloodt, M. M. A. H., & Paredis, E. (2024). Business model innovation in circular start-ups: Overcoming barriers in the circular plastics economy. *International Small Business Journal, 42*(4), 506v550. https://doi.org/ DOI: 10.1177/02662426231217954

Lund, M., & Nielsen, C. (2018). The concept of business model scalability. *Journal of Business Model, 6*(1), 1–18.

Miles, M. B., Huberman, A. M., & Saldaña, J. (2014). *Qualitative data analysis: A methods sourcebook.*

Petri, J. M., de Francisco, A. C., Martins de Souza, A., de Lima, J. D., & Trojan, F. (2025). How do start-ups develop circular business models? A systematic literature review. *Sustainability, 17*, 1246. DOI: 10.3390/su17031246

Saldaña, J. (2021). The coding manual for qualitative researchers. *Sage (Atlanta, Ga.).*

Stahel, W. (1997). *The performance economy*. Palgrave Macmillan.

StartupBlink. (2025). The Global Startup Ecosystem Index Report 2025. https://lp.startupblink.com/report/

Van Opstal, W., & Borms, L. (2023). Startups and circular economy strategies: Profile differences, barriers and enablers. *Journal of Cleaner Production, 396*, 136510. DOI: 10.1016/j.jclepro.2023.136510

Yin, R. K. (2018). *Case study research and applications: Design and methods* (6th ed.). Sage.

# Chapter 5
# Circular Economy and Ethical Compliance:
## Assessing the European Green Deal Policy Framework in Driving Sustainability

**Md. Tota Miah**
https://orcid.org/0000-0001-8536-8565
*Doctoral School of Economic and Regional Sciences, Hungarian University of Agriculture and Life Sciences, Hungary & Department of Business Administration, Varendra University, Bangladesh*

**Md. Faisal-E-Alam**
http://orcid.org/0000-0002-7341-7945
*Begum Rokeya University, Bangladesh*

**Anna A. Kredina**
https://orcid.org/0000-0002-7682-2727
*K. Sagadiyev University of International Business, Kazakhstan*

## ABSTRACT

*This chapter examines the relationship between Circular Economy (CE) practices and Ethical Compliance (EC) within the EGD policy framework to reduce waste, promote resource efficiency, and foster sustainable economic growth. A critical assessment of EU documents and literature highlights the need to integrate ethical compliance, legal values, and sustainability to ensure an inclusive green transition. The analysis reveals key challenges, including labor rights, social justice, fair resource allocation, and access to technology in adopting the policy framework. It further emphasizes the importance of developing countries aligning national*

DOI: 10.4018/979-8-3373-2454-8.ch005

*programs with the framework's vision, leveraging available funding, and fostering multi-stakeholder cooperation. Moreover, the long-term success of the EGD requires a comprehensive strategy that incorporates investment in green technologies, regulatory simplification, and infrastructure improvement across the energy, agriculture, industry, and construction sectors. Overall, the EU Green Deal offers a robust framework for policy development, serving governments, academics, and researchers engaged in sustainable development.*

## 1 INTRODUCTION

The world is facing multifaceted socio-environmental and ethical challenges, including environmental degradation, climate change, biodiversity loss, resource depletion, economic inequality, and job displacement—many of which stem from the traditional economic model (Hartley et al., 2024). In this context, the Circular Economy (CE) has emerged as both an essential component and a strategic priority for the European Union (EU) in pursuing the UN's Sustainable Development Goals (SDGs), adopted in 2015 (Castro-Lopez et al., 2025). CE emphasizes resource reuse in closed-loop systems, product redesign to extend life span, and the reduction of adverse environmental impacts (Nobre & Tavares, 2017). The adoption of CE practices can significantly contribute to climate and sustainability objectives by minimizing industrial waste, prolonging product life cycles, and reducing greenhouse gas emissions. Beyond environmental gains, CE also offers numerous social benefits, including job creation, enhanced social equity and well-being, and community empowerment. In 2019, the EU introduced the European Green Deal (EGD) policy framework, with the primary goals of protecting the environment, advancing social equity, and achieving climate neutrality by 2050 (Alessi et al., 2024). At the core of the EGD framework and CE principles lies a commitment to sustainability and human well-being as central elements of economic policy. Achieving these objectives requires the active involvement of stakeholders across multiple sectors, including energy, transport, agriculture, and construction (Koundouri et al., 2024). If implemented as planned, the framework is expected to influence nearly every area of human activity, stimulating sustainable and labor-intensive economic growth while driving innovation and technological advancement (Trane et al., 2025). Nevertheless, several ethical considerations must be carefully addressed, such as the fair utilization of resources, environmental awareness, and the quality of employment across industries (Ghormare et al., 2024). By design, the circular approach seeks to eliminate waste and optimize resource use, yet concerns remain regarding fairness and equity for all stakeholders (Wen et al., 2025). EU member states are aligning with EGD goals, such as reducing greenhouse gas emissions, enhancing biodiversity, and promoting sustainable agriculture. For instance, Poland has adopted precision farming techniques to reduce chemical inputs, while Bulgaria has expanded organic agriculture—both yielding economic and environmental benefits. Central and Eastern European countries are taking climate-smart strategies to tackle growing climate threats (Koundouri et al., 2024). Therefore, the EGD has become a necessity in response to ongoing environmental and emerging human-caused threats that require large-scale, coordinated actions both ethically and legally (Sapaloglu et al., 2025). Communications and cooperation among interested parties, updating the existing legal framework, adopting new technologies, and substantial financial investment should be equally prioritized (Sandri et al., 2025). The policy framework focuses on three major goals: 1) reducing greenhouse gas emissions, 2) increasing resource efficiency, and 3) boosting economic growth without increasing natural resource consumption (Dubský et al., 2025). Moreover, CE, as a critical operational pillar under the EGD policy, might offer a compelling vision not only for Europe's ecological and economic future but also a way forward for the world. Despite technological innovation, institutional reform, behavioral change, and international coordination are required to implement the policy in practice (Kuyer et al., 2025). Further exploration and structural changes are necessary to ensure social, political, and economic interests (Hereu-Morales et al., 2024). Therefore, academic research should be conducted to guide businesses and governments in developing long-term plans and initiatives with a wide range of potential applications. Thus, the chapter primarily aims to critically assess CE practices and ethical compliance (EC) under the EGD framework alongside the following research objectives:

1) To examine the interconnections between the circular economy, ethical compliance, and the European Green Deal (EGD) framework.
2) To analyze how the EGD integrates circular economy principles and ethical compliance into its sustainability agenda.
3) To explore future industry and business applications, along with challenges and opportunities, arising from the EGD framework.

The chapter also ensures that CE aligns with ethical compliance by assessing the regulatory framework to develop an action plan and implement sustainable reporting. Moreover, embedding CE practices with ethical compliance under the framework involves a multifaceted approach that combines the SDGs, robust regulatory frameworks, practical monitoring tools, stakeholder engagement, and technological innovation. Within the framework, ethical issues—such as environmental implementation, legal, regulatory, and economic implications of the EGD—are discussed. Additionally, potential ethical challenges and opportunities are addressed. Ultimately, this chapter aims to provide insights into how regulatory frameworks can motivate businesses to adopt both circular innovation and ethical accountability, contributing to a fair, green, and resilient European economy in the long term.

## 2 CIRCULAR ECONOMY (CE), ETHICAL COMPLIANCE (EC), AND EUROPEAN GREEN DEAL (EGD): AN OVERVIEW AND RELATIONSHIP

### 2.1 Circular Economy (CE)

The essence of the Circular Economy (CE) lies in the efficient use of renewable resources, energy conservation, and waste reduction, aiming to improve environmental, social, and economic outcomes (Yang et al., 2023). It is considered an innovative economic model and concept actively supported by the European Union (EU), the United Kingdom, China, Japan, and companies worldwide (Liu et al., 2021). The primary objective is to maximize the value derived from resources by maintaining their utility for extended periods through renovation, recycling, and reuse. CE focuses on closing the loop of product lifecycles, in contrast to the conventional linear model of "take-make-dispose" (Daga et al., 2025). In the digital age, CE seeks to minimize waste, promote efficient resource management, and foster sustainability. Despite its potential, new challenges arise in the digital circular economy, such as data privacy and security issues, and regulatory frameworks often require integration with digital innovation and efficient management (Karim et al., 2025). The advent of digital technologies in the fourth industrial revolution (Industry 4.0), such as Artificial Intelligence (AI), the Internet of Things (IoT), Blockchain Technology (BT), and human-digital twins (HDTs), offers significant potential for promoting sustainable development (Roshid et al., 2024). This transformation aims not only to build on the technological progress achieved in Industry 4.0 but also to ensure that these advancements lead to outcomes that are more sustainable, inclusive, and socially responsible through the further integration of the Industry 5.0 framework (Hariyani et al., 2024). According to Roshid et al. (2025, 2026), the synergy of these technologies with CE principles enables organizations to enhance resource optimization, reduce waste, and strengthen operational resilience while advancing sustainability goals. The key principles of CE include: 1) designing products to minimize waste and environmental impact, 2) maintaining the utility of products and materials through practices such as maintenance, repair, reuse, remanufacturing, refurbishing, recycling, and upcycling, and 3) ensuring the replenishment of natural resources and the restoration of ecosystems (Suárez-Eiroa et al., 2019). In this model, products and materials are designed for longevity, repairability, and recyclability while minimizing waste (Schroeder et al., 2019). By encouraging innovation in product design, business models, and supply chains, CE has the potential to stimulate economic growth, create new jobs, and enhance living standards, maintaining a dual focus on environmental conservation and economic development (Wolf et al., 2021).

## 2.2 Ethical Compliance (EC)

Ethical Compliance (EC) is often defined as adherence to moral principles and standards in business conduct. It embodies values such as transparency, accountability, fairness, and social responsibility (Bag et al., 2024). These principles ensure that organizational actions not only serve economic interests but also benefit society and the environment. EC provides a critical ethical foundation for the Circular Economy. Essentially, ethical compliance plays a crucial role in reinforcing the CE model by addressing the moral shortcomings of the linear economy and providing guidance for sustainable decision-making (Ali et al., 2021). For example, a company may comply with all data privacy laws but still engage in unethical practices in collecting and handling user data, thereby creating socio-ethical concerns during the transition to a CE model (Zorpas et al., 2025). The principle of equitable distribution plays a crucial role in this transition. Particularly in marginalized and underdeveloped regions, an ethical circular economy must ensure that the benefits and burdens of the transition are shared fairly. This includes safeguarding workers' health and safety, promoting fair labor practices, and ensuring that communities are not adversely affected by industrial changes (Bashynska, 2025). However, ethical values remain a significant challenge in underdeveloped regions, where limited infrastructure, weak legal systems, and unequal distribution of benefits often place a disproportionate burden on local populations (Oehring & Gunasekera, 2024). Consequently, ethical concerns—such as worker safety, environmental justice, and social equity—are key considerations when implementing CE principles. Ethical gaps remain underexplored, particularly in policy development and in taking effective actions to enhance transparency, accountability, traceability, and resource efficiency (Gunawardhana, 2025).

Furthermore, the recent industrial paradigm poses challenges regarding the ethical use of technologies, including data privacy, digital inequality, fair resource distribution, social inequalities, and biased algorithms, all of which affect the implementation of CE practices (Shabur et al., 2025). To build trust and mitigate risks, stakeholders must be engaged through open communication, regular reporting, and participatory decision-making processes (Veyssière, 2021). Monitoring, evaluation, and adaptive learning are essential to sustain ethical outcomes over time. Moreover, the role of ethical leadership and management cannot be overstated. Leaders who demonstrate ethical behavior and implement strong management control systems are crucial to advancing circular practices within organizations. As a result of their influence, ethical values are embedded throughout operations, and sustainability goals are not only discussed but also realized. Furthermore, a broader societal transformation is necessary to transition to a CE model. Stakeholders, particularly consumers, must reimagine their roles and responsibilities by engaging in ethical practices daily (Bucur, 2023). This includes making conscious purchasing decisions, supporting sustainable brands, and adopting lifestyles that align with one's values. Such changes may also challenge producers to consider the full life cycle of their products (Awan & Sroufe, 2022). For example, from raw material sourcing to end-of-life disposal, consumers should be aware of and encourage industries' responsible behaviors, such as reducing, reusing, and recycling. These actions reinforce ethical commitments to environmental stewardship and moral responsibility for future generations. Therefore, EC is not merely an ethical aspect of CE; it constitutes a moral foundation for an inclusive and resilient sustainability transition.

## 2.3 European Green Deal (EGD) Policy Framework

In 2019, the European Commission (EC) launched the European Green Deal (EGD) policy framework, aiming to address climate change and promote a "just and inclusive" transition to a sustainable economy (Szpilko & Ejdys, 2022). In response to ongoing climate change and environmental degradation, the EU has undertaken its most ambitious plan in history to tackle the existential threat to Europe and the world (Buzogány et al., 2025). The EGD is a growth strategy designed to transform the EU into a modern, resource-efficient, and competitive economy. To achieve these objectives, it aims to reach net-zero greenhouse gas emissions by 2050 and decouple economic growth from resource use, ensuring that no individual or region is left behind. Furthermore, the majority of Europeans consider protecting the environment important (95%), and nearly eight in ten Europeans (77%) believe that environmental protection can stimulate economic growth (European Commission, 2019). According to the Commission, "EGD is a roadmap for making the EU's economy sustainable. It sets out how to make Europe the first climate-neutral continent by 2050, boosting the economy, improving people's health and quality of life, caring for nature, and leaving no one behind." The EGD is not a law in itself but rather a general policy strategy that outlines ambitions and goals across various policy sectors (Fetting, 2020). For its implementation, existing regulations and standards will be revised over the coming years, and new laws and directives will be developed and enforced (Pollex, 2025). Eight key areas comprise the 'Green Deal,' which are presented in Figure 1:

*Figure 1. The EGD policy framework. Source: (European Commission, 2020)*

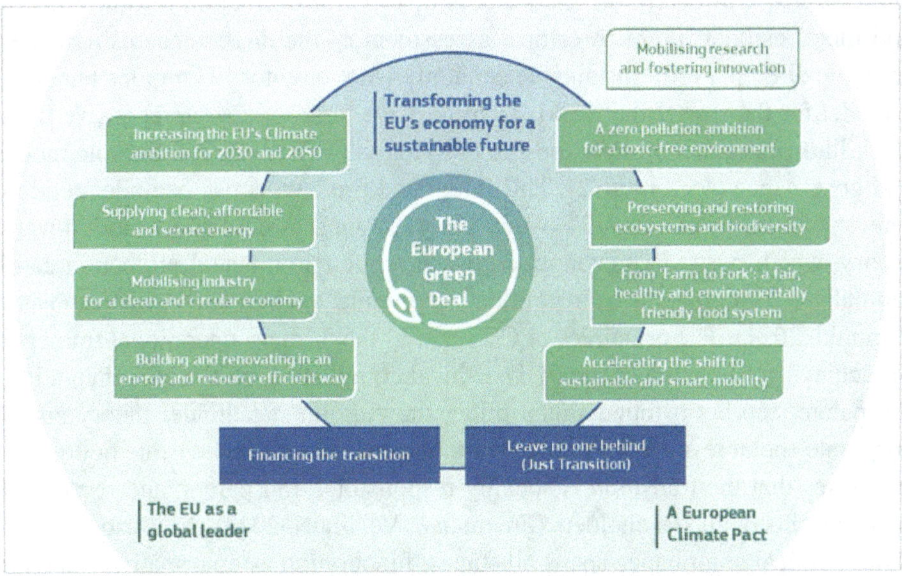


The success of the overall EGD will depend on global leadership in climate action and sustainable development. Implementing this ambitious goal and comprehensive vision requires addressing a wide range of strategies and regulations across the eight key areas (Arta et al., 2021). In some areas, the EGD faces challenges related to resource adequacy, implementation disparities, and potential impacts on entrepreneurship, sustainable mobility, sustainable food and farming, or biodiversity, where European institutions need to elaborate or revise legislation to achieve the objectives (Pianta & Lucchese, 2020).

However, the realization of the EGD and the CE encounters considerable obstacles. Securing financing remains a significant challenge, as the projected one trillion euros over the next decade will cover only about one-third of the EU's estimated investment requirements to meet its climate objectives (Sikora, 2021). Additional hurdles include gaps in innovation capacity, particularly in less developed regions, as well as the absence of precise policy instruments to encourage compliance by businesses and governments. Social and regional inequalities may widen if these challenges are not addressed equitably in regions that rely on fossil fuels or lack the necessary resources for green investments (Hafner & Raimondi, 2020). Careful management and support for affected workers and communities should be prioritized during the technological transformation to avoid labor market disruptions. To sum up, the EGD and CE collectively represent an ambitious and holistic strategy for achieving sustainability, economic growth, and social welfare in Europe. The practical and ongoing commitment of all stakeholders, including governments, businesses, and citizens, will determine the success of this transformation toward a greener and more resilient future. A critical assessment of the three concepts indicates a strategic direction for a sustainable economy—not only for EU regions but also as a model for the rest of the world. Therefore, we propose a comprehensive framework illustrating the interrelationship between the EGD vision, economic models, and governance (see Figure 2). This integrated framework can be understood as a direct application of various existing theories, such as ecological modernization theory, which posits that economic growth and environmental protection can be mutually reinforcing through political, economic, and technological innovation (Leipold, 2021). The principles of CE serve as the primary operational framework for achieving the vision of the EGD. Ethical compliance, on the other hand, is not a separate step but a foundational pillar, drawing on stakeholder theory and the corporate social responsibility (CSR) framework, including the triple bottom line. It ensures that the transition is socially responsible, transparent, and respects the human rights of all stakeholders (Freeman & Velamuri, 2023). As illustrated in the figure, ethical compliance spans all stages, functioning as a normative safeguard. The interdependence of these elements is crucial for implementing a sustainable and equitable future.

*Figure 2. Interrelationship between Circular Economy, Ethical Compliance, and the EU Green Deal framework (Source: authors).*

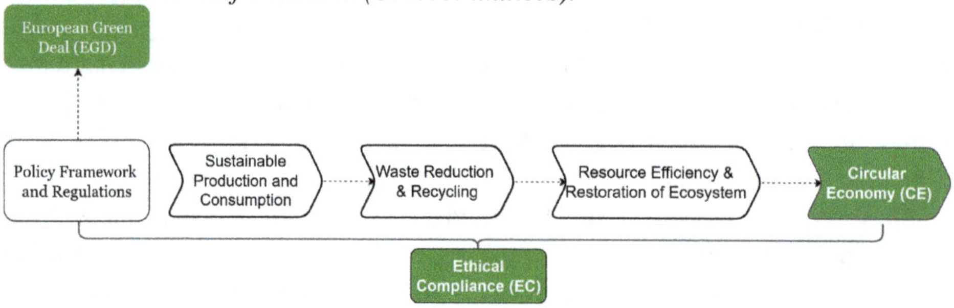

## 3. ETHICAL COMPLIANCES (EC) IN THE CONTEXT OF THE EUROPEAN GREEN DEAL (EGD)

The EU Green Deal links three dimensions of responsibility that correspond to the concepts of ethics, compliance, and sustainability. The basic concept of "ethics" sets the value guidelines for a company's behavior, creating expectations of fairness and mutual trust between all stakeholders (Valentinov, 2024). Moreover, business ethics is a valuable resource for a company. There are two primary perspectives on the relationship between compliance and sustainability. One group of scholars treats compliance as a distinct construct defined as the operational framework of rules and procedures that ensures adherence to external regulations and internal policies. The effectiveness of such frameworks depends on the extent to which legal requirements are translated and understood at every organizational level (Karlsson, 2019). On the other hand, compliance is considered a component of the broader notion of sustainability, which broadens the scope of corporate sustainability (Li & van Rooij, 2022). It incorporates environmental, social, and economic criteria across corporate decision-making, reframes the organizational objective as the pursuit of net positive environmental and social impact. Nevertheless, managerial decisions ultimately remain oriented toward profit generation (Adolph & Beckmann, 2024). The relationship among these three terms can be explained as follows: ethics forms the value foundation, compliance translates values into verifiable procedures, and sustainability sets the direction for development, attracting future generations and the planet as beneficiaries. Without ethical leadership, compliance risks become a formality, and without compliance, sustainability remains a declaration only (Brown & Treviño, 2006).

Since the publication of the Brundtland Report and its institutionalization in Rio 92, the normative paradigm of sustainable development has shifted from the "three pillars" to the idea of systematic transformation. The decisive turn is reflected in the 2019 Global Sustainable Development Report. In this document, the concept of systemic change is a prerequisite for achieving the SDGs (United Nations, 2019; Scoones et al., 2020). On the other hand, evolution is enshrined in the "Do No Significant Harm" principle, as legalized in Article 17 of Regulation (EU) 2020/852 on a taxonomy that sets technical criteria for green investments (European Parliament & Council, 2020). At the corporate level, "ethical compliance" consists of a combination of mandatory standards, voluntary codes, and a culture of values. The effectiveness of such a system increases when value motivation (ethics) is transformed into verifiable procedures (compliance) and integrated into a long-term sustainability strategy.

Regulation EU2021/1119 (the European Climate Law) sets out the legally binding EU targets. The target is to reach at least a 55% net reduction in GHG emissions by 2030. A complementary goal is to achieve climate neutrality by 2050 (Kulovesi et al., 2024). The REPowerEU plan introduced an "ethic of urgency": accelerated investments in renewable energy and energy market reform simultaneously reduce dependence on fossil fuel imports and mitigate price pressure on households (Schramm & Terranova, 2024). At the corporate level, the core of mandatory corporate sustainability reporting is primarily governed by the Corporate Sustainability Reporting Directive (CSRD), the Sustainable Finance Disclosure Regulation (SFDR), and the Corporate Sustainability Due Diligence Directive (CSDDD). CSRD establishes the European Sustainability Reporting Standards, requires external verification, and gradually extends coverage to SMEs. These standards enable investors, banks, and regulators to quickly compare the performance of companies across industries and countries, which is critical for redirecting capital to support the EU Green Deal (Kosi & Relard, 2024). SFDR outlines the Principal Adverse Impacts indicators and categorizes products based on their sustainability. A standardized list of indicators enables all market participants to measure and compare the damage that investments may cause to sustainable factors (Busch, 2023). Finally, the regulatory pressures for the CSDDD expand the mandatory companies. The standard is to achieve a balance of more than 3,000 employees or €450 million in turnover in a company. Thus, a high threshold concentrates obligations where there are the most excellent management and financial resources (Ciacchi, 2024). The EU Taxonomy outlines six environmental targets and the Do No Significant Harm (DNSH) criterion, which links long-term investments to economic sustainability (Bassen et al., 2025). External alignment is achieved through conformity with the 1998 Aarhus Convention and the 2015 Paris Agreement (Bernardo et al., 2021; Barritt, 2024). The EGD framework distributes tasks among five key blocks, each corresponding to a set of SDGs and an ESG dimension (see Table 1).

*Table 1. Key blocks for achieving EGD*

| **Key directions of the EU Green Deal** | **Achievement tool** | **SDG goals** | **ESG measurement indicators** | |
|---|---|---|---|---|
| Energy & Climate | European Climate Law, Fit for 55, REPowerEU | 7, 9, 11, 12, 13 | E | |
| Circular Economy | Ostojic Circular Economy Action Plan | 9, 11, 12 | E | |
| Just Transition & Social Inclusion | Just Transition Mechanism, Social Climate Fund | 8, 10 | S | |
| Corporate Transparency & Due Diligence | Corporate Sustainability Reporting Directive, Corporate Sustainability Due Diligence Directive | All goals | G | |
| Biodiversity & Oceans | EU Biodiversity Strategy 2030, Farm-to-Fork | 14, 15 | E, S | |

- The first block, Energy and Climate, directly advances SDGs 7, 9, 11, 12, and 13 through the European Climate Law, the Fit for 55 package, and the REPowerEU initiative. The targets set out in Regulation EU2021/1119 make progress verifiable and align this block with the environmental objectives of the European Taxonomy.
- The next block, Circular Economy, contributes to SDGs 9, 11, and 12. The "Do No Significant Harm" principle in the EU Taxonomy redirects capital to resource-efficient chains, turning environmental benefits into access to financing. This is achieved through the implementation of the Circular Economy Action Plan. Just Transition and Social Inclusion covers SDGs 8 and 10.
- The Just Transition Mechanism and Social Climate Fund programs compensate households for price shocks and finance worker retraining, constituting key instruments under the social pillar of ESG.
- Corporate Transparency and Due Diligence is transversal across all SDGs and relates to components, e.g., CSRD, ESRD, etc.
- The final block, Biodiversity and Oceans, formally covers SDGs 14 and 15, but remains under-reported. The EU Biodiversity Strategy for 2030 and the Farm-to-Fork strategy for terrestrial and marine ecosystems do not yet have strict financial metrics, which limits investment.

However, ethical credibility is the foundational prerequisite of the long-term legitimacy of the EGD. Analysis of political discourse reveals that when principles of fairness and 'do no harm' are embedded in goals and instruments, public support remains stable even as costs rise. In contrast, neglecting cross-border impacts raises questions about the long-term implementation (da Silva Hyldmo, Rye, & Vela-Almeida, 2024). In reality, ethical standards cannot remain static. Technologies, data governance frameworks, and social expectations change faster than regulations. A recent study of 82 co-designed priorities in pan-European workshops demonstrates that regular revision of reporting criteria and open participation formats improve decision quality and stakeholder trust (Fejzic & Usher, 2025). When companies can demonstrate that their products meet the EU's strict socio-ecological criteria, they gain access to premium segments of the global market. In other words, the stronger the ethical foundation of EGD, the more sustainable the European business model. The preliminary analysis demonstrates that a robust ethical foundation is a necessary condition for the successful implementation of the EGD implications. The task arises to identify the factors that impede the operationalization of the stated principles.

## 4. CHALLENGES IN ENSURING ETHICAL COMPLIANCE (EC) WITHIN THE EUROPEAN GREEN DEAL (EGD)

There are ongoing debates and diverse challenges that need to be minimized, e.g., food losses, the strain on arable land, the increase in nitrogen inputs, and the decline in crop yields (Aicja Sikora, 2021). Fundamentally, the EU green deal frames sustainability as a normative, ethical, and institutional change in addition to an environmental goal. This strategic initiative formulates sustainability as both a new development paradigm and a call for systemic institutional, economic, and lifestyle changes. Agriculture is central to the EGD but faces serious challenges. Declining yields, growing land demand, and nitrogen overuse are pressing concerns. Ethical accountability in this context extends beyond EU legal compliance to include corporate and institutional responsibility toward society (Siddi, 2020). In practice, sectoral strategies operationalize this ethical agenda by combining biodiversity conservation and emissions reduction with guarantees of a decent income for farmers, bridging agroecological practices and sustainable intensification. Similarly, the energy pillar prioritizes large-scale solar photovoltaic deployment, job creation, and reducing energy poverty, all of which rely on fair investment distribution. These priorities work together to create a normative program that turns the EGD into a forum for social renewal rather than just an environmental initiative. A successful implementation can encourage investment and innovation throughout the economy. Additionally, the EGD harmonizes climate objectives with human rights, legal protections, and open governance practices (Szpilko & Ejdys, 2022). The following are the main challenges that have been extracted by analyzing the previous literature and EGD reports, and communications.

### 4.1 Corporate Accountability and ESG Integration

The need for funding is a significant obstacle to achieving this ambitious goal. It is estimated that the Green Transition will cost at least €260 billion a year. Only €7.5, or 3% of the necessary funds, is currently provided by the Just Transition Mechanism. Because capital-intensive projects continue to dominate investment flows, low-income families are disproportionately affected by this disparity (European Commission, 2019). Additionally, businesses must audit their supply chains, disclose emissions data, and eliminate unethical behavior. Global firms are increasingly aligning themselves with these moral principles, particularly those that are directly incorporated into the EGD. EU banks are now refusing loans to businesses that do not disclose important ESG (Environmental, Social, and Governance) risks because the European Commission has set clear expectations (Islam et al., 2025; Maris & Flouros, 2021).

## 4.2 Regional Disparities

Geographic inequality in Green Deal representation is a notable concern. Eastern European countries are underrepresented in EU investment case studies, which reference them in only 10% of a 27,000-word report. In contrast, Western Europe receives significantly more visibility. Meanwhile, environmental burdens are unequally distributed: in 2019 alone, Asian ports received 1.6 million tons of European plastic waste, with half of it from Germany and the Netherlands (Eckert & Kovalevska, 2021). This imbalance in responsibility further widens the ethical gap between EU regions.

## 4.3 Labor Market Transitions

Regional labor markets are already changing as a result of energy system restructuring. For example, about 240,000 jobs could be lost as a result of mine closures, 60,000 in the power generation industry, and 180,000 in the mining industry. About 225,000 construction jobs could be created each year by repurposing these sites for solar energy infrastructure, but this shift requires prompt and efficient retraining. For instance, under the EU4Climate initiative, one Eastern European nation aims to reduce emissions by 65% by 2030; however, enabling legislation has not yet been passed, and reskilling initiatives remain primarily focused on energy-related topics (Kopytsia, 2021). Furthermore, fewer than 25% of green digital jobs in the ICT sector are held by women. Gender and generational disparities in access to the developing green economy present a significant challenge.

## 4.4 Inequitable Access to Subsidies and Technologies

Higher-income households disproportionately utilize home insulation and electric vehicle subsidies. Out of over 20 million passenger cars in Poland, only 119,297 are registered as electric vehicles (Bartosik et al., 2025). The main obstacles mentioned by respondents are a lack of infrastructure and cost. Due to their greater availability, 64% of consumers still opt for gasoline-powered vehicles (Sandri et al., 2025). Green incentives, therefore, benefit the richest quintile the most. With the help of tax credits, high-income households are also the first to install solar panels. Renters and low-income families often struggle to afford the co-financing required by programs like the EU Renovation Wave, as they lack ownership rights or have limited financial resources (Maunuksela, 2024). These populations consequently continue to use fossil fuels and pay greater taxes for oil and petrol.

## 4.5 Public Backlash and the Rise of Populism

While the EGD promotes economic growth and competitiveness, the interests of workers, migrants, young people, and rural communities are often overlooked. Public dissatisfaction increases when farmers, miners, and other stakeholders in traditional industries demonstrate against what they perceive as neglect. The closures of mines and power plants have already resulted in job losses in Germany. A study by Vrontisi et al. (2024) found that regions like Silesia (Poland) and Western Macedonia (Greece) are particularly vulnerable, with the energy sector accounting for 6.5% of regional employment. These regions already have high unemployment rates and low incomes (Buzogány et al., 2025). Despite being portrayed as technologically neutral and scientifically based, EGD decisions are usually made by technocrats with little to no public participation (Nagy et al., 2025).

## 4.6 Emergence of Green Elitism and Greenwashing

Although the EGD aims to promote genuine change, some businesses exaggerate or fabricate their environmental credentials. Greenwashing is a problem that deceives customers and damages public confidence (Cavoski, 2022). Overzealous or deceptive sustainability claims skew investor confidence and consumer choices. This phenomenon has contributed to the emergence of what critics refer to as "green elitism," in which the wealthy and powerful disproportionately benefit from climate action while externalizing the costs to vulnerable groups. The EGD is viewed as elitist or unrelated to common issues in many nations, which feeds right-wing populist narratives and incites farmer protests (Żuk, 2025).

## 4.7 Technocratic Design and Symbolic Compliance

Other ethical issues arise from the EGD's technocratic structure. Since it is often referred to as an expert-driven framework, operational objectives and resources are frequently concealed from the general public. There are criticisms, such as the fact of "sustainable" is widely used and many outdated business practices persist. Citizens are increasingly considered responsible consumers, and environmental responsibility has become more individualized. This approach restricts the public from decision-making. EGD implementations can also be hindered by limited administrative capacity and insufficient capital inflows throughout Eastern and Southern Europe.

## 5. INDUSTRY IMPLICATIONS AND BUSINESS IMPLEMENTATION

The EGD is a blueprint for the future industry and business toward shifting to green practices. There are some important sectors, such as energy, transport, and manufacturing, where more investment is needed for renewable energy and low-carbon technologies. On the business side, new regulations, alignment with EGD strategies should be emphasized in increasing sustainability-driven markets. However, the adoption of these strategies poses various ethical challenges that must be carefully managed in business implementation.

### 5.1 Ethical Considerations in CE Practices

The CE strategies contribute to environmental sustainability, but ethical concerns remain in the case of equitable distribution of resources. For example, as businesses optimize resource use and increase recycling efforts. It is important to ensure that these practices do not aggravate existing inequalities in less developed regions. The benefits of CE practices must be accessible to all, avoiding the perpetuation of social and economic disparities (Kirchherr et al., 2017). Another concern is the quality of employment within circular industries. During the transition, businesses must prioritize job creation that is both sustainable and equitable. This includes ensuring fair wages, safe working conditions, and opportunities for workers displaced by traditional economic models (Waaje et al., 2024). Moreover, workers should have access to retraining programs to help them transition to new roles in the evolving green economy. The shift to a circular economy also requires careful consideration of the natural environment. While reusing materials can reduce environmental impact, it can create new challenges in waste management and resource extraction. Ethical concerns regarding the environmental integrity of certain recycling practices for the poorly managed materials must be addressed (Ghisellini et al., 2016).

### 5.2 Regulatory Framework for Ethical Compliance

To align the CE with ethical compliance, a comprehensive policy framework is necessary. The EGD integrates initiatives, e.g., the CE action plan, the sustainable products initiative, and the EU Taxonomy for sustainable activities to ensure social equity and environmental protection (Fetting, 2020). These regulatory frameworks extend beyond compliance by integrating social responsibility, justice, and fairness into a company's business practices. Businesses must adhere not only to environmental regulations but also to social dimensions. The goal is to foster business models that promote economic growth, environmental sustainability, and social equity simultaneously (Geissdoerfer et al., 2017).

## 5.3 Measuring Ethical Compliance and Circular Economy Performance

To ensure businesses adhere to ethical standards while transitioning to a circular economy, robust monitoring and reporting mechanisms are essential. Transparent and accountable reporting frameworks, such as the CSRD, are pivotal in assessing the environmental and social impacts of CE practices. These frameworks provide businesses with clear guidelines for reporting their sustainability efforts. It allows stakeholders to monitor progress and ensure accountability (European Commission, 2021). According to the World Economic Forum (2025), European companies have decreased their Scope 1 and Scope 2 emissions by 28% between 2019 and 2023, but Scope 3 emissions have remained relatively stable. Meaning that fully decarbonizing supply chains needs further attention. Companies continue to struggle with establishing supply chain-related targets due to the fragmented regulations and the lack of globally accepted standards. Stakeholder engagement is another important element of ethical compliance. Businesses, civil society organizations, and affected communities can ensure that the transition to a CE is inclusive, transparent, and responsive to the needs of all parties.

## 5.4 Technological Innovation and Ethical Challenges

Technological innovation plays a crucial role in enabling CE practices, such as more efficient resource management and advanced recycling technologies. The rapid pace of technological change can raise ethical concerns about the accessibility and equity of these innovations. Technological solutions must be accessible to all stakeholders to prevent exacerbating existing disparities. Furthermore, it must align with the core values of sustainability, ensuring both positive environmental and social outcomes. Businesses should assess the ethical implications of emerging technologies to ensure they do not worsen social inequality or environmental degradation. Inclusive and equitable access to new technologies is vital for maintaining the ethical integrity of the EGD framework (Maunuksela, 2024).

Despite the challenges, the EU green deal framework has been embraced by different industries and sectors. For example, in the construction and building sector, they are responsible for 40% of energy consumption and 36% of greenhouse gas emissions. Through eco-design and comprehensive life cycle assessments, the industry is working toward reducing its environmental footprint. Evidence shows that concrete recycling stands out as a notable practice to minimize waste and reuse materials effectively (Bonoli et al., 2021). On the contrary, the Dutch Green Deals offer another successful example of transition to a circular economy. The policy instruments approach brings different stakeholders together, such as firms, NGOs, universities, and multiple levels of government, in voluntary agreements with the national government. These partnerships aim to identify and remove barriers that hinder innovation related to CE initiatives in the biobased economy and construction sectors, where innovation and collaboration have opened new pathways for sustainable development (van Langen & Passaro, 2021). In Romania and, Czech Republic, they are focusing on decarbonizing the energy system by replacing coal energy sources with renewable sources of electric energy as part of the Paris agreement and the EU green deal (Lapčík et al., 2022). However, these applications offer valuable lessons for policymakers, businesses, and communities striving to make the transition toward more sustainable and resilient economic models.

## 6. FUTURE DIRECTIONS AND OPPORTUNITIES

Further academic and industrial research can be conducted on the EGD framework. As digital technologies are becoming more prevalent, researchers can explore how AI, IoT, and BT can be utilized to improve resource flow tracking, reduce waste, and enhance the fairness of circular economies. EGD's future path is based on an equitable, environmentally friendly, and sustainable Europe. The energy transition is a key pillar of EGD. There should be a stronger emphasis on decarbonization and bolstering energy policies that enable cleaner, more resilient energy systems. With intensified efforts to maintain biodiversity and support nature-based solutions that restore ecosystems and mitigate climate change, environmental protection is equally vital (Oberer & Erkollar, 2024). As this green transition must be inclusive, economic and social inclusion will be crucial. Participation in decision-making, education, and training should be provided to the citizens to develop their skills and information to make society more sustainable. Financial and regulatory tools, e.g., creative financing schemes and sharp tax laws, can support these initiatives to promote environmental investment. The EGD relies on technical innovation, public-private partnerships, and the development of green technologies across a range of industries (Taneja et al., 2024). Additional focus should be given on important sectors, e.g., the building, transportation, and maritime industries, and energy to assess the positive impact on the economy and the environment.

## 7. CONCLUSION

The European Green Deal (EGD) is an innovative policy framework primarily aimed at strengthening the European economy while emphasizing the importance of sustainable development goals. Researchers, decision-makers, and practitioners can use this framework as a policy guide to facilitate the integration of environmental, social, and regulatory considerations. Developing and underdeveloped regions can also benefit significantly from the suggestions and guidelines discussed in this chapter. The chapter argues that the success of the circular economy (CE) under the EGD framework depends on a strong ethical and legal foundation. Active stakeholder engagement, combined with public–private partnerships and government and corporate sustainability initiatives, should place greater emphasis on regulatory and ethical compliance, including fair labor standards, equal access to resources, and protection for vulnerable groups. Although regulatory tools—such as the EU Taxonomy, Corporate Sustainability Reporting Directive (CSRD), and the Sustainable Finance Disclosure Regulation (SFDR)—provide a foundation, their effectiveness depends on technological development and regional integration. Our assessment also suggests that developing countries should establish national programs that align with EGD goals and fully leverage available funding. Collaboration among stakeholders, including governments, businesses, and communities, is essential to building awareness and support for sustainable practices. Investment in green technologies and innovation will help overcome existing challenges and create new opportunities. Finally, simplifying regulations, removing logistical barriers, and improving infrastructure will facilitate the adoption of CE principles by industries. By focusing on these actions, the EGD framework can drive a sustainable future while ensuring socio-economic and environmental benefits.

## REFERENCES


Adolph, J., & Beckmann, M. (2024). Corporate sustainability strategy: From definitional ambiguity toward conceptual clarification. *Business Strategy and the Environment, 33*(5), 4708–4729. DOI: 10.1002/bse.3722

Aicja Sikora (2021). European Green Deal – legal and financial challenges of climate change.

Alessi, L., Cojoianu, T., Hoepner, A. G. F., & Michelon, G. (2024). Accounting for the EU Green Taxonomy: Exploring its concept, data, and analytics. *Accounting Forum, 48*(3), 365–373. DOI: 10.1080/01559982.2024.2369343

Ali, R. F., Dominic, P. D. D., Ali, S. E. A., Rehman, M., & Sohail, A. (2021). Information security behavior and information security policy compliance: A systematic literature review for identifying the transformation process from noncompliance to compliance. *Applied Sciences (Basel, Switzerland), 11*(8), 3383.

Arta, K., Fogarassy, C., & Arta, K. (2021). European Green Deal Policy for the Circular Economy: Opportunities and Challenges. *Hungarian Agricultural Engineering, 39*, 65–73. DOI: 10.17676/HAE.2021.39.65

Awan, U., & Sroufe, R. (2022). Sustainability in the Circular Economy: Insights and Dynamics of Designing Circular Business Models. *Applied Sciences (Basel, Switzerland), 12*(3), 1521. DOI: 10.3390/app12031521

Bag, S., Srivastava, G., Gupta, S., Sivarajah, U., & Wilmot, N. V. (2024). The effect of corporate ethical responsibility on social and environmental performance: An empirical study. *Industrial Marketing Management, 117*, 356–370.

Barritt, E. (2024). The Aarhus Convention and the latent right to a healthy environment. *Journal of Environmental Law, 36*(1), 67–85. DOI: 10.1093/jel/eqae003

Bartosik, A. S., Manev, N., Pyk, A., & Pendaroska, M. (2025). Factors Influencing Powertrain Technology Choice and Barriers to Electric Vehicle Adoption in Poland: Insights into Consumer Preferences and Challenges in the Green Transition. Preprint. https://doi.org/DOI: 10.20944/preprints202504.1376.v1

Bashynska, I. (2025). Ethical Aspects of AI Use in the Circular Economy. *AI & Society*. Advance online publication. DOI: 10.1007/s00146-025-02436-1

Bassen, A., Kordsachia, O., Lopatta, K., & Tan, W. (2025). Revenue alignment with the EU Taxonomy Regulation in developed markets. *Journal of Banking & Finance, 170*, 107339. Advance online publication. DOI: 10.1016/j.jbankfin.2024.107339

Bernardo, C., Wang, L., Vasca, F., Hong, Y., Shi, G., & Altafini, C. (2021). Achieving consensus in multilateral international negotiations: The case study of the 2015 Paris Agreement on climate change. *Science Advances*, *7*(51), eabg8068. Advance online publication. DOI: 10.1126/sciadv.abg8068 PMID: 34910525

Bonoli, A., Zanni, S., & Serrano-Bernardo, F. (2021). Sustainability in Building and Construction within the Framework of Circular Cities and European New Green Deal. The Contribution of Concrete Recycling. *Sustainability*, *13*(4), 2139. DOI: 10.3390/su13042139

Brown, M. E., & Treviño, L. K. (2006). Ethical leadership: A review and future directions. *The Leadership Quarterly*, *17*(6), 595–616. DOI: 10.1016/j.leaqua.2006.10.004

Bucur, L. M. (2023). Exploring the role of consumers in promoting a circular economy: Increasing awareness and engagement. In *Proceedings of the international conference on business excellence* (Vol. 17, No. 1, pp. 38-47).

Busch, D. (2023). EU Sustainable Finance Disclosure Regulation. *Capital Markets Law Journal*, *18*(3), 303–328. DOI: 10.1093/cmlj/kmad005

Buzogány, A., Parks, L., & Torney, D. (2025). Democracy and the European Green Deal. *Journal of European Integration*, *47*(2), 135–154. DOI: 10.1080/07036337.2025.2459859

Castro-Lopez, A., Entrialgo, M., Liao, C. T., & Santos-Vijande, M. L. (2025). Towards circular economy through innovation: The role of entrepreneurial orientation and human resource management. *The International Entrepreneurship and Management Journal*, *21*(1), 36.

Cavoski, A. (2022). The European Green Deal and Technological Determinism. *Environmental Law Review*, *24*(3), 201–213. DOI: 10.1177/14614529221104558

Ciacchi, S. (2024). The newly-adopted Corporate Sustainability Due Diligence Directive: an overview of the lawmaking process and analysis of the final text. ERA Forum, 25, 29–48. https://doi.org/DOI: 10.1007/s12027-024-00791-y

Crespy, A., & Munta, M. (2023). Lost in transition? Social justice and the politics of the EU green transition. *Transfer: European Review of Labour and Research*, *29*(2), 235–251. DOI: 10.1177/10242589231173072

da Silva Hyldmo, H., Rye, S. A., & Vela-Almeida, D. (2024). A globally just and inclusive transition? Questioning policy representations of the European Green Deal. *Global Environmental Change*, *89*, 102946. DOI: 10.1016/j.gloenvcha.2024.102946

Daga, S., Yadav, K., Singh, P., & Mishra, V. (2025). Beyond the Take-Make-Dispose Model—Unlocking the Power of Circular Economy for an Environmentally Resilient Future. In *Circular Economy and Environmental Resilience: Solutions for a Sustainable Tomorrow* (Vol. 1, pp. 1–11). Springer Nature Switzerland.

Dubský, Z., Votoupalová, M., & Vítková, A. (2025). Framing the European Green Deal: Political and media energy discourses in Czechia and Sweden. *Humanities & Social Sciences Communications*, *12*(1), 1–13. DOI: 10.1057/s41599-025-04557-w

Eckert E., Kovalevska O. (2021). Sustainability in the European Union: Analyzing the Discourse of the European Green Deal. Journal of Risk and Financial Management, 14(2), 80. https://doi.org/. ERA Foruma, 21, 681-697.DOI: 10.3390/jrfm14020080

European Commission. (2019). Communication from the commission to the European parliament, the european council, the council, the European economic and social committee and the committee of the regions. European Commission: Brussels, Belgium. Available online: https://ec.europa.eu/commission/presscorner/detail/en/ip_19_6691 (accessed on July 17, 2025).

European Commission (2020). Sustainable Europe Investment Plan/European Green Deal investment plan. Communication from the commission to the European Parliament, the European Council, the Council, the European Economic and Social Committee and the Committee of the Regions 21.

European Commission. (2021). Corporate Sustainability Reporting Directive (CSRD) - Proposal. Retrieved from https://ec.europa.eu/info/business-economy-euro/banking-and-finance/sustainable-finance_en

European Commission. (2021). Regulation (EU) 2021/1056 of 24 June 2021 establishing the Just Transition Fund. *Official Journal of the European Union, L, 231*, 1–20.

European Parliament & Council. (2020). Regulation (EU) 2020/852 of 18 June 2020 on the establishment of a framework to facilitate sustainable investment. *Official Journal of the European Union, L, 198*, 13–43.

Fejzic, E., & Usher, W. (2025). Stakeholder engagement for co-designing European climate and energy research priorities. *Renewable & Sustainable Energy Reviews*, *215*, 115574. DOI: 10.1016/j.rser.2025.115574

Fetting, C. (2020). The European green deal. ESDN report, December, 2(9), 53.

Freeman, R. E., & Velamuri, S. R. (2023). A new approach to CSR: Company stakeholder responsibility. In *R. Edward Freeman's Selected Works on Stakeholder Theory and Business Ethics* (pp. 251–264). Springer International Publishing.

Geissdoerfer, M., Savaget, P., Bocken, N. M., & Hultink, E. J. (2017). The Circular Economy–A new sustainability paradigm? *Journal of Cleaner Production, 143*, 757–768.

Ghisellini, P., Cialani, C., & Ulgiati, S. (2016). A review of circular economy: the expected transition to a balanced interplay of environmental and economic systems. In Sustainability, 8(2), 17-28. https://doi.org/DOI: 10.1016/j.jclepro.2015.09.007

Ghormare, R. R., Fatima, S., Grover, P., Phutela, N., Kandpal, V., & Gonzalez, E. D. S. (2024). Exploring the paradigm shift towards sustainability: A systematic literature review on circular economy and eco-innovation. *AIMS Environmental Science*, *11*(6), 940–959.

Gunawardhana, S. (2025). Ethical Gaps in the Circular Economy Execution Process Undercutting the Benefits: Special Focus on Underdeveloped Countries. *Circular Economy and Sustainability*, *5*(1), 719–728.

Gunawardhana, S. (2025). Ethical Gaps in the Circular Economy Execution Process Undercutting the Benefits: Special Focus on Underdeveloped Countries. *Circular Economy and Sustainability*, *5*(1), 719–728. DOI: 10.1007/s43615-024-00446-7

Hafner, M., & Raimondi, P. P. (2020). Priorities and challenges of the EU energy transition: From the European Green Package to the new Green Deal. *Russian Journal of Economics*, *6*(4), 374–389.

Hariyani, D., Hariyani, P., Mishra, S., & Sharma, M. K. (2024). Leveraging digital technologies for advancing circular economy practices and enhancing life cycle analysis: A systematic literature review. *Waste Management Bulletin*, *2*(3), 69–83.

Hartley, K., Baldassarre, B., & Kirchherr, J. (2024). Circular economy as crisis response: A primer. *Journal of Cleaner Production*, *434*, 140140.

Hereu-Morales, J., Segarra, A., & Valderrama, C. (2024). The European (Green?) Deal: A systematic analysis of environmental sustainability. *Sustainable Development*, *32*(1), 647–661.

Islam, S., & Roshid, Md. M., Chandra Bhowmik, R., Dhar, B. K., Raihan, A., & Karim, R. (2025). Policy pathways for renewable energy, health, and sustainability in sub-saharan Africa: An empirical assessment of energy access and life expectancy. *Energy Policy*, *206*, 114801. DOI: 10.1016/j.enpol.2025.114801

Karim, R., Waaje, A., & Roshid, M. M. (2025). Digital Circular Economy for Attaining Sustainable Development Goals: Technologies for Global Sustainability. In Effects of Digitalization and Circular Economy on Sustainable Policy and Climate Change Prevention (pp. 211-240). IGI Global Scientific Publishing.

Karlsson, N. P. (2019). Business models and business cases for financial sustainability: Insights on corporate sustainability in the Swedish farm-based biogas industry. *Sustainable Production and Consumption, 18*, 115–129. DOI: 10.1016/j.spc.2019.01.005

Kirchherr, J., Reike, D., & Hekkert, M. (2017). Conceptualizing the circular economy: An analysis of 114 definitions. *Resources, Conservation and Recycling, 127*, 221–232. DOI: 10.1016/j.resconrec.2017.09.005

Kopytsia, I. (2021). The legal regulation of climate change in Ukraine: Issues and prospects. *Journal of Environmental Law and Policy, 1*, 105–125. DOI: 10.33002/jelp001.05

Kosi, U., & Relard, P. (2024). Are firms (getting) ready for the Corporate Sustainability Reporting Directive? *Sustainability Nexus Forum, 32*(5), 1–14. DOI: 10.1007/s00550-024-00541-1

Koundouri, P., Alamanos, A., Plataniotis, A., Stavridis, C., Perifanos, K., & Devves, S. (2024). Assessing the sustainability of the European Green Deal and its interlin kages with the SDGs. *Npj Climate Action, 3*(1), 23.

Kulovesi, K., Oberthür, S., van Asselt, H., & Savaresi, A. (2024). The European Climate Law: Strengthening EU procedural climate governance? *Journal of Environmental Law, 36*(1), 23–42. DOI: 10.1093/jel/eqad034

Kuyer, P., Shukla, S., O'Brolcháin, F., Doncel, V. R., & Gordijn, B. (2025). The ethics of digital nudging for sustainable energy consumption–a closer look at the EU green deal. *Technology in Society, 102990*. Advance online publication. DOI: 10.1016/j.techsoc.2025.102990

Lapčík, V., Lapčík, M., & Lapčík, V.Jr. (2022). Current aspects of decarbonisation in the Czech Republic and possibilities of replacement of coal energy sources by renewable sources of electric energy. *Inżynieria Mineralna*, (1), 87–96.

Leipold, S. (2021). Transforming ecological modernization 'from within' or perpetuating it? The circular economy as EU environmental policy narrative. *Environmental Politics, 30*(6), 1045–1067.

Li, N., & van Rooij, B. (2022). Law lost, compliance found: A frontline understanding of the non-linear nature of business and employee responses to law. *Journal of Business Ethics, 178*, 715–734. DOI: 10.1007/s10551-021-04751-1

Liu, Z., Liu, J., & Osmani, M. (2021). Integration of Digital Economy and Circular Economy: Current Status and Future Directions. *Sustainability, 13*(13), 7217. DOI: 10.3390/su13137217

Maris, G., & Flouros, F. (2021). The Green Deal, National Energy and Climate Plans in Europe: Member States' Compliance and Strategies. *Administrative Sciences, 11*(3), 75. DOI: 10.3390/admsci11030075

Maunuksela, O. (2024). Social justice and the European Green Deal: Green transition in a well-ordered society (Master's thesis, University of Turku).

Moesker, K., & Pesch, U. (2022). The just transition fund–Did the European Union learn from Europe's past transition experiences? *Energy Research & Social Science, 91*, 102750. DOI: 10.1016/j.erss.2022.102750

Nagy, G., Heiner, S. Á., & Kovács, Z. (2025). Exploring the Presence and Absence of Academic Discourse on Public Participation in the European Green Deal: A Central and Eastern European Perspective. *Societies (Basel, Switzerland), 15*(3), 49. DOI: 10.3390/soc15030049

Nobre, G. C., & Tavares, E. (2017). Scientific literature analysis on big data and internet of things applications on circular economy: A bibliometric study. *Scientometrics, 111*(1), 463–492.

Oberer, B., & Erkollar, A. (2024). The European Green Deal: A Roadmap to a Sustainable Future. In The International Symposium for Production Research (pp. 69-81). Cham: Springer Nature Switzerland.

Oehring, D., & Gunasekera, P. (2024). Ethical frameworks and Global Health: A narrative review of the "leave no one behind" principle. INQUIRY. *The Journal of Health Care Organization, Provision, and Financing, 61*, 00469580241288346. PMID: 39385394

Ostojic, S., Simone, L., Edler, M., & Traverso, M. (2024). How practically applicable are the EU taxonomy criteria for corporates? *Sustainability, 16*(4), 1575. DOI: 10.3390/su16041575

Pianta, M., & Lucchese, M. (2020). Rethinking the European Green Deal: An Industrial Policy for a Just Transition in Europe. *The Review of Radical Political Economics, 52*(4), 633–641. DOI: 10.1177/0486613420938207

Pollex, J. (2025). Is it all about distribution? – Debating the Green Deal in the European Parliament. *Journal of European Integration, 47*(2), 217–236. DOI: 10.1080/07036337.2025.2459874

Roshid, M. M., Waaje, A., Karim, R., Irtisham, M. R., & Kaswan, M. S. (2025). Industry 5.0 and organizational sustainability: Integrating cutting-edge technologies and human ingenuity. In Industry 5.0 (pp. 50-61). CRC Press.

Roshid, M. M., Waaje, A., Meem, T. N., & Sarkar, A. (2024). Logistics 4.0: A comprehensive literature review of technological integration, challenges, and future prospects of implementation of Industry 4.0 technologies. *International Journal of Technology, Knowledge and Society*, *20*(1), 65. DOI: 10.18848/1832-3669/CGP/v20i01/65-85

Roshid, M. M., Yeamin, M. B., Waaje, A., Karim, R., & Kaswan, M. S. (2026). Understanding the Nexus of Industry 5.0 and Circular Economy for Enriched Lean and Environment Sustainability: Nexus of Industry 5.0 and Circular Economy. In J. de Sá, S. Sharma, S. Kumar, & M. B.R. (Eds.), Transformative Lean Six Sigma Techniques for the Quality 5.0 Paradigm (pp. 349-380). IGI Global Scientific Publishing. https://doi.org/DOI: 10.4018/979-8-3373-0943-9.ch013

Sandri, S., Hussein, H., Alshyab, N., & Sagatowski, J. (2025). The European Green Deal: Challenges and opportunities for the Southern Mediterranean. *Mediterranean Politics*, *30*(1), 196–207. DOI: 10.1080/13629395.2023.2237295

Sandri, S., Hussein, H., Alshyab, N., & Sagatowski, J. (2025). The European green deal: Challenges and opportunities for the Southern Mediterranean. *Mediterranean Politics*, *30*(1), 196–207.

Sapaloglu, I., Ajas, İ. İ., & Yıldırım, N. (2025). A Strategic Roadmap to Circularization of the Turkish Textile and Ready-To-Wear Industry Within the Context of the European Green Deal. *Business Strategy & Development*, *8*(3), e70153. DOI: 10.1002/bsd2.70153

Schlacke, S., Wentzien, H., Thierjung, E. M., & Köster, M. (2022). Implementing the EU Climate Law via the 'Fit for 55'package. *Oxford Open Energy*, *1*, oiab002. Advance online publication. DOI: 10.1093/ooenergy/oiab002

Schramm, L., & Terranova, C. (2024). From NGEU to REPowerEU: Policy steering and budgetary innovation in the EU. *Journal of European Integration*, *46*(6), 943–961. DOI: 10.1080/07036337.2024.2353791

Schroeder, P., Anggraeni, K., & Weber, U. (2019). The relevance of circular economy practices to the sustainable development goals. *Journal of Industrial Ecology*, *23*(1), 77–95.

Scoones, I., Stirling, A., Abrol, D., Atela, J., Charli-Joseph, L., Eakin, H., Ely, A., Olsson, P., Pereira, L., Priya, R., van Zwanenberg, P., & Yang, L. (2020). Transformations to sustainability: Combining structural, systemic and enabling approaches. *Current Opinion in Environmental Sustainability*, *42*, 65–75. DOI: 10.1016/j.cosust.2019.12.004

Shabur, M. A., Shahriar, A., & Ara, M. (2025). From automation to collaboration: Exploring the impact of industry 5.0 on sustainable manufacturing. *Discover Sustainability*, *6*(1), 1–23.

Siddi, M. (2020). The European Green Deal: Assessing its current state and future implementation (FIIA Working Paper No. 114). Finnish Institute of International Affairs.

Sikora, A. (2021, January). European Green Deal–legal and financial challenges of the climate change. In Era forum (Vol. 21, No. 4, pp. 681-697). Berlin/Heidelberg: Springer Berlin Heidelberg.

Suárez-Eiroa, B., Fernández, E., Méndez-Martínez, G., & Soto-Oñate, D. (2019). Operational principles of circular economy for sustainable development: Linking theory and practice. *Journal of Cleaner Production*, *214*, 952–961.

Szpilko, D., & Ejdys, J. (2022). European Green Deal — research directions: A systematic literature review. Ekonomia i Środowisko, 2(81), 8-38. https://doi.org/ DOI: 10.34659/eis.2022.81.2.455

Taneja, S., Bansal, N., & Ozen, E. (2024). What was, is and will be: A bibliometric analysis of European Green Deal (EGD). *International Journal of Technology Management & Sustainable Development*, *23*(1), 85–104.

Trane, M., Borchardt, S., Marelli, L., Siragusa, A., Pollo, R., & Lombardi, P. (2025). What Role for Cities in the EU Green Deal? Analysis of Policy Framework, Interlinkages and Contribution to the 2030 Agenda. Sustainable Development.

United Nations. (2019). *Global Sustainable Development Report 2019: The Future is Now — Science for Achieving Sustainable Development.* UN DESA., DOI: 10.18356/5d04ad97-

Valentinov, V. (2024). Three problems of business addressed by stakeholder theory: Insights from the institutional economics of John R. Commons. *Journal of Business Ethics*, •••, 1–17. DOI: 10.1007/s10551-024-05854-1

van Langen, S. K., & Passaro, R. (2021). The Dutch Green Deals Policy and Its Applicability to Circular Economy Policies. *Sustainability*, *13*(21), 11683. DOI: 10.3390/su132111683

Veyssière, S. (2021). Circularity–The circular economy as an innovative process. Innovation Economics, Engineering and Management Handbook 2: Special Themes, 75-83.

Vrontisi, Z., Charalampidis, I., Fragkiadakis, K., & Florou, A. (2024). Towards a just transition: Identifying EU regions at a socioeconomic risk of the low-carbon transition. *Energy and Climate Change*, *5*, 100129. DOI: 10.1016/j.egycc.2024.100129

Waaje, A., Karim, R., & Roshid, Md. M., & Meem, T. N. (2024). Minimizing Waste in the Hospitality Industry: The Opportunities and Challenges of Lean Tourism. In R. Sucheran (Ed.), *Sustainable Waste Management in the Tourism and Hospitality Sectors* (pp. 125–166). IGI Global Scientific Publishing. https://doi.org/DOI: 10.4018/979-8-3693-6110-8.ch006

Wen, B., Zhang, G., Zhan, C., Chen, C., & Yi, H. (2025). The 2024 revision of the Declaration of Helsinki: A modern ethical framework for medical research. *Postgraduate Medical Journal*, *101*(1194), 371–382. PMID: 39671397

Wesseler, J. (2022). The EU's farm-to-fork strategy: An assessment from the perspective of agricultural economics. *Applied Economic Perspectives and Policy*, *44*(4), 1826–1843. DOI: 10.1002/aepp.13239

Wolf, S., Teitge, J., Mielke, J., Schütze, F., & Jaeger, C. (2021). The European Green Deal—More than climate neutrality. *Inter Economics*, *56*(2), 99–107. PMID: 33840826

World Economic Forum. (2025). Delivering on the European Green Deal Objectives: A Private Sector Perspective – Second Edition.

Yang, M., Chen, L., Wang, J., Msigwa, G., Osman, A. I., Fawzy, S., & Yap, P. S. (2023). Circular economy strategies for combating climate change and other environmental issues. *Environmental Chemistry Letters*, *21*(1), 55–80.

Zorpas, A. A., Naddeo, V., Voukkali, I., Vraka, E., & Bockreis, A. (2025). Environmental ethics considerations in circular economy and waste management. *Waste Management & Research*, *43*(2), 133–136. DOI: 10.1177/0734242X241299891 PMID: 39523494

Żuk, P. (2025). The European Green Deal and the peasant cause: Class frustration, cultural backlash and right-wing nationalist populism in farmers' protests in Poland. *Journal of Rural Studies*, *119*, 103708.

# Chapter 6
# Advancing Sustainability Through the Social Business Model in the Circular Economy of Developing Countries

**Forhad Hossain**
https://orcid.org/0000-0002-3593-0860
*Sophia University, Japan*

**Paulina Permatasari**
https://orcid.org/0000-0002-2457-2252
*Parahyangan Catholic University, Indonesia*

**Johanna Renny Octavia**
https://orcid.org/0000-0002-5581-2634
*Parahyangan Catholic University, Indonesia*

**Theresia Edward Busagara**
*University of Dar es Salaam, Tanzania*

**Shailaja Kasaju**
*Kathmandu Model College, Nepal*

## ABSTRACT

*The chapter explores the operational and structural challenges in implementing Circular Economy (CE) practices in developing countries, with a specific focus on Indonesia, Nepal, and Tanzania. Using a mixed method approach grounded in ex-*

DOI: 10.4018/979-8-3373-2454-8.ch006

*pert survey data and supported by PESTLE and Triple Bottom Line frameworks, the chapter investigates how the Social Business model can act as a strategic enabler of CE. Key findings highlight limited access to finance, policy enforcement gaps, and weak consumer engagement as significant barriers. Despite these challenges, expert insights reveal high potential for CE adoption if targeted support mechanisms are introduced. The chapter discusses real-world social businesses such as Grameen Group and Waste Concern that demonstrate how mission-driven, locally embedded organizations can advance circularity while maintaining sustainability. The chapter concludes with recommendations for integrating Social Business into CE strategies and outlines future research directions to deepen practice-informed knowledge.*

## INTRODUCTION

The circular economy (CE) is increasingly gaining global attention as a model for creating sustainable and resource-efficient systems. The fundamental premise of the circular economy is a shift from the traditional "take-make-dispose" model to one that emphasizes reusing, recycling, and reducing waste. By rethinking how products are designed, manufactured, used, and disposed of, the circular economy aims to benefit both society and the environment. In contrast to the linear economy, which reduces limited resources and generates excess waste, the circular economy seeks to establish a closed-loop system where products, materials, and resources are kept in use for as long as possible, extracting maximum value from them before recovering and regenerating materials at the end of each service life.

Developing countries, which often face dual pressures of poverty alleviation and environmental protection, stand to benefit considerably from the adoption of circular economy practices (Schroeder et al., 2019; Waaje et al., 2024; Roshid et al., 2025a). However, the implementation of CE in these contexts is fraught with challenges. These include inadequate infrastructure, limited technological capabilities, fragmented policy frameworks, insufficient financing mechanisms, and a general lack of public and institutional awareness regarding sustainable practices (UNEP, 2018; Kirchherr et al., 2018; Roshid et al., 2025b). Consequently, many CE initiatives in these regions are small-scale, donor-dependent, and struggle with long-term viability. Frequently, they are mischaracterized as charitable or welfare-driven ventures rather than commercially viable models that can operate independently and scale sustainability.

In response to these challenges, the Social Business model defined by Nobel Peace Prize laureate Muhammad Yunus offers a compelling solution. A social Business is defined as a business created and designed to address a social problem financially and sustainably, without distributing profits to shareholders or owners

(Yunus, 2010). Unlike traditional profit-maximizing firms, social businesses reinvest their profit to expand their impact, making them well-suited to tackle systemic issues in developing contexts. By combining the efficiency of market mechanisms with the mission of social welfare, social business can address core CE challenges such as financial sustainability, operational scalability, and community engagement.

Integrating CE with the Social Business model could create a robust framework for sustainable development in developing countries. Several grassroots initiatives have already begun to reflect this synergy. For example, in Indonesia, organizations like Gringgo Tech are leveraging digital tools to formalize informal waste management sectors, creating inclusive CE ecosystems (ASEAN-Japan Centre, 2024). Similarly, in Nepal, local cooperatives and SMEs have been instrumental in promoting recycling and reuse practices, albeit with limited policy support. These case studies offer valuable insights into how Social Business principles can facilitate circular solutions that are tailored to local contexts.

This chapter aims to explore the intersection of the circular economy and Social Business in the context of developing countries. Specifically, it examines how Social Business models can overcome systemic barriers to CE implementation and foster sustainable, scalable, and socially inclusive practices. The primary focus is on three countries: Indonesia, Nepal, and Tanzania, which serve as representative examples of developing countries.

To guide this inquiry, the chapter addresses the following research questions:

a) What are the key challenges in implementing circular economy practices in developing countries, specifically related to operations and sustainability?
b) How can Social Business models address these challenges, particularly in terms of ensuring long-term sustainability and scalability?
c) What recommendations can be made for applying Social Business models in CE projects to ensure their success and replication?

To address these questions, this study employs a mixed-methods approach, combining literature reviews, case study analysis, policy examination, and stakeholder interviews. By synthesizing insights from diverse data sources, the research aims to construct a holistic understanding of how Social Business can serve as an enabling platform for CE transformation in resource-constrained environments. The structure of this chapter is as follows: after the introduction and background, section 3 presents the literature review, followed by the methodology in section 4. Section 5 outlines the findings, which are analyzed in Section 6. The chapter concludes with a final reflection in section 7.

# BACKGROUND AND THEORETICAL FOUNDATION

## 1. The Concept and Evolution of the Circular Economy

The circular economy (CE) is an emerging paradigm that seeks to address the environmental and social limitations of the traditional linear economic model, which is characterized by a one-way flow of resources: take, make, and dispose. The CE model proposes a regenerative system designed to minimize waste and maximize the value of products, materials, and resources by extending their lifecycle (Geissdoerfer et al., 2017). It emphasizes activities such as reuse, refurbishment, remanufacturing, recycling, and closed-loop production systems.

The origins of the circular economy are interdisciplinary, drawing from several schools of thought such as industrial ecology, cradle-to-cradle design, performance economy, and biomimicry (Murray et al., 2017). These concepts share a common thread: they challenge the notion of waste and seek to reframe materials and outputs as inputs in continuous cycles. Notably, the Ellen MacArthur Foundation (2021), a key proponent of CE globally, identifies three foundational principles: (1) design out waste and pollution, (2) keep products and materials in use, and (3) regenerate natural systems.

In contrast to conventional waste management or recycling, the CE is not solely a downstream activity but one that involves systemic changes in product design, business models, and supply chains (Korhonen et al., 2018). This systemic nature makes CE particularly suitable for addressing the complex sustainability challenges faced by both developed and developing economies.

## 2. Circular Economy in Developing Countries

The adoption of CE in developing countries presents both significant opportunities and complex challenges. These regions are disproportionately affected by environmental degradation, waste mismanagement, and resource scarcity, making CE an appealing strategy for inclusive and sustainable development (Schroeder et al., 2019; Karim et al., 2025; Sultana et al., 2025). The integration of CE principles could reduce dependency on finite resources, promote job creation in waste management and repair sectors, and foster environmental resilience. However, several structural and institutional barriers hinder CE adoption in developing contexts. These include:

**Inadequate Infrastructure:** Many developing countries lack the necessary logistics, collection, and recycling infrastructure to implement CE practices effectively (Singh & Ordonez, 2016).

**Policy and Regulatory Gaps:** There is often an absence of clear regulatory frameworks and incentives to support circular initiatives, coupled with fragmented environmental governance (Cattaneo et al., 2010).

**Limited Financial and technological Capacity:** Insufficient access to capital, innovation, and CE-compatible technologies impedes the scaling of circular initiatives (Saidani et al., 2017).

**Low Public Awareness:** Many communities are unfamiliar with CE principles and lack the education or incentives to engage in circular behaviours (Kirchherr et al., 2018).

Despite these challenges, informal and community-driven CE activities already exist in many developing nations, particularly in sectors such as repair, reuse, and upcycling. These grassroots efforts can serve as a foundation upon which scalable circular models can be built, provided they are supported by inclusive, sustainable, and financially viable mechanisms.

## 3. The Social Business Model: Principles and Practice

Social Business, as articulated by Muhammad Yunus (2010), is a model that addresses social problems through financially self-sustaining businesses that do not aim to maximize profits for shareholders. In contrast to traditional for-profit enterprises or donor-dependent non-profits, Social Businesses operate with a double bottom line: Social impact and financial sustainability. Any surplus generated is reinvested into the business to expand its outreach and effectiveness rather than being distributed as dividends. The seven principles of social business include (Seven Principles of Social Business, 2009):

I. The business objective will be to overcome poverty, or one or more problems (such as education, health, technology access, and environment) which threaten people and society; not profit maximization.
II. Financial and economic sustainability.
III. Investors get back their investment amount only. No dividend is given beyond investment money.
IV. When the investment amount is paid back, the company's profits are reinvested for expansion and improvement.
V. Gender sensitive and environmentally conscious.
VI. The workforce gets a market wage with better working conditions.
VII. …do it with joy.

Social Business is particularly relevant in contexts where traditional market solutions fail to deliver public goods effectively. It offers a model through which underserved communities can access essential services such as health, education, and sanitation through market-based yet socially oriented mechanisms (Battilana & Lee, 2014). Social Businesses have been applied globally in sectors like agriculture, waste management, and health care. One of the most prominent examples is Grameen Danone Foods, a joint venture aimed at combating child malnutrition in Bangladesh through affordable fortified yogurt, demonstrating how a business can be mission-driven yet self-sustaining (Yunus et al., 2015).

## 4. Integrating Social Business with Circular Economy

The intersection between CE and Social Business offers a promising synergy, particularly for developing countries where both environmental sustainability and social equity are high priorities. Circular solutions often face hurdles in financial sustainability and operational scalability challenges that the Social Business model is uniquely designed to address. By embedding CE practices into the operational structure of a Social Business, entrepreneurs can develop solutions that are not only environmentally regenerative but also socially inclusive and economically viable.

For instance, in waste management, one of the primary CE sectors, Social Businesses can formalize informal workers, improve labour conditions, and generate income while providing essential environmental services (Wilson et al., 2006).

## 5. Theoretical Frameworks Supporting CE and Social Business Integration

Two major theoretical frameworks underpin the integration of CE and Social Business:

**The Triple Bottom Line (TBL):** Introduced by Elkington (1997), the TBL framework evaluates business success based on three dimensions: economic, social, and environmental performance. Both CE and Social Business align well with this model. CE focuses on minimizing environmental impact and maximizing resource efficiency, while Social Business emphasizes social equity and economic inclusion. Together, they present a comprehensive strategy for achieving all three pillars of sustainable development.

**PESTLE Framework:** To design the questionnaire, this study also considered the PESTLE framework, which provides a structured lens for analyzing the external factors that influence organizational strategy and implementation. It has six key dimensions: Political, Economic, Social, Technological, Legal, and Environmental, which consider sustainability pressure, ecological impacts, and climate resilience (Yüksel, 2012). In the context of CE implementation, PESTLE helps identify systemic enablers and constraints across multiple domains, particularly relevant for understanding the complex operating environments of developing countries.

**Systems Thinking and Stakeholder Theory:** Systems thinking emphasizes interconnections between different components in a system, which is central to CE. Meanwhile, stakeholder theory posits that businesses should account for the interests of all stakeholders, not just shareholders (Elias, 2017). Social Business, by design, serves marginalized stakeholders such as low-income communities and informal worker groups that are critical in circular systems, especially in waste and material recovery sectors.

Combining these theoretical frameworks provides a holistic lens for understanding and implementing circular solutions that are socially responsible, environmentally sustainable, and economically viable. It also guides policy design, organizational structure, and impact evaluation.

## LITERATURE REVIEW

A growing body of literature underscores the compatibility between CE principles and social entrepreneurship. Social businesses aim to address societal problems with financially sustainable, mission-driven operations (Yunus, 2010). This approach is particularly suitable in developing countries, where traditional CE business models may not align with the social realities of poverty, informality, and weak governance (Schroeder et al., 2019). However, Kirchherr et al. (2018) argue that CE research is dominated by techno-centric and western-centric perspectives, with insufficient empirical grounding in low-and middle-income countries. Similarly, Cattaneo et al. (2010) emphasize the Lack of contextualized strategies for CE implementation in Sub-Saharan Africa and South Asia, where informal sectors dominate waste collection and recycling. This study responds to these gaps by focusing on three developing countries - Nepal, Indonesia, and Tanzania- and evaluating how Social Business models can support CE transitions in resource-constrained settings.

### 1. Insights from Nepal

Nepal is experiencing rapid urbanization, resulting in escalating waste generation and resource inefficiency. According to Subedi et al. (2023), Kathmandu alone produces over 766 tons of solid waste per day, most of which ends up in open landfills. Informal waste pickers manage 15% of household waste but remain unrecognized in official systems (Cataldo et al., 2024). Despite the adoption of pilot projects like source segregation and biogas generation, the country still lacks integrated solid waste management and CE-oriented policy frameworks. Social businesses such as

Doko Recyclers, Blue Waste to Value, and Khaalisisi have created digital platforms to incentivize household-level segregation and recycling. However, Melles et al. (2025) highlight that these initiatives operate without strong institutional support, limiting their scalability. Agriculture, a significant economic sector, produces substantial organic waste, yet circular farming practices remain underexplored and underfunded. Key constraints include a Lack of policy harmonization across local and federal levels; Minimal integration of the informal sector (Melles et al., 2025); Insufficient infrastructure for recycling and composting (Subedi et al., 2023); and Financial and digital limitations of social enterprises (Sodhi & Tang, 2011). Despite these challenges, the use of indigenous knowledge, traditional farming, and circular textile models (e.g., reuse of baby cloths and denim) has proven culturally resonant and effective (Shrestha, Shakya, & Shrestha, 2025).

## 2. Insights from Tanzania

Tanzania faces a similar trajectory, with massive volumes of plastic and organic waste dominating its urban and rural landscapes. In 2021, plastic waste made up 76% of all waste streams (Nipe Fagio, 2021; Isibika et al., 2023). Despite regulatory efforts, including a ban on thin plastic bags and the 2024 launch of the National Plastic Action Partnership, compliance and enforcement remain weak (Kinyondo & Huggins, 2021). Studies highlight the agricultural sector's potential in CE. For example, only 2% of Tanzania's sisal is used, and the rest becomes waste (Msuya & Temu, 2025). Similar inefficiencies exist in coffee, fruit, and food processing sectors. However, initiatives in composting and biofertilizer production are gaining traction through agribusinesses. Waste Picker Organizations (WPOs) are vital grassroots actors. They demonstrate innovation in collecting, segregating, and repurposing waste, often relying on trust-based systems and community networks (Kain et al., 2022). Charles (2019) reports that despite their adaptive strategies, WPOs face poor working conditions, unstable income, and marginalization. Furthermore, recycling in Dar es Salaam and other cities remains informal, mainly, with private sector players controlling 90% of waste logistics (Kirama & Mayo, 2016). Public awareness, infrastructure deficits, and inadequate municipal funding further undermine CE progress.

## 3. Insights from Indonesia

Indonesia provides a valuable case for understanding how circular economy (CE) practices can align with social business strategies. The Country's Circular Economy Roadmap 2025-2045, developed with support from UNDP and the EU, promotes inclusive, low-carbon business models aimed at integrating informal workers and supporting sustainable innovation (Ministry of National Development Planning/

Bappenas, 2024). Social enterprises like Gringoo and Waste4Change have introduced digital platforms using AI and mobile apps to support waste pickers and optimize recycling logistics (Agyabeng-Mensah et al., 2022). Despite these advancements, implementation challenges remain. Municipal capacity for integrating informal workers is limited, and many CE initiatives remain donor-dependent or fragmented (Meidiana & Gamse, 2010).

Additionally, rural areas face a lack of recycling infrastructure and public awareness, hindering widespread adoption (Sutomo, Mukhlis, & Ahadiat, 2023). Research also highlights low consumer participation and weak institutional coordination as persistent barriers. Nevertheless, Indonesia's experience demonstrates the value of community engagement, policy alignment, and digital innovation principles that resonate strongly with the Social Business Model. The country illustrates how CE strategies can be scaled when supported by inclusive entrepreneurship and technology, while also understanding the need for sustainable funding, institutional coherence, and grassroots integration.

Despite the growing global attention to the circular economy and its promising alignment with sustainable development goals, a significant research gap persists in understanding how the circular economy can be effectively operationalized in developing countries through socially oriented business models. While many studies emphasize technological innovations, policy reforms, and industrial applications of CE, there is limited empirical exploration into how Social Business Models prioritize financial sustainability alongside social and environmental impact. Can bridge critical implementation challenges such as informal sector exclusion, weak infrastructure, and regulatory fragmentation. Most of the existing literature is geographically skewed toward high-income contexts, offering limited relevance to economies characterized by informal labor markets, donor dependency, and governance limitations. Furthermore, much of the circular economy discourse remains conceptual, lacking a grounded analysis of real-world initiatives that integrate inclusive employment, community engagement, and localized resource cycles. The roles of grassroots entrepreneurs, women, youth, and informal workers who are often central to waste recovery and reuse practices are typically underrepresented in CE strategy and academic inquiry. This study addresses these deficiencies by focusing on how Social Business models can provide scalable and sustainable solutions for CE implementation in three diverse but similarly constrained contexts: Nepal, Indonesia, and Tanzania. It aims to contribute actionable insights into how financially self-sustaining, socially driven enterprises can overcome systemic barriers and build inclusive, circular ecosystems in developing economies, an area still underexplored in CE research.

## METHODOLOGY

This research employed a mixed-methods approach integrating qualitative and quantitative strategies. This design enabled the triangulation of diverse data sources, offering a comprehensive understanding of the challenges, enablers, and contextual nuances of CE transitions. The study focused on three countries -Indonesia, Nepal, and Tanzania, randomly chosen for their emerging CE activity, socio-economic diversity, and varied institutional frameworks. The research also considers relevant insights from Bangladesh as part of broader comparative reflections. The research methodology consisted of the following key components: literature review, case study analysis, policy review, and quantitative data collection via a structured survey. The design aimed to bridge existing research gaps identified in the literature, particularly the limited empirical grounding of CE studies in the Global South and the underexplored role of social business in this context.

A systematic review of academic publications and policy reports was undertaken to develop a robust conceptual foundation. Databases such as JSTOR, Scopus, and Google Scholar were used to identify peer-reviewed literature from the last 15 years focused on circular economy principles, the Social Business model, and their integration in developing countries. Grey literature from global institutions such as UNEP and the World Bank supplemented this review. Thematic synthesis highlighted key focus areas: the potential of CE in low- and middle-income contexts, the inclusive and regenerative dimensions of social enterprises, and the limitations of current CE frameworks in addressing informality, finance, and scalability. A synthesis matrix was used to map findings, extract theoretical models, and highlight methodological gaps that informed the design of the primary research tools.

This study also collected primary data through a structured expert survey, designed using insights from the TBL, PESTLE framework, and Systems Thinking and Stakeholder Theory. The survey included both closed and open-ended questions to capture expert perspectives on economic, institutional, and policy-related barriers to CE implementation and to assess the perceived effectiveness of social business models as potential solutions. The questionnaire was piloted for clarity and revised before dissemination. Survey distribution targeted stakeholders actively engaged in circular economy initiatives for sustainable policy, such as social entrepreneurs, business leaders, NGO representatives, policymakers, and sustainability experts. A total of 22 responses were collected from participants representing Indonesia (13), Nepal (5), and Tanzania (4). Roles included social entrepreneurs (36%), business leaders (36%), NGO representatives (14%), and a mix of policy and sustainability professionals. The questionnaire was structured around key dimensions such as: Barriers to CE adoption (e.g., funding, technology, awareness, policy); Availability and effectiveness of government support; Business incentives and public perception;

Policy frameworks and regulatory enforcement; Consumer willingness and market behavior. Each of these dimensions was aligned with either TBL or PESTLE categories. For instance, economic barriers were assessed through questions on policy support and regulatory clarity. The quantitative survey data were analyzed using descriptive statistical techniques. Key metrics included mean scores on Likert scale questions and frequency counts of selected multiple-choice responses. Qualitative responses to open-ended questions were coded thematically, enabling the identification of recurring issues across countries and roles. The research concluded with a cross-country comparative analysis to synthesize findings and draw context-sensitive recommendations.

## FINDINGS

This section presents the findings derived from structured expert survey data collected across three countries, Indonesia, Nepal, and Tanzania, to assess the barriers, opportunities, and institutional readiness for implementing CE practices through social business models. The survey, guided by Triple Bottom Line (TBL) and PESTLE analytical frameworks, targeted experts actively involved in sustainability, policy, and enterprise development. Their responses offer critical insights into the multi-dimensional challenges facing CE adoption in developing country contexts.

### 1. Operational and Structural Challenges in CE Implementation

One of the most significant insights from the study lies in understanding the operational and structural challenges associated with transitioning from a linear to a circular economy in developing countries. Survey data from 22 experts across Indonesia, Nepal, and Tanzania indicate that this transition is widely perceived as difficult. The mean difficulty rating was 3.77 out of 5, where five indicates that the transition is "extremely challenging". This high average reflects persistent systemic obstacles that prevent widespread CE adoption. To gain comparative insight, the study analyzed how respondents in each country rated the transition challenges. Figure 1 shows that Indonesia has the widest distribution of ratings, indicating a spectrum of experiences and potentially uneven regional capacities. Nepal and Tanzania show tighter clusters near the high end (scores of 4 & 5), suggesting more uniformly perceived difficulty across regions. The variance reflects differences in CE awareness, available infrastructure, and access to enabling policies or technologies. While Indonesian stakeholders may benefit from better urban CE support or pilot programs, Nepalese and Tanzanian respondents likely face similar challenges nationwide, especially in rural and under-resourced communities.

*Figure 1. Distribution of perceived difficulty in transitioning to circular economy models by country*

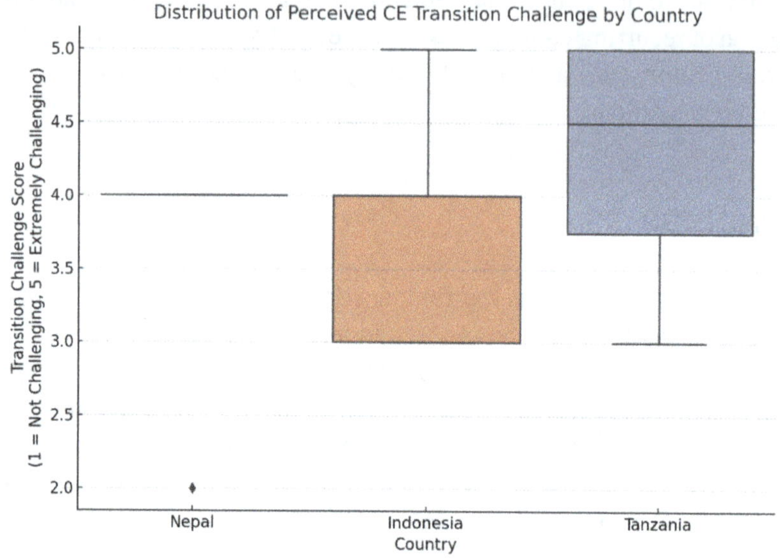


Statistical analysis also reveals a moderate negative correlation (r=0.45) between policy effectiveness and transition difficulty. This means that respondents who rated their country's policies as more effective were more likely to report lower implementation challenges. This supports the intuitive link between enabling institutions and the ease of CE adoption. However, no strong correlations were observed between funding difficulty and transition challenges, implying that financial barriers, while crucial, are just one part of a more complex implementation puzzle.

A content analysis of open-ended responses, along with multiple-choice selections, identified several recurring operational themes that obstruct CE transitions:

**Financial Constraints:** Nine respondents specifically selected or mentioned the Lack of investment and funding as the leading challenge. This includes access to seed capital, microcredit, or CE-focused grants. Several noted that high up-front costs for circular equipment, processing facilities, or certification made CE models economically feasible without external support. *"There are no dedicated investment channels for CE here-everyone's still talking about plastic bans, but not about building alternatives"- Indonesia.*

**Market Pressure from Linear Businesses:** Eight respondents indicated that CE enterprises struggle to compete with traditional market actors. These conventional businesses often have lower costs due to cheaper raw materials, tax avoidance, or regulatory loopholes. *"Recycled goods cost more to produce and less to sell. Without consumer incentives or support schemes, it is hard to break even,"-Tanzania.*

**Supply Chain and Technology Gaps:** Three respondents highlighted difficulty in sourcing sustainable materials, tools, and circular inputs. This is particularly acute in rural areas, where even basic composting tools or eco-packaging are often unaffordable.

**Regulatory and Compliance Costs:** Two responses referenced the burden of complex or mismatched regulations, for instance, high fees for business registration, conflicting waste management bylaws, or unclear environmental standards.

**Skills and Knowledge Deficit:** Though less frequently stated directly, several open comments pointed to the need for training, awareness, and institutional knowledge to implement CE in practice. This includes how to measure circularity, assess life cycle impacts, or manage reverse logistics.

## 2. Policy Gaps and Governance Challenges

Another central theme that emerged concerns weak policy effectiveness. While some countries, particularly Indonesia, have national CE strategies or pilot initiatives in place, most respondents reported limited policy support in practical terms. Out of 22 experts surveyed:

- 68% indicated "some support, but not widely accessible"
- 27% reported "no significant policy or financial support"
- 5% were "not sure" about the policy environment in their country

The average policy effectiveness score across countries was 2.45/5, indicating moderate to low satisfaction, as shown in Table 1.

*Table 1. Policy supports and effectiveness by country*

| Country | Respondents | % No/Unclear Policy Support | Avg. Policy Effectiveness (1–5) |
|---|---|---|---|
| Indonesia | 13 | 30.8% | 2.54 |
| Nepal | 5 | 40.0% | 1.80 |
| Tanzania | 4 | 25.0% | 3.00 |
| **Overall** | **22** | **27.3%** | **2.45** |

Commonly cited governance issues included:

- Poor coordination between ministries responsible for the environment, trade, and small business
- Lack of financial incentives, such as tax credits or subsidies for circular activities
- Regulatory frameworks are inaccessible to informal or community-based enterprises

As one Indonesian respondent noted: *"Even where CE policies exist, implementation is centralized and does not trickle down to small businesses".*

## Linking Policy to CE Adoption:

To examine the link between governance and adoption, a simple linear regression was conducted with **policy effectiveness** as the predictor and the **likelihood of CE adoption if better support existed** as the outcome variable. The results revealed a statistically significant positive relationship ($\beta = 1.00, p = 0.006$), with policy effectiveness explaining 32.5% of the variance in adoption likelihood. This indicates that even moderate improvements in policy effectiveness could substantially increase the likelihood of CE adoption.

Figure 2 illustrates this relationship, showing that higher policy effectiveness scores are generally associated with a greater reported likelihood of CE adoption. While these findings are based on a small sample, they align with broader evidence in the literature that strong governance and targeted incentives are critical enablers of CE scaling.

*Figure 2. Relationship between policy effectiveness and CE adoption likelihood.*

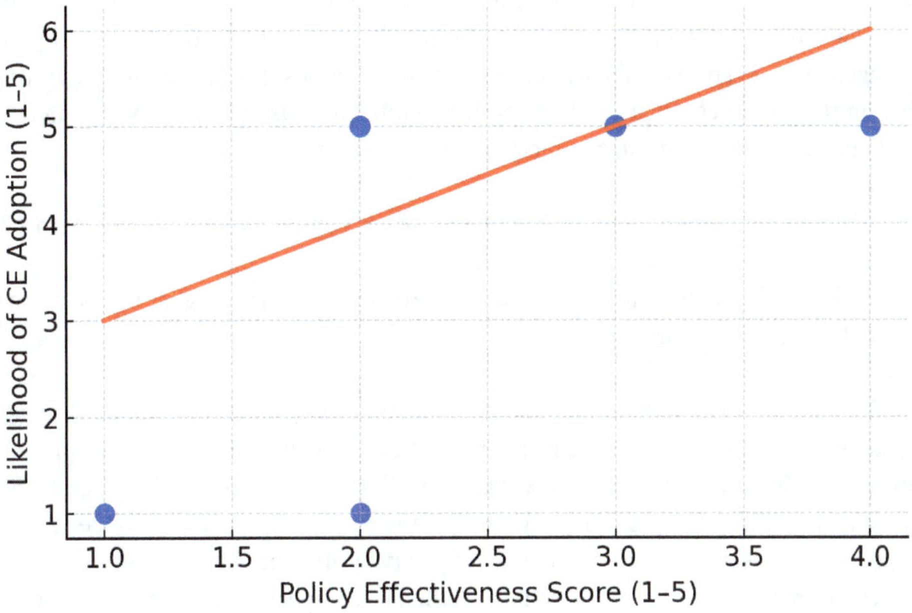


## 3. Market Dynamics and Consumer Readiness

Consumer behaviour and market demand emerged as critical determinants for scaling Circular Economy (CE) practices. Survey responses revealed moderate overall willingness, averaging 2.73/5 across all respondents, with notable variation by stakeholder role, as shown in Table 2.

*Table 2. Average consumer willingness to adopt circular economy products from the observations of stakeholders*

| Stakeholder Role | Respondents (n) | Mean Willingness (1–5) |
|---|---|---|
| NGO Representative | 3 | 3.33 |
| Policymaker / Government Official | 1 | 3.00 |
| Standard Setter on Sustainability Disclosures | 1 | 3.00 |
| Business Leader | 8 | 2.63 |
| Social Entrepreneur | 8 | 2.63 |
| Manager (Corporate Organization) | 1 | 2.00 |

Interestingly, NGO representatives and Policy Actors reported slightly higher willingness scores (≥3.0), often citing increased awareness within specific community segments. In contrast, business leaders and social entrepreneurs, those most directly engaged in market transactions, rated consumer willingness lower (~2.63), suggesting that on-the-ground purchasing behaviour remains cautious despite growing interest.

Qualitative feedback highlighted two recurring barriers:

1. Limited awareness, especially outside urban centres, where exposure to CE concepts is minimal.
2. Affordability constraints-sustainable products often carry higher upfront costs, deterring low-income consumers.

Respondents also noted a feedback loop where CE projects face funding shortages, businesses have reduced capacity for marketing, consumer education, or product innovation, leading to low visibility and slow adoption. This dynamic suggests that consumer readiness is not solely a demand-side issue, but is heavily influenced by supply-side investment, outreach, and supportive policy environments.

To address these market-readiness challenges, targeted awareness campaigns, price-reduction strategies (e.g., subsidies, cooperative buying), and integrating CE principles into mainstream retail channels should help break the cycle of low demand and low investment.

## 4. Social Business as a Strategic Enabler

Findings from the expert survey consistently highlight an enduring lack of finance, policy backing, market access, and inclusion gaps that the Social Business model is often well-positioned to address. Although Social Business was not directly surveyed, its structures centered on social impact and sustainability align with many catalysts needed to overcome CE implementation challenges. A defining feature of Social Business is its mission-driven orientation, prioritizing environmental and societal benefits over profit maximization far beyond typical commercial cycles. Equally important is how Social Businesses are embedded within local ecosystems, mobilizing low-income and informal workers from waste pickers to rural farmers into circular systems. Examples include Grameen Shakti in Bangladesh, which has installed over one million solar home systems using local technicians, and Waste Concern, which processes over 700 tons of organic waste daily into compost while formalizing livelihoods in the informal sector (Barua, 2011; Enayetullah & Hashmi, 2006). In healthcare, the Micro Healthcare Entrepreneurship (MHE) model employs local health workers and digital tools to deliver affordable, preventive services, reinvesting revenue into community-based improvements (Hossain et al., 2024).

One survey respondent from Tanzania emphasized this resilience: “Because we are a social enterprise, we can survive on narrow margins and grow slowly without losing sight of our environmental mission”. The hybrid revenue strategies common to Social Business, combining sales, services, training, and advocacy, enable simultaneous funding of CE activities and local capacity building. Social Business adoption in the circular economy is not commonly practiced. In fact, many people are not aware of the social business concept properly, but many successful examples of Social Business have been established to solve social issues. By comparing with the characteristics of Social Business and CE implementation challenges, we can have an idea about its impact as mentioned in Table 3.

*Table 3. Circular Economy challenges and possible contribution by Social Business*

| Typical Barrier in CE Implementation | Social Business Feature that Addresses It |
|---|---|
| High startup costs, Lack of affordable finance | Uses long-term funding and reinvests profits locally |
| Weak or unclear policies | Advocates for change and pilot community models |
| Limited market access for small producers | Builds local networks and niche market links |
| Low public awareness of CE | Includes education in business activities |
| Exclusion of marginalized groups | Involves informal workers and rural communities |

However, the evidence also suggests limitations that policymakers and practitioners should anticipate:

- **Scalability Constraints:** Localized, labour-intensive models may face challenges expanding beyond niche markets without compromising quality or mission.
- **Funding Dependencies:** Many social businesses rely on grants and subsidies for initial capital. Entrepreneurs should focus on sustainable income from the beginning or start with their own capital to run until they break even. You should not entirely rely on a third party for initial capital.
- **Mission Drift Risks:** As they diversify revenue streams or seek scale, some enterprises risk diluting their social and environmental priorities.
- **Policy and Market Barriers:** Without enabling regulation or access to mainstream markets, Social Businesses may remain marginal players in large CE ecosystems.

Recognizing both strengths and vulnerabilities positions Social Business not merely as participants but as strategic, though not sufficient on its own, architects of inclusive and sustainable CE ecosystems, particularly in low-resource contexts.

## 5. Correlation Analysis of Implementation Factors

To explore interdependencies between key CE implementation barriers, a correlation matrix was computed using four primary survey variables:

Q5_Funding: "On a scale of 1 to 5, how would you rate the difficulty of obtaining funding for CE projects in your country? (1=Very Easy 5=Extremely Difficult)"

Q6_Transition: "How challenging is it for businesses in your country to transition from a linear economy (take-make-dispose) to a CE? (1=Not Challenging, 5=Extremely Challenging)"

Q10_Policy: "On a scale of 1 to 5, how would you rate the effectiveness of your country's policies in promoting CE practices? (1=Not Effective, 5=Highly Effective)"

Q13_Consumers: "In your experience, how willing are consumers to adopt sustainable and circular products in your country? (1=Not Willing, 5=Willing)"

In Figure 3, the analysis revealed a moderate negative correlation between Q6_Transition and Q13_Consumers ($r \approx -0.46$). This indicates that when consumers are more willing to adopt circular products, businesses perceive fewer challenges in transitioning from linear models. It suggests that public awareness and consumer engagement can play a critical role in easing the CE adoption journey for businesses. Other pairwise relationships, such as those between Q5_Funding, Q10_Policy, and Q13_Consumers, were weak or negligible, indicating no strong linear association in this dataset. These findings imply that while financial and policy-related barriers remain important, consumer readiness may be a more immediate level for reducing transition difficulty. Overall, the results show the value of public engagement strategies in supporting CE implementation, especially in developing country contexts where both policy effectiveness and funding access may be limited or unevenly distributed.

*Figure 3. Correlation matrix of key circular economy implementation factors*

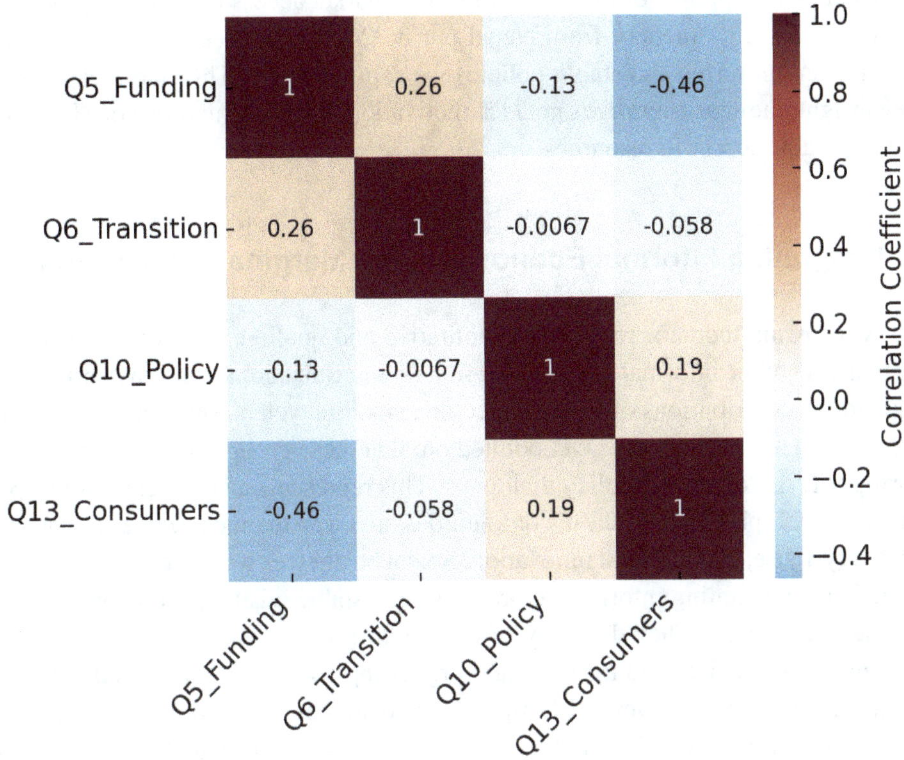


## 6. Implementation Potential and Support Conditions

To assess the practical viability of CE expansion, the survey asked how likely businesses would be to implement circular models if provided with improved financial and policy support. The response was overwhelmingly positive: 86% of participants indicated "Very Likely", highlighting strong latent capacity and motivation within the ecosystem. Only three respondents selected "Not Likely", citing persistent structural issues such as weak governance, fragmented markets, or cultural resistance to sustainable alternatives. These results indicate that CE readiness is not limited by ideology or awareness but instead constrained by a lack of enabling infrastructure and support mechanisms. To delve deeper into what kind of support is most needed, open-ended responses were thematically analyzed. The most frequently cited support needs included clear and stable policy frameworks, dedicated financial tools such as low-interest loans and green grants, and capacity-building programs for both technical and managerial skills. Other commonly mentioned needs included access to sustainable suppliers or tools and improved market access for circular products. As one social entrepreneur from Nepal put it, "We do not need charity, we need finance on fair terms and stable policies we can build on". This quote captures a key insight: the ecosystem does not lack motivation; it lacks structure, coordination, and equitable access to resources.

## 7. Integrating Informal Economies and Marginalized Groups

A recurring theme across both quantitative and qualitative responses was the critical role of the informal sector in CE transitions. Stakeholders consistently highlighted the contributions of waste collectors, small recyclers, rural producers, and women-led microenterprises, yet pointed out that these groups are often left out of formal CE strategies and policy dialogues. This represents a significant blind spot in current CE planning, as these communities are often the most active in material recovery, reuse, and low-cost innovation. Social businesses were seen as a powerful vehicle for integrating informal actors into more stable, visible, and economically viable CE systems. They do so by offering fair employment, providing training, linking small producers to markets, and organising cooperative models that legitimate informal work. However, Multiple respondents cautioned that current policies do not explicitly address informality, and without targeted legal and financial tools, these actors may remain excluded from the benefits of CE scaling. Thus, to build inclusive circular systems, national and regional CE frameworks must recognize and institutionalize the role of informality, not as a barrier, but as a foundation.

## 8. Final Insights and Cross-Country Comparison

Combining the findings across all three countries, several patterns emerge that help contextualize both the opportunities and the challenges of CE adoption in the Global South.

- **Indonesia** demonstrates leadership in CE-related innovation, especially in urban recycling systems and digital waste platforms. However, significant urban-rural disparities remain, with rural and informal sectors lacking the infrastructure and incentives available in cities.
- **Nepal** shows strong grassroots CE activity, composting, textile reuse, and cooperative recycling driven by community ingenuity. However, fragment-

ed governance, donor dependency, and limited national coordination hinder scaling. Technical capacity is concentrated in urban centres.

- **Tanzania** holds strong potential in eco-tourism, agro-waste valorization, and informal waste networks, with clear opportunities for a circular bioeconomy. The main obstacles include weak regulatory enforcement, scarce CE-specific funding, and isolated project implementation.

A consistent cross-country theme is the emerging catalytic role of Social Businesses in bridging institutional voids, piloting low-cost CE models, and integrating marginalized communities into circular systems. However, to scale and sustain these efforts, enabling policies, financial access, and multi-sector partnerships are essential.

**Action Roadmap for CE Scaling:** Based on the analysis in this study, the following actionable recommendations (Table 4) are proposed. Each recommendation directly addresses one or more of the persistent gaps observed across contexts—such as policy incoherence, financing bottlenecks, market fragmentation, and awareness deficits. These measures can help policymakers, development partners, and social business entrepreneurs translate their strategic goals into the effective implementation of CE initiatives.

*Table 4. Actionable roadmap for scaling circular economy implementation in developing countries*

| Priority Action | Timeframe | Responsible Actors | Funding Mechanisms | Key Performance Indicators (KPIs) |
|---|---|---|---|---|
| Develop national CE strategies that integrate multiple sectors (Waste, Agriculture, Renewables, Water, Tourism, Manufacturing, etc.) | Mid Term (3-5 Years) | Ministries or Government Specialized Departments | National budget; donor grants | CE strategy adopted; number of sectors covered |
| Establish urban-rural CE linkage programs to share infrastructure, training, and markets. | Mid Term (3-5 Years) | Local governments, social business support organizations, NGOs, and Local business alliances | Public-private partnerships; grants; community support | % of rural CE enterprises with market access; infrastructure gap reduction |
| Create CE finance facilities for MSMEs, Startups | Short Term (1-3 Years) | Government, Donors, Big Corporations | Startup fund; collateral free loan; low/zero-interest loan; grants; CSR funds | Total CE funding disbursed; repayment rates |

continued on following page

*Table 4. Continued*

| Priority Action | Timeframe | Responsible Actors | Funding Mechanisms | Key Performance Indicators (KPIs) |
|---|---|---|---|---|
| Strengthen regulatory frameworks with enforcement capacity | Mid Term (3-5 Years) | National environmental agencies; local authorities | Government enforcement budgets; international technical assistance | Compliance rate with CE-related regulations |
| Build national CE innovation hubs linking universities, entrepreneurs, and communities. | Long Term (5-10 years) | Universities, innovation agencies, and chambers of commerce | Innovation funds; corporate CSR programs | Number of prototypes piloted; patents or commercialized solutions |
| Mainstream CE awareness in public education and media | Short Term (1-3 Years) | Government; educational institutes; community groups; social organizations/ clubs; media agencies | Government budgets, organizational budgets, and sponsorships | % of population with basic CE literacy |

## DISCUSSION

This study examined the challenges and opportunities in implementing CE models in developing countries through the lens of Social Business, with a specific focus on Indonesia, Nepal, and Tanzania. Through expert surveys and analytical frameworks grounded in the Triple Bottom Line and PESTLE models, we address three guiding questions: The operational challenges of CE, the role of Social Business in overcoming these challenges, and the enabling conditions required for scalable and sustainable CE transitions.

Findings from this research confirm and extend existing literature on the structural and operational barriers to CE adoption in the Global South. Respondents consistently identified Lack of financial resources, weak policy enforcement, and limited consumer demand as primary impediments, findings that resonate with studies by Kirchher et al. (2018), which note that CE transitions in developing economies are hindered by poor regulatory coherence and infrastructure. The survey also confirmed that in countries like Nepal and Tanzania, CE readiness is high in motivation but low in operational capacity, echoing the observations of Nasir et al. (2017) regarding the technology-financing gap in low-income regions. Interestingly, our data shows a moderate negative correlation between perceived policy effectiveness and transition difficulty, which supports the broader CE literature stressing the importance of strong institutions and multi-stakeholder governance (Geissdoerfer et al., 2017). This confirms the assertion that policy clarity and consistency are not merely supportive features but foundational enablers of CE transitions. While existing frameworks in Indonesia demonstrate incremental progress, challenges in coordination and localized enforcement persist, illustrating that CE frameworks need both top-down coherence and bottom-up adaptability.

In response to the second research question, the study reveals that Social Business models offer a uniquely appropriate vehicle for implementing CE in developing countries. Experts praised social enterprises for their mission alignment with sustainability goals, ability to engage informal and marginalized groups, and their financial adaptability, often combining multiple revenue streams with community reinvestment. This supports and extends Yunus' (2010) conceptualization of Social Business as a non-dividend, impact-first structure capable of addressing market failures. More recent empirical studies have highlighted the success of social enterprises in implementing circular models, particularly in sectors like waste management, recycling, and bioeconomy (Bocken et al., 2016). Our findings corroborate these patterns. Social entrepreneurs in our survey emphasized that their ability to operate on thin margins and prioritize long-term goals made them more resilient and more attuned to the complexities of CE compared to traditional firms. This adds empirical weight to emerging claims that Social Business can act as an institutional bridge, linking policy frameworks, community needs, and CE objectives in ways that conventional firms may struggle to replicate. Moreover, respondents from Nepal and Tanzania specifically cited the role of social enterprises in organizing informal waste pickers into cooperatives and facilitating circular agri-value chains. Therefore, inclusive CE models that integrate the informal economy as an asset rather than an obstacle. However, as our findings show, policy frameworks have yet to formally recognize social enterprises, leaving them in a precarious position when it comes to accessing finance, license, or CE-specific incentives.

The third research question asked what enabling conditions are required for Social Business to scale CE transitions. The most compelling insight from the survey was that 86% of respondents believed CE adoption would be "Very Likely" if appropriate financial and policy support were provided. This aligns with findings from Ghisellini et al. (2016), who emphasize the importance of financial innovation and policy orchestration in driving CE forward. Thematic analysis of open responses identified five key areas for intervention: Policy clarity, access to finance, capacity building, supplier infrastructure, and market access. These support areas align well with recommendations in the Ellen MacArthur Foundation's (2021) CE policy toolkit and echo empirical insights from Lazarevic and Valve (2017), who advocate for multi-layered governance and sectoral customization in CE programs. Another notable finding relates to the third mild negative relationship between funding difficulty and customer willingness. While not statistically wrong, this trend highlights a feedback loop often overlooked in CE literature: insufficient funding limits outreach and innovation, which then weakens consumer demand and market visibility. In this sense, consumer engagement is not purely a behavioral issue but deeply linked to financial and institutional access, especially for early-stage or social enterprises that lack economies of scale.

This study contributes new empirical knowledge in several ways. First, it offers one of the few comparative assessments of CE implementation across three developing economies, using standardized metrics and grounded survey data. Second, it extends the literature on Social Business and CE by showing how these models perform in a real-world context, not just theoretically, but particularly under conditions of regulatory fragility and financial scarcity. Third, it underscores the importance of policy design that is not only circular but inclusive, especially of the informal sector and grassroots innovators. For policymakers, the implications are clear: if CE is to be viable in the Global South, strategies must move beyond top-down mandates and engage with the ecosystems where real work happens. This includes formally recognizing social enterprises, offering blended finance instruments, and ensuring policy incentives are accessible to small-scale actors. For donors and development institutions, supporting Social Business ecosystems may offer one of the most cost-effective pathways to scalable and sustainable CE transitions.

**Limitations and Future Works:** This study has several limitations that should be acknowledged. First, the sample size of 22 respondents across three countries is too small to support substantive cross-country statistical comparisons. While the qualitative richness of the expert insights offers valuable depth, the absence of a larger and more diverse respondent pool, including direct operators of social business, informal sector workers, and grassroots community stakeholders, limits the representativeness. While this study focuses on major CE implementation industries, there are many other industries that can be discussed in future studies. Likewise, the study does not explicitly compare urban and rural contexts, which could have highlighted important differences in infrastructure availability, consumer awareness, and market readiness. Integrating country-level indicators, for example, GDP per capita, waste generation rates, or environmental policy enforcement indexes, could further validate the relationships identified in the survey. Future research should therefore expand the sample size and stakeholder diversity, include broader coverage, apply integrated macroeconomic and policy indicators, and incorporate field-based implementation studies and impact evaluations.

## CONCLUSION

This study set out to explore the implementation of CE practices in developing countries through the lens of Social Business models, focusing on three primary research questions: the key challenges in CE implementation, the potential of Social Business to address those challenges, and the enabling conditions for success and replication. Drawing from expert survey data across Indonesia, Nepal, and Tanzania, supported by correlation analysis, thematic analysis, and comparative literature, the findings offer meaningful insights into the realities and opportunities of circular transitions in low-resource settings. The study confirms that operational and structural barriers, including Lack of finance, weak policy enforcement, inadequate infrastructure, and limited consumer engagement, persist across all three countries. These challenges are compounded by fragmented regulatory frameworks and the absence of scalable, inclusive models that can translate CE theory into practice. However, it also reveals that the majority of respondents expressed a strong willingness to implement CE models if supported by enabling financial and policy environments. The strong alignment between the identified implementation gaps and the structural strengths of Social Business models positions these enterprises as strategic enablers of inclusive and sustainable CE ecosystems.

In conclusion, the pathway to sustainable and inclusive CE in developing countries is complex, yet increasingly navigable. By recognizing the unique capabilities of Social Business models and aligning them with supportive ecosystems of finance, policy, and knowledge, policymakers and practitioners can unlock transformative, community-centered solutions. This study offers a foundational step in identifying those leverage points and calls for deeper, more collaborative efforts to scale them equitably and effectively.

## REFERENCES


Agyabeng-Mensah, Y., Afum, E., & Ahenkorah, E. (2022). Exploring financial performance and green logistics management practices: Examining the mediating influences of market, environmental and social performances. *Journal of Cleaner Production, 258*, 120613. DOI: 10.1016/j.jclepro.2020.120613

ASEAN-Japan Centre. (2024, December 24). *From trash for tech to tech for good: The Gringgo journey with Febriadi Pratama*. ASEAN-Japan Centre – ASEAN Promotion Centre on Trade, Investment and Tourism. https://www.asean.or.jp/en/impact-asean/20240607/

Barua, D. C. (2011). Strategy for promotions and development of renewable technologies in Bangladesh: Experience from Grameen Shakti. *Renewable Energy, 22*(1–3), 205–210. DOI: 10.1016/S0960-1481(00)00012-4

Battilana, J., & Lee, M. (2014). Advancing research on hybrid organizing – Insights from the study of social enterprises. *The Academy of Management Annals, 8*(1), 397–441. DOI: 10.5465/19416520.2014.893615

Bocken, N. M. P., Short, S. W., Rana, P., & Evans, S. (2016). A literature and practice review to develop sustainable business model archetypes. *Journal of Cleaner Production, 65*, 42–56. DOI: 10.1016/j.jclepro.2013.11.039

Cataldo, N. L., Oyinlola, M., Sigdel, S., Nguyen, D., & Gallego-Schmid, A. (2024). Waste management in Nepal: Characterization and challenges to promote a circular economy. *Circular Economy and Sustainability, 4*(1), 439–457. DOI: 10.1007/s43615-023-00283-0

Cattaneo, O., Engman, M., Saez, S., & Stern, R. M. (Eds.). (2010). *International trade in services: New trends and opportunities for developing countries*. World Bank., https://documents.worldbank.org/curated/en/464591468158719636

Charles, G. (2019). Sustainability of social enterprises involved in waste collection and recycling activities: Lessons from Tanzania. *Journal of Social Entrepreneurship, 12*(2), 219–237. DOI: 10.1080/19420676.2019.1686712

Elias, A. A. (2017). Systems thinking and modelling for stakeholder management. *IIM Kozhikode Society & Management Review, 6*(2), 123–131. DOI: 10.1177/2277975216681105

Elkington, J. (1997). *Cannibals with forks: The triple bottom line of 21st century business*. Capstone.

Ellen MacArthur Foundation. (2021). *Universal circular economy policy goals: Enabling the transition to scale.* https://ellenmacarthurfoundation.org

Enayetullah, I., & Hashmi, Q. S. I. (2006, October 30–November 1). *Community based solid waste management through public-private-community partnerships: Experience of Waste Concern in Bangladesh.* 3R Asia Conference, Tokyo, Japan. https://www.env.go.jp/recycle/3r/en/asia/02_03-3/06.pdf

Geissdoerfer, M., Savaget, P., Bocken, N. M. P., & Hultink, E. J. (2017). The circular economy – A new sustainability paradigm? *Journal of Cleaner Production, 143*, 757–768. DOI: 10.1016/j.jclepro.2016.12.048

Ghisellini, P., Cialani, C., & Ulgiati, S. (2016). A review on circular economy: The expected transition to a balanced interplay of environmental and economic systems. *Journal of Cleaner Production, 114*, 11–32. DOI: 10.1016/j.jclepro.2015.09.007

Hossain, F., Islam, R., Osugi, T., Shah, F., Mine, T., Nakashima, N., & Ahmed, A. (2024). Concept of Micro Healthcare Entrepreneurship (MHE) to facilitate Universal Health Coverage (UHC): Prospects and challenges. *Sustainability, 16*(6), 2268. DOI: 10.3390/su16062268

Isibika, A., Simha, P., Vinnerås, B., Zurbrügg, C., Kibazohi, O., & Lalander, C. (2023). Food industry waste – An opportunity for black soldier fly larvae protein production in Tanzania. *The Science of the Total Environment, 858*(Part 3), 159985. DOI: 10.1016/j.scitotenv.2022.159985 PMID: 36368404

Kain, J. H., Zapata, P., de Azevedo, A. M. M., Carenzo, S., & Charles, G. (2022). Characteristics, challenges and innovations of waste picker organizations: A comparative perspective between Latin American and East African countries. *PLoS One, 17*(7), e0265889. DOI: 10.1371/journal.pone.0265889 PMID: 35905045

Karim, R., Waaje, A., & Roshid, Md. M. (2025). Digital Circular Economy for Attaining Sustainable Development Goals. In Durmuş Çağrı Yıldırım, Seda Yıldırım, & Vinay Kandpal (Eds.), *Effects of Digitalization and Circular Economy on Sustainable Policy and Climate Change Prevention* (pp. 211–240). IGI Global. https://doi.org/DOI: 10.4018/979-8-3693-9909-5.ch008

Kinyondo, A., & Huggins, C. (2021). State-led efforts to reduce environmental impacts of artisanal and small-scale mining in Tanzania: Implications for fulfilment of the sustainable development goals. *Environmental Science & Policy, 120*, 157–164. DOI: 10.1016/j.envsci.2021.02.017

Kirama, A., & Mayo, A. W. (2016). Challenges and prospects of private sector participation in solid waste management in Dar es Salaam City, Tanzania. *Habitat International*, *53*, 195–205. DOI: 10.1016/j.habitatint.2015.11.014

Kirchherr, J., Reike, D., & Hekkert, M. (2018). Conceptualizing the circular economy: An analysis of 114 definitions. *Resources, Conservation and Recycling*, *127*, 221–232. DOI: 10.1016/j.resconrec.2017.09.005

Korhonen, J., Honkasalo, A., & Seppälä, J. (2018). Circular economy: The concept and its limitations. *Ecological Economics*, *143*, 37–46. DOI: 10.1016/j.ecolecon.2017.06.041

Lazarevic, D., & Valve, H. (2017). Narrating expectations for the circular economy: Towards a common and contested European transition. *Energy Research & Social Science*, *31*, 60–69. DOI: 10.1016/j.erss.2017.05.006

Meidiana, C., & Gamse, T. (2010). Development of waste management practices in Indonesia. *European Journal of Scientific Research*, *40*, 199–210.

Melles, G., Gautam, S., & Shrestha, R. (2025). Circular economy for Nepal's sustainable development ambitions. *Challenges*, *16*(1), 8. DOI: 10.3390/challe16010008

Ministry of National Development Planning/Bappenas. (2024). *Circular economy national roadmap & action plan: Indonesia 2025–2045*. https://www.un-pageindonesia.org/assets/uploads/4c832-en-roadmap-action-plan-ce-18-09-2024-.pdf

Msuya, N., & Temu, A. (2025). From waste to wealth: Sisal by-product opportunities in Tanzania. *Tanzania Journal of Science*, *51*(1), 182–195. DOI: 10.4314/tjs.v51i1.13

Murray, A., Skene, K., & Haynes, K. (2017). The circular economy: An interdisciplinary exploration of the concept and application in a global context. *Journal of Business Ethics*, *140*(3), 369–380. DOI: 10.1007/s10551-015-2693-2

Nasir, M. H. A., Genovese, A., Acquaye, A. A., Koh, S. C. L., & Yamoah, F. (2017). Comparing linear and circular supply chains: A case study from the construction industry. *International Journal of Production Economics*, *183*(Part B), 443–457. https://doi.org/DOI: 10.1016/j.ijpe.2016.06.008

Nipe Fagio. (2021). *Nipe Fagio national report 2021*. https://brandaudit.breakfreefromplastic.org/wp-content/uploads/2022/11/LDI_National_Report_2021_v2C.pdf

Roshid, Md. M., Waaje, A., Karim, R., Irtisham, M. R., & Kaswan, M. S. (2025a). Industry 5.0 and Organizational Sustainability: Integrating Cutting-Edge Technologies and Human Ingenuity. In Mahender Singh Kaswan, Rajeev Rathi, & Jose Arturo Garza-Reyes (Eds.), Industry 5.0: Concepts and Strategies for Digital Transformation (pp. 50–61). CRC Press. https://doi.org/DOI: 10.1201/9781003535188-7

Roshid, Md. M., Waaje, A., Karim, R., Irtisham, M. R., & Kaswan, M. S. (2025b). Industry 5.0 and Organizational Sustainability: Integrating Cutting-Edge Technologies and Human Ingenuity. In Mahender Singh Kaswan, Rajeev Rathi, & Jose Arturo Garza-Reyes (Eds.), *Industry 5.0: Concepts and Strategies for Digital Transformation* (pp. 50–61). CRC Press. https://doi.org/DOI: 10.1201/9781003535188-7

Saidani, M., Yannou, B., Leroy, Y., & Cluzel, F. (2017). How to assess product performance in the circular economy? Proposed requirements for the design of a circularity measurement framework. *Recycling*, *2*(1), 6. DOI: 10.3390/recycling2010006

Schroeder, P., Anggraeni, K., & Weber, U. (2019). The relevance of circular economy practices to the Sustainable Development Goals. *Journal of Industrial Ecology*, *23*(1), 77–95. DOI: 10.1111/jiec.12732

Seven Principles of Social Business. (2009, June 22). Muhammad Yunus. Retrieved May 28, 2025, from https://www.muhammadyunus.org/post/363/seven-principles-of-social-business

Shrestha, R., Shakya, R., & Shrestha, A. (2025). Circular economy and its impact on economic, environmental and societal aspects: A survey among fashion industry businesses in Lalitpur, Nepal. *Apex Journal of Business and Management, 4*(1), 27–40. https://nepjol.info/index.php/ajbm/article/view/76937

Singh, J., & Ordoñez, I. (2016). Resource recovery from post-consumer waste: Important lessons for the upcoming circular economy. *Journal of Cleaner Production*, *134*, 342–353. DOI: 10.1016/j.jclepro.2015.12.020

Sodhi, M. S., & Tang, C. S. (2011). Social enterprises as supply-chain enablers for the poor. *Socio-Economic Planning Sciences*, *45*(4), 146–153. DOI: 10.1016/j.seps.2011.04.001

Subedi, M., Pandey, S., & Khanal, A. (2023). Integrated solid waste management for the circular economy: Challenges and opportunities for Nepal. *Journal of Multidisciplinary Research Advancements*, *1*(1), 21–26. DOI: 10.3126/jomra.v1i1.55100

Sultana, N., Rahman, K. S., Karim, R., & Ahmad, S. (2025). Entrepreneurship: A rising way leading to economic development in Bangladesh. *Procedia Computer Science*, *259*, 1947–1954. DOI: 10.1016/j.procs.2025.04.150

Sutomo, B., Mukhlis, M., & Ahadiat, A. (2023). A circular economy, waste management, and sustainable development: A case study of a transmigration rural area on the Indonesian island of Sumatra. *Quality - Access to Success, 24*(192). Advance online publication. DOI: 10.47750/QAS/24.192.04

UNEP. (2018). *Single-use plastics: A roadmap for sustainability.* United Nations Environment Programme. https://www.unep.org/ietc/ja/node/53

Waaje, A., Karim, R., & Roshid, Md. M., & Meem, T. N. (2024). Minimizing Waste in the Hospitality Industry: The Opportunities and Challenges of Lean Tourism. In R. Sucheran (Ed.), *Sustainable Waste Management in the Tourism and Hospitality Sectors* (pp. 125–166). IGI Global Scientific Publishing. https://doi.org/DOI: 10.4018/979-8-3693-6110-8.ch006

Wilson, D. C., Velis, C., & Cheeseman, C. (2006). Role of informal sector recycling in waste management in developing countries. *Habitat International, 30*(4), 797–808. DOI: 10.1016/j.habitatint.2005.09.005

Yüksel, İ. (2012). Developing a multi-criteria decision-making model for PESTEL analysis. *International Journal of Business and Management, 7*(24), 52–66. DOI: 10.5539/ijbm.v7n24p52

Yunus, M. (2010). *Building social business: The new kind of capitalism that serves humanity's most pressing needs*. PublicAffairs.

Yunus, M., Moingeon, B., & Lehmann-Ortega, L. (2015). Building social business models: Lessons from the Grameen experience. *Long Range Planning, 43*(2–3), 308–325. DOI: 10.1016/j.lrp.2009.12.005

# Chapter 7
# Stitching Sustainability in Fashion:
## Strengthening Circular Economy Drivers in the Textile Industry

**Bablu Kumar Dhar**
https://orcid.org/0000-0001-8768-8634
*Mahidol University, Thailand*

**Rejaul Karim**
https://orcid.org/0000-0001-6175-8825
*Varendra University, Bangladesh*

**Md. Soleman Mollik**
https://orcid.org/0009-0003-5703-0606
*Bangladesh University of Professionals, Bangladesh*

## ABSTRACT

*This chapter discusses the pressing need for the textile and fashion industry to embrace Circular Economy (CE) principles with reference to their environmental impacts. The textile industry — a significant economic driver on its own — is also one of the biggest polluters out there (accounting for almost 10% of global greenhouse gas emissions and heavily contributing to widespread water pollution through waste effluents). In light of the above, this chapter showcases that moving from a linear "take-make-dispose" model to a circular economy is crucial to tackling waste generation, improving resource efficiency, and mitigating environmental degradation. The textile sector serves as a case of Circular Economy with key drivers including sustainable materials innovation, product life extension, reverse logistics, and business model innovation in this chapter. The chapter also discusses*

DOI: 10.4018/979-8-3373-2454-8.ch007

*the obstacles to CE adoption, such as restrictions in terms of technology, economic barriers, and issues with markets and consumer behavior. It also underlines the importance of a collaborative ecosystem in combination with enabling policy to work across the value chain and engage the consumer for successful circular adoption. This chapter, illustrated with real-world case studies and strategies of overcoming challenges, presents how businesses can integrate CE to minimize waste, maximize resource utilization, and ensure long-term sustainability. Key to furthering apparel's circular revolution are future trends such as regenerative fashion, AI for production optimization, and localized manufacturing. The chapter highlights that embedding CE principles will enhance the triple bottom line by improving social well-being, environmental sustainability, and financial performance in the textile and fashion sectors.*

## 1. INTRODUCTION

Industrialization is widely considered to have begun in the late 18th century in Britain. It began with the textile and clothing (TC) industry (Farhana et al., 2022), playing a central role in driving mechanization, factory production, and economic transformation. This historical foundation highlights the textile sector's long-standing influence on industrial development and provides context for its contemporary role in sustainability transitions. It has significantly benefited many emerging economies by boosting export earnings, creating employment opportunities, and enhancing living standards. In 2019, global trade in the TC sector reached a value of approximately USD 807 billion, providing jobs to over 70 million people worldwide across its extensive supply chains (WTO, 2019). Countries such as Bangladesh, India, and Vietnam are particularly reliant on this industry (Saha et al., 2022). For instance, garment exports account for 18% of Vietnam's total export revenue (Vietnam Briefing, 2020), and an impressive 80% for Bangladesh (BGMEA, 2020). Beyond its economic impact, the industry has made notable social contributions, particularly by advancing women's empowerment (Naqvi et al., 2024).

The term "textile" refers to underlying materials made from cotton, jute, silk, wool, nylon, acrylic, polyethylene, polyester, and similar fibers (Schumacher & Forster, 2022). To transform these materials into garments, clothing, and various other products (Table 1) (Kumar et al., 2020), a range of processes are applied, including weaving and non-woven techniques such as stitching. These textile materials can be categorized as plant- or animal-based, cellulosic, organic, synthetic or semi-synthetic, and inorganic types (Fig. 1) (Piribauer and Bartl, 2019).

*Table 1. Textile market globally*

| Category | Description |
|---|---|
| Product Type | Fibers, yarn, fabrics, thread, home furnishings and floor coverings, textile and fabric finishing, and fabric coating mills. |
| Process | Woven, non-woven. |
| Textile Application Outlook | Broad-woven fabrics, narrow fabric mills and Schiffli machine embroidery, non-woven fabric, knitted fabrics, home furnishings, floor coverings, textile and fabric finishing mills, fabric coating mills. |
| Raw Material | Natural Fibers (e.g., Cotton, Wool, Silk, Linen, Jute, Hemp), Synthetic Fibers (e.g., Polyester, Nylon, Rayon, Acrylic, Polypropylene, Spandex), Blended Textiles, and Recycled Fibers. |
| Application/End-Use Industry | Fashion & Apparel (e.g., clothing, accessories), Household & Home Textiles (e.g., bedding, floor coverings, upholstery), Technical/Industrial Textiles (e.g., automotive, medical, construction, protective, agriculture). |

Source: Authors' creation

Over recent decades, the demand for cotton, wool, and cellulosic products has stayed almost the same. Meanwhile, the demand for synthetic materials, particularly polyester, has increased significantly due to their superior performance and lower cost (Kabir et al., 2019; Maity et al., 2025). Currently, synthetic fibers make up 60% of clothing and 70% of household textiles, and this share is likely to keep rising. Cotton accounts for 24%, and wool makes up 14% (Akter et al., 2022). Today, 60% of the world's fiber production goes to the fashion industry (Schumacher & Forster, 2022). The rest is used for domestic, industrial, geo-, agro-, and hygienic textiles (Niinimäki et al., 2020).

In recent years, the textile industry has experienced significant growth, making it a vital part of the global economy. According to Market Research Future (2024), the worldwide textile industry increased from approximately USD 573.2 billion in 2022 to around USD 611 billion in 2023, reflecting a compound annual growth rate (CAGR) of about 6.6%. The market is projected to keep expanding and may reach roughly USD 755 billion by 2027, with a CAGR of 5.5%. In 2022, the Asia-Pacific region led the global textile market, followed by Western Europe as the second-largest market (Statista, 2024). Furthermore, global textile production is expected to surpass 99 million tons annually (Ütebay et al., 2020). These figures highlight the industry's significance, not only economically but also in terms of its environmental footprint.

Different fiber types used in textile production can be broadly grouped into natural fibers (such as cotton, wool, silk, jute) and synthetic or semi-synthetic fibers (like polyester, nylon, acrylic, and viscose).

Despite its economic and social benefits, the textile and clothing (TC) industry's environmental impact is deeply concerning. It is considered the world's second most polluting industry after oil (UN, 2019), accounting for around 10% of global greenhouse gas (GHG) emissions. Effluents from dyeing processes have severely contaminated rivers near TC factories, leading to the destruction of aquatic ecosystems (Masum et al., 2025). Additionally, this sector creates global industrial water pollution by 20% (approximately) and discharges untreated dyes into water bodies by almost 200,000 tons (Sustain Your Style, 2020; Uddin, 2021). Consequently, groundwater and air are severely polluted, posing serious health risks to humans and animals (Al-Tohamy et al., 2022). The recent rapid expansion of the fashion and textile industry has brought about profound environmental costs. This growth results in mass production and consumption, fueling significant pollution of air, water, and soil (Niinimäki et al., 2020). Moreover, emerging as energy-intensive processes, the sector emits huge amounts of greenhouse gases (Schumacher & Forster, 2022). Furthermore, textile manufacturing practices, including the extraction of raw materials, dyeing, and finishing, have led to significant resource depletion, threatening biodiversity and disrupting natural systems (Ütebay et al., 2020).

The textile industry is witnessing significant and diverse environmental impacts. The environmental impacts contain substantial water consumption, particularly for cotton cultivation and wet processing, and chemical pollution from dyes and finishing agents (Kabir et al., 2019). Besides, producing and washing synthetic fibers begets microplastics, hampering food chains and ecosystems (Maity et al., 2025). Indeed, the textile's linear "take-make-dispose" model consumes enormous resources and generates huge waste, signifying the application of urgent circular economy (CE) and sustainable options.

*Figure 1. Environmental impact of the textile industry*

*Source: Authors' creation*

*Figure 2. Infographic illustrating the environmental effects of the textile and apparel industry.*

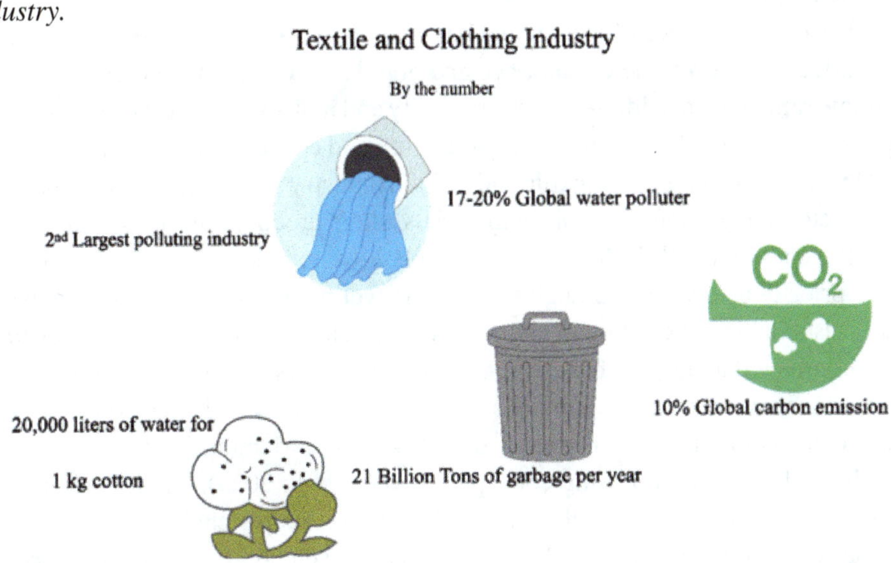


*Source: Textile Today Bangladesh*

Standing at a critical juncture, the textile industry needs to transition from a linear "take-make-dispose" model to a CE model that emphasizes waste reduction, resource efficiency, and environmental sustainability (Aus et al., 2021; Chen et al., 2021). This transition drives a shift regarding how apparels are designed, produced, consumed, disposed of, and managed throughout its entire life cycle (D'Itria & Aus, 2023). The textile industry's profound environmental impacts demand an innovative solution, which lies in CE principles (Musová et al., 2021). Besides, CE implementation in the textile industry promotes sustainable economic growth, reduces hazardous environmental impacts, and brings innovation in businesses (Jia et al., 2020). Most of the textile industries still use a linear model, resulting in substantial losses of materials, economic values, and environmental ecosystems (Schumacher & Forster, 2022). CE can address maximum requirements to ensure businesses and the environment are sustainable for a longer time (Lawrence et al., 2023).

Sustainability has become a major concern for the textile industry due to its significant environmental impact, resource intensity, and social implications. It encompasses the responsible management of natural resources, reduction of waste and emissions, and promotion of social well-being throughout the production and consumption lifecycle. CE principles offer a framework for this sustainability transition by emphasizing resource efficiency, product life extension, recycling, and innovative business models. Integrating these practices enables the textile sector to move from linear, resource-intensive operations toward a more regenerative and resilient system.

The exact evidence of the circular economy (CE) concept is missing (Ahmed et al., 2022). However, the concept became popular in China in the early 1990s as a policy framework to address resource scarcity and industrial waste, before spreading globally as a model for sustainable production and consumption (Brunner & Morf, 2025). The relevance to the textile and fashion industry soon became apparent, given the sector's dependence on virgin materials, short product lifecycles, and growing waste generation. With the expanding global awareness of CE, textiles emerged as a priority area where strategies such as recycling, product lifecycle extensions, and sustainable material innovation could translate CE principles into measurable environmental and social benefits. Additionally, diverse global fields are adopting this concept widely in their domains to achieve efficiency and effectiveness in their entire business functions (Karim et al., 2024; Ooi & Memon, 2025).

Indeed, CE is a promising concept (Rhaouti et al., 2025), and it contains sub-concepts, including industrial ecology, closed-loop supply chain (CSC), and so on (Gharibi & Abdollahzadeh, 2025). In particular, CE emerges as an economic system that eliminates waste and continually reuses resources to develop a closed-loop system (Badran & Lazarov, 2025; Karim et al., 2025). Designing CE model prioritizes remanufacturing, recycling, reusing, refurbishment, and sustainable consumption in addition to reducing resource input and waste generation (Gautam et al., 2025; Mehta et al., 2025). Unlike the traditional model, CE focuses on the continuous usage of resources, preservation of their lifetime values, and repurposing of used materials for new product development (Shirvanimoghaddam et al., 2020). The textile industry contributes substantial amounts of plastic microfibers and carbon to the environment every year. Applying CE principles, the apparel industry can reduce or eliminate environmental damage, enhance reutilization of resources, and construct resilient and sustainable business models (Damayanti et al., 2021). The European Commission finds CE practices best suited for the textile industry (Coscieme et al., 2022). Thus, CE can stitch sustainability into the textile industry through waste management, resource recycling, and environmental responsibility.

Unlike the linear economy (which follows the "take-make-dispose" model), the circular economy emphasizes restorative and regenerative approaches, ensuring that materials and products retain their value for as long as possible (Provensi et al., 2025). It aligns with sustainability goals by reducing environmental impact, promoting efficient resource use, and fostering economic resilience (Verma et al., 2025). The concepts of industrial ecology, biomimicry, bioprinting, and industrial symbiosis are all convergent in the concept of CE (Genta et al., 2025). These include extending the life of the product, lengthening its life cycle, optimizing material value, and reducing energy, waste, emissions, and input materials (Grover et al., 2025).

A few scholars have identified the motivational factors and challenges of CE adoption in diverse domains. However, there prevails a significant knowledge gap regarding how to strengthen CE drivers and weaken its challenges to produce sustainability in the textile industry. Molla et al. (2022) urged scholarly work on strategies to reinforce the CE drivers and challenges in myriad sectors. For this reason, this chapter aims to cement this scholarly work. The proposed chapter will potentially cover the following areas:

This chapter aims to explore how circular economy (CE) drivers can transform and strengthen sustainability practices in the textile and fashion industry. The objective is to clarify the vital role these drivers play in shifting the industry from a resource-intensive, linear "take–make–dispose" model toward a regenerative, circular model that minimizes waste, conserves resources, and reduces environmental impact. By identifying and analyzing these key drivers, the chapter seeks to show how businesses, policymakers, and stakeholders can integrate circular principles to foster long-term sustainability in textile production, distribution, and consumption.

Additionally, the chapter aims to present a deeper understanding of the strategic, operational, and policy-related measures so that the textile stakeholders can move closer to the circular economy. Furthermore, offering practical insights in this chapter will help textile industry leaders, researchers, and policymakers design actionable roadmaps for sustainable transformation into a circular economy from the linear take-make-dispose model.

The chapter begins with the title "Stitching Sustainability in Fashion: Strengthening Circular Economy Drivers in the Textile Industry," highlighting how the fashion and textile industry can be sustainable by identifying, strengthening, and applying CE drivers. The following section sheds light on the key CE drivers in the textile industry that help enable CE transformation. Next, the chapter discusses CE challenges in the textile industry. The subsequent sections present ways to strengthen CE drivers, share case studies, strategic recommendations, and related real-world examples. Finally, key findings and implications, future trends and opportunities, and conclusions of CE are discussed. This structure essentially intends to guide the readers from the basic understanding to the applied strategies of the circular economy in the textile industry. It also posits a critical reflection on the future of CE adoption in the textile industry.

The chapter mainly eyes on the textile and fashion industry in developing countries. It examines how CE principles can enhance sustainability, social well-being, and financial performance. Additionally, the chapter shows how the CE drivers can be strengthened. While global trends are briefly discussed for the context, the primary perspective and examples are drawn from emerging economies, where resource constraints and social challenges make circular practices particularly impactful.

## 2. UNDERSTANDING CIRCULAR ECONOMY IN THE CONTEXT OF FASHION

Circular economy appears as an alternative to the linear paradigm, fostering the utilization of resources on a circular basis, maintenance of their values throughout their lifecycle, and the repurposing of required materials for the creation of new products (Shirvanimoghaddam et al., 2020). Reducing environmental impacts and cultivating more robust and sustainable business models in the textile industry requires implementing the circular economy (Karim et al., 2025). The textile industry has already presented the limitations of the linear economic model (take-make-dispose) by creating plastic microfibers and emitting enormous carbon into the environment (Koszewska, 2018). Addressing these problems necessitates the implementation of the circular economy in the textile industry (Coscieme et al., 2022).

### 2.1 Challenges of Implementing CE in Fashion

Circular fashion, which applies circular economy (CE) principles to the textile and apparel sector, faces a number of significant barriers that hinder its progress. While the main goal is to extend garment lifespans through recycling, reuse, and waste reduction, many of the toughest challenges arise around recycling (Biyada & Urbonavičius, 2025; Sultana et al., 2025). For instance, most garments are not designed with recyclability in mind, and the dominance of polyester–cotton blends makes material separation technically difficult. On top of this, there are few systems that allow consumers to deposit used clothing, and large-scale collection and recycling remain largely unmanaged due to the absence of dedicated organizations. These problems are compounded by uncertainty over how nationwide implementation could work, the limited availability of suitable technologies, and the demand for specialized skills across the entire lifecycle of circular fashion operations.

These issues can be understood within broader, interconnected categories—cultural, technological, market-related, and regulatory barriers—that often overlap and influence one another. Neglecting even one of these dimensions' risks undermining the success of circular fashion initiatives. For this reason, businesses are encouraged to carefully review each type of barrier, along with its subcategories, in the context of their own business models. Such an approach helps companies identify the root causes of potential failures when moving from linear to circular systems, allowing them to directly address challenges and build a stronger foundation for a successful transition (Biyada & Urbonavičius, 2025)

### 2.1.1 Cultural barriers

Cultural barriers are among the most significant obstacles to adopting circular fashion in the apparel industry. Key issues include consumers' limited interest and awareness of sustainable products and the entrenched linear mindset within company operations. Many consumers hesitate to pay higher prices for eco-friendly items, as most business models still rely on high-volume sales that push prices lower (Waaje et al., 2025a). Even though younger generations are generally aware of social and environmental concerns, this awareness rarely translates into actual purchasing decisions (Alam et al., 2025). Contributing factors include limited availability and affordability of sustainable options, lack of clear product information, and perceptions that eco-friendly garments are less stylish. Moreover, consumers are influenced by habits tied to disposable fashion, preferring to replace rather than repair clothing.

On the company side, cultural resistance also stems from traditional linear production systems. Many manufacturers stick to established processes and show reluctance toward adopting circular practices. Implementing a truly circular approach requires cooperation across the entire supply chain, but it is often challenging to find partners willing to change their models as well. Manufacturers might not fully know the origins of smaller components like buttons or trims, making it difficult to guarantee that products are completely circular or free from unsustainable practices.

These cultural barriers are closely linked to other challenges, such as technological limitations, market dynamics, and policy gaps, as shown in Figure 2.10 (Biyada & Urbonavičius, 2025). Together, they can significantly slow the transition to circular fashion. Businesses are therefore encouraged to analyze these interconnected barriers in detail to identify and address the underlying causes, ultimately supporting a shift from linear to more circular business models.

### 2.1.2 Technological barriers

Technological barriers present significant challenges to adopting circular practices in the textile and apparel industry (Roshid et al., 2025). While the ambition to produce fully circular products is growing, practical and systemic limitations such as gaps in technology availability, slow innovation processes, and difficulties in managing data and materials across complex supply chains continue to hinder progress. Ensuring the right technology is available at each production stage is essential, yet it remains uncertain whether these technologies are accessible across all market sectors (Waaje et al., 2025b). Even when firms are motivated to introduce innovative designs and production methods, implementation often proves slow and time-consuming. Access to historical data is also critical for researchers and scientists seeking to develop improved processes, while the quality of recycled materials must be enhanced through targeted upgrades to current manufacturing technologies (Ahmad et al., 2024). In addition, effective systems for collecting used garments—beginning locally at the municipal level and scaling nationally—are needed, alongside standardized technologies for sorting textiles into streams suitable for high-quality raw material production. Finally, challenges such as limited information-sharing systems, inadequate training and expertise, and gaps in knowledge transfer across businesses and suppliers, compounded by data fragmentation, further restrict cooperation and innovation throughout the supply chain.

### 2.1.3 Market barriers

Although interest in circular fashion is increasing, these products remain costly and primarily attract a niche group of consumers who prioritize environmental and social responsibility (Firdaus et al., 2025). The market for circular fashion is still relatively small and complex to distinguish from the broader apparel industry, consisting of only a limited number of consumers, businesses, and organizations. Customers in this segment often have specific preferences and expectations, and while they may be willing to pay extra, this adds complexity for producers and retailers. Additionally, higher prices for virgin materials can raise overall production costs, making circular products more expensive than conventional alternatives. Consumers' purchasing decisions are also shaped by their financial capacity, meaning those with greater economic means are better positioned to afford such premium items. On top of this, fulfilling the specialized demands of circular fashion requires capable suppliers and tailored operations, but in many cases, few—if any—suppliers are able to consistently meet these requirements.

### 2.1.4 Barriers related to government regulations

Most government policies and regulations are still built around the traditional linear economy, which emphasizes disposal over reuse, and this often limits support for circular practices while slowing the transition to a circular economy. As Firdaus et al. (2025) note, legislation creates several barriers, including vague or unclear regulatory messages, complex and frequently changing rules that overlap, limited enforcement combined with weak regulatory pressure, and restrictions that reduce opportunities for innovation.

In many places, there are no clear policies for collecting waste separately, leading to mixed waste streams that lower the quality and raise recycling costs. Due to strict quality standards, regulations sometimes also prohibit the use of recycled materials. In existing waste laws, success is primarily measured by quantity, such as weight-based recycling targets, rather than on the quality and reuse potential of materials.

Circular solutions such as prevention, reuse, and high-quality recycling may be supported by combining better regulatory frameworks with economic incentives.

### 2.1.5 Economic/Financial Barriers

Economic and financial barriers also hinder circularity in fashion, as high upfront costs for recycling technologies and durable designs, along with a lack of economic incentives for both businesses and consumers, make adoption difficult (Roshid et al., 2025). Implementing circular systems requires significant initial investment, and without sufficient financial support, many businesses are hesitant to transition. The reverse logistics process, which includes collecting, sorting, and transporting used garments, further adds to the cost burden. At the same time, consumer price sensitivity poses another challenge, since circular products are often more expensive and struggle to compete with cheaper, conventionally produced alternatives.

### 2.1.6 Infrastructure Barriers

Infrastructure barriers also limit the shift toward circular fashion, particularly due to the absence of large-scale collection systems for post-consumer textile waste and the lack of advanced sorting facilities needed to separate textiles by fiber type, color, and quality. Without efficient collection and sorting infrastructure, recycling efforts are significantly constrained. In addition, the limited number of mechanical and chemical recycling plants makes it difficult for the industry to process large volumes of textile waste effectively, further restricting progress toward circularity.

### 2.1.7 Consumer Behavior Barriers

Consumer behavior barriers add another layer of difficulty, as many people lack awareness or education about the environmental impact of fashion and how to participate in circular initiatives. Distrust in sustainability claims and recycling programs, along with a strong preference for novelty and trends, discourages reuse and durability. In addition, circular activities are often seen as inconvenient since they require extra effort, such as repairing items or dropping off clothes for recycling. These barriers take several forms, including limited knowledge of how fashion choices affect the environment, doubts about the credibility of sustainability initiatives, the ingrained desire to keep up with ever-changing trends, and the perception that circular practices reduce convenience.

### 2.1.8 Supply Chain Complexity and Transparency Barriers

The highly globalized and fragmented nature of the textile supply chain creates major challenges for material tracing and the implementation of circular processes. With numerous stakeholders spread across many countries, maintaining transparency and applying circular practices becomes difficult. This complexity is further compounded by limited collaboration and information sharing between designers, manufacturers, retailers, recyclers, and consumers, which slows systemic change and hinders progress toward a more sustainable and circular fashion system.

### 2.1.9 Technical & Material Composition Barriers

Technical and material composition barriers pose significant challenges to circularity in the textile industry, largely due to the widespread use of mixed fiber blends such as cotton-polyester, which are difficult and costly to recycle mechanically or chemically. Added complications come from dyes, coatings, and chemical finishes that can contaminate recycling streams or hinder reprocessing. Mechanical recycling also often leads to fiber quality degradation, restricting the range of applications for recycled materials. While chemical recycling offers promise, it remains in the early stages of scaling. Overcoming these barriers is essential for the successful integration of circular economy principles into the textile sector.

## 3. KEY CIRCULAR ECONOMY DRIVERS IN THE TEXTILE INDUSTRY

### 3.1. Sustainable Materials Innovation

Sustainable material innovation injects circularity into the textile industry. It requires replacing conventional resource-intensive fabrics with bio-based fibers such as hemp, organic cotton, or bamboo (Pavan et al., 2025). Besides, dependency on virgin resources can be reduced by integrating recycled textiles into production lines. Further, adopting low-impact dyes helps minimize water pollution and chemical waste, assuring that garments are aesthetically appealing as well as environmentally responsible (Karim et al., 2024; Sahu & Rath, 2025). A truly circular and resilient fashion system can be built on the foundation of this shift in material sourcing.

### 3.2. Extended Product Life Strategies

Circular fashion or garments prosper with a longer circular design. By establishing repair services, reuse initiatives, and upcycling programs, consumers can make their clothes last longer while needing less new production. Applying principles such as design for durability to factors like garment construction (e.g., sturdy seams), style that does not date, or modular construction will lend lasting functionality and appeal over multiple life cycles of that item. Stitching sustainability through integration of the textile industry can integrate sustainability from the origin of the product to its lifecycle.

### 3.3. Reverse Logistics and Take-back Systems

Retrieving clothes or garments at the end of their life is a cornerstone for any robust textile circular economy. Recovery systems are a potential solution to this challenge and can be used in reverse logistics, take-back schemes, which will collect post-consumer waste so that companies can take it back, sort it out, and recycle it for new products. Incorporating a closed loop with a textile-to-textile recycling promise, it will simultaneously help to cut down on landfill waste and resource extraction, while also establishing processable protocols for other mills interested in following the same footsteps. Fashion brands that run such systems not just reveal their eco-consciousness but also reaffirm the narrative of regeneration and stitching sustainability in a fashion cycle.

### 3.4. Business Model Innovation

Moving away from the traditional "take-make-waste" or "buy-and-dispose" linear economy, the textile industry is exploring new business models that extend product use and seek to make better use of the planet's resources. In addition, platforms based on renting, leasing, resale, and services are enabling consumers to embrace fashion without being owners. They allow value to be obtained repeatedly from the same product, reduce overproduction, and motivate consumption.

### 3.5. Digitalization and Technology Adoption

Emerging technologies are revolutionizing the way businesses can make fashion more circular. Blockchain guarantees transparency within the supply chain (Karim et al., 2025; Mollik et al., 2024). This is a big step in ensuring that consumers will be able to verify that the sources maintain sustainability and ethical production practices. AI can predict demand, reducing waste linked to overproduction; 3D printing lets companies make items when needed, using fewer materials. It is in response to these challenges that advanced technologies, offering digital-based and transparent solutions, enhance operational efficiency while improving trust between brands and consumers, so it should ideally be a crucial thread woven into the tapestry of a sustainable textile future.

### 3.6. Consumer Engagement and Awareness

Consumers must come into play if the concept of circular fashion has to blossom fully. There is a direct connection between the wider trend of ethical consumerism, which relates to greater knowledge of environmental and social footprints made possible by eco-labels, as well as brand stories and certification schemes (Karim et al., 2025; Hossain & Mollik, 2022; Roshid et al., 2025a). Brands can inspire consumers to buy, care for, and discard clothing more sustainably by conveying the life cycle and how good it is. This shift in societal habits, from fast fashion to conscious fashion and purchasing practices, sustains sustainability beyond just the outward fibers of clothes into the fabric of society.

Table 2 highlights several drivers that replace the textile industry in the circular economy model from the traditional model. The table shows cases where several companies use circular textile design to become cost-effective, along with environmentally and sustainability-focused.

*Table 2. Drivers of the circular economy transition in the textiles industry*

| Company/Area of action | Aim(s) | Action | Results |
|---|---|---|---|
| BIONIC®<br><br>Recycling technologies<br><br>(Jiang et al., 2024) | To apply advanced technologies that convert recycled plastics into high-quality raw materials and textiles, setting new benchmarks for design, performance, and functionality.<br><br>To deliver fully traceable, premium textiles and polymers—produced from coastal and marine plastic waste—to both consumer and industrial markets. | BIONIC® Polymer is a hybrid made from recovered marine and coastal plastics, available as Polyethene Terephthalate (PET) and High-Density Polyethene (HDPE). It is designed for injection and blow molding with customizable pellet colors and physical properties. Additionally, three types of fibers—DPX®, HLX®, and FLX®—are produced, each containing between 40% and 100% recycled plastic, featuring different yarn structures and recycled content levels. | Bionic Yarn estimates that over three years, they have collected 7 million plastic bottles from shorelines worldwide to produce yarn, contributing to the protection of marine ecosystems. |
| Tonlé<br><br>Zero-waste design and production model<br><br>(Ofori et al., 2025) | To transform all pre-consumer textile waste into fashionable clothing | Tonlé creates zero-waste fashion collections using surplus fabric from larger manufacturers, who typically discard about 11% due to inefficient cutting. Through innovative pattern-making and reworking leftover fabric into new garments, Tonlé utilizes over 97% of the materials it sources, with the remaining scraps turned into paper instead of being thrown away. Additionally, workers receive wages well above the minimum and work in safe, comfortable conditions. | In 2014, Tonlé successfully kept 10 tons of textiles out of landfills. Its collections are made from 90% pre-consumer textile waste, with the remaining 10% consisting of upcycled local garment waste. By choosing recycled over virgin materials, Tonlé reduces environmental impact.<br>As of 2025, Tonlé continues this commitment to sustainability, further refining its zero-waste design process and expanding its impact in the fashion industry. |
| Mud Jeans International BV<br><br>Recycling technologies based on a "lease a jean" consumption model<br><br>(Koszewska, 2018) | To encourage use rather than ownership and support the shift toward a circular fashion economy by offering jeans for lease and recycling or upcycling materials. | After leasing jeans for a year, customers can choose to keep them, exchange them for a new pair, or return them for reuse or recycling. Mud Jeans then sells the returned items as vintage pieces or recycles the material into new products. The entire production process is socially responsible, ensuring fair treatment and respect for workers' rights. | The clothing production process consumes significantly less water and avoids the use of chemicals compared to conventional jeans manufacturing. |

continued on following page

*Table 2. Continued*

| Company/Area of action | Aim(s) | Action | Results |
|---|---|---|---|
| G-Star and Circle Economy pilot project with Wieland Textiles and Recover<br><br>Waste collection & recycling technologies<br><br>(Choudhary & Kumar, 2025) | To bring denim items returned to G-Star back into circulation and develop new denim fabrics that match virgin cotton denim in cost, quality, and appearance.<br><br>The goal is also to demonstrate both the business viability and environmental benefits of high-value textile-to-textile denim recycling. | G-Star chose one of its best-selling denim fabrics and worked to include recycled content in its production. The aim was not only to launch a limited capsule collection but also to integrate recycled denim into the brand's long-term sourcing strategy for lasting impact. | Recycled denim fabric costs about 12.5% more than virgin denim. To maintain the required strength for weaving and finishing, the recycled yarn can include up to 30% recycled fibers. Even with just 12% recycled content, the environmental benefits are notable: water use drops by 9.8%, energy consumption by 4.2%, and $CO_2$ emissions by 3.8% compared to virgin denim. |
| ReShare – a division of the Salvation Army: Project partnered with Circle Economy (Recycling technologies)<br><br>(Das et al., 2025) | To transform discarded Dutch military workwear into new textile products, demonstrating that used workwear can be repurposed into high-quality items while delivering substantial environmental benefits. | A sustainable and safe solution was developed for around 600 tons of discarded military workwear provided by the Dutch Ministry of Defence. The used garments, averaging a 50:50 cotton/polyester blend, were combined with virgin PET fibers and mechanically recycled into new yarn, which was then used to produce blankets for humanitarian aid. | Several tons of old Dutch navy and army uniforms were recycled into new yarns used to make humanitarian aid blankets. A Life Cycle Assessment of yarns containing 80% recycled military uniforms revealed an 87% reduction in water usage, 42% less energy consumption, and a 33% decrease in CO2 emissions compared to non-recycled yarn. |
| ReBlend, a Dutch circular fashion & textiles agency<br><br>Product design and recycling technologies<br><br>(Singh, 2025) | To demonstrate that mixed post-consumer textiles can be recycled into new, high-quality products, supporting the case for a closed-loop textile system. Ask ChatGPT | Over 70% of post-consumer textile waste, which was no longer wearable, was transformed into yarn and textiles through an eco-friendly process that avoided water use, added chemicals, and dyeing, minimizing environmental impact. The project produced 100% recycled yarns used in a new collection of knitted and woven fashion and upholstery products. | In the project, nearly 7 tons of post-consumer garments were processed to produce 6 tons of 100% recycled yarns. Four yarn colors were created using 70% recycled post-consumer garments and 30% recycled PET (RPET). Life Cycle Assessment of one recycled yarn showed a 33% reduction in energy use, 62% less water consumption, and an 18% decrease in greenhouse gas emissions compared to a virgin yarn with similar composition. |

Source: Authors' creation

## 4. THEMATIC CONFIGURATION OF CIRCULAR ECONOMY IN THE TEXTILE INDUSTRY

Figure 3 displays a thematic framework that entails a dynamic interplay between the enablers, challenges, and technological factors, influencing sustainability in the textile industry. The textile industry's sustainability is promoted by fostering resource efficiency, innovative business models material reuse, and strengthened with the presence of CE drivers. In contrast, CE challenges impede progress by creating barriers, including infrastructure gaps, financial constraints, and consumer resistance. Eventually, supportive policies command the industry toward circularity by setting regulatory standards and offering incentives. Additionally, digital technologies correlate with sustainability by enabling efficiency, transparency, and waste reduction.

*Figure 3. Thematic analysis of CE in the Textile industry*

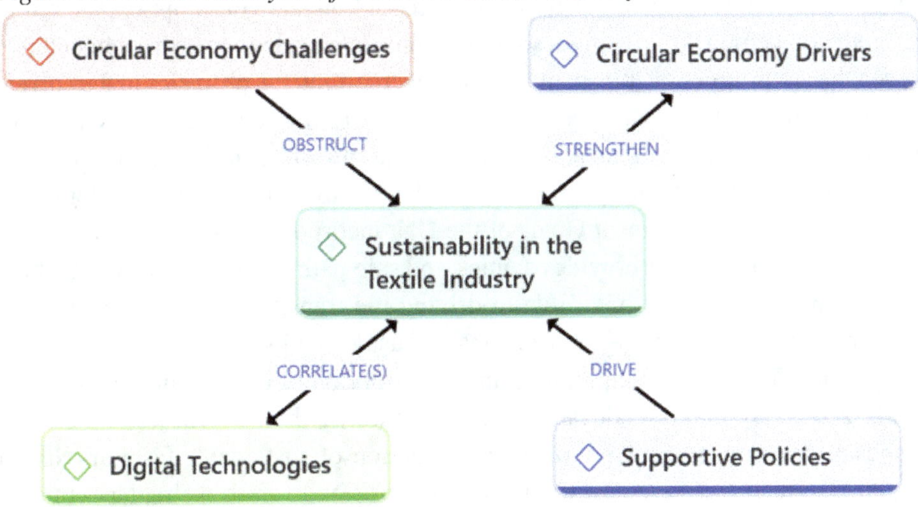


*Source: Authors' creation*

The CE challenges function as pressure points that obstruct the transition, while CE drivers counterbalance the pressures through innovation and consumer engagement. Thus, supportive policies offer the structural push for adopting circular strategies; digital technologies reinforce the policies by operationalizing data-driven solutions. Textile industry's sustainability emerges as the central outcome of these interactions. It reflects the balance between forces that either promote or hinder circular transformation.

## 5. STRENGTHENING THE CIRCULAR ECONOMY DRIVERS: STRATEGIES AND BEST PRACTICES

A multifaceted strategy that tackles every phase of the fashion value chain, from design and production to consumption and end-of-life management, is needed to strengthen the circular economy drivers in the textile sector. Implementing the circular economy entails recycling and recovery, product life extension, sharing economies, circular supply chains, and product-service systems (Karim et al., 2024). Through creative business models, these concepts give companies a foundation for redesigning their operations, cutting waste, and generating new revenue streams (Roshid et al., 2025b). The textile circularity is mostly determined by its design, which places a strong emphasis on using sustainable materials, designing for durability and recyclability, and minimizing production waste. In order to ensure sustainable development and long-term prosperity through waste reduction, recycling, reuse, and material recovery during production and consumption, a circular economy that emphasizes using resources in multiple cycles can be implemented to lessen the environmental impact of fashion. Overcoming the obstacles in the current textiles recovery system and promoting the shift to a circular economy requires cooperation, system harmonization, and data and information sharing. In line with European laws like the European Green Deal, the Circular Economy Strategy, and the Sustainable Development Goals of the UN, incorporating the Circular Economy into the textile sector provides a quick and safe path to sustainability (Mustafa, 2025). Consumers play a vital role in driving the transition to a circular fashion economy, with their purchasing decisions and usage habits significantly impacting the demand for sustainable products and the adoption of circular practices (Alam et al., 2025). By making educated choices regarding their purchases and clothes disposal, consumers can support the consumption of sustainable fashion (Nuh et al., 2023). Businesses are now motivated to transition towards a circular economy by minimizing raw material extraction, manufacturing more sustainable and eco-friendly products, and utilizing more recycled materials (Pranta et al, 2025). New business models are emerging that prioritize rental, repair, and resale, extending the lifespan of garments and reducing waste (Wicaksono et al., 2024). Business models can potentially influence both production and consumption practices as they define how a company conducts business and shapes the company-consumer relationship (Tunn et al., 2018). Circular business models, emphasizing reuse, repair, and recycling, are crucial for minimizing waste and maximizing resource utilization. Ideally, waste is prevented by changing consumer buying behavior, wearing clothes for a longer period, implementing other business models in the fashion industry, such as textile rentals, or by applying design for longevity principles (Harmsen et al., 2021). Furthermore, companies are beginning to focus on creating a sustainable ecosystem where products and services are not just sold to end customers, but there is a bottom-up approach to engaging them in business activities and building market competitiveness (Raj & Aithal, 2020).

The fashion sector exemplifies circular economy models, incorporating consumer behavior as a crucial element of the product life cycle (Ly, 2021). Stakeholder collaboration, involving brands, manufacturers, retailers, consumers, and policymakers, is essential for establishing a circular fashion ecosystem. This involves sharing knowledge, coordinating efforts, and developing common standards and targets for circularity. The fashion industry is beginning to respond to growing worries about unethical labor practices and environmental damage by implementing new business strategies, technologies, and materials (Kumar & Sahu, 2024). By reinventing value propositions and modifying different business model components, CE principles, such as reduce, repair, reuse, and recycle, can be integrated into current business models. A three-level framework is created that can assist the fashion industry in transitioning from LE to CE, and this study will help fashion companies understand how they can collaborate with various stakeholders to create sustainable business models (Mishra et al., 2020).

## 5.1 Policy and Regulatory Frameworks

Robust policy and regulatory frameworks will be key drivers of the progression to a circular textile economy. For example, measures such as Extended Producer Responsibility (EPR) make the fashion industry accountable for the environmental impact of their products at every stage of a garment's lifecycle, not to mention providing impetus to brands to create more durable and recyclable clothing. The overall trend is towards textile policies at international and national levels that increasingly push waste reduction, resource efficiency, and green production initiatives, with green public procurement schemes showing the way through a preference for sustainable textiles in institutional purchasing. Not only do these regulations ensure compliance, but they also level out the competition, which will eventually lead to a widespread circular economy approach within the fashion industry.

## 5.2 Collaboration Across the Value Chain

The textile industry is so intricate that it can only be circular if every step of the value chain participates. It takes each of these players, including brands, suppliers, recyclers, and non-governmental organizations (NGOs), to close material loops at scale with sustainable solutions. Collaborative projects like the development of common recycling infrastructure, supply chain visibility measures, and sustainability learning programs can bridge technical, logistical, and financial gaps. Together, stakeholders can sew a stronger, more circular, and much fairer fashion ecosystem rather than diminishing the planet and all its inhabitants.

### 5.3 Investment in Research and Development

Without long-term R&D investments, it will be difficult to bring about the necessary textile innovations needed for more circular economic ambitions. Investing in the development of sustainable fibers, advanced recycling technology, and more environmentally friendly manufacturing processes will help steer the industry away from its current resource-intensive and wasteful model. Research and development are also crucial in getting from great ideas to practical, scalable solutions, such as true chemical textile recycling or fully biodegradable fabrics. In striving for innovation, the fashion industry can evolve its toolset for sustainability, thereby reinforcing and interweaving a web of circular components that are more robust and agile in withstanding the heft of tomorrow's obstacles.

### 5.4 Education and Skill Development

Designing circularity in fashion relies on training designers, technicians, merchandisers, and retail teams on low-impact material choices; designing for durability/repair; the ability to disassemble for recycling; and aligning with consumers. Training delivers highly applicable curricula on pattern engineering for repairability, fiber identification and sorting, care-and-repair methods, and digital skills for traceability to equip the workforce to waste less and recycle products more. By supplementing their internal training with public-facing education (care guides to instruct customers on how to use and care for a product; take-back instructions to show consumers how they can return products once they reach end of life), businesses create a culture of stewardship that strengthens policy goals (like EPR or green procurement) and enables circular business models to work operationally from end-to-end in the value chain.

## 6. FUTURE TRENDS AND OPPORTUNITIES

### 6.1 Rise of Regenerative Fashion

Regenerative fashion is regarded as the next step of sustainability in textiles since businesses have an obligation to the planet to not only minimize harm but also restore ecosystems. This pathway is incorporating an alternative for regenerative agriculture, i.e., farming organic cotton that helps soil health, biodiversity, and water retention in the process of supply chain. Brands that source from farms that sequester carbon, rebuild degraded land, and support rural livelihoods will contribute to reversing environmental degradation caused by decades of extractive production. It uses the language of climate and ecology to shift fashion away from being an extractive industry that relies solely on biomimicry to make products less bad, towards being a regenerative force able to sew the very seeds of its recovery into the clothes. Consequently, the stakeholder may benefit further.

## 6.2 Role of Artificial Intelligence and Big Data

AI and Big Data are revolutionizing the way fashion companies can predict demand, optimize production, and tackle waste. For sustainable fashion, overproduction represents one of the biggest hurdles in the industry to predictive analytics that can be used to reveal purchasing trends and seasonal preferences, as well as production numbers necessary for this, thus significantly minimizing such. Material Selection, Defect Detection, and Supply Chain Optimization by adopting Machine Learning algorithms helps to achieve circularity effectively. When fashion brands combine AI with up-to-the-minute market data, they can quickly adopt a structured and evidence-based approach that will not only cater to the consumer but will also prove highly beneficial for the planet.

## 6.3 Emergence of Bio-fabricated Textiles

Additionally, new horizons are being presented by bio-fabricated textiles like cultured leather and mycelium-sourced materials. These are high-performance materials of the future that offer the look, feel, and qualities of leather and fabrics, but with a significantly lower amount or no land footprint, water consumption, or chemical inputs. Further, they allow for the possibility of complete biodegradability and, therefore, no end-of-life waste issues. With the increasing sophistication of technology and decreasing cost, these bio-fabricated alternatives could potentially co-opt existing supply chains to provide scalable solutions that are fully integrated with the circular frameworks of fashion or textiles.

## 6.4 Localized Production and On-Demand Manufacturing

The body part of cloth production is reshaping as a result of localized production and delivery, which reduces distant emissions, thereby lowering unsold inventories. Manufacturing closer to the end customer, and also in small runs (producing items only when somebody orders them), means brands can perfectly synchronize supply with actual demand - rather than overproducing and creating waste through huge markdowns. Responsive production is also enhancing the connection between local communities and fashion producers, delivering supply chains that are transparent, traceable, and resilient, in line with circular economy objectives.

## 7. CONCLUSION

For the garments and textile sector, a shift towards a circular economy is more than a challenge; it embodies an unprecedented potential for turning economic growth into environmental protection alongside social value. This will allow the sector to shift from the linear "take-make-dispose" model by incorporating both regenerative production and consumption systems that not just avoid waste in the first place but truly contribute towards changing ecological wear. This chapter has considered the fundamental enablers of circularity as a means to ensure these novel concepts, from sustainable material innovation to lifespan extension strategies, reverse logistics, and turning old business models on their heads, digitalization, or engaging consumers, are more than just abstractions. It also marks responsibilities for creating the conditions to enable these drivers to be effective and scalable, enabling policy and regulatory frameworks, cross-sector collaboration, targeted investment in R&D, education, and skill building on circular design and production. Moving forward, new trends like regenerative fashion, AI-fueled predictive analytics, bio-fabricated textiles, and on-demand manufacturing with small carbon footprints are prepared to reshape the industry's sustainability conversation. While these can allow material loops to close more efficiently, they might also reduce overproduction and hence emissions, both directly from production (as a byproduct of simply making less) and indirectly due to influencing the type of consumer. To capture this potential, utilities will need to continuously innovate, communicate openly, and hold the value chain accountable together.

Ultimately, circularity in the textile sector is not so much a destination as it is an iterative journey towards switching to systemic design that simultaneously grows the economy and regenerates nature. This chapter described the drivers (why), enabling mechanisms (how), and future opportunities (where) in their interaction for building a resilient, just sustainable fashion ecosystem by underpinning developments that help brands, policymakers, suppliers, innovators, and consumers to come together.

## REFERENCES


Ahmad, S., Saxena, C., Islam, S., & Karim, R. (2024). Impact of Insur-Tech on the Premium Performance of Insurance Business. *SN Computer Science*, *5*(1), 138. DOI: 10.1007/s42979-023-02462-0

Ahmed, Z., Mahmud, S., & Acet, H. (2022). Circular economy model for developing countries: Evidence from Bangladesh. *Heliyon*, *8*(5). Advance online publication. DOI: 10.1016/j.heliyon.2022.e09530 PMID: 35663755

Akter, M. M. K., Haq, U. N., Islam, M. M., & Uddin, M. A. (2022). Textile-apparel manufacturing and material waste management in the circular economy: A conceptual model to achieve sustainable development goal (SDG) 12 for Bangladesh. *Cleaner Environmental Systems*, *4*, 100070. DOI: 10.1016/j.cesys.2022.100070

Al-Tohamy, R., Ali, S. S., Li, F., Okasha, K. M., Mahmoud, Y. A. G., Elsamahy, T., & Sun, J. (2022). A critical review on the treatment of dye-containing wastewater: Ecotoxicological and health concerns of textile dyes and possible remediation approaches for environmental safety. *Ecotoxicology and Environmental Safety*, *231*, 113160. DOI: 10.1016/j.ecoenv.2021.113160 PMID: 35026583

Alam, S. S., Ahsan, M. N., Kokash, H. A., Ahmed, S., & Di, W. (2025). Remanufactured consumer goods buying intention in circular economy: Insight of value-belief-norm theory, self-identity theory. *Journal of Remanufacturing*, *15*(1), 179–206.

Aus, R., Moora, H., Vihma, M., Unt, R., Kiisa, M., & Kapur, S. (2021). Designing for circular fashion: Integrating upcycling into conventional garment manufacturing processes. *Fashion and Textiles*, *8*(1). Advance online publication. DOI: 10.1186/s40691-021-00262-9

Badran, G., & Lazarov, V. K. (2025, February). From Waste to Resource: Exploring the Current Challenges and Future Directions of Photovoltaic Solar Cell Recycling. In *Solar* (Vol. 5, No. 1, p. 4). MDPI.

BGMEA. (2020). Trade information. Available from https://www.bgmea.com.bd/home/pages/TradeInformation

Biyada, S., & Urbonavičius, J. (2025). Circularity in textile waste: Challenges and pathways to sustainability. *Cleaner Engineering and Technology*, •••, 100905.

Brunner, P. H., & Morf, L. S. (2025). Waste to energy, indispensable cornerstone for circular economy: A mini-review. *Waste Management & Research*, *43*(1), 26–38. PMID: 38282576

Chen, X., Memon, H. A., Wang, Y., Marriam, I., & Tebyetekerwa, M. (2021). Circular Economy and Sustainability of the Clothing and Textile Industry. *Materials Circular Economy*, *3*(1). Advance online publication. DOI: 10.1007/s42824-021-00026-2

Choudhary, A. K., & Kumar, P. (2025). Challenges in Handling and Systematically Organizing Textile Waste. *Nanotechnology-Assisted Recycling of Textile Waste: Sustainable Tools for Textiles*, 29-58. https://doi.org/DOI: 10.1002/9781394175017.ch2

Coscieme, L., Manshoven, S., Gillabel, J., Grossi, F., & Mortensen, L. F. (2022). A framework of circular business models for fashion and textiles: The role of business-model, technical, and social innovation. *Sustainability Science Practice and Policy*, *18*(1), 451. DOI: 10.1080/15487733.2022.2083792

D'Itria, E., & Aus, R. (2023). Circular fashion: Evolving practices in a changing industry. *Sustainability Science Practice and Policy*, *19*(1). Advance online publication. DOI: 10.1080/15487733.2023.2220592

Damayanti, D., Wulandari, L. A., Bagaskoro, A., Rianjanu, A., & Wu, H. (2021). Possibility Routes for Textile Recycling Technology [Review of Possibility Routes for Textile Recycling Technology]. Polymers, 13(21), 3834. Multidisciplinary Digital Publishing Institute. https://doi.org/DOI: 10.3390/polym13213834

Das, A. K., Hossain, M. F., Khan, B. U., Rahman, M. M., Asad, M. A. Z., & Akter, M. (2025). Circular economy: A sustainable model for waste reduction and wealth creation in the textile supply chain. *SPE Polymers*, *6*(1), e10171. DOI: 10.1002/pls2.10171

Farhana, K., Mahamude, A. S. F., & Mica, M. T. (2022). The scenario of textile industry in Malaysia: A review for potentiality. *Materials Circular Economy*, *4*(1), 20. DOI: 10.1007/s42824-022-00063-5

Firdaus, F., Setyono, P., Gravitiani, E., & Liquiddanu, E. (2025). Key Drivers and Barriers to Circular Economy Practices in the Global Textile and Fashion Industries: Sustainable Strategies for the Indonesian Batik Industry. *International Journal of Sustainable Development and Planning*, *20*(1).

Gautam, P., Salunke, D., Lad, D., & Gautam, A. (2025). Convergent synergy of carbon capture within the circular economy paradigm: A nexus for realizing multi-faceted sustainable development goals. *Carbon Research*, *4*(1), 3.

Genta, C., Sanyé-Mengual, E., Lombardi, P., & Sala, S. (2025). A local analysis of circular economy through a stakeholders' lens: From definitions and collaborative efforts to metrics for monitoring. The case of Turin (Italy). *Environmental Impact Assessment Review*, *112*, 107736.

Gharibi, K., & Abdollahzadeh, S. (2025). A mixed-integer linear programming approach for circular economy-led closed-loop supply chains in green reverse logistics network design under uncertainty. *Journal of Enterprise Information Management*, *38*(1), 1–31.

Grover, T., Wang, L., Nayak, R., Padhye, R., & Khandual, A. (2025). Sustainable strategies for the valorization of cotton and cotton blended waste: Pathways to circular economy. *Cellulose (London, England)*, •••, 1–24.

Harmsen, P. F. H., Scheffer, M., & Bos, H. L. (2021). Textiles for Circular Fashion: The Logic behind Recycling Options. *Sustainability*, *13*(17), 9714. DOI: 10.3390/su13179714

Hossain, M. K., & Mollik, M. S. (2022). Exploring the impact of Social Media Platforms on Instigating Tourist's Emotions and Behavioral Intentions to Co-Create and Visit Tourist Spots-Bangladesh Perspective. https://doi.org/DOI: 10.58753/jbspust.3.1.2022.13

Jia, F., Yin, S., Chen, L., & Chen, X. (2020). The circular economy in the textile and apparel industry: A systematic literature review. *Journal of Cleaner Production*, *259*, 120728. DOI: 10.1016/j.jclepro.2020.120728

Jiang, M., Deng, W., & Lin, H. (2024). Sustainability through biomimicry: A comprehensive review of bionic design applications. *Biomimetics*, *9*(9), 507. DOI: 10.3390/biomimetics9090507 PMID: 39329529

Kabir, S. M. F., Chakraborty, S., Hoque, S. M. A., & Mathur, K. (2019). Sustainability Assessment of Cotton-Based Textile wet processing. *Cleanroom Technology*, *1*(1), 232–246. DOI: 10.3390/cleantechnol1010016

Karim, R., Roshid, M., & Waaje, A. (2024). Circular economy and supply chain sustainability. In *Strategic Innovations for Dynamic Supply Chains* (pp. 1–30). IGI Global., DOI: 10.4018/979-8-3693-3575-8.ch001

Karim, R., Waaje, A., & Roshid, M. M. (2025). Digital Circular Economy for Attaining Sustainable Development Goals: Technologies for Global Sustainability. In *Effects of Digitalization and Circular Economy on Sustainable Policy and Climate Change Prevention* (pp. 211-240). IGI Global Scientific Publishing. https://doi.org/DOI: 10.4018/979-8-3693-9909-5.ch008

Karim, R., Waaje, A., Roshid, M. M., & Yeamin, M. B. (2025). Turning the waste into wealth: Progressing toward global sustainability through the circular economy in waste management. In *Sustainable waste management in the tourism and hospitality sectors* (pp. 507–552). IGI Global Scientific Publishing., DOI: 10.4018/979-8-3693-6110-8.ch019

Koszewska, M. (2018). Circular Economy — Challenges for the Textile and Clothing Industry. *AUTEX Research Journal*, *18*(4), 337. DOI: 10.1515/aut-2018-0023

Kumar, A., & Sahu, R. (2024). Entrepreneurship of Eco-friendly Fashion Startups: Pioneering Sustainable Practices in the Fashion Industry. SSRN Electronic Journal. https://doi.org/DOI: 10.2139/ssrn.4926224

Kumar, P., Samuchiwal, S., & Malik, A. (2020). Anaerobic digestion of textile industries wastes for biogas production. *Biomass Conversion and Biorefinery*, *10*(3), 715–724. DOI: 10.1007/s13399-020-00601-8

Lawrence, J., Bettinetti, R., Torretta, V., & Rada, E. C. (2023). Investigating the effectiveness of clothes recovery programs in promoting a circular economy: A review [Review of Investigating the effectiveness of clothes recovery programs in promoting a circular economy: A review]. E3S Web of Conferences, 408, 1002. EDP Sciences. https://doi.org/DOI: 10.1051/e3sconf/202340801002

Ly, B. (2021). Competitive advantage and internationalization of a circular economy model in apparel multinationals. *Cogent Business & Management*, *8*(1). Advance online publication. DOI: 10.1080/23311975.2021.1944012

Maity, D., Borkar, A., Ghosh, S., & Sabnis, A. S. (2025). Overview of Textile Recycling (Solid Waste). *Nanotechnology-Assisted Recycling of Textile Waste: Sustainable Tools for Textiles*, 59-96. https://doi.org/DOI: 10.1002/9781394175017.ch3

Market Research Future. (2024). Textile market research report: Forecast to 2027. Market Research Future. https://www.marketresearchfuture.com/reports/textile-market-9271

Masum, M. M. H., Islam, M. B., Anowar, M. R., Arsh, K., & Alam, S. (2025). Assessment of textile industry effluents and their impact on local water bodies in an urban setting of Bangladesh. *Discover Environment*, *3*(1), 59. DOI: 10.1007/s44274-025-00245-3

Mehta, J., Dilbaghi, N., Deep, A., Hai, F. I., Hassan, A. A., Kaushik, A., & Kumar, S. (2025). Plastic waste upcycling into carbon nanomaterials in circular economy: Synthesis, applications, and environmental aspects. *Carbon*, *234*, 119969.

Mishra, S., Jain, S., & Malhotra, G. (2020). The anatomy of circular economy transition in the fashion industry. *Social Responsibility Journal, 17*(4), 524. DOI: 10.1108/srj-06-2019-0216

Molla, A. H., Shams, H., Harun, Z., Ab Rahman, M. N., & Hishamuddin, H. (2022). An assessment of drivers and barriers to implementation of circular economy in the end-of-life vehicle recycling sector in India. *Sustainability, 14*(20), 13084.

Mollik, M. S., Rahman, S. M., Rahat, M. R., Kulsum, C. U., & Sagir, S. A. M. (2024). DOES TRUSTWORTHINESS INFLUENCE TRAVEL SERVICE USE INTENTIONS AT AN ONLINE TRAVEL AGENCY? A STUDY ON THE DIGITALIZATION OF THE TOURISM SECTOR IN BANGLADESH. *Geo Journal of Tourism and Geosites, 52*(1), 30–40. DOI: 10.30892/gtg.52103-1180

Musová, Z., Musa, H., Drugdova, J., Lăzăroiu, G., & Alayasa, J. (2021). Consumer Attitudes Towards New Circular Models in the Fashion Industry. *Journal of Competitiveness, 13*(3), 111. DOI: 10.7441/joc.2021.03.07

Mustafa, M. (2025). Integrating Life Cycle Analyses and Circular Economy: A Pathway Towards Sustainable Development in Bangladesh. In *The Contribution of Life Cycle Analyses and Circular Economy to the Sustainable Development Goals* (pp. 313–338). Springer Nature Switzerland.

Naqvi, S. L. H., Nadeem, M., Ayub, F., Yasar, A., Naqvi, S. H. Z., & Tanveer, R. (2024). Social and environmental impacts in textile production. In *Dye pollution from textile industry: Challenges and opportunities for sustainable development* (pp. 423–453). Springer Nature Singapore., DOI: 10.1007/978-981-97-5341-3_19

Niinimäki, K., Peters, G., Dahlbo, H., Perry, P., Rissanen, T., & Gwilt, A. (2020). The environmental price of fast fashion. *Nature Reviews. Earth & Environment, 1*(4), 189–200. DOI: 10.1038/s43017-020-0039-9

Nuh, A., Munir, M. M., & Muhibban, M. (2023). Fashion engagement and pro-environmental attitudes: Drivers of sustainable fashion consumption in Indonesia. *Journal of Enterprise and Development, 5*(3), 460. DOI: 10.20414/jed.v5i3.7292

Ofori, E. A., Akweley, B. D., Eghan, B., Seidu, R. K., & Acquaye, R. (2025). A review on upcycling as a strategic marketing tool for sustainability in the fashion industry. *Journal of Responsible Production and Consumption, 2*(1), 1–24. DOI: 10.1108/JRPC-02-2024-0007

Ooi, S. K., & Memon, K. R. (2025). Responsible Innovation, Circular Economy and Organisational Resilience: Investigating Multiple Pathways Towards Sustainability Performance. *Business Strategy and the Environment.*

Pavan, M., Samant, L., Nageshkumar, T., Samreen, S., & Yashaswini, N. G. (2025). Eco-Production of Apparel: Energy-and Water-Efficient Luxury Textiles. In *Crafting Sustainability in Luxury Textiles for a Zero-Waste Future* (pp. 195–219). Springer Nature Switzerland., DOI: 10.1007/978-3-031-86773-6_13

Piribauer, B., & Bartl, A. (2019). Textile recycling processes, state of the art and current developments: A mini review. *Waste Management & Research, 37*(2), 112–119. DOI: 10.1177/0734242X18819277 PMID: 30632932

Pranta, A. D., Rahaman, M. T., & Repon, M. R. (2025). Sustainable Waste Management of Textiles and Apparel in a Circular Economy: Recent Advances and Future Prospects of Life Cycle Assessment. *Materials Circular Economy, 7*(1), 28.

Provensi, T., Marcon, M. L., Sehnem, S., Campos, L. M., & Queiroz, A. F. S. D. (2025). Exploring ESG and circular economy in Brazilian companies: The role of stakeholder engagement. *Benchmarking*.

Raj, K., & Aithal, P. S. (2020). A New Concept of Circular Model of Management for Achieving Sustainable Success and Growth. *International Journal of Management Technology and Social Sciences, 1*. Advance online publication. DOI: 10.47992/ijmts.2581.6012.0079

Rhaouti, Y., Taha, Y., & Benzaazoua, M. (2025). Life cycle assessment and industrial synergy for carbon reduction: A circular economy approach. *The Science of the Total Environment, 964*, 178592. PMID: 39862508

Roshid, Md. M., Waaje, A., Karim, R., Irtisham, M. R., & Kaswan, M. S. (2025a). Industry 5.0 and Organizational Sustainability: Integrating Cutting-Edge Technologies and Human Ingenuity. In Mahender Singh Kaswan, Rajeev Rathi, & Jose Arturo Garza-Reyes (Eds.), *Industry 5.0: Concepts and Strategies for Digital Transformation* (pp. 50–61). CRC Press. https://doi.org/DOI: 10.1201/9781003535188-7

Roshid, Md. M., Saiyed, S., Karim, R., Haque, T., Siddique, Md. S. R., Haque, Md. A., Kumar, V., & Habib, S. Bin. (2025b). The Role of Green Entrepreneurship in Driving Sustainable Innovation in Bangladesh's Startup Ecosystem: Insights from a Belt and Road Emerging Market. *Environment. Innovation and Management, 01*, 2550010. DOI: 10.1142/S3060901125500103

Saha, K., Dey, P. K., & Papagiannaki, E. (2022). Implementing circular economy in the textile and clothing industry. In *Supply chain sustainability in small and medium sized enterprises* (pp. 239–276). Routledge., DOI: 10.4324/9781003018551-10

Sahu, D. K., & Rath, G. (2025). Innovative Approaches to Recover Waste in the Textile Sector. *Nanotechnology-Assisted Recycling of Textile Waste: Sustainable Tools for Textiles*, 129-169. DOI:DOI: 10.1002/9781394175017.ch5

Schumacher, K. A., & Forster, A. L. (2022). Textiles in a circular economy: An assessment of the current landscape, challenges, and opportunities in the United States. *Frontiers in Sustainability, 3*. Advance online publication. DOI: 10.3389/frsus.2022.1038323

Shirvanimoghaddam, K., Motamed, B., Ramakrishna, S., & Naebe, M. (2020). Death by waste: Fashion and textile circular economy case. *The Science of the Total Environment, 718*, 137317. DOI: 10.1016/j.scitotenv.2020.137317 PMID: 32088483

Singh, N. (2025). Circular economy in textile and apparel supply chains. In *Innovation and Sustainability Through Circular Economy in Businesses* (pp. 25–64). IGI Global Scientific Publishing., DOI: 10.4018/979-8-3693-4123-0.ch002

Statista. (2024). Textile industry worldwide – statistics & facts. Statista. https://www.statista.com/topics/9614/textile-industry-worldwide/

Sultana, N., Rahman, K. S., Karim, R., & Ahmad, S. (2025). Entrepreneurship: A rising way leading to economic development in Bangladesh. *Procedia Computer Science, 259*, 1947–1954. DOI: 10.1016/j.procs.2025.04.150

Sustain your style (2020). The fashion industry is the second largest polluter in the world. Available from https://www.sustainyourstyle.org/old-environmental-impacts #anchor-link-soils

systematic and bibliometric analysis. *EMIDWORLD 2nd International Congress on Economics Public Finance Business & Social Sciences*. (pp. 784–802). Altınbas University. www.emidworld.com

Tunn, V., Bocken, N., van den Hende, E. A., & Schoormans, J. (2018). Business models for sustainable consumption in the circular economy: An expert study. *Journal of Cleaner Production, 212*, 324. DOI: 10.1016/j.jclepro.2018.11.290

Uddin, F. (2021). Environmental hazard in textile dyeing wastewater from local textile industry. *Cellulose (London, England), 28*(17), 10715–10739. DOI: 10.1007/s10570-021-04228-4

UN. (2019). UN launches drive to highlight environmental cost of staying fashionable. Available from https://news.un.org/en/story/2019/03/1035161

Ütebay, B., Çelik, P., & Çay, A. (2020). Textile Wastes: Status and Perspectives. In IntechOpen eBooks. https://doi.org/DOI: 10.5772/intechopen.92234

Verma, S., Bhalla, P., & Dutta, T. (2025). Happiness and Circular Economy: A New Framework for Economic and Social Well-Being. In *Innovation and the Sharing Economy* (pp. 211-229). Routledge.

Vietnam Briefing. (2020). Seizing Investment Opportunities in Vietnam's Garment and Textile Industry. Available from https://www.vietnam-briefing.com/news/seizing-investment-opportunities-vietnams-textile-garment-industry.html/

Waaje, A., Karim, R., Roshid, M. M., & Meem, T. N. (2025a). *Minimizing Waste in the Hospitality Industry: The Opportunities and Challenges of Lean Tourism. Sustainable Waste Management in the Tourism and Hospitality Sectors*. IGI Global Scientific Publishing., DOI: 10.4018/979-8-3693-6110-8.ch006

Waaje, A., Karim, R., Roshid, M. M., Yeamin, B., & Meem, T. N. (2025b). Integration of industry 4.0 with personalized medicine: A shift of the paradigm in healthcare. In Chemingui, H., & Lamloumi, M. (Eds.), *AI-Driven Personalized Healthcare Solutions* (pp. 27–69). IGI Global., DOI: 10.4018/979-8-3693-7858-8.ch002

Wicaksono, P. A., Sari, D. P., Azzahra, F., & Sa'adati, N. A. (2024). Analysis of Consumer Behaviour in Purchasing Second-Hand Fashion Products: An Extended Theory of Planned Behaviour Model. *International Journal of Sustainable Development and Planning, 19*(8), 2955. DOI: 10.18280/ijsdp.190813

WTO. (2019). WTO DATA available from https://data.wto.org/. https://www.ilo.org/global/about-the-ilo/newsroom/news/WCMS_008075/lang--en/index.htm

# Chapter 8
# Integrating Circular Economy Principles Into Responsible Consumption and Production

**Mohammad Shahab Uddin**
https://orcid.org/0000-0003-3591-7284
*University of Chiattagong, Bangladesh*

**Md. Soleman Mollik**
https://orcid.org/0009-0003-5703-0606
*Bangladesh University of Professionals, Bangladesh*

**Amir Mohammad Nasrullah**
*University of Chittagong, Bangladesh*

## ABSTRACT

*Integrating circular economy (CE) principles into responsible consumption and production (RCP) drives long-term sustainability for businesses, consumers, society, and the environment by addressing environmental challenges and resource constraints of the 21st century. Despite their growing relevance, research on CE-RCP integration remains limited. Against this backdrop, this chapter aims to analyze existing literature to bridge the research gap, using qualitative methodology. The findings reveal that the integration of CE-RCP enhances resource efficiency, advances sustainable consumption and production (SDG12), and facilitates waste reduction. However, key challenges relate to substantial financial requirements, limited circular*

DOI: 10.4018/979-8-3373-2454-8.ch008

*supply chains, and consumer behavioral barriers. The chapter guides stakeholders toward a sustainable and resilient future by offering managerial, policy, societal, and scholarly implications.*

## 1. INTRODUCTION

Circular economy (CE) appears to be an emerging economic model. It strives to redefine social and industrial growth based on multipurpose benefits (Mubarik et al., 2024; Sami & Khan, 2025). CE functions according to key principles, such as reuse, remanufacture, reduce, recycle, or refurbish (Jain et al., 2025; Otasowie et al., 2025). It remains differentiated from the traditional linear economy model (take-make-dispose) (Karim et al., 2025). The circular economy helps extend product life cycles (Steinbiß, 2025), promote resource efficiency (Singh, 2025; Velenturf et al., 2019), and facilitate the reuse and recycling of materials (Nademi & Kalmarzi, 2025). Additionally, CE lessens environmental costs by lowering greenhouse gas emissions, protecting biodiversity, and restraining environmental degradation (Xiao, 2025). Therefore, CE addresses environmental challenges while ensuring economic growth, responsible consumption and production, and diminishing environmental costs (Munuhwa & Mawonde, 2025).

The worsening climate change, biodiversity loss, and resource depletion globally have led the United Nations to establish Sustainable Development Goals (SDGs) in 2015 as a universal call to action (Vogel et al., 2025). The SDGs outline 17 shared goals, aiming to promote prosperity, social equity, and environmental protection (Ivanov et al., 2025; Morshed, 2025). SDG 12 (responsible consumption and production) remains one of them. It urges reducing the use of natural resources and minimizing waste and pollutants across the consumption and production process (Dhaigude et al., 2025).

The integration of CE and RCP clings to SDG 12. CE promotes sustainable production and consumption through stimulating behavioral shifts of producers and consumers. Consumers' behavioral shifts contain reuse, recycling, repair, and remanufacturing, while responsible production folds innovative designs, efficient and effective supply chains, and industrial symbiosis. These CE-RCP shifts advance social and economic well-being (Gautam et al., 2025; Parvathy et al., 2025). Thus, implementing CE principles into RCP helps achieve broader sustainable development objectives.

The circular economy has become a popular alternative to the linear economic model (Sibanda et al., 2025). Here, the reasons contain regenerative natural systems, recycling products and materials, and minimizing waste at the best possible level

(Islam & Zheng, 2025). Moreover, the integration between CE and SDG 12 provides a stronger route to reach long-term sustainable development (Gautam et al., 2025).

The rapid growth of industry, urbanization, and population creates a significant increase in resource and energy use, as well as waste generation. Consequently, the traditional linear economic model proves unsustainable (Roshid et al., 2025c) as it depends on finite natural resources and contributes to severe environmental pollution (Kandasamy et al., 2025). Thus, the transition towards sustainable production and consumption remains urgent to prolong product life cycles, abate waste, and reduce environmental costs. Emerging as an alternative to the linear model (Sudusinghe & Seuring, 2025), CE tends to strengthen industrial and economic growth, using finite resources (Schultz & Pies, 2024). Since the CE promotes reuse, recycling, remanufacturing, and regenerative design (Balakrishnan et al., 2025; Madaan et al., 2024), it results in closed-loop materials and reduces waste (Bourdin & Jacquet, 2025). In the meantime, RCP (SDG 12) urges the sustainable use of finite natural resources across product life cycles.

Although the circular economy and responsible consumption and production target sustainability, there exists a significant gap regarding how CE principles integrate RCP practices to ensure sustainability. Therefore, the examination of CE-RCP integration remains urgently needed (Islam & Zheng, 2025) to enhance consumer awareness, address climate challenges, and promote more efficient management of limited resources.

The chapter mainly aims to discover how the integration of CE-RCP can benefit organizations, consumers, society, and the environment. The objective is then broken down into specific variants, including conceptualizing the nexus between CE and RCP in the context of sustainable development, identifying the drivers and barriers of CE-RCP integration in the existing business models, critically examining business cases and extract the best practices illustrating successful CE-RCP integration, and proposing a framework or roadmap for policymakers, industries, and consumers to guide the transition towards circular and responsible production-consumption systems.

The CE-RCP integration entails the potential to advance entire sustainable development. Therefore, the transition needs concrete knowledge about the effective ways of CE-RCP integration in real-world settings, which influence consumers and producers.

Responsible production- which emphasizes minimizing environmental impact, maximizing resource efficiency, and rethinking product lifecycles- is central to achieving Sustainable Development Goal 12. While CE offers theoretical strategies for sustainable production, there is a need to explore how these principles can be translated into practical applications across industries.

This chapter generates significant contributions to the academic, practical, and social domains. Academically, it enriches the interdisciplinary discourse on sustainability by bridging two influential paradigms: Circular Economy (CE) and Responsible Production and Consumption (RPC), which have often been treated separately. By examining how CE principles can be strategically aligned with SDG 12 objectives, this study contributes to the theoretical development of integrated sustainability models. It offers a nuanced understanding of the institutional, behavioral, and structural dynamics that influence the adoption of circular practices. Furthermore, the study fills a gap in the literature regarding the operationalization of CE in RPC contexts, particularly in emerging and resource-constrained economies, providing empirical insights that can inspire further research in this growing field.

Practically, the study provides a clear roadmap for stakeholders, including governments, businesses, and civil society organizations, to design systems, products, and behaviors that are not only environmentally sustainable but also economically viable and socially inclusive. By identifying key barriers and enablers to CE-RPC integration, the research offers actionable strategies for mainstreaming circularity into supply chains, business models, consumer engagement, and product-service systems.

For emerging economies, where rapid industrial growth is frequently accompanied by resource depletion, pollution, and social inequality, this integration presents a compelling model for sustainable industrialization. The study demonstrates how CE can enable resource decoupling, reduce environmental burdens, and stimulate green entrepreneurship, particularly through innovation in product design, packaging, and reverse logistics. It also outlines how circular value chains can be fostered through supportive policy environments, cross-sectoral collaboration, and investment in green technologies.

Policy-wise, the study offers evidence-based recommendations that can inform national sustainability strategies, waste management frameworks, and climate action plans. Moreover, it assists policymakers in aligning their practices with the SDGs, according to the Paris Agreement. As countries seek post-pandemic recovery models, these insights remain significantly relevant for building resilient, inclusive, and climate-smart pathways.

The chapter discloses new opportunities for future research, including cross-cultural comparisons of CE-RCP integration and their sector-specific analyses. Furthermore, it will encourage scholars to explore how CE can be properly scaled, adapted, and sustained in different socio-economic contexts.

The remaining sections of the chapter are distributed as follows. Section two reviews existing and relevant literature. The chapter then illustrates the methodology and presents the findings and discussions. It concludes with implications, limitations, and future research directions.

## 2. CIRCULAR ECONOMY

The circular economy represents an economic system where the resources are reused, materials are recycled, and disposed items are remanufactured. This regenerative system contrasts with the traditional linear economic model (take-make-dispose) (Kanda et al., 2021; Mahdikhani & Karbasi, 2025). CE aims to extend the value of products, materials, and resources across the production and consumption domains by fostering closed-loop cycles (Suppipat & Intrachooto, 2025). This economic system encompasses innovative designs, a transitional business model (De Martino et al., 2024), and consumer participation (Cerchione, 2025) to promote waste elimination, product and material circulation, and natural regeneration (Stahel & MacArthur, 2019; Espuny et al., 2025). Although CE integration in business domains faces significant challenges in developing countries, the concept remains well-established in developed nations. The extent of CE integration varies across sectors (Murray et al., 2017). For instance, the Netherlands is actively exploring effective approaches to implement circular economy practices across diverse sectors, particularly medical devices, while the United States and Australia remain cautious about this transition. However, CE promotes sustainability in challenging environments (Karim et al., 2024).

Although most scholars have viewed CE as an effective solution to unsustainable consumption and production (Karim et al., 2025b; Waaje et al., 2025a), Korhonen et al. (2018) unraveled it as an "Oxymoron." They have questioned CE's conceptual robustness due to the inherent conflict between continuous economic growth and finite ecological resources. On the other hand, Geissdoerfer et al. (2017) described CE as a holistic sustainability model, arguing that CE fosters economic growth by optimizing resource use through innovation and design. These contrasting viewpoints shed light on the need for contextual and critical application of CE principles.

### 2.1 Responsible Production (RP)

Responsible production (RP) involves integrating environmental consciousness throughout the life cycle of products, from design and raw material extraction to manufacturing, distribution, and disposal, while prioritizing ecological considerations (Maldonado-Romo et al., 2024). RP emphasizes abating environmental costs or impacts, conserving natural resources, and promoting economic and social sustainability. Moreover, it appears to be the cornerstone for achieving SDG 12 (UNCTAD, 2023; UNEP, 2023). Unlike the traditional take-make-dispose linear approach, RP clings to CE, resulting in the creation of closed-loop systems to reduce waste and pollution (UNEP, 2015; Waaje et al., 2025b).

In a responsible production system, the circular economy functions as a catalyst. CE favors RP by interpreting how products should be designed, materials should be sourced, and waste should be managed. CE also strengthens closed-loop systems by converting waste into new input materials for new production. This practice reduces full dependence on virgin materials. RP and CE advocate long-term sustainability by eco-design, life cycle thinking, modular manufacturing, and industrial symbiosis (Kanda et al., 2021).

Research indicates that integrating CE-RP enables businesses to minimize operating costs and maximize competitive advantages by satisfying consumer demand for sustainability. Furthermore, modern technologies facilitate the integration of the circular economy and responsible production (De Jesus & Mendonça, 2018), which ensures sustainable business practices (Hossain & Mollik, 2022; Mollik et al., 2024).

## 2.2. Responsible Consumption (RC)

Responsible consumption refers to the ways of selecting and consuming goods and services that ensure economic sustainability, foster social equity, and minimize environmental degradation. It urges sustainable consumption and production patterns. RC encourages consumers to protect the environment by lessening irresponsible consumption, selecting ethical products, avoiding waste, and promoting the sustainable use of resources (UNEP, 2023). CE and RC complement each other based on the 4Rs (Reduce, Recycle, Reuse, Repair) and closed-loop systems. The integration of CE with RC empowers consumers to become responsible and active participants in contributing to a regenerative economy (White et al., 2019).

Prior research claimed that CE made consumers aware of responsible consumption and environmental sustainability (Mugge et al., 2017). Consumer awareness favors RC, whereas limited knowledge, high costs, and perceived inconvenience regarding CE hinder sustainability and transition towards RC (Joshi & Rahman, 2015; Vermeir & Verbeke, 2006). Removing these barriers requires CE-based tools, including digital tracking apps, eco-labels, and reverse logistics systems, to transform consumer behavior into a sustainable one (Katiyar & Gedam, 2025).

The chapter aims to promote responsible consumption and production, along with strengthening the critical success factors of the circular economy transition. When consumers demand circular products, businesses respond with responsible production systems, creating a feedback loop that reinforces sustainable practices across the entire value chain. Thus, embedding CE principles into consumption behaviors is crucial for achieving a balanced, circular, and sustainable economy.

Responsible Production and Consumption refer to the sustainable management and efficient use of natural resources throughout the lifecycle of goods and services (United Nations, 2015). RPC emphasizes reducing negative environmental and social impacts while maximizing economic value. In practice, RCP encompasses cleaner production technologies, eco-design, sustainable sourcing, ethical labor practices, and promotion of sustainable consumption behavior (Mont & Plepys, 2008; Roshid et al., 2025b).

The origins of SDG 12 can be traced back to 1972, when the Club of Rome conducted a computer-based study to simulate a world with limited natural resources. This model evaluated the implications of economic and population growth on planetary sustainability. It predicted that, without changes to growth patterns, the world would face a sharp and irreversible decline in population and industrial output by the year 2072. The findings, published in *The Limits to Growth (LTG)* report, were later validated with real-world data from 1972 to 2000 (Hughes, 2020). Over time, the concept of responsible consumption and production gained global attention and was supported by major international organizations. The release of the "Our Common Future" report in 1987 by the Brundtland Commission was a critical moment at that time. The report urged narrowing the global poverty and consumption gap between poorer and wealthier populations to control global consumption habits (Dhaigude et al., 2025). Furthermore, the United Nations Conference on Environment and Development (Earth Summit) emphasized the same issue in 1992. The World Summit on Sustainable Development, held in Johannesburg, unveiled 10-year frameworks on actions to track and drive progress in this domain. At its 2012 conference, the UN reinforced the protection of natural resources by adopting sustainable consumption and production as a pathway to achieving the Sustainable Development Goals (Dhaigude et al., 2025).

## 2.3 Behavioral Interventions (Nudge Techniques for Responsible Consumption)

The circular economy contributes to responsible consumption through fostering behavioral change among consumers. *Amsterdam's Repair Café* appears as an example in this context. It pertains to a movement where communities host open workshops to repair broken furniture, electronics, and clothes (Repair Café International Foundation, 2025). These initiatives are likely to reduce virgin consumption, prolong product life, and promote repair literacy, complying with CE principles. Behavioral economics suggests that reducing friction in repair, such as offering free service and social events, enhances participation (Thaler & Sunstein, 2009).

## 2.4 Key Targets of Responsible Consumption and Production

SDG 12 calls for sustainable consumption and production. This sustainability requires efficient use of finite natural resources alongside heightened awareness of environmental well-being. SDG 12 outlines RCP targets (Table 1) to achieve sustainability in business, environmental management, and society by creating the framework of RCP. Achieving these targets will benefit all stakeholders. The targets are planned to be attained by 2030. Table 1 illustrates the targets below.

*Table 1. Key targets of SDG 12 related to responsible consumption and production*

| Targets | Explanations |
|---|---|
| 1. Target 12.1: Implementing the 10-year "Sustainable Consumption and Production Patterns" framework programs | This target urges the adoption of the 10-year "Sustainable Consumption and Production Patterns" framework programs. It promotes sustainability by strengthening public procurement, lifestyles, food systems, and tourism through global cooperation. |
| 2. Target 12.2: Achieving sustainable management and efficient natural resource usage by 2030 | The target advances closed-loop systems to ensure sustainable management and efficient utilization of natural resources by 2030. It discourages the linear take-make-dispose approach in production and consumption. |
| 3. Target 12.3: Minimizing losses and per capita food waste in the total supply chain by 2030 | The target emphasizes reducing food waste across agriculture, production, retail, and supply chains by making consumers and businesses responsible for consumption and production. |
| 4. Target 12.4: Reducing harmful releases and protecting health and the environment through sound chemicals and waste management by 2020 | This target aims to protect humans and the environment from toxic chemicals and waste through a proper chemical and waste management plan. |
| 5. Target 12.5: Reducing global waste substantially through recycling, reuse, reduction, and prevention by 2030 | This target aims to reduce global waste through the circular economic principles by 2030. The target aligns well with the integration of CE with RCP. |
| 6. Target 12.6: Motivating companies to adopt sustainable business practices in operations and reports | The target encourages organizations to adopt sustainable business practices in operations. It also urges the disclosure of environmental data in reports that are related to responsible production. |
| 7. Target 12.7: Fostering sustainable public procurement and national priorities to strengthen green markets | This target urges the government and authorities to produce and supply eco-friendly goods and services across the entire supply chain. |
| 8. Target 12.8: Ensuring global access to information for stakeholders and making them aware of bringing sustainability by 2030 | This target encourages stakeholders to be aware of responsible production and consumption to achieve sustainability by 2030. It further urges the organizations to ensure access to the required information for stakeholders. |
| 9. Target 12. a: Supporting developing nations to excel in science and technology, essential for sustainable production and consumption | This target calls for strengthening scientific capacity in developing nations to bring sustainable development in production and consumption by creating diverse competitive advantages. |

continued on following page

*Table 1. Continued*

| Targets | Explanations |
|---|---|
| 10. Target 12. b: Achieving sustainable development in terms of developing metrics and tools that will monitor tourism practices, create jobs, and foster local culture | This target aims to excel in tourism consumption through sustainable tourism practices among all stakeholders, considering environmental and social well-being. |
| 11. Target 12. c: Reducing waste and supporting environmental sustainability by tracing inefficient fossil fuel subsidies | This target aims to protect developing countries from the adverse impacts of waste and environmental costs produced by developed nations. |

Source: Authors' compilation (Adapted)

SDG 12 targets provide comprehensive guidance for achieving sustainable consumption and production across stakeholder groups (Shafik, 2025). Stitching these targets into business and consumption will foster long-term sustainability in business, consumption, society, and the environment.

## 2.5 Digital Technologies as Enablers of the Circular Economy

The transformation of a company into a circular economy model necessitates the incorporation of digital technologies in design and manufacturing. Digital technologies entail Artificial Intelligence (AI), Robotics, Collaborative (cobots), Digital Twins (DT), Simulation technologies, and Blockchains. These technologies facilitate demand forecasting (Mollik et al., 2024; Hossain & Mollik, 2022), predictive maintenance, and optimization of the product lifecycle. Additionally, technology reduces and recycles tons of waste, contributing to the utilization of finite resources and the protection of the environment. The personnel of the organization become more skilled and competitive in managing waste. Moreover, being adapted to technology enables workers with disabilities to improve the quality of their lives, leading to the achievement of SDG 9. Technology enables industries such as plastics, electronics, healthcare, textiles, and automobiles to trace the provenance of materials, allowing downstream actors to verify recycled content and assess sustainability claims (Ellen MacArthur Foundation, 2021).

## 2.6 Circular Economy in the Global South (Role of Informal Sector in Circular Systems)

The informal sector in many developing countries plays a critical role in resource recovery. Brazil's *catadores,* informal waste pickers, collect profuse recyclables and feed them into industrial loops. However, their contributions often lack deserving recognition, fair pay, and legal protection. Properly integrating them into formal CE strategies (as done through cooperatives like Movimento Nacional dos Catadores de Materiais Recicláveis) remains essential for inclusive and circular transitions (Dias, 2016; WIEGO, 2025).

Additionally, India forms the backbone of urban recycling systems by hosting thousands of waste picker cooperatives. SWaCH cooperative in Pune collaborates with municipal authorities to collect waste from over 850,000 households. Thus, the organization practices CE significantly. Formalizing informal labor and ensuring satisfactory wages ultimately lead to social equity in the Global South (Kirchherr et al., 2018; WIEGO, 2025).

## 2.7 Circular Economy in Emerging Economies

### Case: Philippines' Sachet Economy Upcycling Innovations

The Philippines encounters a mounting plastic crisis, driven by its "sachet economy." They mostly rely on small, single-use packages of shampoo, coffee, and other convenience goods for affordability. This practice yields over 160 million plastic sachets daily (GAIA, 2022), overwhelming municipal waste systems.

Grassroots enterprises and social ventures have pioneered upcycling solutions to address plastic pollution. A Manila-based startup, *Plastic Flamingo*, collects plastic sachets and transforms them into eco lumber for construction. Likewise, *"The Sustainable Sagada Project"* empowers local women by providing them with an opportunity to weave discarded sachets into handbags and homeware. In addition to reducing waste, these initiatives provide livelihood opportunities.

These upcycling models urge circular solutions in emerging economies to be localized, low-cost, and socially inclusive. The Philippines' example presents the potential for community-led CE initiatives to address both socio-economic development and environmental degradation.

## 2.8 Policy Enforcement – Fitness versus Incentives (European Union's EPR, versus Rwanda Plastic Ban)

Different countries adopt different regulatory strategies to enact circularity. The Extended Producer Responsibility (EPR) of the EU urges financing the end-of-life collection and processing of products. Further, it incentivizes eco-design. Rwanda, in contrast, implemented one of the world's strictest plastic bans, backed by fines and imprisonment. Ultimately, both approaches demonstrate that regulation can accelerate the adoption of CE, depending on cultural and political contexts.

As shown in Figure 1, the global demand for virgin materials continues to climb.

*Figure 1. Projected virgin material demand vs. planetary boundaries (2025-2040)*

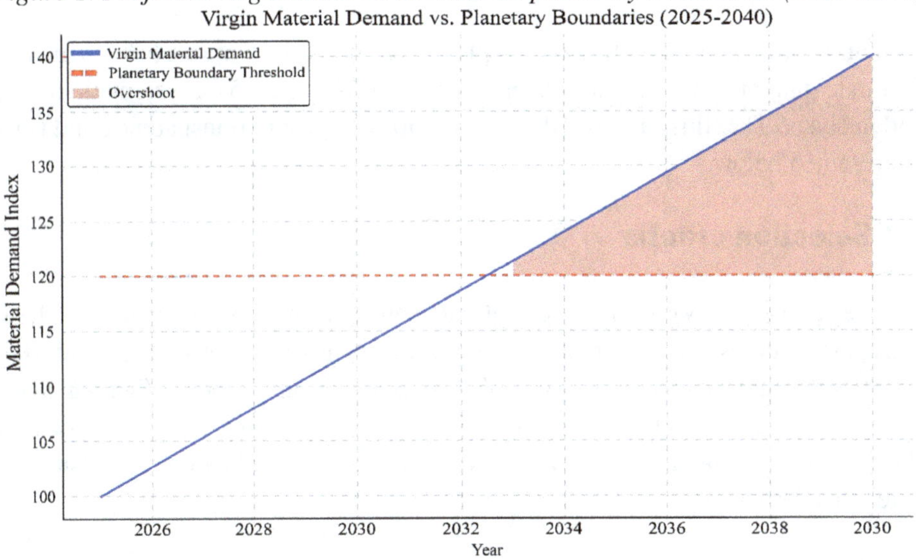


*Source: Authors' compilation based on the Circular Economy Gap Report 2024*

The figure reports that virgin material demand is expected to mount linearly from an index value of 100 in 2025 to 140 by 2040, driven by unsustainable consumption and production patterns. The Circle Economy (2024) depicts that material extraction could rise by 60% without a significant transition to CE models. Consequently, irreversible environmental degradation will occur. The planetary boundary threshold in the graph shares a critical insight, set at a material demand index of 120. The red dashed horizontal line represents this threshold. This boundary typically marks the ecological limits beyond which material extraction and usage become unsustainable, signaling serious stakes to biodiversity, climate stability, and ecosystem resilience. Further, the shaded red area labeled "Overshoot" fully visualizes the ecological transgression projected to happen from 2033 (approximately) onward. Furthermore, the gap between actual demand and sustainable threshold broadens significantly between 2035 and 2040. This picture indicates that environmental degradation will accelerate if the existing linear economic model persists. These trends signify an urgent global call for systematic transitions toward responsible consumption and production, integrating "CE principles."

## 3. METHODOLOGY: DATA COLLECTION PROCESS

This chapter analyzes literature employing a qualitative approach. This method facilitates mapping themes and concepts, using ATLAS.ti 2025 for systematic coding and analysis. The illustration of the methodology to ensure transparency and rigor is presented below.

### 3.1 Selection criteria

The chapter analyzes three types of data sources. They include academic literature, case studies, strategic reports, and policy documents related to CE and RCP. Databases searched entail Scopus, Web of Science, and Google Scholar using keywords such as "Circular Economy," "Responsible Consumption," "Responsible Production," "Sustainable Consumption," "Sustainable Production," and "sustainable manufacturing."

### 3.2 Coding process and theme derivation

Data from the relevant literature of CE and RCP were imported into ATLAS. ti 2025 to identify initial codes reflecting CE and RCP concepts. Next, codes were grouped into axial codes (e.g., circular economy enablers, policy implications, barriers, and opportunities). Finally, selective coding identified higher-level themes from the literature that structured the narrative of this chapter, including:

- The synergy of CE and RCP.
- Responsible consumption and production as the enablers of circularity.
- Policy frameworks supporting sustainable and effective transformation.
- Challenges and strategic opportunities of CE-RCP integration.

### 3.3 Validation techniques

To ensure the chapter's robustness:

i. A second author independently coded a portion (~50%) of the data to evaluate inter-coder reliability. Cohen's kappa was calculated and found to be acceptable (>0.75), indicating strong agreement.
ii. Regular and critical team discussions resolved discrepancies and refined code captions to enhance reliability.
iii. Data triangulation from academic, industry, and policy sources further increased validity.

# 4. BARRIERS, ENABLERS, AND CONSEQUENCES OF INTEGRATING CE WITH RCP

## 4.1 Barriers and Challenges

The integration of CE with RCP clings to several barriers as follows:

i. Institutional barriers: Complex regulatory frameworks, weak governance, and vulnerable enforcement impede effective integration of CE with the proposed sustainable domain (Milios, 2018).
ii. Organizational barriers: When organizations lack the required capabilities, financial resources, strategic vision, or adaptive culture, they cannot integrate CE practices with sustainable consumption and production (Ritzén & Sandström, 2017).
iii. Behavioral constraints: Consumer-side circularity gets hampered by the lack of awareness, resistance to change, and habitual consumption patterns among consumers (De Chiara et al., 2024).
iv. Technological constraints: Digital technologies foster the transition of CE with RCP. However, the adoption of technologies becomes costly and complex for small and medium-sized enterprises.

## 4.2 CE implementation heatmap (Sectoral barrier severity index)

Small and medium-sized enterprises face several additional barriers to adopting CE practices. The circular economy implementation heatmap (Figure 2) illustrates specific barriers faced by small and medium-sized enterprises (SMEs) across five major industries, including food, textiles, electronics, construction, and automotive, in their adoption of circular economy practices."

*Figure 2. CE implementation heatmap (Sectoral barrier severity index)*

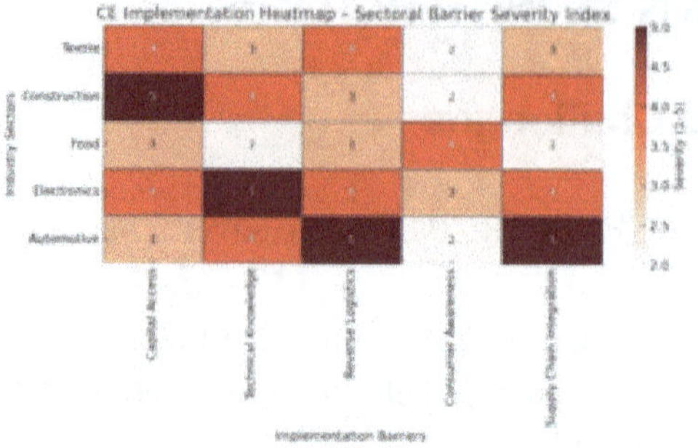


*Source: Authors' compilation*

Each barrier has been rated on a scale from 1 (low severity) to 5 (high severity). The heatmap highlights a comparative barrier severity index across five critical dimensions, including capital access, reverse logistics, technical knowledge, consumer awareness, and supply chain integration, based on insights from the Ellen MacArthur Foundation (2023) and the Circle Economy (2024). Capital access and technical knowledge rank as two major barriers to construction and electronics. Reverse logistics meet critical challenges for the automotive and textile industries. Consumer awareness poses higher challenges in the food sector, where changing their behavior remains more critical. Furthermore, supply chain integration becomes a widespread issue for the construction and automotive industries. The findings support experts as follows.

Professor Adewale Olufemi, renowned environmental economist at the University of Lagos, emphasized that *"the lack of access to affordable reverse logistics systems is the biggest pain point for African SMEs. Eventually, circularity often ends at the recycling stage due to poor infrastructure and weak municipal partnerships."*

Dr. Nisha Ravichandran, Sustainability Advisor, TERI (The Energy and Resource Institute), commented that *"The uncertainty in the policy usually kills momentum. Circular Economy startups working with textile recycling often face bottlenecks in certification and export due to outdated linear policies in India. Thus, Circular design essentially needs legislative clarity, not just innovation."*

## 4.3 Enablers or Success Factors and Consequences of CE-RCP integration

There exist some critical success factors that facilitate the integration between CE and RCP. The factors facilitating the integration of CE-RPC are as follows:

i. Leadership and Vision: Strong top-level commitment is essential for driving sustainability strategies (Lüdeke-Freund et al., 2018).
ii. Collaborative Networks: Partnerships across industries, academia, and government foster knowledge exchange and innovation (Etzkowitz & Zhou, 2017)
iii. Consumer Engagement: Transparency, trust, and participation are key to encouraging circular consumer behavior (Peattie & Peattie, 2009).
iv. Education and Awareness: Embedding CE concepts into curricula and professional training builds capacity.

Drawing from an extensive body of literature, this study synthesizes the theoretical and empirical insights surrounding responsible production and responsible consumption to develop comprehensive conceptual models (see Fig. 3 and Fig. 4). Numerous scholarly contributions emphasize the growing importance of aligning production and consumption with sustainability principles to meet global environmental, social, and economic challenges. By analyzing peer-reviewed studies, policy frameworks (such as the UN SDG 12), and industry reports, the antecedents, attributes, and consequences of responsible practices were systematically identified. Key drivers such as regulatory pressure, organizational leadership, consumer awareness, and technological advancement repeatedly emerged as central elements. These insights laid the foundation for constructing two integrative models—one for responsible production and the other for responsible consumption—each illustrating a structured relationship between influencing factors, core attributes, and resulting outcomes.

Additionally, the comparison of circular economy policy levers presents how the policy levers critically impact SMEs (see Table 2). The insights from the comparison illustrate that a balanced policy mix of incentives and penalties encourages and facilitates the CE transition. Subsidies and green procurement, as incentive mechanisms, promote initial investment and encourage enduring, sustainable demand for circular economy adoption.

Several penalties like landfill tax, EPR, and raw material levy drive CE transition by increasing linear disposal costs, shifting responsibility to producers, and discouraging virgin resource use, respectively.

*Table 2. Comparison of CE Policy Levers*

| Policy Levers | Type | Impact on SMEs |
|---|---|---|
| Tax Breaks | Incentive | Reduces the cost of transition |
| Subsidies for CE | Incentive | Supports initial investments |
| Green Public Procurement | Incentive | Creates stable CE demand |
| Landfill Tax | Penalty | Increases linear disposal costs |

continued on following page

*Table 2. Continued*

| Policy Levers | Type | Impact on SMEs |
|---|---|---|
| Extended Producer Responsibility | Penalty | Shift responsibility to producers |
| Raw Material Levy | Penalty | Discourages virgin resource use |

Source: Authors' compilation

*Figure 3. Antecedents, attributes, and consequences of responsible production*

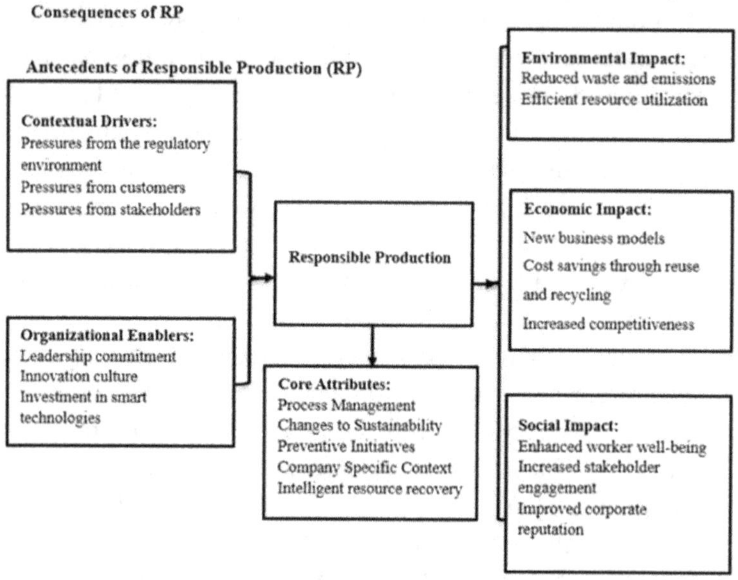


*Source: Authors' compilation*

The models of responsible production and responsible consumption presented in this study offer a holistic framework that maps the interconnected variables influencing sustainable practices across both supply and demand perspectives. The "responsible production model" illustrates how contextual pressures and internal firm characteristics drive sustainable production practices, which in turn lead to favorable outcomes at the customer, firm, and employee levels.

*Figure 4. Antecedents, attributes, and consequences of responsible consumption*

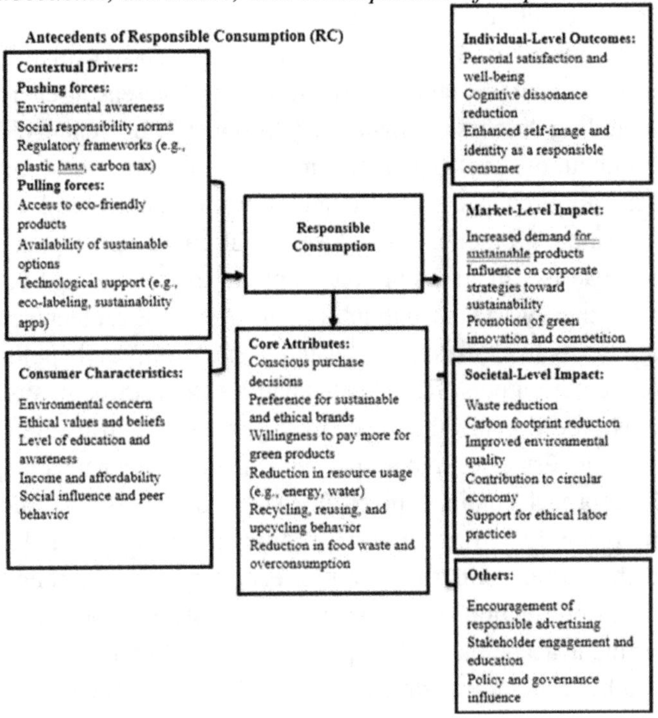


*Source: Authors' compilation*

In the meantime, the RC model shapes consumer behavior and values in a responsible way. It generates positive impacts on individual consumers, markets, society, and the environment. This model strengthens the theoretical understanding of sustainable production and consumption. Additionally, it provides insights into practical implications for scholars, businesses, and policymakers who seek to promote a circular and responsible economic atmosphere.

## 4.4 Concept Analysis

The integration of CE with RCP converts the linear approach into a closed-loop system where resources, materials, and products are recycled, reused, and regenerated. The concept figure (Fig. 5) encompasses interrelated keywords that span economic, operational, and environmental domains. The word "sustainability" in the figure denotes the goal of CE and RCP, which urges to balance economic development, social equity, and environmental protection. Additional terms, including "consumption," "production," "resource," "waste," and "management," remain significant to understand how C functions in practice.

CE urges production based on durability, modularity, and recyclability. It necessitates reducing waste, utilizing renewable materials, and increasing energy efficiency. The words, including "process," "material," "product," and "energy," prioritize an eco-friendly and innovative manufacturing. Responsible consumption signifies awareness, behavioral changes, and the adoption of responsible lifestyles. Terms like "impact," "value," "practice," and "use" highlight responsible and sustainable decision-making across all stages of consumption, from sourcing to disposal.

CE tends to reduce waste through its strategies, such as remanufacturing, reuse, and recycling. Words like "material," "water," "emission," and "pollution" encapsulate environmental outputs. CE tries to manage them through sustainable production systems. Additionally, the word "management" signifies strategic planning and coordinated governance. Integrating CE with RCP demands a system-based approach. The approach entails policy frameworks, supply chains, and stakeholder collaboration. Terms like "firm, industry, company, system, and chain" echo the approach. Furthermore, scaling the circular economy model through responsible consumption and production (RCP) necessitates collaboration across sectors, involving government, academia, and the private sector. Although integrating CE with RCP requires a high initial investment, it offers long-term financial gains, creates new markets, and promotes innovation. Terms like "cost, economy, market, and value" highlight this issue.

*Figure 5. Concept analysis of CE-RCP integration*

Figure 4. Concept analysis of CE integration for RPC (Source: Authors' creation)

*Source: Authors' compilation*

The concept figure concludes that integrating CE into RCP requires a multi-dimensional effort. It signifies technological innovation, adapted consumer behavior, business models, and a supportive regulatory environment. These must be economically viable, socially responsible, and environmentally sound.

## 4.5 Sectoral Case Studies

Table 3 presents the business models and use cases of five companies leveraging emerging technologies to create value aligned with Sustainable Development Goal 12 (SDG 12). It's important to note that a single use case may contribute to multiple SDGs and incorporate various emerging technologies. The left-hand column displays the company name, country of origin, relevant SDGs, and the emerging technologies applied. The companies and use cases are organized in alphabetical order.

*Table 3. Companies and cases*

| Company Information | Value Proposal | Value Creation | Value Capture |
|---|---|---|---|
| 1. AMP Robotics, USA, SDG 12 (AMP Robotics, 2024) | Recycling materials using robotic systems equipped with computer vision and deep learning technologies. | They employ a robotic system to automate the detection and separation of recyclable materials from mixed waste streams. The AI-powered platform is self-learning, continually improving its ability to recognize various characteristics such as color, texture, shape, size, pattern, and even brand labels to determine material types and their recyclability. This enables the robotic system to sort materials with high precision- down to specific types of plastic- at a rate exceeding 80 items per minute. | Revenue is generated by enhancing the value of recyclable materials through improved separation, increased purity, and the discovery of new end markets for recycling and reuse. |
| 2. Biopipe, Turkey, SDG 6, 12 Recycling (Küfeoğlu, 2022) | The company created a patented and proprietary wastewater treatment system that operates entirely within the pipeline. | Biopipe is an innovative biological wastewater treatment technology where the complete treatment process takes place inside a network of pipes. Specialized microorganisms within the system break down and eliminate impurities in the wastewater. The design leverages pressure differentials to naturally draw in air, fostering an optimal environment for bacterial growth and accelerating aerobic digestion. | Households, hotels, residential complexes, and government entities benefit financially from the wastewater treatment system. Thanks to its compact design and high adaptability, Biopipe can be installed at any location where wastewater is generated, regardless of the scale or capacity. |
| 3. Effa, Ukraine, SDG 3, 12 Bioplastics, biotechnology and biomanufacturing, and recycling (Yaremova et al., 2024) | The company aims to minimize plastic use in toothbrush production and ensure that its products are fully recyclable | The toothbrush handle is made from paper, while the head is composed of cornstarch and Nylon11. Designed for easy separation, the head and handle can be disposed of in appropriate recycling bins. The bristles are integrated with the head as a single unit and remain securely attached, ensuring they do not detach from the product. | The company earns revenue from its toothbrush product, which lowers CO2 emissions by 60% and cuts plastic usage by 70%. This product helps establish an eco-friendly user profile. |

continued on following page

*Table 3. Continued*

| Company Information | Value Proposal | Value Creation | Value Capture |
|---|---|---|---|
| 4. Lignin, Sweden, SDG 12, advanced materials, bioplastics, biotech & biomanufacturing, and recycling (Hassegawa et al., 2022) | The company is a biotechnology firm that transforms lignin to make it compatible with both virgin and recycled thermoplastics, enabling the creation of innovative bioplastics. | This method reduces the boiling point of the lignin polymer without requiring extra pressure, making the process highly energy-efficient and free from toxic emissions or hazardous by-products. Within minutes, one ton of lignin can be converted into one ton of RENOL bio-granulates, which can then be blended into various types of bioplastics. | The company reduces environmental harm by preventing lignin from being burned for energy. Instead, it generates revenue by improving the production of biodegradable bioplastics. |

Source: Authors' compilation

Therefore, the circular economy finds RCP significant for the present situation in the business world.

## 4.6 Critical Reflections on CE: Exploring Degrowth-CE tensions and synergistic possibilities

While the circular economy remains highlighted as compatible with green growth, degrowth scholars challenge this notion. They argue that CE, when pursued within a growth-centric framework, might replicate extractive and unequal systems (Cosme et al., 2017). Though circular strategies often signify efficiency, they overlook sufficiency, failing to address overconsumption and material rebound effects. Therefore, Korhonen et al. (2018) argued that CE alone cannot avert ecological overshoot unless absolute decoupling takes place. This critique suggests that CE practices must be installed with broader socio-economic transitions that prioritize sufficiency, not merely efficiency.

# 5. DISCUSSION ON CE INTEGRATION FOR RPC

The integration of Circular Economy (CE) principles into Responsible Production and Consumption (RPC) systems represents a paradigm shift in how societies manage resources, design products, and engage with consumption. This chapter examines the theoretical foundations, empirical insights, and conceptual models that support CE-RPC integration, with the overarching goal of promoting Sustainable Development Goal 12 (SDG 12). In this section, authors critically reflect on the implications of the proposed framework, examine the interactions between core variables, and discuss the practical relevance for different stakeholder groups.

## 5.1 Interlinkages between CE and RCP

CE and RCP appear to be an effective and productive combination. The combination favors the stakeholders from diverse perspectives. CE strategies, such as eco-design, industrial symbiosis, and circular logistics, minimize waste and maximize resource efficiency significantly (McDonough & Braungart, 2010; Kanda et al., 2021). These practices remain economically viable and environmentally beneficial due to integrating with CE that fosters innovation and long-term value creation (Lüdeke-Freund et al., 2018)

Moreover, CE and RC encourage consumers' behavioral transition toward reuse, recycling, repair, remanufacturing, and shared use, unlike the linear approach (Mugge et al., 2017). It requires trust, social acceptance, and behavioral transition of consumers toward the transition of CE with RCP (Vermeir & Verbeke, 2006). Additionally, consumer access to sustainable alternatives drives the transition. Thus, the integration between CE and RCP serves as a bidirectional system. Herein, responsible production influences responsible consumption, offering sustainable product design and availability. Besides, consumer demand for circular and sustainable products, along with their feedback, affects organizations to provide sustainable offerings.

### CE-RCP Maturity Assessment Tool: Readiness, Practice, and Circularity

A five-dimensional scoreboard (0-5 scale each) (Table 4) assessing an organization's readiness and maturity is as follows:

R: Resource Optimization (Use, Reuse, Waste Minimization)
P: Process Circularity (remanufacturing, reverse, logistics)
C: Cultural Readiness (leadership, training, stakeholder buy-in)
Governance Integration (certification, policy alignment)
Innovation Index (data tools, design for CE)

*Table 4. CE-RCP Maturity Assessment scoreboard- Evaluating readiness, practice, and circularity dimensions*

| Dimension | Score (0-5) | Description |
|---|---|---|
| Resource Optimization | 3 | Recycling present, limited reuse strategy |
| Process Circularity | 2 | Reverse logistics underdeveloped |
| Cultural Readiness | 4 | Employees trained and equipped, CE mission in place |
| Governance Integration | 2 | No CE KPIs in strategic policy |
| Innovation Index | 1 | Minimal digital CE tools applied |

Source: Authors' compilation

This CE-RCP scoreboard enables SMEs and corporations by assessing their positions and prioritizing interventions in the CE journey. It replaces generic models with practical tools designed for economies with finite resources.

## 5.2 Systemic Enablers: Role of Institutions, Culture, and Technology

While CE and RPC align theoretically, systemic enablers are necessary to operationalize this alignment. Regulatory and institutional frameworks, including Extended Producer Responsibility (EPR) and green procurement policies, play a critical role in incentivizing sustainable actions by firms and consumers (Milios, 2018; OECD, 2022). However, implementation remains uneven, especially in developing economies where regulatory enforcement and infrastructure are limited.

Supportive culture and digital technologies remain significant for attaining sustainability (Roshid et al., 2025a; White et al., 2019). Therefore, emerging technologies such as Artificial Intelligence (AI) and Digital Twins (DT) can be leveraged to support sustainability integration. Digital technologies facilitate organizational shifts toward circular economy and responsible production models, improving efficiency and quality. Consequently, the alignment of interests among institutions, consumers, regulatory bodies, and other stakeholders is essential for advancing sustainability via circular approaches. Digital technologies can act as a moderating element in this transition (Milios, 2018).

## 5.3 Thematic Analysis

Figure 6 unravels the thematic analysis of CE-RCP integration. The figure presents a coherent framework that connects the core components of sustainable economic development. The CE integration, central to the figure, functions as the core driver for ensuring social and environmental sustainability. It integrates interconnected components of CE and RCP. These practices encourage the effective use of resources, the adoption of renewable inputs, and the reduction of waste. Responsible consumption enforces responsible production. Consequently, responsible production remains attached to CE principles.

Businesses and society can benefit from the integration of CE-RCP. Businesses can innovate new and cost-sensitive products, enhance resource efficiency, and promote long-term value generation. Society receives positive outcomes of fostering inclusivity and environmental sustainability. Additionally, integration empowers consumers with wider information. Consumers become more informed and active with circular and responsible practices. Empowered consumers demand eco-friendly and ethical products, leading to strengthening the CE and RCP model.

*Figure 6. Thematic analysis of integrating CE with SDG 12*

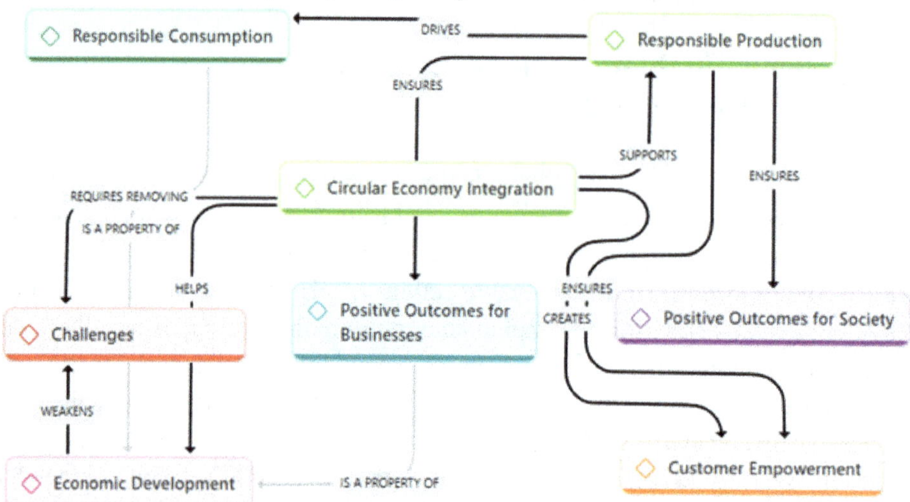


Figure 9: Thematic analysis of integrating CE with SDG 12
Source: Authors' compilation

*Source: Authors' compilation*

However, the integration of CE-RCP has some limitations. There exist some challenges that must be addressed with critical consideration. Otherwise, they would weaken the integration from different perspectives. Overall, the figure highlights the potential of CE-RCP integration, considering wider benefits.

## 6. IMPLICATIONS FOR STAKEHOLDERS

The conceptual model proposed in this chapter carries significant implications for multiple stakeholder groups, as outlined below:

- Policymakers should design integrative and adaptive regulatory frameworks that support CE-RPC alignment across sectors. This includes harmonizing waste management laws, incentivizing circular business practices, and investing in circular infrastructure.
- Businesses and Industry Leaders are encouraged to adopt systems thinking and life cycle perspectives in product and service design. Firms can gain competitive advantages by innovating around circular business models and responding to shifting consumer expectations.
- Consumers and Civil Society play a vital role in reinforcing the demand side of circularity. Advocacy, responsible purchasing, and participation in community-based sharing initiatives are key actions that align individual behavior with systemic sustainability goals.
- Academics and Researchers can contribute by developing metrics, conducting longitudinal studies, and exploring cross-cultural applications of CE-RPC integration. Interdisciplinary research is particularly valuable in unpacking the complexity of circular transitions.

## 7. THEORETICAL AND PRACTICAL CONTRIBUTIONS

From a theoretical standpoint, this chapter advances sustainability research by bridging two parallel streams: CE and RPC. This is achieved through the application of multi-level theories, including Institutional Theory (Zucker, 1987), the Resource-Based View (Barney, 1991), and Paradox Theory (Smith & Lewis, 2011). These frameworks provide lenses to understand organizational behavior, strategic capability development, and tensions in sustainability transitions.

The conceptual framework offers a holistic model to guide empirical studies and policy interventions. It integrates production-side and consumption-side variables with enablers and outcomes, making it adaptable for future quantitative or case-based validation.

Practically, the chapter provides actionable guidance for key stakeholder groups:

- Governments can use the model to design comprehensive circular economy roadmaps.
- Firms can align CE strategies with sustainability performance metrics.

- Consumers and NGOs can use the findings to advocate for better product transparency, access, and participation in circular systems.

## 8. LIMITATIONS AND OPPORTUNITIES

Despite the strengths of the model, certain limitations must be acknowledged. First, the model is conceptual and has not yet been empirically tested. Future research could apply the framework to specific sectors or regions to validate its components and interrelationships. Second, the model does not explicitly address informal economies, which are critical in many developing countries for activities such as recycling and repair. Incorporating informal actors could enhance the model's inclusivity and real-world relevance.

Moreover, the dynamic nature of CE-RPC integration suggests the need for adaptive governance frameworks. As technology evolves and social values shift, institutions must remain flexible and responsive. For example, the rise of AI-generated product designs or decentralized digital marketplaces could transform existing circular systems in ways that are not yet fully understood.

Finally, there is a need for cross-SDG alignment. While this model focuses on SDG 12, circular practices also contribute to SDG 13 (Climate Action), SDG 9 (Industry, Innovation, and Infrastructure), and SDG 11 (Sustainable Cities and Communities). Future frameworks should incorporate these interlinkages to reflect the holistic nature of sustainable development. While the chapter contributes significantly to the discourse on CE-RPC integration, it is not without limitations. First, the conceptual model has not been empirically tested. Its application to real-world settings—such as different industries, geographic regions, or institutional environments—requires further research.

Second, the model may be limited in scope by excluding informal economies, especially relevant in Global South contexts where recycling, repair, and waste collection are often managed informally. Integrating these actors into the formal CE system remains a major gap.

Third, the chapter provides a broad framework rather than sector-specific solutions. While some case examples were used (e.g., textiles, electronics, food), more in-depth sectoral studies are required to operationalize CE-RPC linkages effectively.

## 9. CONCLUSION AND FUTURE RESEARCH DIRECTIONS

The global transition to sustainability demands profound changes in how we produce, consume, and manage resources. The integration of Circular Economy

(CE) principles into Responsible Production and Consumption (RPC) systems offers a promising path toward meeting these demands, especially in alignment with Sustainable Development Goal 12 (SDG 12). This chapter presents a comprehensive analysis of CE-RPC integration, grounded in theory, supported by literature, and structured through a conceptual model. The goal has been to provide both academic and practical insights into how circular strategies can transform linear consumption and production patterns into sustainable, regenerative systems.

To build upon the foundation established in this chapter, several promising avenues for future research are proposed:

i. Empirical Validation of the Model

Future studies should apply the conceptual framework in specific contexts to test its relevance and robustness. Sector-specific or country-level analyses can provide more grounded insights.

ii. Longitudinal Studies

The CE-RPC transition is an evolving process. Longitudinal research can capture how CE adoption influences production and consumption patterns over time and how systemic enablers evolve in response.

iii. Consumer Behavior Research

There is a pressing need to explore behavioral interventions that bridge the attitude–behavior gap. Future research could investigate the effectiveness of nudging, gamification, and digital transparency tools in promoting responsible consumption.

iv. Circular Metrics and Measurement

Developing and standardizing metrics for CE-RPC outcomes is critical. Studies can explore how existing tools like the Circularity Gap Report, CTI (Circular Transition Indicators), or Material Circularity Indicator (MCI) can be improved and integrated into corporate and national reporting systems.

v. Cross-SDG Synergies

Future frameworks should explore the co-benefits of CE-RPC integration for other SDGs, such as SDG 9 (Industry & Innovation), SDG 11 (Sustainable Cities), and SDG 13 (Climate Action). A systems-thinking approach is essential for understanding trade-offs and synergies.

vi. Digitalization and CE Integration

With the rise of Industry 4.0 technologies, there is growing potential to leverage data analytics, IoT, and blockchain for enhancing CE efficiency. Further exploration of these technologies in real-world RCP systems is warranted.

## REFERENCES


Balakrishnan, A. S., Ramanathan, R., Ramanathan, U., & Srinivasan, R. R. (2025). Responsible battery disposal management practices: An investigation in the electric vehicle industry from a circular economy perspective. *Journal of Environmental Management, 383*, 125488. DOI: 10.1016/j.jenvman.2025.125488 PMID: 40273794

Barney, J. (1991). Firm resources and sustained competitive advantage. *Journal of Management, 17*(1), 99–120. DOI: 10.1177/014920639101700108

Bourdin, S., & Jacquet, N. (2025). Closing the loop at the local scale: Investigating the drivers of and barriers to the implementation of the circular economy in cities and regions. *Ecological Economics, 231*, 108542. DOI: 10.1016/j.ecolecon.2025.108542

Cerchione, R. (2025). Design and evaluation of a blockchain-based system for increasing customer engagement in circular economy. *Corporate Social Responsibility and Environmental Management*. Advance online publication. DOI: 10.1002/csr.2948

Circle Economy. (2024). The Circularity Gap Report 2024. http://www.circularity-gap.world

Cosme, I., Santos, R., & O'Neill, D. W. (2017). Assessing the degrowth discourse: A review and analysis of academic degrowth policy proposals. *Journal of Cleaner Production, 149*, 321–334.

De Chiara, A., Gallo, M., & Simonacci, V. (2024). Re-examining consumer engagement in the circular economy. *Journal of Consumer Marketing*. Advance online publication. DOI: 10.1108/jcm-03-2023-5918

De Jesus, A., & Mendonça, S. (2018). Lost in transition? Drivers and barriers in the Eco-Innovation Road to the Circular economy. SSRN *Electronic Journal*. https://doi.org/DOI: 10.2139/ssrn.3038887

De Martino, M., Apicerni, V., & Gravagnuolo, A. (2024). Sustainable hospitality and tourism in the Anthropocene era: The need for a more radical shift of the current circular economy models. *International Journal of Contemporary Hospitality Management*. Advance online publication. DOI: 10.1108/ijchm-06-2023-0854

Dhaigude, A. S., Verma, A., & Nayak, G. (2025). Sustainable production and consumption: A bibliometric analysis of SDG-12 literature through a financial management lens. *Cogent Economics & Finance, 13*(1). Advance online publication. DOI: 10.1080/23322039.2025.2467882

Dias, S. M. (2016). Waste pickers and cities. *Environment and Urbanization, 28*(2), 375–390. DOI: 10.1177/0956247816657302

Ellen MacArthur Foundation. (2021). Link (https://www.ellenmacarthurfoundation.org/)

Ellen MacArthur Foundation. (2021). *Case Study: Renault remanufacturing*. https://ellenmacarthurfoundation.org/circular-examples/renault

Ellen MacArthur Foundation. (2023). *Circular economy in practice: Industry transitions.* https://ellenmacarthurfoundation.org

Espuny, M., Da Mota Reis, J. S., Giupponi, E. C. B., Rocha, A. B. T., Costa, A. C. F., Poltronieri, C. F., & De Oliveira, O. J. (2025). The role of the Triple Helix Model in Promoting the circular Economy: Government-Led Integration Strategies and Practical application. *Recycling, 10*(2), 50. DOI: 10.3390/recycling10020050

Etzkowitz, H., & Zhou, C. (2017). *The triple helix: University–Industry–Government innovation and Entrepreneurship.*

Gautam, P., Salunke, D., Lad, D., & Gautam, A. (2025). Convergent synergy of carbon capture within the circular economy paradigm: A nexus for realizing multifaceted sustainable development goals. *Carbon Research, 4*(1). Advance online publication. DOI: 10.1007/s44246-024-00178-1

Geissdoerfer, M., Savaget, P., Bocken, N. M., & Hultink, E. J. (2017). The Circular Economy – A new sustainability paradigm? *Journal of Cleaner Production, 143*, 757–768. DOI: 10.1016/j.jclepro.2016.12.048

Global Alliance for Incinerator Alternatives (GAIA). (2022). *Sachet Economy in the Philippines: Environmental and Social Impacts*. https://www.no-burn.org/sachet-economy-philippines/

Hassegawa, M., Karlberg, A., Hertzberg, M., & Verkerk, P. J. (2022). Innovative forest products in the circular bioeconomy. *Open Research Europe, 2*, 19. PMID: 37645287

Hossain, M. K., & Mollik, M. S. (2022). Exploring the impact of Social Media Platforms on Instigating Tourist's Emotions and Behavioral Intentions to Co-Create and Visit Tourist Spots-Bangladesh Perspective. https://doi.org/DOI: 10.58753/jbspust.3.1.2022.13

Hughes, C. (2020). *SDG 12: responsible consumption and production. Jean Monnet Sustainable Development Goals Network Policy Brief Series*. RMIT University Melbourne.

Islam, M. Z., & Zheng, L. (2025). Why is it necessary to integrate circular economy practices for agri-food sustainability from a global perspective? *Sustainable Development*. Advance online publication. DOI: 10.1002/sd.3135

Ivanov, S., Seyitoğlu, F., & Webster, C. (2025). Tourism, automation and responsible consumption and production: A horizon 2050 paper. *Tourism Review*. Advance online publication. DOI: 10.1108/tr-12-2023-0898

Jain, V. K., Kumari, A., Singh, S., & Tyagi, V. (2025). Electric vehicles as a pathway to decarbonisation in India: Assessing the environmental benefits and challenges of EV adoption for Sustainable Tomorrow. *Circular Economy and Sustainability*. Advance online publication. DOI: 10.1007/s43615-025-00619-y

Joshi, Y., & Rahman, Z. (2015). Factors affecting green purchase behaviour and future research directions. *International Strategic Management Review*, *3*(1–2), 128–143. DOI: 10.1016/j.ism.2015.04.001

Kanda, W., Geissdoerfer, M., & Hjelm, O. (2021). From circular business models to circular business ecosystems. *Business Strategy and the Environment*, *30*(6), 2814–2829.

Kandasamy, J. KEK, V., Ethirajan, M., & Murali, N. (2025). Role of Industry 4.0 in the Future Circular Supply Chain Management. In *Circular Supply Chain Management: A Sustainability Approach* (pp. 143-178). Cham: Springer Nature Switzerland.

Karim, R., Roshid, M., & Waaje, A. (2024). Circular economy and supply chain sustainability. In *Strategic Innovations for Dynamic Supply Chains* (pp. 1–30). IGI Global., DOI: 10.4018/979-8-3693-3575-8.ch001

Karim, R., Waaje, A., & Roshid, M. M. (2025). Digital Circular Economy for Attaining Sustainable Development Goals: Technologies for Global Sustainability. In *Effects of Digitalization and Circular Economy on Sustainable Policy and Climate Change Prevention* (pp. 211-240). IGI Global Scientific Publishing. https://doi.org/ DOI: 10.4018/979-8-3693-9909-5.ch008

Karim, R., Waaje, A., Roshid, M. M., & Yeamin, M. B. (2025b). Turning the waste into wealth: Progressing toward global sustainability through the circular economy in waste management. In *Sustainable waste management in the tourism and hospitality sectors* (pp. 507–552). IGI Global Scientific Publishing., DOI: 10.4018/979-8-3693-6110-8.ch019

Katiyar, A., & Gedam, V. V. (2025). The circular economy and fertilizer industry: A systematic review of principal measuring tool. *Environment, Development and Sustainability*, *27*(6), 12437–12475.

Kirchherr, J., Piscicelli, L., Bour, R., Kostense-Smit, E., Muller, J., Huibrechtse-Truijens, A., & Hekkert, M. (2018). Barriers to the circular Economy: Evidence from the European Union (EU). *Ecological Economics*, *150*, 264–272. DOI: 10.1016/j.ecolecon.2018.04.028

Korhonen, J., Honkasalo, A., & Seppälä, J. (2018). Circular Economy: The Concept and its Limitations. *Ecological Economics*, *143*, 37–46. DOI: 10.1016/j.ecolecon.2017.06.041

Küfeoğlu, S. (2022). SDG-12: Responsible consumption and production. In *Emerging technologies: value creation for sustainable development* (pp. 409–428). Springer International Publishing.

Kushwaha, D., Talib, F., & Asjad, M. (2025). The intellectual structure of research on environmental sustainability 4.0 in manufacturing: A bibliometric analysis and future research directions. *Environment, Development and Sustainability*. Advance online publication. DOI: 10.1007/s10668-025-06495-8

Lüdeke-Freund, F., Carroux, S., Joyce, A., Massa, L., & Breuer, H. (2018). The sustainable business model pattern taxonomy—45 patterns to support sustainability-oriented business model innovation. *Sustainable Production and Consumption*, *15*, 145–162. DOI: 10.1016/j.spc.2018.06.004

Madaan, G., Singh, A., Mittal, A., & Shahare, P. (2024). Reduce, reuse, recycle: Circular economic principles, sustainability and entrepreneurship in developing ecosystems. *Journal of Small Business and Enterprise Development*, *31*(6), 1041–1066. DOI: 10.1108/jsbed-01-2023-0009

Mahdikhani, M., & Karbasi, A. (2025). Integration of TQM and Circular Economy practices in supply chain: Insights from industry perspective. *Total Quality Management & Business Excellence*, •••, 1–18. DOI: 10.1080/14783363.2025.2477168

Maldonado-Romo, J., Ahmad, R., Ponce, P., Mendez, J. I., Mata, O., Rojas, M., Montesinos, L., & Molina, A. (2024). Advancing sustainable manufacturing: A case study on plastic recycling. *Production & Manufacturing Research*, *12*(1). Advance online publication. DOI: 10.1080/21693277.2024.2425672

McDonough, W., & Braungart, M. (2010). Cradle to cradle: Remaking the way we make things. *Choice Reviews Online*, *40*(02), 40–0914. DOI: 10.5860/choice.40-0914

Milios, L. (2018). Advancing to a Circular Economy: Three essential ingredients for a comprehensive policy mix. *Sustainability Science*, *13*(3), 861–878. PMID: 30147792

Mollik, M. S., Rahman, S. M., Rahat, M. R., Kulsum, C. U., & Sagir, S. A. M. (2024). DOES TRUSTWORTHINESS INFLUENCE TRAVEL SERVICE USE INTENTIONS AT AN ONLINE TRAVEL AGENCY? A STUDY ON THE DIGITALIZATION OF THE TOURISM SECTOR IN BANGLADESH. *Geo Journal of Tourism and Geosites*, *52*(1), 30–40. DOI: 10.30892/gtg.52103-1180

Mont, O., & Plepys, A. (2008). Sustainable consumption progress: Should we be proud or alarmed? *Journal of Cleaner Production*, *16*(4), 531–537. DOI: 10.1016/j.jclepro.2007.01.009

Morshed, A. (2025). Ethical challenges in designing sustainable business models for responsible consumption and production: case studies from Jordan. *Management & Sustainability an Arab Review*. https://doi.org/DOI: 10.1108/msar-09-2024-0131

Mubarik, M. S., Kontoleon, A., & Shahbaz, M. (2024). Beyond the hurdles: Exploring policy obstacles in the path to circular economy adoption. *Journal of Environmental Management*, *370*, 122667. DOI: 10.1016/j.jenvman.2024.122667 PMID: 39369529

Mugge, R., Jockin, B., & Bocken, N. (2017). How to sell refurbished smartphones? An investigation of different customer groups and appropriate incentives. *Journal of Cleaner Production*, *147*, 284–296. DOI: 10.1016/j.jclepro.2017.01.111

Munuhwa, S., & Mawonde, D. (2025). Green Reverse Logistics as a Sustainable Strategy towards Circular Economies in Africa. In *Palgrave studies of marketing in emerging economies* (pp. 183–208). https://doi.org/DOI: 10.1007/978-3-031-81627-7_10

Murray, A., Skene, K., & Haynes, K. (2017). The Circular Economy: An interdisciplinary exploration of the concept and application in a global context. *Journal of Business Ethics*, *140*(3), 369–380. DOI: 10.1007/s10551-015-2693-2

Nademi, Y., & Kalmarzi, H. S. (2025). Breaking the unemployment cycle Using Circular Economy: Sustainable Jobs for Sustainable Futures. *Journal of Cleaner Production*, *488*, 144655. DOI: 10.1016/j.jclepro.2025.144655

OECD. (2022). Global materials resource outlook to 2060: Economic drivers and environmental consequences. https://www.oecd.org/en/publications/global-material-resources-outlook-to-2060_9789264307452-en.html

Otasowie, O., Aigbavboa, C. O., & Oke, A. E. (2025). Theoretical perspectives on circular economy in organisations and policies. In Green energy and technology (pp. 33–70). https://doi.org/DOI: 10.1007/978-3-031-88322-4_3

Parvathy, S. U., Kolil, V. K., Raman, R., Vinuesa, R., & Achuthan, K. (2025). Integrating sustainable development goals into life cycle thinking: A multidimensional approach for advancing sustainability. *Environment, Development and Sustainability*. Advance online publication. DOI: 10.1007/s10668-024-05810-z

Peattie, K., & Peattie, S. (2009). Social marketing: A pathway to consumption reduction? *Journal of Business Research*, *62*(2), 260–268. DOI: 10.1016/j.jbusres.2008.01.033

Repait Café International Foundation. (2025). Repair Café-About us. Retrieved July 29, 2025, from https://www.repaircafe.org/en/

Ritzén, S., & Sandström, G. Ö. (2017). Barriers to the circular economy – integration of perspectives and domains. *Procedia CIRP*, *64*, 7–12. DOI: 10.1016/j.procir.2017.03.005

Robotics, A. M. P. (2024). AMP Robotics Secures $91 Million to Transform Recycling and Waste Management. https://www.causeartist.com/amp-robotics-secures-91-million-to-transform-recycling-and-waste-management/

Roshid, M. M., Karim, R., Waaje, A., Islam, S., & Bhowmik, R. C. (2025c). Mapping the carbon finance research: A bibliometric exploration of innovation in sustainable finance. *Development and Sustainability in Economics and Finance, 100048*. Advance online publication. DOI: 10.1016/j.dsef.2025.100048

Roshid, M. M., Saiyed, S., Karim, R., Haque, T., Siddique, M. S. R., Haque, M. A., Kumar, V., & Habib, S. B. (2025b). The Role of Green Entrepreneurship in Driving Sustainable Innovation in Bangladesh's Startup Ecosystem: Insights from a Belt and Road Emerging Market. *Environment. Innovation and Management*, *1*, 2550010. DOI: 10.1142/S3060901125500103

Roshid, M. M., Waaje, A., Karim, R., Irtisham, M. R., & Kaswan, M. S. (2025a). Industry 5.0 and organizational sustainability: Integrating cutting-edge technologies and human ingenuity. In *Industry 5.0* (pp. 50-61). CRC Press. https://doi.org/DOI: 10.1201/9781003535188-7

Sami, F., & Khan, I. H. (2025). From Waste to Resource: Embracing the Circular Economy. In *Adaptive Strategies for Green Economy and Sustainability Policies* (pp. 581-600). IGI Global Scientific Publishing.

Schultz, F. C., & Pies, I. (2024). The circular economy growth machine: A critical perspective on "post-growth" and "pro-growth" circularity approaches. *Journal of Industrial Ecology*, *28*(1), 17–24. DOI: 10.1111/jiec.13450

Shafik, W. (2025). SDG 12: Responsible Consumption and Production—The Circular Economy. In *Factoring Technology in Global Sustainability: A Focus on the Sustainable Development Goals* (pp. 365–390). Springer Nature Singapore.

Sibanda, V., Mhlanga, J., & Munuhwa, S. (2025). Circular Economy practices in supply chain management. In *Advances in logistics, operations, and management science book series* (pp. 13–38). https://doi.org/DOI: 10.4018/979-8-3693-6175-7.ch002

Singh, A. (2025). AI-Driven Innovations for Enabling a Circular Economy: Optimizing Resource Efficiency and Sustainability. In *Innovating Sustainability Through Digital Circular Economy* (pp. 47–64). IGI Global Scientific Publishing.

Smith, W. K., & Lewis, M. W. (2011). TOWARD a THEORY OF PARADOX: A DYNAMIC EQUILIBRIUM MODEL OF ORGANIZING. *Academy of Management Review*, *36*(2), 381–403. DOI: 10.5465/amr.2011.59330958

Stahel, W. R., & MacArthur, E. (2019). *The Circular Economy: A User's Guide.*

Steinbiß, K. (2025). The principle of repair in circular economy: extending the lifespan of products. In *CSR, sustainability, ethics & governance* (pp. 91–96). https://doi.org/DOI: 10.1007/978-3-031-70749-0_9

Sudusinghe, J. I., & Seuring, S. (2025). The practitioner perspective is more complete: Analysing supply chain collaboration for the circular economy. *Production Planning and Control*, •••, 1–24. DOI: 10.1080/09537287.2025.2484556

Suppipat, S., & Intrachooto, S. (2025). Bio-Circular-Green Economic Model in Thailand: A Case of Upcycle Circular Economy Certification for Products. In *The Palgrave Handbook of Practical Sustainability* (pp. 329–350). Springer Nature Switzerland.

UNEP. Sustainable Consumption and Production.(2015). https://sustainabledevelopment.un.org/content/documents/1951Sustainable%20Consumption.pdf. Accessed June 2025.

Thaler, R. H., & Sunstein, C. R. (2009). *Nudge: Improving decisions about health, wealth, and happiness.* Penguin.

UNCTAD (2023). *SDG 12 Progress: Global South Diagnostics.*

UNEP. (2023). Global Environment Outlook 7.

United Nations. (2015). Transforming our world: The 2030 Agenda for Sustainable Development. https://sdgs.un.org/2030agenda

Velenturf, A. P., Archer, S. A., Gomes, H. I., Christgen, B., Lag-Brotons, A. J., & Purnell, P. (2019). Circular economy and the matter of integrated resources. *The Science of the Total Environment*, *689*, 963–969. DOI: 10.1016/j.scitotenv.2019.06.449 PMID: 31280177

Vermeir, I., & Verbeke, W. (2006). Sustainable Food Consumption: Exploring the consumer "Attitude – Behavioral Intention" gap. *Journal of Agricultural & Environmental Ethics*, *19*(2), 169–194. DOI: 10.1007/s10806-005-5485-3

Vogel, R., De Vass, T., Croy, G., & Sohal, A. (2025). Circular Economy: Contested definitions and an evolving concept. In *CSR, sustainability, ethics & governance* (pp. 181–199). https://doi.org/DOI: 10.1007/978-3-031-70749-0_16

Waaje, A., Karim, R., Roshid, M. M., & Meem, T. N. (2025b). Minimizing waste in the hospitality industry: The opportunities and challenges of lean tourism. *Sustainable waste management in the tourism and hospitality sectors*, 125-166. https://doi.org/DOI: 10.4018/979-8-3693-6110-8.ch006

Waaje, A., Karim, R., Roshid, M. M., Yeamin, B., & Meem, T. N. (2025a). Integration of Industry 4.0 With Personalized Medicine: A Shift of the Paradigm in Healthcare. In *AI-driven personalized healthcare solutions* (pp. 27–70). IGI Global Scientific Publishing., DOI: 10.4018/979-8-3693-7858-8.ch002

White, K., Habib, R., & Hardisty, D. J. (2019). How to SHIFT consumer behaviors to be more sustainable: A literature review and guiding framework. *Journal of Marketing*, *83*(3), 22–49.

Women in Informal Employment. Globalizing and Organizing (WIEGO). (2025). Waste pickers. Retrieved July 30, 2025, from https://www.wiego.org/

Xiao, D. (2025). Evaluating and prioritizing strategies to reduce carbon emissions in the circular economy for environmental sustainability. *Journal of Environmental Management*, *373*, 123446. DOI: 10.1016/j.jenvman.2024.123446 PMID: 39591714

Yaremova, M., Tarasovych, L., Kilnitska, O., Buluy, O., & Kravchuk, N. (2024). Global trends in the development of a sustainable bioeconomy for rural growth in Ukraine.

Zucker, L. G. (1987). Institutional theories of organization. *Annual Review of Sociology*, *13*, 443–464.

# Chapter 9
# Circular Economy and Sustainability in the Wine Sector:
## An Analysis on the Reuse of By-Products in the Context of SDGs

**Mariagrazia Provenzano**
https://orcid.org/0000-0003-0737-1907
*Tuscia University, Italy*

**Francesco Pacchera**
https://orcid.org/0000-0002-5809-392X
*Tuscia University, Italy*

**Cecilia Silvestri**
https://orcid.org/0000-0003-2528-601X
*Tuscia University, Italy*

**Alessandro Ruggieri**
*Tuscia University, Italy*

## ABSTRACT

*The wine sector is a strategic component of European agribusiness, mainly because of its environmental impact. This study aims to contribute to the scientific debate on the sustainability of the supply chain through a systematic review of the potential for recovery and valorization of its by-products. Accordingly, the objective of the research is twofold. The first concerns identifying and analysing products from wine by-products reused in agriculture, aiming to close the production cycle. The*

DOI: 10.4018/979-8-3373-2454-8.ch009

*second focuses on critically assessing the environmental, economic and social impacts of such practices about the Sustainable Development Goals (SDGs). The review process identified three main application areas: organic fertilizers, animal feed, and wastewater treatment. These practices are found to be consistent with the SDGs: fertilizers, goals 2, 12, and 15; animal feed, goal 12; and water treatment, goals 6 and 12.*

## 1. INTRODUCTION

A crucial alternative to the linear economy, characterized by waste and overproduction, is the circular economy. The ability of circular systems to guarantee that products retain their added value by converting every waste product into a resource is a crucial characteristic that sets them apart from the other two economic models (Abbate et al., 2024). Thus, waste treatment reduces the strain on natural resources by becoming an opportunity for recovery and value-adding rather than only a disposal activity (Provenzano et al., 2024).

The shift to a circular economy is crucial for strengthening the EU's industrial base, decoupling economic growth from resource use, and reaching climate neutrality by 2050, as highlighted in the European Commission's Circular Economy Action Plan (2020). To create closed-loop production and consumption systems, the strategy emphasizes techniques including product durability, reparability, reuse, and the phase-out of single-use products. Additionally, it gives priority to industries like electronics, textiles, construction, and food that have a high resource consumption and potential for circularity.

In keeping with the global goal established by the Sustainable Development Goals (SDGs), circular economy practices provide a route to a more resilient and sustainable future by combining economic, environmental, and social factors (Berg et al., 2018; Mehta, 2023).

The SDGs and the circular economy are related ideas that enhance one another. According to Ortiz-de-Montellano et al. (2023), the two strategies aim to promote sustainability in a comparable way.

By focusing on the extraction, production, use, and disposal of resources, the circular economy seeks to offer an alternative to the conventional linear economy (Bocken et al., 2014; Scarpellini, 2022). Economic, social, and environmental aspects are all included in the SDGs' holistic vision of sustainable development (Brodny & Tutak, 2023; Khaskheli et al., 2020). In this regard, manufacturing chains can implement circular business practices, including greener production, a sharing economy, product recycling, and material reuse (Gallardo-Vázquez et al., 2024).

According to Xu et al. (2024), the use of circular techniques in agricultural systems can result in a notable decrease in greenhouse gas emissions by substituting bio-based products made from agricultural waste for conventional ones.

Considering the above, approaching agribusiness with a circular mindset leads to reduced resource waste and greater economic savings.

Given its importance for primary economic production and international trade, the wine industry is one of the most analyzed agri-food systems from the environmental sustainability perspective (Alessandri et al., 2024; Pratt, 2012). The 2023 report of the International Organisation of Vine and Wine (The International Organization of Vine and Wine, n.d.) estimates world wine production in 2022 at 258 million hectoliters. Viticulture and winemaking represent a significant cultural heritage both economically and socially. However, they affect the environment like any other economic activity (Santos et al., 2023). The wine production process generates about 9 million tons of waste, which has a negative impact on the environment due to the disposal of this waste. Several studies (Abbate et al., 2024; Constantin et al., 2024; Gabzdylova et al., 2009) have demonstrated the significant environmental impact of the wine supply chain due to the disposal and composting of waste outside of vineyards. These impacts, which lead to vegetation clearance, soil salinity, and erosion, are most noticeable in soil quality. The atmosphere is also degrading due to the usage of pesticides, chemical sprays, greenhouse gas emissions, and the large carbon footprint of transportation (Abbate et al., 2024).

The wine business must radically reevaluate its production techniques in light of growing market and environmental constraints, gradually shifting toward more resource-efficient and sustainable methods.

The methods used to handle waste and byproducts from the many phases of the winemaking process have changed significantly in recent years. In a circular economy perspective, more businesses in the industry are implementing recycling and reuse-based solutions that seek to turn what was once thought of as waste into actual secondary raw resources that can be used to create high-value products (Gabur et al., 2024).

In real terms, repurposing agricultural waste is an essential tactic for tackling numerous global issues in accordance with the SDGs. Reusing agricultural biomass allows for the development of new sustainable production chains and the reduction of waste generated by the agriculture industry (Kobayashi & Nakajima, 2021).

In addition to being a tangible answer to the demands of environmental sustainability, this shift to creative wine by-product recovery and valorization techniques is also having a good economic and social impact (Garrido et al., 2024; Morone et al., 2015). In fact, waste disposal is a major expense item in business budgets, particularly for small and medium-sized producers, and wineries can drastically cut operating costs related to this expense. By creating new industrial chains based on by-products with minimal environmental impact, waste valorization also makes it possible to diversify sources of revenue. Important examples include livestock feed created from lees (Câmara et al., 2020; Gabur et al., 2024) or biofertilizers made from pomace (Abbate et al., 2024; Gabur et al., 2024; Sousa et al., 2025a).

In this instance, the sustainable management of wine by-products is becoming increasingly recognized as a strategic chance to develop new business strategies and boost industry competitiveness by fusing environmental preservation with enhanced financial results. Furthermore, these approaches directly address the needs of contemporary consumers, who are becoming more conscious of sustainability issues and aware of the source and effects of the goods they purchase.

According to Abbate et al. (2024), these practices are therefore consistent with the SDGs of Agenda 2030 since they encourage sustainable and circular patterns and help to lower emissions related to waste management and the use of low-impact alternative resources.

However, as was mentioned, scientific research still has gaps that require filling. The lack of a thorough examination of the primary uses of wine by-products in agriculture and their effects on sustainability objectives is a crucial issue. A methodical study could find new ways to reuse wine by-products, completing the circular economic cycle, in addition to naturally gathering and analyzing the body of existing literature.

This type of in-depth study is essential to promote sustainable solutions that reduce emissions associated with waste management and use alternative low-impact resources. Ultimately, the goal is to integrate strategies that improve wineries' efficiency and contribute to a more sustainable and environmentally friendly future.

For this reason, the study aims to achieve two main objectives: to fill current gaps in the scientific literature and make a concrete contribution to the advancement of knowledge in the fields of sustainability and circular economy applied to the wine sector.

The first objective is to identify and systematically analyze the main new products obtained from the recovery and valorization of wine by-products, with a focus on those destined for the agricultural sector. In fact, using wine by-products in agriculture allows the cycle to be closed within the same production system, promoting a closed-loop circular economy. In this way, the interconnection between complementary production sectors is strengthened, helping to create virtuous synergies between environmental sustainability, technological innovation and social development. Promoting the agricultural use of wine by-products thus means not only reducing environmental impacts related to waste disposal but also actively supporting the ecological transition of the agricultural sector, turning it into a central node of circular economy strategies on a territorial scale.

The study's second objective is to critically and articulately assess the main environmental, economic, and social impacts associated with such recovery solutions, paying particular attention to their consistency with the SDGs promoted by the United Nations. In this context, the study aims to explore how the use of wine by-products in the agricultural sector can contribute to achieving the SDGs. This integrated assessment will highlight not only the technical and economic potential of valorized by-products but also their strategic role in the ecological transition of the agricultural and agribusiness sector as a whole.

For this reason, the study aims to answer the following research questions:

> RQ1. What are the main new products from winery waste applied in the agricultural sector from a circular economy perspective?
> RQ2. What are the main impacts of new products from winery waste according to the SDGs?

The article is structured as follows: the methodology is analyzed in the next section. Sections 3 and 4 present the main findings and discussion. Finally, Section 5 describes the main conclusions.

## 2. METHODOLOGY

The study used the systematic literature review methodology to meet the proposed aims. This methodology originated in the medical field (Saade et al., 2020) and has since been adopted across various disciplines, including social sciences, engineering (Bastas & Liyanage, 2018; Sassanelli et al., 2019), economics, and business (Aquilani et al., 2017; Colicchia & Strozzi, 2012; Merli et al., 2018; Silvestri et al., 2021). This approach is a systematic method "to identify, evaluate, and interpret the entire corpus of research relevant to a specific issue, subject area, or phenomenon of interest" (Kitchenham et al., 2007).

In order to answer research questions and accomplish study goals, a systematic literature review combines qualitative and quantitative evaluations (Denyer & Tranfield, 2009). By using transparent and replicable methodologies to thoroughly examine and map primary studies (Greenhalgh, 1997), it allows for a thorough grasp of a considerable body of literature (Çimen, 2021), highlighting research gaps and prospective future developments (Abbate et al., 2024).

According to Denyer and Tranfield (2009), the systematic literature review is a rigorous process that is divided into four distinct stages: (1) identifying existing studies; (2) selecting and evaluating contributions; (3) analyzing and synthesizing data; and (4) articulating results to produce conclusive conclusions. Furthermore, the four sequential steps of the procedural model put out by Mayring (2004) are (1) material collection, (2) descriptive analysis, (3) category selection, and (4) material evaluation.

In accordance with the guidelines proposed by Denyer & Tranfield (2009) and Mayring (2004), the present review process has been structured into four distinct steps, as illustrated in Figure 1.

*Figure 1. Review process*

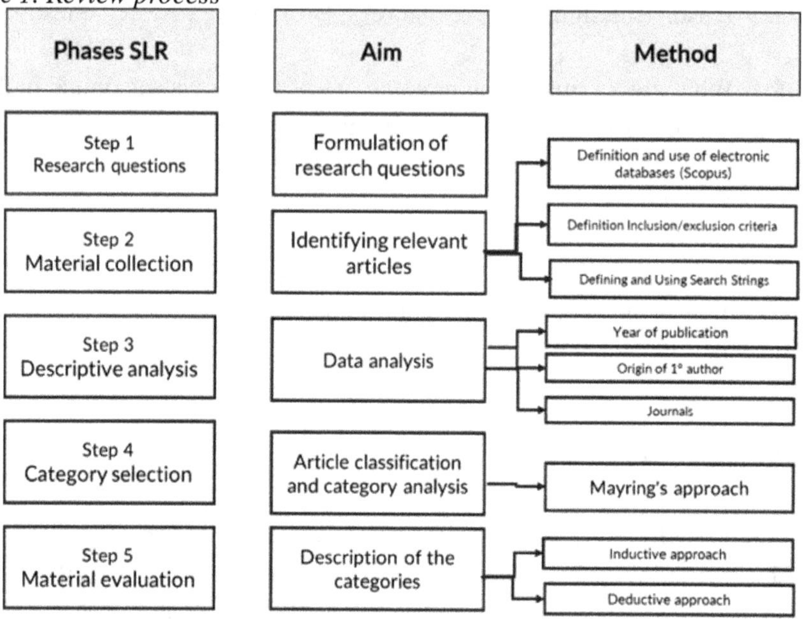


*Source: Authors' elaboration.*

## 2.1. Material collection

The study refers to the Elsevier Scopus database, which was selected based on its comprehensive coverage, as a "primary reference database with over 50 million records from approximately 5,000 publishers." (De Mauro et al., 2016; Merli et al., 2018; Sassanelli et al., 2019). The search for relevant articles was conducted on March 19, 2025, and the string used on Scopus was "Wine" AND "Waste" AND "Circular Economy." In the Scopus search, the research criteria were "Title, Keywords, Abstract." The initial number of articles identified was 152. Subsequently, we applied the C-I-M-O framework to establish the inclusion/exclusion criteria (Silvestri et al., 2024) as follows: Inclusion criteria: all papers analyzed in this study focused on the waste valorization in the wine sector according to a circular economy approach, with a particular emphasis on the potential for reuse in the agricultural sector.

Exclusion criteria: all papers that do not focus on waste valorization in the wine sector and with reuse in the agricultural sector, although according to a circular economy approach. Furthermore, limiting the search to scientific articles written in English, the number of articles was 95. By analyzing the entire article, the final sample consisted of 21 articles. Figure 2 shows the process followed for the selection and analysis of articles.

*Figure 2. Item selection process*

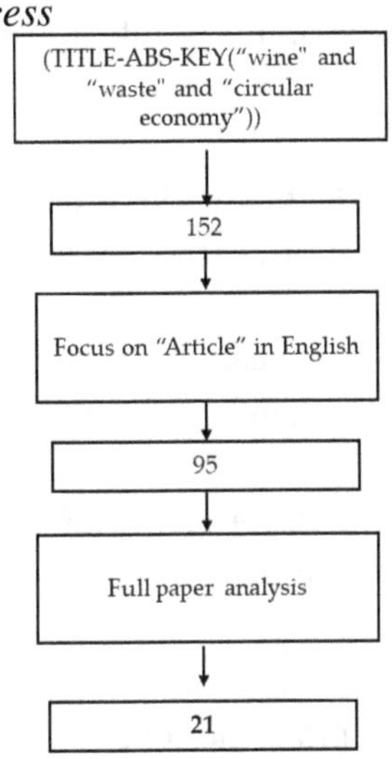


*Source: Authors' elaboration.*

## 2.2. Category selection

In accordance with the findings of (Mayring, 2004) and consideration of the research questions, a few structural dimensions were identified. Subsequently, analytical categories were identified to evaluate the material. The categorization process employs a combination of inductive and deductive approaches (Mayring, 2004; Merli et al., 2018; Seuring & Müller, 2008). The deductive approach defines dimensions and categories based on a specific theory or an existing model (Polit, 2004). These factors were therefore established prior to the material's study (Seuring & Müller, 2008). The inductive technique moves the emphasis from the specific to the general by determining dimensions and categories through material analysis (Silvestri et al., 2021).

Based on content analysis, dimensions and associated categories were found using the inductive approach (Tranfield et al., 2003). The following phases are part of the iterative approach that was used as a result. In the initial phase, the researchers need to become aware of with the chosen sample of papers (Dixon-Woods, 2011).

Second, they must refrain from picking preset themes and instead let topics grow and evolve naturally. After that, the data was grouped into categories (Prajapati et al., 2019). The dimensions of trash and newly created products in the agricultural sector were addressed using this method.

Specifically, the waste dimension is divided into four analytical categories as follows:

- vine pruning, waste from the first stage of the wine production chain.
- grape pomace and stems, waste from harvesting, destemming, aging, and bottling.
- wine wastewater and lees, fermentation phase waste.
- general residues from every step in the production process.

The same process was followed for clustering new products generated by the valorization of wine waste in the agricultural sector:

- animal feed, products from agricultural waste.
- fertilizer, substances to improve the soil, obtained from organic waste.
- treatment of wastewater, processes to purify and reuse wastewater.

In the deductive approach, the identification of dimensions and related categories was based on analyzing models and frameworks in literature. For the United Nations Sustainable Development Goals (SDGs) classification, we referred to the 2030 Agenda for Sustainable Development, adopted by all United Nations Member States in 2015 (United Nations, 2015).

By identifying the dimensions and their categories, it becomes easier to analyze the articles from many angles, which improves understanding of the topic and, in turn, results in a more thorough response to the research questions of this article.

The categorization process enables us to (1) classify the waste from the wine supply chain and study its valorization from a circular economy perspective in the agricultural sector and (2) analyze the impact of these new products through the lens of sustainability dimensions and their connection to the SDGs.

Before analysing the three structural dimensions, a general description of the sample was conducted.

This research defined two structural dimensions using the inductive approach, while one dimension was defined using the deductive approach (see Table 1).

*Table 1 - Approach, structural dimensions, and analytical categories*

| Approach | Structural dimensions | Analytical categories |
|---|---|---|
| Inductive | Waste | Vine pruning<br>Grape pomace and stems<br>Wine wastewater and lees<br>General residues |
| Inductive | New products | Animal feed<br>Fertilizer<br>Treatment of wastewater |
| Deductive | United Nations Sustainable Development Goals (SDGs) | 17 Sustainable Development Goals (SDGs) |

Source: Authors' elaboration.

## 3. RESULTS

This section presents the results of the study, structured into two parts: a descriptive analysis of the selected literature and a material evaluation of the valorization solutions identified.

### 3.1. Descriptive sample analysis

The descriptive sample analysis classifies articles according to the following criteria: (1) date of publication, (2) journal, and (3) origin of 1st authors.

The first element analyzed is the year of publication of the selected articles. Most of the articles were published in 2023 (6 articles, 29% of the sample), a considerable portion were published in 2024 (5 articles, 24% of the sample), followed by the articles published in 2021 (4 articles, 19% of the sample) (Figure 3).

*Figure 3. Papers distribution for years*

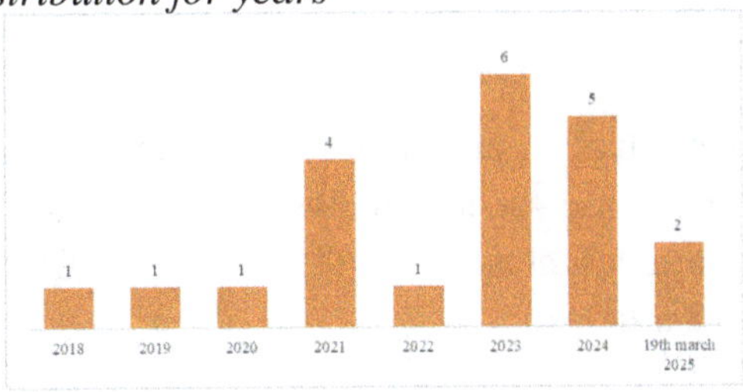


*Source: Authors' elaboration.*

Another analysis involved classifying the papers concerning journals. Table 2 shows the list of selected journals, and "Agriculture" is the journal with the highest number of papers (10% of the sample). Indeed, this journal aims to publish research on postharvest handling of produce, management of the natural resource base for agricultural production (land, soil, and water), and agriculture in changing environments.

*Table 2. List of selected journals*

| *Journals* | *f* | % |
|---|---|---|
| Agriculture (Switzerland) | 2 | 10% |
| Agronomy | 1 | 5% |
| Algal Research | 1 | 5% |
| Applied Sciences (Switzerland) | 1 | 5% |
| Biochar | 1 | 5% |
| Ecotoxicology and Environmental Safety | 1 | 5% |
| Environmental Science and Pollution Research | 1 | 5% |
| Fuel | 1 | 5% |
| Journal of Applied Phycology | 1 | 5% |
| Journal of Environmental Management | 1 | 5% |
| Mediterranean Fruits Bio-wastes: Chemistry, Functionality and Technological Applications | 1 | 5% |
| Microchemical Journal | 1 | 5% |
| Nematology | 1 | 5% |
| Process Safety and Environmental Protection | 1 | 5% |
| Science of the Total Environment | 1 | 5% |
| Sustainability (Switzerland) | 1 | 5% |
| Waste Management | 1 | 5% |
| Waste Management and Research | 1 | 5% |
| Water (Switzerland) | 1 | 5% |
| Water Science and Technology: Water Supply | 1 | 5% |
| *Total* | 21 | 100% |

Source: Authors' elaboration.

The geographical origin of the articles was ascertained by examining the country of the primary author (Figure 4). The analysis reveals that Spain has the highest number of articles (9 articles, constituting 43% of the total), followed by Portugal (6 articles) and Italy (2 articles).

*Figure 4. Number of papers by country*

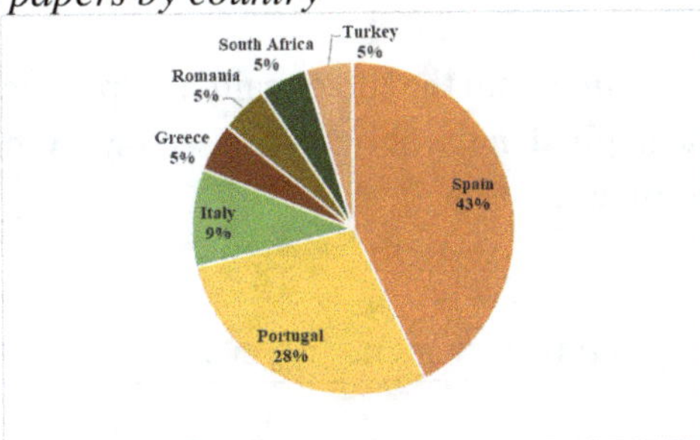


*Source: Authors' elaboration.*

## 3.2. Material evaluation

This section reports the results, first presenting the structural dimensions identified through both inductive and deductive methods, followed by an in-depth analysis of the valorization solutions. The findings aim to provide a comprehensive understanding of the circular opportunities within the wine sector, highlighting environmental, economic, and social implications.

### 4.2.1 Waste analysis

The papers' analysis shows that general residues are the largest waste from the supply chain used in a circular economy perspective in the agricultural sector (accounting for 43% of the total sample), followed by grape pomace and stems (33%) (Table 3).

*Table 3. The main waste from wine supply chain.*

| Waste | *f* | % |
|---|---|---|
| General residues | 9 | 43% |
| Grape pomace and stems | 7 | 33% |
| Wine wastewater and lees | 4 | 19% |
| Vine pruning | 1 | 5% |
| *Total* | 21 | 100% |

Source: Authors' elaboration.

### 4.2.2 New products analysis

Table 4 divides the sample into the three main new products. Fertilizer, derived from winery waste and applied in the agricultural sector, represents the biggest new product (71% of the sample).

*Table 4. New products from winery waste*

| New Products | *f* | % |
|---|---|---|
| Fertilizer | 15 | 71% |
| Animal feed | 4 | 19% |
| Treatment of wastewater | 2 | 10% |
| *Total* | 21 | 100% |

Source: Authors' elaboration.

### 4.2.3 New products and SDGs

The 17 SDGs are statutes adopted globally to address the planet's grand challenges by 2030. They cover the three key dimensions of sustainability: 1) environment, through action against climate change, protection of biodiversity and responsible use of resources; 2) economy, through sustainable growth, decent work and innovative practices; and 3) society, providing for the fight against poverty and inequality, access to education and health, and gender equality (United Nations, 2015).

The valorization and transformation of waste into new products is a concrete way to achieve the goals of the 2030 Agenda. Table 5 shows the impact of new products on the three dimensions of sustainability and their connection with the SDGs.

*Table 5. Impact of new products from winery waste on SDGs goals.*

| *Waste* | *New Products* | *SDGs Target* |
|---|---|---|
| General Residues<br>Grape Pomace<br>Vine Pruning<br>Wine Lees<br>Wine Wastewater | Fertilizer | SDGs 2, 12 & 15 |
| General Residues<br>Grape Stems | Animal Feed | SDG 12 |
| Wine Wastewater | Treatment of wastewater | SDGs 6 & 12 |

Source: Authors' elaboration.

The results show how the processing of wine waste into animal feed contributes to the achievement of Goal 12 “Ensure sustainable production and consumption patterns,” a goal also shared by the processing of wine waste into fertilizer and treated wastewater to be reused for agricultural purposes.

Wastes processed into fertilizer also contribute to the achievement of Goals 2 “End hunger, achieve food security, improve nutrition and promote sustainable agriculture” and 15 “Protect, restore and foster sustainable use of the Terrestrial ecosystem.” Treated wastewater impacts Goals 6 “Ensure the availability and sustainable management of water and sanitation for all” and 11 “Make cities and human settlements inclusive, safe, durable and sustainable.”

## 4. DISCUSSION

The increasing focus on environmental sustainability, food security, and resource efficiency has brought to the forefront the need to rethink traditional production models, especially in agriculture. As a result, by-products from wine production cease to be considered a problem mainly related to their disposal but rather as a resource that can be valorized within heterogeneous production processes (Cancelli, n.d.; Lisičar Vukušić et al., 2023). The circular economy enables companies to optimize the use of these wastes, generating competitive advantages, reducing operating costs and contributing to the achievement of sustainability goals, in line with the growing market demand for low environmental impact solutions (Gabur et al., 2024). By virtue of this, the integration of wine by-products into a circular economy model can contribute across the board to achieving the Sustainable Development Goals.

RQ1. What are the main new products from winery waste applied in the agricultural sector from a circular economy perspective?

The wine industry generates a significant amount of organic waste and wastewater throughout its production cycle. In a circular economy perspective, these by-products represent valuable resources rather than disposal problems. Their transformation into new products not only contributes to environmental sustainability but also brings economic and social benefits by reducing production costs, creating new value chains, and promoting innovation. Three major avenues of value creation are examined in this section: fertilizer, animal feed, and wastewater reuse. Their potential to address sustainability issues and promote the shift to regenerative agricultural systems is highlighted.

## New Product: Fertilizer

A key element of agriculture, soil serves as the foundation for all chemical and physical processes. The natural soil foundation accessible for food production has been diminished in recent decades due to insufficient replacement of soil nutrients (Manniello, Statuto, Di Pasquale, Giuratrabocchetti, et al., 2020). Long-term usage of chemical fertilizers limits the development and production of plants by acidifying, degrading, and compacting the soil (Y. Zhang et al., 2022). A shift from a fossil fuel-dependent economy to a bio-based one that is based on the effective extraction of nutrients from waste streams is feasible if the necessity of addressing these issues is acknowledged (Nascimento-Gonçalves et al., 2024).

Because it improves the amount of organic matter and is a low-impact, technically and financially advantageous procedure, this solution aims to solve conventional agriculture's over-reliance on chemical fertilizers (Abbate et al., 2024). There are also financial benefits because using leftovers is less expensive than using traditional materials (Arvanitoyannis et al., 2006).

One of the most widely used methods for turning organic waste into soil conditioners is composting. Wine industry waste is one of the organic wastes that can be valorized as compost since it contains more polyphenols than any other kind of agro-industrial waste (Abbate et al., 2024).

According to Gabur et al. (2024), cellar trash can be converted into organic soil conditioner due to its high content of organic matter and macronutrients.

To make vineyard waste appropriate for broader commercial application, new technologies are essential (Niculescu & Ionete, 2023). Cutting-edge methods for treating wine waste are crucial for enhancing the industry's sustainability, generating revenue, and lessening its negative effects on the environment (Beres et al., 2017). In order to enhance the recovery of bioactive chemicals and lessen the environmental impact through more sustainable methods, several cutting-edge technologies have recently been developed (Shahabi, Mohammadabadi et al., 2022).

In this case, cellar waste is utilized in anaerobic digestion, which is a process that occurs without oxygen and permits organic matter to break down, resulting in digestate. Spreading digestate from winery waste on vineyards could be an experimental approach to recycling nutrients and organic matter, creating a circular economy (Da Ros et al., 2018).

Cellar waste can be transformed into valuable products such as biochar, reducing the waste management problem and contributing to the circular economy. Biochar obtained from cellar waste could have potential as a slow-release fertilizer, gradually releasing nutrients as the soil is irrigated, acting as a sink. In addition, biochar, used as a soil conditioner, can absorb and capture pollutants, sequestering carbon in the soil, and effectively controls pests that feed on host plants that pose a serious threat to agriculture, thus again helping to mitigate climate change (Nunes et al., 2021).

Grape marc is nutrient-rich, biologically active, and free of polyphenols, which can be a proper soil conditioner if treated properly and sustainably making it suitable for treatment by vermicomposting (Gabur et al., 2024).

In fact, vermicomposting is another new technology, inexpensive and rapid, that effectively recycles organic waste and converts it into nutritious soil for agricultural production. Vermicomposting is a natural process that relies on using earthworms with endogenous microorganisms present in the waste following the decomposition of organic matter. Vermicompost is a nutrient-rich material characterized by high porosity, high water retention capacity, and low carbon/nitrogen ratio, which, from a perspective of promoting sustainable zero-carbon agriculture, underscores its sustainability as an alternative for economical treatment of organic waste (Lazcano & Domínguez, 2011). From a circular economy perspective, vermicompost, derived from grape pomace and used for vine treatment, has shown increased grape production and better organoleptic properties in the final wine (Nascimento-Gonçalves et al., 2024).

Grape marc was also used as a product for soil biodisinfestation. The crops tested consisted of a rotation of tomato and beet under greenhouse conditions. The effects of biodisinfestants on soil fertility, nematode community and crop yield were evaluated. Results showed the immediate impact, after treatment application, of nematode reduction. Additionally, it was discovered that soil fertility has improved (López-Pérez & Sánchez-Moreno, 2021).

On the other hand, winery wastewater has been utilized as a sustainable addition to microalgae farming, which is consistent with the ideas of the circular economy. The macronutrient composition of the microalgae produced by this kind of fertilizer demonstrates its potential as a bio-additive for soil, especially as a source of phosphorus and nitrogen (Sartori et al., 2023). In particular, the microalgae grew more than the control media without this water when percentages of winery wastewater were added, which decreased the amount of polyphenols present (Lindner & Pleissner, 2022; Spennatia et al., 2019).

Furthermore, by lowering reliance on synthetic fertilizers, this kind of nutrient-rich biomass can be used as a bio-additive in viticulture to increase soil fertility, encourage vine growth, and support sustainable agricultural practices (Sousa et al., 2025b).

Each of these approaches has distinct benefits, as can be seen when comparing them: technologies such as anaerobic digestion and the production of biochar require more infrastructure but provide wider environmental benefits, while composting and vermicomposting are affordable and appropriate for small to medium-sized operations. When combined, they serve as complimentary instruments for creating a circular, regenerative agricultural paradigm based on reduced foreign inputs and local waste valuation.

## New Product: Animal Feed

The conversion of wine by-products into animal feed is another important option for circular innovation in the wine industry, and it has attracted a lot of attention lately. This topic is covered in about 19% of the reviewed publications, indicating an increasing interest in maximizing the economic and nutritional worth of residues derived from grapes.

The volatile composition of winery wastes can improve the quality of aquatic feeds in terms of flavor and provide health benefits to aquaculture livestock. Esters and terpenoids positively contributed to flavor with fruity notes, playing an essential role as a feed attractant. Other compounds, however, impact odor, stimulating a slight anti-stress effect on animals (Câmara et al., 2020). Cellar waste, moreover, can be used directly or processed to create protein-rich biomass that can be added to animal diets (pigs, cattle, lambs), in which it has successfully replaced oat bran and wheat bran (Constantin et al., 2024).

Grape marc has also been shown to be a possible feed ingredient, particularly for Ruminants as up to 3% of the total marc can be used in animal feed (Dwyer et al., 2014). In Europe, for example, dry grape extract is authorized as a potential feed additive for most animal species.

In pig feed, incorporation of fermented grape marc improves meat color stability and total polyunsaturated fatty acids in adipose (subcutaneous) fat while reducing lipid peroxidation (Marin et al., 2020). In addition, a specific amount of grape pomace, when included in feed, influences pig productivity by improving several antioxidant systems in body tissues and blood (Kafantaris et al., 2018). Grape pomace added in broiler diets also improved immunity and antioxidants and decreased diet cost per kg live weight (Ebrahimzadeh et al., 2018; Gabur et al., 2024).

Grape stems, which are rich in fiber and polyphenols, prove to be an attractive addition to animal diets. In this regard, rabbit breeding emerges as a target species for breeding, as fiber content is essential in rabbits' diets to prevent digestive disorders, and polyphenols are associated with better performance in animals due to their antimicrobial and antioxidant activities. The grape stalk ingredient is proposed as a secondary feed for rabbits, reducing dependence on other fiber sources (San Martin et al., 2024).

Grape stems, properly processed, also prove beneficial in the diet of ruminants. Although grape stems contain lignin, which limits microbial enzymatic access to cellulose (Atuhaire et al., 2016), it affects digestibility (Tucker et al., 2003). Therefore, hydrolysis emerges as a strategy with great potential to increase fiber digestibility since it is a chemical reaction to break the bonds of a particular substance to obtain smaller, more digestible fractions (San Martin et al., 2023). Altogether, these diverse applications illustrate how the wine industry can not only reduce waste but also create new revenue streams in the animal nutrition sector. By reintroducing by-products into the food chain through livestock feed, producers can reduce feed imports, cut emissions, and support more resilient agri-food systems rooted in local circular practices.

## New Product: Treatment of Wastewater

Finally, wastewater reuse is emerging as a vital strategy for resource efficiency and sustainability in the wine sector. As climate change intensifies water scarcity—particularly in the Mediterranean basin — there is growing urgency to identify alternative water sources that reduce pressure on freshwater reserves (Oliveira et al., 2019), promoting the circular economy and avoiding the use of raw materials. From an environmental point of view, wastewater can be reused for irrigation, just as treated sludge is used as fertilizer or soil conditioner in vineyards. From a socioeconomic perspective, wastewater treated and used for agricultural purposes can dramatically reduce the costs (up to 60 times) of winery wastewater treatment and management compared to conventional treatment and management (Flores et al., 2023).

This management and reuse model could be an advantage in wastewater management, particularly in the arid areas of the Mediterranean, where demand for new water resources is identified as one of the most critical hotspots in future climate change projections for the Mediterranean basin. This approach toward the goal of "closing the loop" by reusing on-site treated industrial wastewater plays a key role in wine production water management (Oliveira et al., 2019).

Across fertilizers, animal feed, and wastewater treatment, wine sector by-products offer a rich source of opportunities for circular innovation. The diverse applications not only reduce environmental impacts but also generate economic value, support rural livelihoods, and promote resource self-sufficiency. When considered together, these solutions present a compelling case for rethinking the wine production model—not as a linear, extractive system, but as a closed-loop, regenerative process aligned with the principles of the circular economy.

RQ2. What are the main impacts of new products from winery waste according to the SDGs?

Waste management is an issue that reflects different aspects of economic growth in all three domains of sustainability: environment, economy and society (Gallardo-Vázquez et al., 2024).

Waste management now encompasses resource optimization and value recovery in addition to disposal in this larger sustainability framework. This is in line with the growing recognition on a worldwide scale that the conventional approach to economic development is no longer sustainable.

The linear economic model is acknowledged by the UN Sustainable Development Goals (SDGs) as being inappropriate for tackling the intricate problems of supply, demand, and long-term growth. There are several threats to ecosystems, economy, and society from this imbalance.

One strategic answer to this is the move toward the circular economy. In order to lessen reliance on virgin materials and its negative effects on the environment, the circular model encourages the thoughtful design, reuse, and reintegration of resources into different production cycles (Gabur et al., 2024). An important advance in sustainable development is this reconsideration of the dynamics of production and consumption.

Furthermore, improper disposal methods or insufficient trash collection not only damage the environment but also raise major health issues for the general public. Because of this, a number of SDGs naturally promote the circular economy by placing a strong emphasis on sustainable resource use (SDG 12) and environmental protection (SDG 15).

By integrating circular economy concepts into waste management systems, it is possible to attain economically resilient and socially inclusive growth, ensuring benefits for present and future generations (Geissdoerfer et al., 2017).

Agriculture is one of the industries most impacted by this shift. For example, maximizing waste from the wine supply chain offers farmers a substantial financial potential. The accomplishment of several SDGs depends on the creation of sustainable farming methods and the decrease of food waste (Gallardo-Vázquez et al., 2024).

Agriculture can support rural communities, generate steady incomes, protect natural ecosystems, and meet the world's food demands if it is managed sustainably. But present farming methods are speeding up biodiversity loss and soil degradation, and climate change is making environmental stresses like floods and droughts worse (United Nations, 2015).

Considering this, agricultural biodiversity becomes an essential instrument for guaranteeing food security, enhancing farmer incomes, and creating robust, sustainable food systems.

## SDG 2: "Zero Hunger"

Implementing resilient agriculture techniques that not only boost productivity but also support ecosystem preservation and gradual soil improvement is the focus of SDG 2 "Zero Hunger" in this context (United Nations, 2015).

The process of turning grape marc, the main solid by-product of winemaking, into fertilizer provides a concrete illustration of this. This method encourages sustainable soil management in addition to providing an easy and environmentally responsible way to value wine industry wastes (Nascimento-Gonçalves et al., 2024).

Given that soil may store up to twice as much carbon as the atmosphere, it is essential for long-term sustainability and climate mitigation (Manniello, Statuto, Di Pasquale, & Picuno, 2020). Thus, using composted grape marc improves soil fertility and aids in environmental preservation and carbon sequestration.

## SDG 6: "Clean Water and Sanitation"

Water is another vital resource for sustainable agriculture, in addition to soil. Particularly in areas that are experiencing drought and water scarcity—which are frequently among the poorest and most vulnerable—SDG 6 "Clean Water and Sanitation" emphasizes the pressing need to enhance water management (United Nations, 2015). Reusing and treating wastewater from wineries is a viable strategy in this area. This approach can greatly lessen the strain on freshwater resources, especially in Mediterranean regions where climate change is predicted to have a considerable impact on water availability (Oliveira et al., 2019). Furthermore, reusing wastewater from wineries has both financial and environmental benefits. It reduces the environmental impact of wine production by avoiding the use of chemicals, requiring low energy inputs, and providing good technical performance (Flores et al., 2023). This makes it a key strategy for water management in arid agricultural areas.

## SDG 12: "Responsible Consumption and Production"

SDG 12, which encourages responsible consumption and production, is in line with these principles. By promoting sustainable infrastructure, equitable job opportunities, and the effective use of energy and resources, this goal embodies the earlier ones (United Nations, 2015).

In order to optimize output and social benefits while reducing resource input and environmental damage, the circular model advocates the idea of "doing more with less" (Lee, 2013; Omer et al., 2025).

Making biochar from grape marc is one example of this, which accomplishes multiple goals at once. It serves as a natural solution for chemical treatments in plant pest control and offers a sustainable alternative to synthetic fertilizers (Fernandes et al., 2024).

This kind of biochar can absorb pollutants and lessen their discharge into the air, water, and soil because of its high adsorption ability, which will ultimately increase ecological health and agricultural output (Nunes et al., 2021).

## SDG 15: "Life on Earth"

SDG 15 "Life on Earth" advocates for the recovery of ecosystems and the sustainable management of natural habitats, and all of these initiatives ultimately lead to this goal. This involves stopping the spread of exotic species that can negatively impact biodiversity and tackling land degradation brought on by drought, floods, and desertification (United Nations, 2015).

Accordingly, horticultural and viticultural by-products from the wine supply chain are being recycled to improve soil quality and shield crops from pests and illnesses (López-Pérez & Sánchez-Moreno, 2021).

When properly applied, these sustainable fertilizers raise soil organic carbon levels, improving fertility, increasing crop yields, and helping the wine industry become carbon neutral (Gabur et al., 2024).

The viability and multifaceted benefits of turning winery waste into new products are amply illustrated by these solutions. For instance, organic wastes like grape pomace and lees can be converted into compost or natural fertilizers, which enhance soil fertility and structure while lowering the need for costly chemical inputs. By encouraging regenerative agriculture and reducing input costs, this strategy directly benefits farmers' bottom lines (Pérez-Bibbins et al., 2015). It also provides opportunities for nearby composting businesses and waste valorization social cooperatives.

Similar to this, winery water, which is frequently highly concentrated with organic matter, can be efficiently processed and repurposed for irrigation of vineyards (Bustamante et al., 2005). This lowers wastewater treatment costs and reduces dependency on freshwater supplies, which is a rising concern in water-scarce places. From a societal perspective, this approach strengthens rural communities' long-term water security and helps small-scale farmers remain resilient to climate change.

Additionally, residues from the production of wine might be helpful components in animal feed. Grape pomace, which is high in fiber and nutrients, can partially replace commercial feedstocks, which are frequently expensive and come from far places (Rondeau et al., 2013). This promotes a more regional and circular agri-food economy by lowering feed costs for livestock farmers and condensing supply chains. This method supports job retention and economic diversification in rural communities, particularly those with mixed farming systems, by enhancing economic synergies between livestock farms and vineyards (Arvanitoyannis et al., 2006).

All three of these strategies operate in conjunction to change the entire production process into a circular, interconnected system. In addition to lowering waste production, completing resource loops, and encouraging environmental stewardship, each step builds on the one before it, while also promoting social inclusion and economic efficiency.

This holistic vision aligns perfectly with several UN Sustainable Development Goals. Using by-products for animal feed contributes to the "Zero Hunger" goal (SDG 2) by enhancing food system resilience and affordability; water treatment and reuse support the "Clean Water and Sanitation" goal (SDG 6), particularly in agricultural contexts; the principle of circularity itself reflects the "Responsible Consumption and Production" goal (SDG 12); and improving soil health and promoting biodiversity through compost use relates directly to the "Life on Land" goal (SDG 15). Additionally, by reducing production costs, creating local jobs, and strengthening rural economies, these practices also contribute to ensure decent work and economic growth, and to create sustainable communities (Figures 5).

*Figure 5. Winery waste valorization process and SDGs' connection*

*Source: Authors' elaboration.*

## 5. CONCLUSION

The transition to a circular economy enables the transformation of environmental constraints into opportunities for innovation, stimulating the development of skills and knowledge that contribute to sustainable and competitive growth. In this context, the valorization of winery by-products offers an attractive path toward sustainable resource use and waste management in the wine industry. By converting waste into useful resources, generating new sources of income, and promoting the creation of eco-friendly products, wineries may support a more circular and ecologically conscious approach.

Investigating the primary new products derived from wine supply chain residues that are used in the agricultural industry was the aim of this study.

The focus was on analyzing the impacts generated by these new products in relation to the Sustainable Development Goals (SDGs) promoted by the United Nations.

The study found that the new products, applied to the agricultural sector, generated by waste from the wine supply chain are fertilizer, feed and wastewater for agricultural use. These three products respond to specific goals of the 2030 Agenda such as SDGs 2 "Zero Hunger," 6 "Sustainable Water Management," 12 "Sustainable Consumption and Production Patterns," and 15 "Protect, Restore and Foster Sustainable Use of the Earth's Ecosystem."

A wide range of stakeholders, including farmers and legislators, must be involved in order to realize these solutions at scale. They also depend on consumer acceptance, infrastructure accessibility, and economic viability. Implementation may be limited particularly in small businesses, by cultural resistance, legislative gaps, and technological complexity. Investment, focused legislative support, and public awareness are needed to remove these obstacles and guarantee that circular processes are socially inclusive and scalable.

From a management standpoint, wine operators have a tangible chance to incorporate the circular economy into their business plans through the value-adding of wine dregs. They can lower disposal costs, provide new value streams, and support sustainable development by turning by-products into valuable agricultural resources.

In addition, adopting sustainable practices enhances corporate reputation, facilitating access to environmentally sensitive markets and paving the way for green certifications (Ngwasiri et al., 2022; Yadav et al., 2024; Q. Zhang & Zhang, 2024).

From a scientific point of view, this review is a contribution in advancing the state of the art on the application of the circular economy in the agribusiness sector, with a specific focus on the valorization of wine residues. The article provides an overview of innovative products derived from wine waste and analyzes their impact with respect to the Sustainable Development Goals, filling a gap in the literature on sustainability and agricultural waste management.

In addition, the work opens new perspectives for interdisciplinary studies that integrate agricultural, environmental and economic-managerial sciences, promoting more resilient and sustainable production models.

The manual search method used may have excluded or omitted potentially relevant articles. Although this approach provides more direct control over source selection, there is a risk that important contributions will be missed. Second, although Scopus is one of the most comprehensive and reliable platforms for scholarly literature, using a single database is another major limitation. Therefore, the next steps in the search include using other databases, such as Web of Science, to expand the sample of articles. In addition, future research developments will focus on developing methods and tools to evaluate the effectiveness of achieving the SDGs through quantitative metrics such as emission reductions, improved waste management, and economic benefits, focusing on return on investment, savings from reduced disposal costs, and the creation of new value streams through product enhancement. To this end, the study of existing case studies related to the application of new agricultural products from waste in the wine sector will be helpful, as they provide concrete examples of applying circular economy practices and achieving sustainable development goals. However, current literature still presents a lack of empirical case studies and an insufficient representation of socio-economic factors, such as stakeholder engagement, employment impact, and social acceptance. Addressing these gaps will be essential to ensure a holistic assessment of circular practices and their alignment with both environmental and social dimensions of sustainability.

## REFERENCES


Abbate, S., Centobelli, P., & Di Gregorio, M. (2024). Wine waste valorization: Crushing the research domain. *Review of Managerial Science*, •••, 1–36.

Alessandri, G., Daddi, T., & Iraldo, F. (2024). Environmental sustainability in the wine industry, a literature review. *Cleaner Production Letters*, *7*, 100067.

Aquilani, B., Silvestri, C., Ruggieri, A., & Gatti, C. (2017). A systematic literature review on total quality management critical success factors and the identification of new avenues of research. *The TQM Journal*, *29*(1), 184–213.

Arvanitoyannis, I. S., Ladas, D., & Mavromatis, A. (2006). Potential uses and applications of treated wine waste: A review. *International Journal of Food Science & Technology*, *41*(5), 475–487.

Atuhaire, A. M., Kabi, F., Okello, S., Mugerwa, S., & Ebong, C. (2016). Optimizing bio-physical conditions and pre-treatment options for breaking lignin barrier of maize stover feed using white rot fungi. *Animal Nutrition*, *2*(4), 361–369. PMID: 29767044

Bastas, A., & Liyanage, K. (2018). Sustainable supply chain quality management: A systematic review. *Journal of Cleaner Production*, *181*, 726–744.

Beres, C., Costa, G. N. S., Cabezudo, I., da Silva-James, N. K., Teles, A. S. C., Cruz, A. P. G., Mellinger-Silva, C., Tonon, R. V., Cabral, L. M. C., & Freitas, S. P. (2017). Towards Integral Utilization of Grape Pomace from Winemaking Process: A Review. *Waste Management (New York, N.Y.)*, *68*, 581–594. PMID: 28734610

Berg, A., Antikainen, R., Hartikainen, E., Kauppi, S., Kautto, P., Lazarevic, D., Piesik, S., & Saikku, L. (2018). Circular economy for sustainable development. *Reports of the Finnish Environment Institute*, *26*, 2018.

Bocken, N. M. P., Short, S. W., Rana, P., & Evans, S. (2014). A literature and practice review to develop sustainable business model archetypes. *Journal of Cleaner Production*, *65*, 42–56.

Brodny, J., & Tutak, M. (2023). Assessing regional implementation of Sustainable Development Goal 9 "Build resilient infrastructure, promote sustainable industrialization and foster innovation" in Poland. *Technological Forecasting and Social Change*, *195*, 122773.

Bustamante, M. A., Paredes, C., Moral, R., Moreno-Caselles, J., Pérez-Espinosa, A., & Pérez-Murcia, M. D. (2005). Uses of winery and distillery effluents in agriculture: Characterisation of nutrient and hazardous components. *Water Science and Technology*, *51*(1), 145–151. PMID: 15771110

Câmara, J. S., Lourenço, S., Silva, C., Lopes, A., Andrade, C., & Perestrelo, R. (2020). Exploring the potential of wine industry by-products as source of additives to improve the quality of aquafeed. *Microchemical Journal*, *155*, 104758.

Cancelli, U. (n.d.). Circular Economy Strategies for Management and Valorization of Some By-Products of the Wine Industry. *On the Developments in the Italian PhD Research on Food Science, Technology and Biotechnology*, 342.

Çimen, Ö. (2021). Construction and built environment in circular economy: A comprehensive literature review. *Journal of Cleaner Production*, *305*, 127180.

Colicchia, C., & Strozzi, F. (2012). Supply chain risk management: A new methodology for a systematic literature review. *Supply Chain Management*, *17*(4), 403–418.

Constantin, O. E., Stoica, F., Ra u, R. N., Stănciuc, N., Bahrim, G. E., & Râpeanu, G. (2024). Bioactive Components, Applications, Extractions, and Health Benefits of Winery By-Products from a Circular Bioeconomy Perspective: A Review. *Antioxidants*, *13*(1), 100. PMID: 38247524

Da Ros, C., Libralato, G., Ghirardini, A. V., Radaelli, M., & Cavinato, C. (2018). Assessing the potential phytotoxicity of digestate from winery wastes. *Ecotoxicology and Environmental Safety*, *150*, 26–33. PMID: 29268111

De Mauro, A., Greco, M., & Grimaldi, M. (2016). A formal definition of Big Data based on its essential features. *Library Review*, *65*(3), 122–135.

Denyer, D., & Tranfield, D. (2009). *Producing a systematic review.*

Dixon-Woods, M. (2011). Using framework-based synthesis for conducting reviews of qualitative studies. *BMC Medicine*, *9*, 1–2. PMID: 21492447

Dwyer, K., Hosseinian, F., & Rod, M. (2014). The market potential of grape waste alternatives. *Journal of Food Research*, *3*(2), 91–106.

Ebrahimzadeh, S. K., Navidshad, B., Farhoomand, P., & Aghjehgheshlagh, F. M. (2018). Effects of grape pomace and vitamin E on performance, antioxidant status, immune response, gut morphology and histopathological responses in broiler chickens. *South African Journal of Animal Science*, *48*(2), 324–336.

European Commission. Directorate-General for Communication. (2020). *Circular economy action plan: for a cleaner and more competitive Europe*. Publications Office of the European Union. https://data.europa.eu/doi/10.2779/05068

Fernandes, R. A., Lopes, S., Ferreira, N., Santos, J., Martins, J. M., & Carvalho, L. H. (2024). Binderless particleboards obtained 100% from winery by-products for the packaging industry. *Frontiers in Food Science and Technology*, *4*, 1376415.

Flores, L., Josa, I., García, J., Pena, R., & Garfí, M. (2023). Constructed wetlands for winery wastewater treatment: A review on the technical, environmental and socio-economic benefits. *The Science of the Total Environment*, *882*, 163547. PMID: 37080316

Gabur, G.-D., Teodosiu, C., Fighir, D., Cotea, V. V., & Gabur, I. (2024). From Waste to Value in Circular Economy: Valorizing Grape Pomace Waste through Vermicomposting. *Agriculture*, *14*(9), 1529.

Gabzdylova, B., Raffensperger, J. F., & Castka, P. (2009). Sustainability in the New Zealand wine industry: Drivers, stakeholders and practices. *Journal of Cleaner Production*, *17*(11), 992–998.

Gallardo-Vázquez, D., Scarpellini, S., Aranda-Usón, A., & Fernández-Bandera, C. (2024). How does the circular economy achieve social change? Assessment in terms of sustainable development goals. *Humanities & Social Sciences Communications*, *11*(1), 1–18.

Garrido, R. A., Manrique, R., Fredes, J., Rodriguez, P., Rodríguez, A., Serafini, D., Mena, M., Masip, Y., & Díaz, I. (2024). Evaluating hydrogen production from grape pomace gasification: Unveiling the potential for Chile's wine industry and its solid waste recovery as energy source. *Renewable Energy*, *223*, 119953.

Geissdoerfer, M., Savaget, P., Bocken, N. M. P., & Hultink, E. J. (2017). The Circular Economy–A new sustainability paradigm? *Journal of Cleaner Production*, *143*, 757–768.

Greenhalgh, T. (1997). How to read a paper: Papers that summarise other papers (systematic reviews and meta-analyses). *BMJ (Clinical Research Ed.)*, *315*(7109), 672–675. PMID: 9310574

Kafantaris, I., Stagos, D., Kotsampasi, B., Hatzis, A., Kypriotakis, A., Gerasopoulos, K., Makri, S., Goutzourelas, N., Mitsagga, C., & Giavasis, I. (2018). Grape pomace improves performance, antioxidant status, fecal microbiota and meat quality of piglets. *Animal*, *12*(2), 246–255. PMID: 28712376

Khaskheli, A., Jiang, Y., Raza, S. A., Qureshi, M. A., Khan, K. A., & Salam, J. (2020). Do CSR activities increase organizational citizenship behavior among employees? Mediating role of affective commitment and job satisfaction. *Corporate Social Responsibility and Environmental Management*, *27*(6), 2941–2955.

Kitchenham, B., Budgen, D., Brereton, P., Turner, M., Charters, S., & Linkman, S. (2007). Large-scale software engineering questions - Expert opinion or empirical evidence? *Software, IET*, *1*, 161–171. DOI: 10.1049/iet-sen:20060052

Kobayashi, T., & Nakajima, L. (2021). Sustainable development goals for advanced materials provided by industrial wastes and biomass sources. *Current Opinion in Green and Sustainable Chemistry*, *28*, 100439.

Lazcano, C., & Domínguez, J. (2011). The use of vermicompost in sustainable agriculture: impact on plant growth and soil fertility. *Soil Nutrients*, *10*(1–23), 187.

Lee, K.-M. (2013). Sustainable products and sustainable production: Introduction. In *Handbook of sustainable engineering* (pp. 261–267). Springer.

Lindner, A. V., & Pleissner, D. (2022). Removal of phenolic compounds from olive mill wastewater by microalgae grown under dark and light conditions. *Waste and Biomass Valorization*, *13*(1), 525–534.

Lisičar Vukušić, J., Millenautzki, T., Reichert, L., Mokhlis Saaid, A., Müller, L., Clavijo, L., Hof, J., Mösche, M., & Barbe, S. (2023). Conversion of Problematic Winery Waste into Valuable Substrate for Baker's Yeast Production and Solid Biofuel: A Circular Economy Approach §. *Food Technology and Biotechnology*, *61*(4), 430–438. PMID: 38205048

López-Pérez, J. A., & Sánchez-Moreno, S. (2021). Soil response to root-knot nematode management with wine vinasse in a solarised horticultural soil under glasshouse conditions. *Nematology*, *24*(4), 431–451.

Manniello, C., Statuto, D., Di Pasquale, A., Giuratrabocchetti, G., & Picuno, P. (2020). Planning the flows of residual biomass produced by wineries for the preservation of the rural landscape. *Sustainability*, *12*(3), 847.

Manniello, C., Statuto, D., Di Pasquale, A., & Picuno, P. (2020). Planning the flows of residual biomass produced by wineries for their valorization in the framework of a circular bioeconomy. *Innovative Biosystems Engineering for Sustainable Agriculture, Forestry and Food Production: International Mid-Term Conference 2019 of the Italian Association of Agricultural Engineering (AIIA)*, 295–303.

Marin, D. E., Bulgaru, C. V., Anghel, C. A., Pistol, G. C., Dore, M. I., Palade, M. L., & Taranu, I. (2020). Grape seed waste counteracts aflatoxin B1 toxicity in piglet mesenteric lymph nodes. *Toxins, 12*(12), 800. PMID: 33333857

Mayring, P. (2004). Qualitative content analysis. *A Companion to Qualitative Research, 1*(2), 159–176.

Mehta, S. (2023). Circular Economy and Sustainable Development Goals. *MET Management Review, 10*(1), 24–29.

Merli, R., Preziosi, M., & Acampora, A. (2018). How do scholars approach the circular economy? A systematic literature review. *Journal of Cleaner Production, 178*, 703–722.

Morone, P., Tartiu, V. E., & Falcone, P. (2015). Assessing the potential of biowaste for bioplastics production through social network analysis. *Journal of Cleaner Production, 90*, 43–54.

Nascimento-Gonçalves, E., Azevedo, T., Lopes, H., Sousa, J. R., Oliveira, P. A., Roboredo, M., Coimbra, A. M., & Morais, M. C. (2024). Vermicomposting as a Valorization Solution to the Winery Sector By-Products. *Agronomy (Basel), 14*(6), 1111.

Ngwasiri, P. N., Ambindei, W. A., Adanmengwi, V. A., Ngwi, P., Mah, A. T., Ngangmou, N. T., Fonmboh, D. J., Ngwabie, N. M., Ngassoum, M. B., & Aba, E. R. (2022). A review paper on agro-food waste and food by-product valorization into value added products for application in the food industry: Opportunities and challenges for Cameroon bioeconomy. *Asian J. Biotechnol. Bioresour. Technol, 8*, 32–61.

Niculescu, V.-C., & Ionete, R.-E. (2023). An Overview on Management and Valorization of Winery Wastes. *Applied Sciences (Basel, Switzerland), 13*, 5063.

Nunes, L. J. R., Rodrigues, A. M., Matias, J. C. O., Ferraz, A. I., & Rodrigues, A. C. (2021). Production of biochar from vine pruning: Waste recovery in the wine industry. *Agriculture, 11*(6), 489.

Oliveira, M., Costa, J. M., Fragoso, R., & Duarte, E. (2019). Challenges for modern wine production in dry areas: Dedicated indicators to preview wastewater flows. *Water Supply, 19*(2), 653–661.

Omer, M. A. E., Ibrahim, A. M. M., Elsheikh, A. H., & Hegab, H. (2025). A Framework for Integrating Sustainable Production Practices Along the Product Life Cycle. *Environmental and Sustainability Indicators*, •••, 100606.

Ortiz-de-Montellano, C. G.-S., Samani, P., & van der Meer, Y. (2023). How can the circular economy support the advancement of the Sustainable Development Goals (SDGs)? A comprehensive analysis. *Sustainable Production and Consumption, 40*, 352–362.

Pérez-Bibbins, B., Torrado-Agrasar, A., Salgado, J. M., de Souza Oliveira, R. P., & Domínguez, J. M. (2015). Potential of lees from wine, beer and cider manufacturing as a source of economic nutrients: An overview. *Waste Management (New York, N.Y.), 40*, 72–81. PMID: 25824282

Polit, D. F. (2004). *Nursing research: Principles and methods*. Lippincott Williams & Wilkins.

Prajapati, H., Kant, R., & Shankar, R. (2019). Bequeath life to death: State-of-art review on reverse logistics. *Journal of Cleaner Production, 211*, 503–520.

Pratt, M. (2012). *Evaluating the Comparison of Sustainability Programs in the Wine Industry*. ITWM.

Provenzano, M., Pacchera, F., Silvestri, C., & Ruggieri, A. (2024). From Vineyard to Value: A Circular Economy Approach to Viticulture Waste. *Resources, 13*(12), 172.ù

Rondeau, P., Gambier, F., Jolibert, F., & Brosse, N. (2013). Compositions and chemical variability of grape pomaces from French vineyard. *Industrial Crops and Products, 43*, 251–254.

Saade, M. R. M., Yahia, A., & Amor, B. (2020). How has LCA been applied to 3D printing? A systematic literature review and recommendations for future studies. *Journal of Cleaner Production, 244*, 118803.

San Martin, D., Ibarruri, J., Gutierrez, M., Ferrer, J., Garcia-Rodriguez, A., Goiri, I., Urkiza, J., Zufía, J., Sáez de Cámara, E., & Iñarra, B. (2024). Valorization of grape stem as an alternative ingredient in rabbit feed. *Waste Management & Research*, 0734242X241259660.

San Martin, D., Ibarruri, J., Luengo, N., Ferrer, J., García-Rodríguez, A., Goiri, I., Atxaerandio, R., Zufía, J., Sáez de Cámara, E., & Iñarra, B. (2023). Evaluation of Valorization Strategies to Improve Grape Stems' Nutritional Value as an Ingredient for Ruminants' Diets. *Sustainability, 15*(15), 11951.

Santos, I. V., Renaud-Gentié, C., Roux, P., Levasseur, A., Bulle, C., Deschênes, L., & Boulay, A.-M. (2023). Prospective life cycle assessment of viticulture under climate change scenarios, application on two case studies in France. *The Science of the Total Environment, 880*, 163288. PMID: 37028673

Sartori, R. B., Padrón, R. A. R., & Mérida, L. G. R. (2023). Current status of the microalgae application in agriculture and aquaculture. In *Microalgae-Current and Potential Applications*. IntechOpen.

Sassanelli, C., Rosa, P., Rocca, R., & Terzi, S. (2019). Circular economy performance assessment methods: A systematic literature review. *Journal of Cleaner Production, 229*, 440–453.

Scarpellini, S. (2022). Social impacts of a circular business model: An approach from a sustainability accounting and reporting perspective. *Corporate Social Responsibility and Environmental Management, 29*(3), 646–656.

Seuring, S., & Müller, M. (2008). From a literature review to a conceptual framework for sustainable supply chain management. *Journal of Cleaner Production, 16*(15), 1699–1710.

Shahabi Mohammadabadi, S., Goli, M., & Naji Tabasi, S. (2022). Optimization of Bioactive Compound Extraction from Eggplant Peel by Response Surface Methodology: Ultrasound-Assisted Solvent Qualitative and Quantitative Effect. *Foods, 11*, 3263. PMID: 37431011

Silvestri, C., Silvestri, L., Forcina, A., Di Bona, G., & Falcone, D. (2021). Green chemistry contribution towards more equitable global sustainability and greater circular economy: A systematic literature review. *Journal of Cleaner Production, 294*, 126137.

Silvestri, C., Silvestri, L., Piccarozzi, M., & Ruggieri, A. (2024). Toward a framework for selecting indicators of measuring sustainability and circular economy in the agri-food sector: A systematic literature review. *The International Journal of Life Cycle Assessment, 29*(8), 1446–1484. DOI: 10.1007/s11367-022-02032-1

Sousa, A. C., Dias, C., Martins, A. R., Gomes, A. G., & Santos, C. A. (2025a). Using winery effluents for cultivating microalgae as bio-additives for vineyards. *Journal of Applied Phycology*, •••, 1–14.

Sousa, A. C., Dias, C., Martins, A. R., Gomes, A. G., & Santos, C. A. (2025b). Using winery effluents for cultivating microalgae as bio-additives for vineyards. *Journal of Applied Phycology*, •••, 1–14.

Spennatia, E., Casazzaa, A. A., Peregoa, P., & Solisioa, C. (2019). Microalgae growth in winery wastewater under dark conditions. *CHEMICAL ENGINEERING, 74*.

The International Organization of Vine and Wine. (n.d.). *OIV Report 2023*.

Tranfield, D., Denyer, D., & Smart, P. (2003). Towards a methodology for developing evidence-informed management knowledge by means of systematic review. *British Journal of Management, 14*(3), 207–222.

Tucker, C. B., Weary, D. M., & Fraser, D. (2003). Effects of three types of free-stall surfaces on preferences and stall usage by dairy cows. *Journal of Dairy Science, 86*(2), 521–529. PMID: 12647958

United Nations. (2015)... *Agenda (Durban, South Africa), 2030*, •••. https://sdgs.un.org/2030agenda

Xu, Q., Zhang, T., Niu, Y., Mukherjee, S., Abou-Elwafa, S. F., Nguyen, N. S. H., Al Aboud, N. M., Wang, Y., Pu, M., & Zhang, Y. (2024). A comprehensive review on agricultural waste utilization through sustainable conversion techniques, with a focus on the additives effect on the fate of phosphorus and toxic elements during composting process. *The Science of the Total Environment*, •••, 173567. PMID: 38848918

Yadav, S., Malik, K., Moore, J. M., Kamboj, B. R., Malik, S., Malik, V. K., Arya, S., Singh, K., Mahanta, S., & Bishnoi, D. K. (2024). Valorization of agri-food waste for bioactive compounds: Recent trends and future sustainable challenges. *Molecules (Basel, Switzerland), 29*(9), 2055. PMID: 38731546

Zhang, Q., & Zhang, H. (2024). Assessing Agri-Food Waste Valorization Challenges and Solutions Considering Smart Technologies: An Integrated Fermatean Fuzzy Multi-Criteria Decision-Making Approach. *Sustainability, 16*(14), 6169.

Zhang, Y., Ye, C., Su, Y., Peng, W., Lu, R., Liu, Y., Huang, H., He, X., Yang, M., & Zhu, S. (2022). Soil Acidification caused by excessive application of nitrogen fertilizer aggravates soil-borne diseases: Evidence from literature review and field trials. *Agriculture, Ecosystems & Environment, 340*, 10817.

# Chapter 10
# Circular Fashion and Recycling:
## A Sustainable Approach to Transform the Fashion and Retail Industry

**Shivangi Kashyap**
https://orcid.org/0009-0002-1685-330X
*University of Hyderabad, India*

**Pravin Kumar**
https://orcid.org/0009-0008-9393-7324
*Brainware University, India*

## ABSTRACT

*The fashion industry is the center of global culture and global economy and has increasingly come under criticism due to its influence on the environment and society in specific ways (particularly mass production, overproduction, waste, and pollution by fast fashion). One such potential solution option will be the circular fashion system, which focuses on recycling, upcycling, closed-loop processes, and increased product lifecycle. Circular economy methods can grow down ecological harm, diminish relation on resources, and generate innovative economic prospects. The present research illuminates the effectiveness of the stakeholders, policies, and technological innovations, enhanced by the brand case studies implementing sustainable practices. It also tackles difficulties in scaling circular models as well as defines directions towards wider levels of application. The chapter concludes that circular fashion is sustainable in its growth because it connects environmental recovery and economic growth and societal well-being.*

DOI: 10.4018/979-8-3373-2454-8.ch010

## INTRODUCTION

The world fashion market is a highly diverse and complex market that has benefited the economies, cultures, and even personal lives in many ways. Fashion is a major source of revenue (estimated to be 2.5 trillion rupees every year), thus not only a creator of economy but also a contributor to culture. Beneath its glitzy face, there is another darker side of the industry, which is contributing to some grave environmental and social problems. It contributes almost 10 percent of global carbon emissions- more than the emissions produced by international flights and maritime shipping. Dyeing alone is a process that pollutes water supplies with toxic dyes, and fast fashion contributes to industrial overproduction, waste, and exploitation, as landfills become littered with cast-offs and supplies dumped in resources, and pressure on the ecosystems they are a part of.

Alternatively, instead of the current take-make-dispose set-up, there is a circular fashion, which presents itself as a promising alternative at this strategic crossroads. The circular fashion is based on the principles of the circular economy that presupposes reducing waste, extending the life of goods, and closing the cycles of the resource cycle. The paradigm shift not only presents an environmental opportunity, but it also presents business model challenges to the industry to rewire their supply chains and rethink consumer behavior (Waaje et al., 2025; Karim et al., 2024a). This way, it tries to find an optimal solution between the cultural and economic value of fashion and the need for sustainability.

The significance and the topicality of this transformation are manifold. Circular fashion responds to international issues of raw material shortages and overconsumption, and follows consumer needs of ethical and ecological products. To policymakers, it provides a guide in which to develop regulations that encourage sustainable practices. To businesses, it brings the potential to drive customer loyalty, provide new revenue sources, and stabilize against resource uncertainty. The circular fashion model gives consumers a chance to make purchasing choices which does not conflict with their personal values, to remain a culture of responsible consumption. The Circular Economy Action Plan (2020) by the EU, which mandates textile manufacturers to recycle their products and increase the life span of their products, is being applied to clothing and circumvented by the EPR policy to bolster the plastic industry in India. The models of regulation inform how a policy framework can be used to accelerate change or become one of the obstacles at the industry level.

### Research Objectives

The purpose of this chapter is to investigate in which way circular fashion can transform the fashion and retail industry by answering the following questions:

- *To determine the various ways that different circular fashion activities contribute to mitigating the impact of the fashion industry on the climate.*
- *To discuss the economic and social impact of circular fashion and its role in community welfare.*
- *To determine whether the long-term stakeholder value generated by the adoption of circular fashion over the long term is worth the cost and risk.*

By means of these objectives, the chapter strives to deliver practical ideas and considerations to stakeholders interested in working in the visitor surfaces of sustainability, profitability, and consumer behavior.

## Methodological Note

The chapter is interpretive and qualitative in approach. The originality of the analysis is anchored in secondary data obtained from academic sources, industry reports, and policy papers to obtain information on the sustainability practices and the challenges of change in the fashion industry. In order to translate theory into practice, the chapter includes three case studies on Patagonia, H&M, and Stella McCartney, focusing on how leading brands already incorporate circular principles into their various business models. A thematic review approach was chosen to detect similarities in the practice of circles, and triangulation with validation literature and evidence in the industry provides rigor to the analysis. It is through this integrated approach that the chapter can relate conceptual frameworks with real-world applications in a circular fashion. The EU has proposed a Circular Economy Action Plan (2020), and it requires textile manufacturers to make their products recyclable and increase product life span, whereas India has an Extended Producer Responsibility (EPR) policy, already in use with plastic, that is being adjusted to work in the textile sector. Those regulatory models demonstrate how certain policy frameworks can expedite or inhibit industry-wide change.

## Chapter Structure

This chapter is structured in the following way: The Literature Review presents a background to the theoretical discourse and recent discussions about circular fashion. The Case Studies will take up the initiatives of Patagonia, H&M, and Stella McCartney to demonstrate practical applications. The Discussion contains a discussion of how technology, in this case, blockchain and advanced recycling systems, can play a part and addresses the obstacles and challenges to scale that are encountered by the industry. Lastly, the Conclusion draws the commonalities of the arguments presented in the above sections with implications that need to be addressed by policymakers, business owners, and consumers, as well as future directions. The EU Circular Economy Action plan (2020) prescribes recycling by textile manufacturers and extending the service life of their products, whereas EPR policy on plastics, currently in use in India, is being employed in adaptation to textiles. These control paradigms demonstrate how a policy framework may either hasten or inhibit industry-wide change.

## Literature Review

Circular fashion has been defined as a response to unsustainable trends in the fast fashion industry (Cachon & Swinney, 2011). It is based on the conceptual thought leadership of the circular economy and suggests addressing the issue of fashion products by changing the way they are designed, produced, consumed, and disposed of (Bocken et al., 2016; Waaje et al., 2025). This is opposite to the linear take-make-dispose mode, which is to prevent losses to sources of resources, to produce no waste, and to create renewable chains of resources where the products and materials can maintain their utility as long as they can (Macarthur, 2017). In this regard, circular fashion is not merely an environmental movement but also an economic and social one and defines how businesses, citizens, and governments approach the subject of sustainability (Ki et al., 2020).

On a fundamental level, circular fashion draws attention to the durability, resource consumption, and recycling of waste and its conversion to resources (Fletcher, 2013). The design of durable, repairable, and recyclable garments is being promoted by designers and manufacturers (Gwilt, 2020). This is a striking difference to the fast fashion industry that creates, burns off its stock of products faster than it can be hoped to stock, and ends up destroying much (Cachon & Swinney, 2011). The fashion industry is breaking this paradigm through circular fashion practices such as recycling, upcycling, and closed-loop production, re-conceptualizing the fashion industry as a mechanism of value creation rather than one of extracting a finite resource base (Bocken et al., 2016).

The ecological impact of the fashion industry is immense, and that is why circularity needs to be implemented as soon as possible (Chen & Burns, 2006). Cotton cultivation is linked to high water consumption, huge energy, and chemical pollution of the environment (Khattab et al., 2020). The synthetic fibers introduce an additional aspect to the issue because they also cause microplastic pollution in the oceans (Chen & Burns, 2006). By seeing these problems and facing them, circular fashion proposes to go further and develop what is available now, more alternatives, such as alternative fibers, renewable materials, and systems to prolong the life of garments (Fletcher, 2013). Recycling in post-consumer clothing uses existing products, diverting them from landfills and converting them into new fibers, whereas upcycling means that designers can up-cycle the old clothes into even better items, thus creating higher value (Gwilt, 2020). Closed-loop systems take this a step further and design garments in a fashion that would allow them to be broken down, and all parts can be used in the production process once again (Bocken et al., 2016).

Circular fashion, in economic terminology, can result in business model innovation, according to the literature (Blau & Kimmig, 2024). New business models, including clothing rental, resale platforms, and repair services, have the potential to create additional revenue opportunities as well as contribute to the needs of the environmentally conscious clientele (Ki et al., 2020). Instead of focusing on mass production and fast movement, fashion companies are starting to realize that they can develop customer loyalty in terms of being transparent and responsible, and participating in circular efforts (Pérez-Bou et al., 2021). Both the large corporations and smaller companies are trying to explore second-hand markets, digital traceability mechanisms, and new materials that coincide with the expectations of consumers on sustainability (Henninger et al., 2016). This transition can be seen as part of a bigger trend to recognise that sustainable profitability is not antithetical to sustainable practices as long as a business redesigns its business model so that it is circular in nature (Bocken et al., 2016).

On the social level, such a change towards circular fashion occurs due to the social concern of the environmental/ethical impact that fashion has (Gardetti & Torres, 2017). Shifting trends toward sustainable fashion can be explained by a growing number of people criticizing the environmental degradation caused by fast fashion among younger generations (Henninger et al., 2016). Cultural norms concerning consumption are also in a stage of transition, with second-hand clothing and clothing hire as well as upcycled goods becoming increasingly mainstream (Fletcher, 2013). These changes are important as previously consumer habits have interfered with fashion business in going green, with prices being generally the biggest factor, and comfort, amongst other choices (Gwilt, 2020). Transparent communications and advocacy work by brands are essential in the attitude reshaping aimed at consumers, integrating sustainability in their daily purchases (Pérez-Bou et al., 2021).

## Context and Importance of Circular Fashion

Garnering over 2.5 trillion in global industry and being a cultural force as well as an economic stimulus, the fashion industry is also one of the largest resource users in the world (Macarthur, 2017). It also emits almost 10 percent of global carbon emissions, which is more than international flights and shipping emissions combined (Chen & Burns, 2006). The contamination of water supply with poisonous dyes, excessive use of freshwaters in manufacturing textiles, and the emergence of fast fashion have also boosted such pressures (Khattab et al., 2020).

Circular fashion is the answer to this problem by changing the system from linear to the system of closed-loop system (Bocken et al., 2016). This concept brings sustainability design, responsible production, conscious consumption, and recycling together (Ki et al., 2020). The three pillars at its heart are material innovation, extended product life-cycles, and reverse logistics that keep the resources in flow as opposed to becoming waste (Macarthur, 2017).

*Figure 1. Transforming Fashion: From Waste to Sustainability*

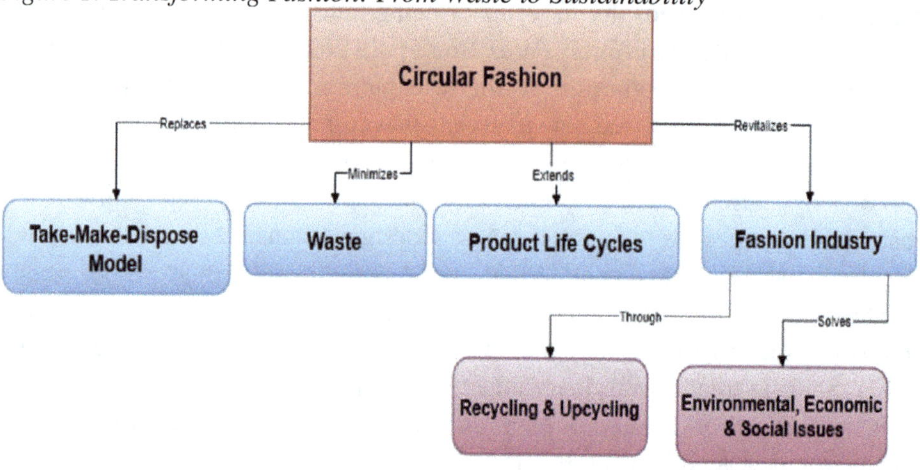


*(Source: Author's Work)*

Recycling is in the middle of this upheaval (Fletcher, 2013). Recycling post-consumer textiles into fibers and fabrics reduces the number of virgin resources used and reduces the use of water and energy (Chen & Burns, 2006). The recycling of inputs also contributes to curbing landfill congestion, transforming fashion to being a resource regenerative one rather than an extractive industry (Gwilt, 2020).

It is not just the ecological advantages of circular fashion (Gardetti & Torres, 2017). It creates new economic possibilities with resale markets, rental business, and repair services, and increases customer loyalty to brands that appeal to the customer's values (Henninger et al., 2016). Socially, it promotes more socially responsible patterns of consumption, and it affords the consumer a means of relating his/her values with buying behaviour (Pérez-Bou et al., 2021). The role of policymakers is of crucial importance as well (Ki et al., 2020). The Circular Economy Action Plan by the EU mandates recyclability and lengthy lifetimes in textile products, whereas India has been finding success applying its Extended Producer Responsibility policy to textiles based on what it has become accustomed to with plastics (Macarthur,

2017). These fruitful examples of regulation explain how its regulation can speed up the transition or, in case of misalignment, slow it down (Bocken et al., 2016).

As a combination of environmental renewal, business innovation, and cultural change, circular fashion transforms into a paradigm of change that could reshape the industry and solve its most pressing problems (Fletcher, 2013).

Nevertheless, numerous impediments to a high adoption rate of the circular models have been identified in the literature (Blau & Kimmig, 2024). Technological issues are not absent, especially when it comes to recycling of mixed fiber blends and producing economical alternatives to virgin resources (Chen & Burns, 2006). The price of circular innovation is often prohibitive, even to small firms, and consumer culture, not yet accustomed to the logistics of sustainable purchases, is unlikely to purchase either rapidly or in large quantities (Cachon & Swinney, 2011). The lack of definition of what makes a product circular is also causing concern, with the problem of greenwashing, with brands promoting circular products and actions without making changes to their systems (Henninger et al., 2016). These barriers reaffirm the difference between systematic enablers, including favourable policy systems, financial, and technological advancements (Gwilt, 2020).

The stakeholders play a major role in promoting circularity (Ki et al., 2020). Designers can affect the sustainability of clothing at the concept design phase, whereas brands and retailers have a say as to how business models will span their processes, which they cost-effectively incorporate practices like take-back schemes, repair shops, or recycling collaborations (Fletcher, 2013). Policymakers can influence the industry by establishing regulations, doling out subsidies, and implementing extended responsibilities where producers are rewarded when practicing sustainable options and punished when doing otherwise (Macarthur, 2017). Consumers, in turn, have the potential to boost the demand of a sustainable fashion industry by adopting decisions that are based on ethical and environmental beliefs (Gardetti & Torres, 2017). Civil society groups and NGOs also have a share in the power through creating awareness, nudging businesses, and accountability monitoring (Pérez-Bou et al., 2021). Combined with the other actors, an ecosystem is created where collaboration is the key to scaling circular fashion beyond individual projects (Bocken et al., 2016).

One more theme in the literature is the technological innovation in the enablement of circularity (Blau & Kimmig, 2024; Roshid et al., 2025). The technological advances in chemical recycling allow recycling more sophisticated fabrics into reusable fibers, and the digitization of supply chains through technologies such as blockchain solutions helps to achieve more transparency and traceability (Khattab et al., 2020). Such innovations are not only coming to meet practical issues related to recycling and reuse, but also in creating confidence in the consumer in the verification of sustainable sources and ethically acceptable labor (Henninger et al.,

2016). With the integration of technology into circular thoughts, fashion brands can address logistical and credibility issues and present themselves as industry leaders regarding sustainable change (Gwilt, 2020).

In spite of advances, there are unmet information needs in academia and industry (Fletcher, 2013). Most of the current literature addresses individual or sporadic actions, such as recycling or rental, without considering a complete picture of the connections between the strategies in a holistic system of cycles (Ki et al., 2020). Likewise, technologies such as blockchain and advanced recycling have often been the subject of discussion, but few studies exist on a large-scale impact (Blau & Kimmig, 2024). In addition, differences in regional policy, consumer culture, and economic potential would make circular approaches to fashion inapplicable everywhere in the industry (Gardetti & Torres, 2017). Such gaps highlight the significance of the case studies that can demonstrate how circular practices work in various contexts and show what circumstances make them thrive or falter (Macarthur, 2017).

In summary, the literature defines circular fashion as a transformational model in dealing with the environmental, economic, and social deficits of the existing fashion system (Bocken et al., 2016). It provides avenues of sustainable innovation but also shows that it faces huge obstacles that have to be mitigated by collaboration, supportive policy-making, technological development, and cultural transformation (Ki et al., 2020). Considerable research has now accumulated to support this practice, but it also leaves unanswered questions regarding scale, integration, and long-term stakeholder benefit (Fletcher, 2013). The existence of these gaps justifies the underlying chapter that tries to connect theory and practice with the help of case studies of other dominant brands and the identification of conditions favorable to a truly circular fashion industry (Gwilt, 2020).

## Gaps in the Current Literature

Although there is increasing interest circularly, several gaps exist. Most of the studies are theoretical in nature and concentrate on a single micro-action to a given solution, like recycling or renting, with little investigation on the interplay of micro-solutions in a holistic system. The atomized view is limiting the aggrandizement of a circular economy perspective.

Another gap relates to differences across business models, markets, and cultural contexts. Most of the research papers regarding sustainability design lump it all together without paying any attention to the variation existing between fast fashion and luxury, and developed and developing economies.

Technological solutions such as blockchain and advanced recycling are also mentioned, but there is little actual evidence as to the effectiveness and scalability of these technological advancements. Although operational data were not available, it is tough to determine their contribution to closing material loops or their industry-wide cluster applications.

There is also an inconsistency in the area of measurement. Consistent metrics to assess circularity are still in their infancy, so it can be difficult to strike a balance between environmental, economic, and social considerations or compare different initiatives with one another. This is associated with the debate of greenwashing, wherein brands distract by virtue of some not even significant initiatives, as a demonstration of sustainability with little accountability.

These lacunae dictate additional research that has the capacity to link theory and practice. This chapter identifies case studies of Patagonia, H&M, and Stella McCartney to help demonstrate how these companies incorporate elements of circular fashion in different contexts, barriers that they continue to face, and how technologies, consumer participation, and policy can collaborate to accelerate the scale of circular fashion.

## Case Studies of Leading Initiatives

Case studies show that circular fashion is workable and demonstrates how far the sector can be changed. They illustrate what can be done about making sustainability an organizational core competence, lowering society's environmental impact, and linking up with environmentally concerned buyers. On earth, 92 million tons of textile waste are produced annually, and only 12 percent of it becomes the object of reprocessing to make something new. The textile waste is projected to be 8 million tons in India by the year 2030.

### Patagonia: The Worn Wear Program

Patagonia's Worn Wear program works circularly. This convinces clients to bring back their used garments for repair, reselling, or reclaiming materials for remaking. On top of that, the Circum–Wear Program not only increases the number of use cycles of clothing but also solidifies the brand Patagonia's focus on being environmentally friendly. The brand can come up with repair services and resale alternatives, all of which would help to prevent textile waste from ending up in the dustbin. Furthermore, most of the time that this company operates is transparent about its practices, making Patagonia one of the symbols of sustainable outdoor apparel.

### H&M: Conscious Collection and Garment Collection Programs

H&M has demonstrated a good attempt at circular process adoption in its functioning. The elements of Conscious Collection include products made of organic and recycled materials to prove that sustainable fashion is quite cheap and trendy. In addition, the brand's garment collection programme encourages consumers to return any unwanted clothes, from any brand, to receive a discount. These actions help to prevent the emission of textile waste and also create interaction of the consumers with the circular economy. Thus, linking sustainable thinking and actions to conventional fashion, H&M shows that rapid scaling of circular models is viable in fast-fashion systems.

### Stella McCartney: Setting the Standard for Sustainable Luxury

The British designer Stella McCartney has always appealed to the concerns of sustainability and has combined aesthetics with respect for the environment. The recycling of polyester is evident in the manufacturing of the products, with the incorporation of organic cotton underpins the company's environmentally sustainable development goals for the high-end fashion market. As for the bull leather controversy, Stella McCartney retains a policy of closed-loop systems, where materials stay in use to prevent being wasted. Apart from that, this approach also serves as an example of being environmentally friendly for established luxury designers, as well as for encouraging other designers to establish effective eco-friendly solutions in the luxury segment. The Circular Economy Action Plan (2020), the EU has compelled the textile producers to make their products recyclable and have a longer product life cycle, whereas India is adopting the Extended Producer Responsibility (EPR) policy it has on plastics to textiles. These regulatory paradigms demonstrate how policy systems will become more rapid or sluggish depending on how an industry transforms.

These examples show that circular fashion can be a starting point relevant to various brands, and its principles can be otherwise incorporated. All of these include the repair and recycling programs, line sustainable collections, and using unique materials, which show that fashion can be circular and effective. Altogether, they are bringing the change that can make the new perspective to the sports industry more sustainable and ethical.

## DISCUSSION

In the chapter, the authors emphasize the importance of considering circular fashion as one of the key pillars of the change in the fashion industry's business model (Pal & Gander, 2018). The transition from the traditional 'taking-making-disposing' pattern to a circular economy approach can be an answer to various worldwide environmental, economic, and social concerns (Niinimäki et al., 2020).

Recycling, which forms the basis of circularity in fashion, indicates the vast potential to serve as a solution to minimize the waste of textile products (Pomponi & Moncaster, 2017). However, chemical recycling still has issues with adaptability due to the use of sophisticated technology and protocols, especially for SMEs. Innovation and creativity give value to upcycling, but it is still a limited concept due to a lack of knowledge and integration in the marketplace (Kozlowski et al., 2018). Typically, expected perfect sustainable systems like closed-loop systems need huge initial capital and strong supply chain systems to initiate mass production.

From an economic perspective, circular fashion represents a dual opportunity: reduction of the inputs needed and the discovery of new methods of generating an equal or higher value of business from recycling or a recycling-based economy (Pal & Gander, 2018). These and similar services, as second-hand marketplaces and repair platforms, meet the increasing demand for sustainability. But it gets to the issue of second-hand clothing, which is traditionally associated with some markets or creating value for the affordability of eco-friendly fabrics (Morgan & Birtwistle, 2009).

From a societal perspective, the move to the circular economy of fashion products requires a radical approach to people's behavioural patterns (Morgan & Birtwistle, 2009). The speed at which fast fashion is presented as cheap and easily accessible remains the major barrier to a sustainable approach (Niinimäki et al., 2020; Karim et al., 2024b). The issue, therefore, requires powerful educational campaigning and unmasked branding so that consumers can be informed appropriately and patronize the products (Pedersen & Andersen, 2015).

The chapter also offers insights into the role of technology, tapping into the possibilities of blockchain for supply chain visibility and the opportunities presented by advanced recycling technologies for material recovery. In practical terms, blockchain has already been piloted in the textile and fashion industry by several brands and consortia. For example, luxury fashion houses have adopted blockchain platforms to certify the authenticity and origin of raw materials such as organic cotton and ethically sourced wool, thereby reducing the risk of counterfeit products and greenwashing. In addition, initiatives such as the TextileGenesis project use blockchain to trace the life cycle of fibers, including polyester, viscose, and recycled materials, allowing manufacturers, retailers, and consumers to verify sustainable sourcing claims. By providing immutable records of production and distribution, blockchain helps ensure that garments entering recycling streams are properly identified for fiber recovery, thereby strengthening closed-loop systems. These applications demonstrate that blockchain is not merely a theoretical tool but a functional mechanism driving accountability, circularity, and trust in the textile industry. Stakeholders, mainly the government, also have a significant responsibility of putting in measures such as tax credits, research grants, and monetary fines for the wastage.

Despite this, the various stakeholders, including the government, business organisations, and consumers, must work together (Pedersen & Andersen, 2015). Sustainable circular fashion, backed by innovation and the prevalence of the systems, can grasp opportunities for economic development with global responsibility as the pioneer of a new trend among industries (Niinimäki, 2013).

## Understanding Circular Fashion

Circular fashion is the set of practices that respond to the linear model of utilization 'take, make, dispose' and are inspired by the circular economy (Pomponi & Moncaster, 2017). By definition, the circular economy does not allow for waste, maximizes the amount of time a material is in use, and seeks to restore the natural system's health (Niinimäki, 2013). In the fashion context, circular fashion refers to fashion goods designed, manufactured ready for reuse, to have low environmental impact, and lastly, ready for recovery and resale. Each year, the textile industry generates 92 million tons of waste, of which only 12 percent is recycled in new products around the globe (WRAP, 2017). In India alone, it is estimated to hit 8 million tons of textile wastes of the garment industry by the year 2030.

The principles guiding circular fashion include:

- *Design for Longevity:* Apparel is designed to last for several wear-and-repair cycles (Kozlowski et al., 2018).
- *Resource Efficiency:* The use of sustainable materials, saving water and energy input, and raw material savings (Niinimäki et al., 2020).
- *Waste Minimization:* Minimizing waste during manufacturing, promoting garment recycling at the consumer level (Morgan & Birtwistle, 2009).
- *Material Recovery:* This would ensure that textiles can be recovered instead of being landfilled for reuse (Pomponi & Moncaster, 2017).

### Core Components

#### *Recycling: Transforming Waste into Reusable Materials*

Recycling remains an important step in circularity for fashion, which is all about repurposing waste materials into new fibers, fabrics, or products (Niinimäki et al., 2020). It cautiously cuts down the utilization of virgin raw materials that, by and large, are resource-intensive, energy and water and water-intensive. For example, recycled PET, which is produced from used plastic bottles, is a more environmentally friendly substitute for virgin PET, which has a small level of carbon footprint and energy consumption.

Scaling circular practices requires systemic enablers, e.g., technology, policy, and collaboration of different stakeholders (Pal & Gander, 2018). Graphically depicted in Figure 2, such enablers contribute to the large-scale use of circular fashion since the recycling process is enhanced, friendly regulations are in place, and intersector collaboration is facilitated. The Circular Economy Action Plan dictates that textile producers should ensure the recyclability of their products, and the Indian EPR policy on plastic is adapted to textiles. These regulatory paradigms demonstrate that policy frameworks can lead to accelerating the speed of change within an industry or slow it down.

*Figure 2. Recycling in Circular Fashion: A Sustainable Approach*

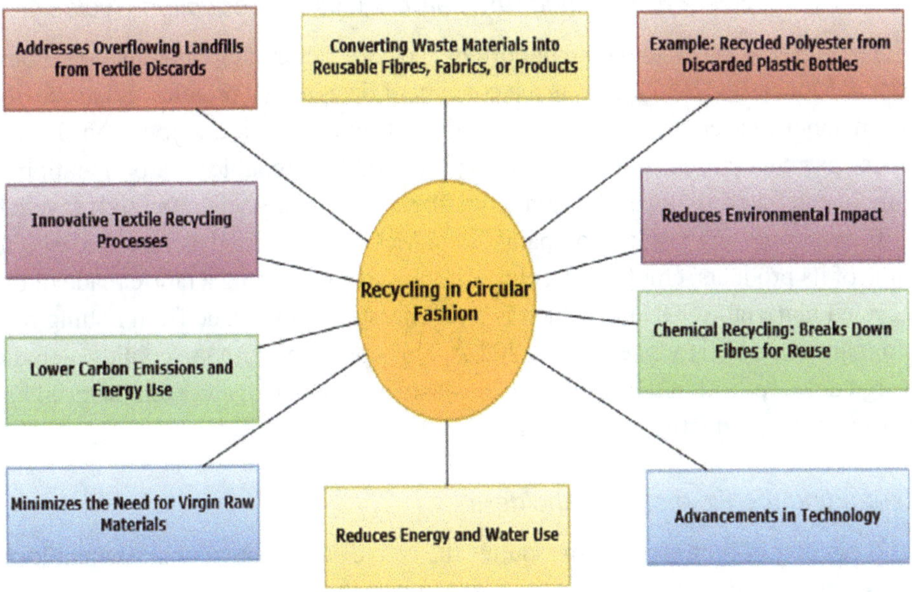


*(Source: Author's illustration)*

In the course of the last few years, chemical recycling has been put forward as a new innovative process of textile recycling that involves the decomposition of fibers into chemical constituents, which are again ready to be reused (Pomponi & Moncaster, 2017). These methods not only minimize waste but also solve the ever-increasing problem of landfills from disposed textiles.

### *Upcycling: Creative Repurposing of Textiles*

This is different from recycling, which can be defined as a process of converting new and waste textiles and garments into products of equal standard (Kozlowski et al., 2018). While recycling products involves changing the material to small pieces or even Tiny particles, upcycling increases the artistic and or practical value of the product. For instance, there can be a reuse of old denim to make handbags or small pieces of fabric to come up with patchwork clothes.

This practice is wise in terms of resource use because it also encourages creativity and differentiation in fashion and gives consumers products that are distinct, special, and, indeed, the one and only. Artisans at the low end of the market alongside luxury brands are turning to up-cycling as a way of combining sustainability with artisanal skill; in so doing, consumers are slowly beginning to redefine what constitutes 'new.'

### *Closed-Loop Systems: Continuous Material Use in Production Cycles*

The closed-loop system is an important and core factor of circular fashion (Pomponi & Moncaster, 2017). This is the concept of designing a product and processes in a manner that ensures that materials do not leave the production loop. In this model, garments are designed to be easily disassembled to enable the aforementioned components, such as zippers, buttons, and fibers, to be easily separated and recycled.

For example, the fashion company Stella McCartney started using circularity in some of its products, employing a closed-loop system by using a fabric made of the recycled material ECONYL, which is a regenerated nylon made from fishing nets and industrial plastic waste (Sipari, 2022). This conforms with the cradle-to-cradle design development in the case in which waste is eliminated, and each material has its role in the production process chain.

### *Expanding the Role of Stakeholders*

A circular fashion ecosystem requires the active participation of all stakeholders (Pedersen & Andersen, 2015):

- **Designers:** Must adopt circular principles in the design phase, focusing on longevity and material recovery (Kozlowski et al., 2018).
- **Consumers:** To behave responsibly as consumers, like choosing products with durability or supporting any recycling/upcycling initiative (Morgan & Birtwistle, 2009).
- **Brands and Retailers:** Business model innovation is needed to integrate circular practices, such as clothing rental or resale platforms (Pal & Gander, 2018).

- **Policymakers:** It is their role to provide regulations and incentives for practicing circular fashion. The Circular Economy Action Plan (2020) in the EU requires textile manufacturers to make their products recyclable and extend their useful life, and the Extended Producer Responsibility (EPR) policy to manage plastics is being replicated to apply to textiles in India. The above regulatory patterns illustrate that policy regimes are capable of accelerating or decelerating the process of change across an industry.

Recycling, upcycling, and closed-loop systems built into the operation mean that fashion can take responsibility not only for immediate environmental concerns but also provide a gold standard for other industries to follow (Niinimäki, 2013).

## The Concept of Circular Fashion

Circular fashion represents a transformative shift in the way the fashion industry operates, departing from the traditional "take-make-dispose" model to embrace sustainability, resource efficiency, and ethical responsibility (Niinimäki et al., 2020). At its core, circular fashion is grounded in the principles of the circular economy, aiming to extend the lifecycle of fashion products while addressing critical environmental, economic, and societal challenges (Pomponi & Moncaster, 2017). This will be achieved by embracing practices involving recycling, upcycling, and closed-loop systems that reduce waste, cut reliance on virgin resources, and drive innovation toward the attainment of sustainable development goals.

### Recycling: Transforming Waste into Resources

The concept of circular fashion differs radically from the conventional linear 'take-make-dispose' model of fashion consumption and production (Niinimäki et al., 2020). In this chapter, circular fashion is defined as fashion that is informed by the circular economy and seeks to decrease the environmental, economic, and social impact of fashion products. Recycling, upcycling, and making the construction industry a closed-loop system. The industry aims to reduce waste, maximize the use of post-consumer recycled material, and encourage innovation in the construction industry that is in line with the Sustainable Development Goals.

Over recent years, various innovations have significantly contributed to making recycling within fashion much easier. Previously, mechanical recycling patterns of fabric, such as tearing them into fibers, have been used alongside chemical recycling techniques. Chemical recycling disassembles multifaceted textile mixtures to their base composition for enhanced fiber regeneration and recycling (Pomponi & Moncaster, 2017). For instance, recycled polyester from used plastic bottles helps save water and energy, besides providing a better solution to the standard polyester fabrication process. This method solves two problems at once, namely, the pollution of the earth by plastic and the disposal of unnecessarily used textiles.

Recycling has many advantages for the environment. Recycling also contributes toward reduced emissions of carbon, the optimal utilization of natural resources, and the reduction of pressure on the overloaded dumpsites (WRAP, 2017). For example, using recycled fibers in clothing decreases water use concern for companies and consumers with limited access to water. Furthermore, using recycled products in manufacturing provides an opportunity for the fashion industry to reduce its impact on the environment and attain a circular economy.

## Upcycling: Adding Value Through Creativity

Upcycling, another foundational approach to circularity, is the innovation of methods towards making new, higher-value products out of old or waste garments (Kozlowski et al., 2018). While recycling reprocesses material, upcycling takes objects and transforms them into something more useful while making the objects more appealing. Besides the benefits of non-renewable resource conservation, it also fosters individuality and innovation in the design of products.

Today, upcycling has become an important philosophy of designers and artisans who seek to combine sustainability with inspiration. Avail of these omnipresent necessities was turned into glorious handbags from old, worn-out jeans and modern, classy clothes from scraps and patches. This approach meets the increasing demand of the market for products that are made individually and are not produced in several identical pieces. Moreover, upcycling provides a way of reusing other materials that may otherwise circulate within the different cycles of production and end up as waste in recycles.

Thus, upcycling has become more popular in luxury fashion segments as a mark of good work and practicing sustainability. Some manufacturers have used a second-life material as part of their collections to show that luxury and sustainability can go hand in hand. This shift is also associated with other social signals that consumers can change their perspective and see sustainable fashion clothing as trendy and stylish.

## Closed-loop Systems: Designing for Continuous Use

There is a bit more to Circularity within Fashion -- Closed-loop systems are the epitome of Fashion Circularity (Pomponi & Moncaster, 2017). These systems are intended to keep materials in constant utilization so that no waste can be produced. A closed-loop garment development model involves designing garments for recycling since their end-of-life design, allowing for easy disassembly and reuse.

For instance, garments designed for easy disassembly permit the decomposition and recycling of different components, including zippers, buttons, in addition to fibers. This chapter argues how closed-loop systems promote circularity and how McCartney's integration of ECONYL regenerated nylon from fishing nets and other industrial plastics supports this cause (Sipari, 2022). The regenerated material, ECONYL, used in the sector, can be recycled continuously without any reduction in quality, which proves the ways materials retain their usefulness in multiple cycles of production.

The cradle-to-grave design is also followed in closed-loop systems where the waste stream does not exist, and each part has its specific function in the production process. The above-discussed integrated Human Resource Management systems hold a promise for making the operations of the fashion industry substantially environmentally friendly.

## Broader Implications of Circular Fashion

Indeed, circular fashion does not limit advantages in economic and social realms to the sum of environmental savings (Pal & Gander, 2018). These are encouraging innovation, developing new sources of income, and encouraging change in the culture of consumption.

## Economic Innovation and Growth

Circular fashion is also a source of significant economic opportunity as businesses adjust, innovate, and seize new markets (Valerie et al., 2022). Reuse and repurpose models create additional revenue streams as a consequence of the demand for sustainable and fair Trade products. Furthermore, new value propositions like second-hand platforms, rental clothes, and platforms for repairing clothes also represent circular fashion as economically sustainable.

While implementing circular approaches, companies tend to have competitive advantages since sustainability remains one of the major customers' motives. For instance, Patagonia's Worn Wear campaign promotes returning worn garments to the company to repair or upcycle to reduce landfill; in the meantime, customers will be loyal to the brand (Syniuk, 2021). Similarly, H&M's case of the Conscious Collection and garment recycling are good examples demonstrating how big brands with top-line business models can apply circular economy models, thereby improving their financial performance and sustainability levels at the same time.

## Societal Transformation

At a societal level, circular fashion is the new way of conducting business through embracing consumed products as a norm. One has to know these changes are happening and that one's actions can make a difference, which is why awareness campaigns and education are part of any sustainable marketing mix. Generation Green: the increased demand for true durability, fair production, and full transparency implies that circular fashion strategies are no longer considered exotic.

However, changing well-established consumer behavior patterns is still a problem. Most people still prefer cheap, throwaway fashion for the same reasons – they are cheap and require no washing. This can only be done through a concerted effort by all the players in the market, which comprises businesses, policymakers, and local and international non-governmental organizations, for the creation of a culture of responsibility and sustainability. Through its Circular Economy Action Plan (2020), the EU requires textile producers to make their products recyclable and work with longer life spans, and India is adapting its EPR initiative on plastics to work on textiles. These paradigms of regulations demonstrate how governmental regulations may boost or halt transformation on an industry-wide scale.

## Challenges and Opportunities

While the potential of circular fashion is undeniable, several challenges must be addressed to realize its full benefits:

- *High Implementation Costs*: Creating new sustainable materials and new recycling technologies could take quite a lot of money and hence can be prohibitive, especially to SMEs.
- *Technological Barriers*: Despite the promising technologies in recycling and upcycling, the question of how these can be scaled to industry-level requirements is still an issue.
- *Behavioral Resistance*: Sustaining awareness creation for consumers to develop and embrace sustainable consumption patterns takes time and resourcefulness. Consumers, governments, NGOs, and multinational retailers are all affected by circularity in different ways. An example is that of the H&M Garment Collecting Program, which recycles more than 20,000 tons of textiles on a yearly basis, yet civil society pressure has questioned whether overproduction is thwarting sustainability targets.
- *Policy Gaps*: Only with supportive regulations and incentives are firms willing to invest in circular practices. In this regard, decisions like providing incentives by special tax treatments for organizations that practice sustain-

ability, as well as punishing companies for unsustainable practices, must be adopted. The Circular Economy Action Plan of the EU (2020) requires textile manufacturers to make their products recyclable and create longer product cycles, and India is adapting its Extended Producer Responsibility (EPR) policy concerning plastics to textile producers. These regulatory models give an example of how policy structures can either hasten transformation within the industry or it may acting as a drag.

All of these ailments require partnerships and coordination between the traditional healthcare system and Complementary and Alternative Medicine (CAM) practitioners. Policymakers, industry, and the general public have a collective responsibility to build a circular fashion economy. New solutions in tech that include Blockchain for supply chain and new recycling methods can unlock present difficulties and speed up the shift to sustainability. The Circular Economy Action Plan (2020) by the EU forces textile manufacturers to make the products recyclable and to extend their life cycles, and India is taking their plastic Extended Producer Responsibility policy as its example, aiming to adjust and use it to make their textile products more sustainable. These regulatory paradigms reveal how systems of policy may hasten or limit industrial-scale change.

## Case Studies: Success Stories in Circular Fashion

Several brands have successfully implemented circular fashion principles, serving as models for the broader industry:

- *Patagonia's Worn Wear Program*: Patagonia has been advising its consumers to bring back their worn-out clothes for repair, to be repurposed, or to be recycled. Most of the time, this helps cut waste while at the same time reinforcing the understanding of brand development toward sustainability.
- *H&M's Conscious Collection*: As shown in the flow above, H&M shows that fast fashion brands are not exempt from incorporating the use of organic and recycled materials. Their garment collection program also helps in the recycling of garments.
- *Stella McCartney's Sustainable Practices*: She, as one of the leaders in sustainable luxury fashion, shows that the concept of the circular economy is plausible in the luxury sector.

## The Path Forward

So, circular fashion is not just a way to go but the only sustainable way forward in the fashion industry. With the use of actions like reuse, recycling, and the systems-closed concepts, the value system can be reconstructed and placed at a new level to embrace scientific change based on sustainability, practical creativity, and ethically acceptable business standards. For circular fashion to function optimally and provide consumers with their desired outcome, stakeholders must work to eliminate barriers.

As for recycling, there will be a lot left to focus on, from developing the technologies to integrate into recycling systems around the world to investing in the politics and propaganda needed to prepare individuals for the change. When what is environmentally desirable is also what is economically rational and culturally preferred, society is ready for a fashion of sustainability. Only if all stakeholders participate circularly can it become a tool that will revolutionize the fashion world while maintaining beauty and sustainability.

## Environmental Benefits of Circular Fashion

The sustainability benefits of circular fashion are vast and encompassing, while it directly tackles major ecological issues in the world and provides indications of change in the rest of the industries. With the help of initiatives like recycling solutions, upcycling techniques, and closed-loop solutions, the fashion industry is moving forward, reducing its impact on the environment and certainly protecting the resources of importance to human life.

## Environmental Gains of Circular Fashion

By examining circular fashion, possibilities for radical changes in fashion consumption can be outlined to face global environmental issues. These aspects of recycling free the fashion industry from a high amount of harm towards the environment by cutting back on emissions of carbon, using less water through virgin material reduction while replacing it with recyclable water, and addressing the issues of waste management through the adoption of upcycling and a closed-loop system. Also, it reduces the emission of microplastics in water through the use of long-lasting, superior-quality products. Raw materials. Combined, these tactics work in a way that provides a roadmap to a sustainable or eco-friendly industry in fashion.

*Reducing Carbon Emissions*: Among the most important benefits, circular fashion can be credited for its contribution towards the reduction of carbon emissions. It is important to understand that conventional textile manufacturing is resource-hungry and plays a major role in the emission of greenhouse gases. Recycling processes, on the other hand, are much more energy-efficient than new processes. Fashion industries reducing emissions of greenhouse gases every time recycled fibers are incorporated into manufacturing processes play their role in the global fight against climate change. Replacing virgin polyester with recycled polyester can significantly reduce emissions, with noteworthy examples of circular practices changing the worst for the better.

*Conserving Water Resources*: The availability of water is becoming a problem in the world, and the fashion industry is one of the largest users of water. The textile industry could be characterized as a water-intensive sector, especially in connection with the growth of fabric materials, which includes the provision of cotton. Circular fashion provides a workable solution to this problem through the use of recycled or renewable materials. On the same note, reused fibers, for instance, require far less water than conventional systems. Employing recycled materials, for instance, polyester from used plastic containers, shows that changes can save millions of liters of water, a resource that is in short supply.

*Minimizing Waste*: Textile waste remains another problem, as billions of tons end up in landfills every year. This issue is solved in a circular fashion, where there is a lesser amount of wastage since it's based on upcycling and a closed loop. Upcycling utilizes end-of-use clothing items and turns them into new and valuable products, thus reducing their likelihood of ending up as waste. Closed-loop systems extend this concept one level further, where no waste is dissipated because all products circulate within the production process. Such practices also minimize impacts on the environment and file flaws that are connected with linear supply chains.

*Reducing Microplastic Pollution*: Synthetic textiles, in particular polyester, rank among the leading contributors to microplastic pollution. These small-sized plastic particles shed off from washing and wear are swept into the water courses, harming aquatic ecosystems. This problem is reduced through a circular fashion since it consists of a few sheds due to durability and the use of high-quality materials. Further, recycling reduces the production of new synthetic fibers into the environment and protects ecosystems from microplastic contamination.

## Holistic Environmental Impact

There are significant environmental implications when circularity practices are integrated into the fold. Adopting circular fashion is the way forward to address environmental factors such as carbon emission, water rationing, and waste reduction, as well as microplastic pollution. These works not only reduce the environmental pressure from fashion industries but also encourage other industries to take similar actions.

***Setting a Standard for Sustainability***: By emphasizing resource management through the recycling and reuse of things, circular fashion provides a roadmap to other environmental objectives. Closed-loop systems, along with other effective recycling technologies, are proving that sustainable ideas can be both feasible and effective. Circularity not only helps brands reduce their environmental impact but also expanding its use by more brands has a snowball effect, which can change the fashion industry for the better. Each year, the textile industry generates 92 million tons of waste, of which only 12 percent is recycled into products. By the year 2030, India will have turned into a 2-billion-ton peak of textile waste.

***Collaborative Solutions***: To attain higher levels of circularity in fashion, players ranging from designers right down to policymakers and consumers need to come on board. These efforts can be supported by the governments through actions such as offering tax credits and exemptions for recycling technologies. We also need consumer education campaigns to increase awareness of the programs and to promote the uptake of recycling and upcycling programs. The EU Circular Economy Action Plan (2020) requires textile manufacturers to make their products recyclable and extend product life cycles, The EU Circular Economy Action Plan (2020) will demand that textile manufacturers make their products recyclable and lengthen product life cycles, and the Indian government is amending its Extended Producer Responsibility policy (on plastics also) to also target textiles. These approaches to regulatory framework demonstrate how these can accelerate or slow down the tempo at which change can occur in an industry. And the Indian government is adapting its Extended Producer Responsibility policy (also on plastics) to apply to textiles. These regulatory models illustrate how policy frameworks can push or halt the pace of change in an industry. Consumers, governments, NGOs, and retailers are getting involved with circularity in countless ways. To give an example, H&M has a program related to the recycling of textiles, recycling more than 20,000 tons of material each year, but the civil society stakeholders have mentioned that there is a risk of overproduction that goes against the idea of sustainability.

The advantages of circular fashion in terms of the environment are enormous and revolutionary. Focusing on the key ecological issues of the industry, circular solutions create the basis for further change. Following innovation, partnership, and, more importantly, stringent adherence to resources, the fashion business has the power to reduce environmental impact drastically and now calls for other industries to embrace change. The textile industry produces wastes in the amount of 92 million tons; 12 percent of the given volume is being reused in further production. In India, as much as 8 million tons of textile waste, by 2030, in India. There has been an understanding that the capacity of textile waste will pivot at an unprecedented scale.

## Economic and Societal Impact

Circulated fashion encompasses multiple economic benefits while initiating social change at the same time. They include cutting the cost of production, establishing new economic value recovery points, and creating sustainability that meets the present standards in the market. Besides, such models enable the creation of a more sustainable economy and promote a change in people's mentality about excessive purchases.

### New Business Models

Different aspects of circular fashion include new types of business concepts for second-hand products, rental services, and repair platforms. Online sales of second-hand clothes take advantage of the increased popularity of second-hand products within the market by providing consumers with an opportunity to purchase sustainable and cheap clothing. Regarding rental services, this provides an opportunity for the clients to wear quality clothes without having to own them and, therefore, eliminates cases of stockpiles that end up remaining unsold. Such websites also prolong the use of garments, which benefits the customer and cuts back on fabric purchases. Altogether, these models help to build a resource-efficient economy and, therefore, are more sustainable.

### Consumer Behavior Shift

Circularly, the consumer steps out of the fast fashion norm and moves from purchasing an ephemeral product to owning a slower, better-quality product. Public sensitization efforts and advocacy in that respect complement the overall efforts and initiatives towards the achievement of sustainable consumption objectives and enable citizens to make consumer decisions that would bear a positive and sustainable impact on the environment or society. The gradually increased focus on sustainability through consumerism has helped to change the culture of fashion and thus open the debate for circular thinking.

### Brand Loyalty

There is also a belief that firms implementing circular strategies have better consumer communications. Transparency and sustainability equations are laudable, therefore creating a loyal customer base for eco-conscious consumers. The Shaping Wonderful DTC communication strategy aligns with the needs of the brands in a competitive market to show consumers that they have a responsibility towards social and environmental issues to gain the customers' trust, which will keep them loyal. This association with consumer values not only enhances the brand image but also supports mega and continuous business development.

Hence, the chapter has demonstrated the socio-economic benefits and costs of circular fashion as a consumption model. This chapter has therefore explained how the industry can attain profitability while at the same time sustaining its business models by changing the consumers' behavior and building a loyal base. These changes make up a small but strategic shift in the right direction concerning equity and the environment.

## Challenges in Implementing Circular Fashion

Nonetheless, circular fashion remains a highly promising solution as a post-linear style of consumption for the textile and fashion industries; however, various critical challenges slow down broader circular fashion consumption all over the world. Solving these challenges is vital to improving the circular practices and their impact to the highest level.

### High Initial Costs

This shows that one of the main challenges of circularity in the fashion industry is the large capital outlay needed for sustainable resources and technologies. However, practices such as internal audit and cost control are quite costly, and SMEs, especially, are financially challenged to undertake such activities. Implementation of such solutions as the use of recycled fibres and biodegradable materials, as well as establishing factories that recycle and upcycle materials, requires significant investment. In addition, training the employees on how to operate with new technologies and processes is another cost that continues to build up. To overcome this barrier, the governments, together with the industry, must make subsidies and incentives, as well as funding available for SMEs to adopt a circular economy.

### Consumer Habits

Another great threat seems to be the changing consumer behavior. Fast fashion currently allows consumers to be constantly enveloped with easy, cheap, and trend-linked products, without regard for sustainability or product endurance. There is a need to interrupt this pattern, hence we must engage in specific sensitisation activities to enlighten the specific consumer on the effects caused to the environment. Another imperative is to make sustainable fashion desirable by creating affordable, kick-ass products that customers want to buy. Cooperation of businesses, schools, and governments remains the only effective way to create a culture of sustainable consumption.

### Scalability

Sustainable fashion at high levels is always a problem. Creating recycling, upcycling, and closed-loop systems all depend on infrastructure that must be substantially (individually) invested in. The most modern technologies of recycling and only the platforms where the products are collected and sorted out demand great logistical support. Additionally, the supply chain is geographically disproportional, which complicates the circumstances of implementing the practices of closeness to the same extent in different countries. As with any nascent market where the goal is to become globally scaled, there must be innovation, a willingness to make a massive investment in R&D, and a need to collaborate across stakeholders. R&D is especially necessary since it is the breakthrough that is needed to overcome incumbent technology and cost restrictions. An example is how R&D can be used to develop sustainable materials (like eco-friendly fibers and lab-grown fabric), which can be used in place of those resource-hungry virgin materials. It also speeds up advancements in the field of textile recycling, including chemical processing that may individually segregate blends of fibers that were once considered non-recyclable. On top of that, R&D contributes to digital innovations such as blockchain, artificial intelligence-based sorting automation, which increase the transparency of supply chains and effectiveness of recycling logistics. R&D can enhance the rate at which circular fashion transforms from multiple small-scale pilots to large-scale commercial operations by lowering the cost of production over time, making materials more economical to reuse, and ensuring the process is transparent. By doing so, the adoption of circular fashion is made possible on the scale of fashion industries, not only technically, but also economically, making it competitive with standard fast fashion. This is a work in progress where the governments, industries, and NGOs are collaborating to put together the enabling infrastructure and regulation.

## Recommendations

### The Role of Policymakers and Stakeholders

Circular fashion demands planning among the businesses, governments, and consumers. A prime example can be seen in the EU action plan on a Circular Economy that is underway, whereby products must be designed to be recycled and have longer product life cycles. In India, its EPR policy has been adopted by the textile industry, and factors such as cost of recycling, likely regulatory limitations, and regulations that restrict consumption, like those imposed in the EU, have shown how regulation can be used to inhibit or drastically increase the speed of industry transformation. A government can start to enact measures in driving a massive change through the provision of incentives, investments into sustainable materials, and sanctions on unsustainable behavior.

Businesses also need to transition their role within the business community through the investment in waste-reduction, recycling, and upcycling operations, and through cooperation with recycling partners the efficiency and innovation of the relevant operations. The sourcing and production processes should be transparent, too, to boost customer confidence.

One of the solutions is the consumers, who should also align their purchases with sustainable principles. Through environmental awareness, people can alter their consumption patterns towards the practices of a true cycle. Coupled together, these stakeholders are the basis for circularizing the fashion system.

### Technological Innovations and the Future of Circular Fashion

Technology will provide essential means of developing circularity. Chemical recycling will help recycle fibers in less supportive existing textile blends, eliminating the use of virgin materials and minimizing environmental damage. Blockchain increases the level of transparency in the supply chain by documenting the entire life cycle of a garment, including sourcing, resell, or recycling on the digital ledger systems. This beefs up accountability, removes fakes, and empowers consumer assurance.

The two innovations work hand in hand: where chemical recycling ends at the circularity of materials, blockchain will keep them both traceable and ethical. Combined, they make up a strong foundation of a fully circular fashion system in which fabrics are developed with an eye towards durability, reuse, and as little waste as possible. As the technologies are becoming more mature, they are going to play a decisive role in transforming the industry to become sustainable.

## FINDINGS

Relevant conclusions highlight the effectiveness of costs, environmental effects in the form of emissions and resource consumption, as well as social costs. The three important concepts of circular fashion, namely recycling, upcycling, and closed-loop systems, best describe the volumes that circular fashion can create. Recycling cuts down on the exploitation of primary resources, curbs energy demand, and the conversion of waste into useful products reduces pollution. Creating a new product out of the old ones is what makes upcycling foster creativity, in that it promotes the development of new products out of waste. Integrated systems, which are crucial for sustainable business, guarantee that all materials being used are cycled or reused – hence, no waste.

Economically, circularity reduces costs for materials, energy, and water and decreases or eliminates emissions of greenhouse gases, toxicity, and microplastics. The economic feasibility study shows potential for fresh sources of income coming from secondary markets, rental and repair services, in addition to encouraging consumption with loyalty programs. Concerning the larger society, circular fashion is a lifestyle change from the unnecessary desire to be dressed in the latest fashion trends to embracing reforms in fashion consumption patterns.

The chapter also reveals some barriers, such as high costs of technologies, changes in consumer behaviours, and small businesses' in implement sustainable measures. These barriers require combined efforts from policymakers, businesses, and consumers to ensure programs and policies are striving forward that will favour the business. Real-life examples are present in campaigns such as Patagonia's Worn Wear and H&M's Conscious Collection, while policy recommendations call for incentives and penalties to be used to encourage sustainability. The EU's Circular Economy Action Plan (2020) mandates textile producers to ensure recyclability and longer product life spans, while India's Extended Producer Responsibility (EPR) policy for plastics is being adapted for textiles. These regulatory models show how policy frameworks can either accelerate or constrain industry-wide transformation.

These fluctuations are helpful for technologies such as chemical recycling and blockchain to manage the difficulties. These tools raise the level of transparency, optimization, and factorability to make the fashion industry's future more sustainable. Finally, the chapter argues that a circular economy of fashion should be embraced as it leads to environmental conservation, contributes to economic development, and addresses social and moral issues that revolve around sustainability.

## CONCLUSION

Circular fashion is a crucial reinvention of the way the fashion and retail business can be geared towards sustainable concepts. It offers an alternative system that does not rely on a traditional linear model of take-make-dispose, but builds in the concepts of durability, resource efficiency, and material regeneration. Through this chapter, the concept has been traversed through literature, brand case studies, and technological views, providing an integrated view of opportunities as well as challenges.

Environmentally, circular fashion results in significant emissions of greenhouse gases and water savings, and waste savings. In a closed-loop process, recycling, and upcycling, the climate-related problems of fast fashion are largely reduced. Some, like Patagonia and its Worn Wear program, illustrate how repair and reuse can prolong the life of products, and how companies can use the benefits of consumer involvement in sustainability. In an analogous vein, H&M, through its Conscious Collection, has demonstrated that even a huge retailer can provide manufacturing-level quantities of recycled materials, although critiques of the necessity of such efforts in relation to fast-fashion production scales persist. Stella McCartney is taking the lead in luxury fashion, becoming more sustainable, with circularity embedded in material design and the use of partners working in biotech and recycled materials. These examples show that circular fashion does not occupy one niche in the market but is relevant to the entire market, starting with mass retail and ending with luxury.

On the one hand, as an economic and business question, circular fashion presupposes different profitability models. Sustainability cannot be reconciled with traditional revenue streams of generating revenue through high volumes of short-cycle production. Rather, opportunities can be realized in resale platforms, rental markets, and repair services where businesses will be able to build long-term customer relationships and profits. Investment in research and development (R&D) is especially significant in stimulating rapid adoption of the scale. The enterprises also focus on technological advancement and achieve material innovation, such as biodegradable fibers, new textile recycling processes, and enhanced traceability systems to develop consumer trust through an R&D process. With a priority shift placed on innovation, the company can eventually reduce costs, make sustainable products more accessible, and keep abreast of the changing market.

Circular fashion builds the culture of accountability. Customers wish to know more about the source of products and where they are. The implementation of the principles of circularity not only meets these expectations but also promotes behavior change through conscious consumption. Policymakers also have a crucial part to play because the concept of circularity aligns with the overall ambitions of society to manage less waste, preserve resources, and ensure green job creation. Policy, norms, and sensitization can contribute to enabling environments in which prudent practices can be comfortable rather than anomalies. As an industry, textiles generate 92 million tons of waste, whereas only 12% of this waste is converted into some new material. It is expected that the capacity to generate 8 million tons of textile waste will be surpassed in India by the year 2030. The European Union, having an Extended Producer Responsibility (EPR) on plastics that will be extended to textiles, in line with the Circular Economy Action Plan (2020), asks producers and importers of their textile products to promote slow fashion by recycling and extending the life span of their products little by little. These frameworks of regulation show how policy can accelerate or limit industrial-level change. The above-mentioned consumer, governments, NGOs, and multinational retailers have unique roles to play in enhancing circularity. As an illustration, H&M's Garment Collecting Program gathers and recycles over 20,000 tons of textiles per year, though civil society organizations have questioned the fact that the current production volumes contradict the aims of sustainability.

Technology use is also a facilitator of the change that has become so inevitable. As an illustration, blockchain can be implemented and used to record immutability that will enable them to track apparel across their lifespan, where they will be able to verify its authenticity, enhance tracing along supply chains, and enable consumers to make better-informed decisions. High-tech recycling has also led to the reuse of fibers by recycling and reintroducing them into production, closing a loop that was previously thought to be impossible to achieve. However, implementation of such tools would involve intense coordination, investment, and inter-industry collaboration.

Circular fashion is, however, not free of difficulties. Sky-high entrance fees, less level infrastructure, and unwillingness to pay high prices are huge barriers that are still there. In addition, greenwashing in companies might result in reputation loss since there are no visible tracks to sustainability-related commitments. To work around these barriers, a systemic shift must be created where businesses go beyond advertising that they are sustainable but meaningfully operate in a sustainable way, where governments put prevention and enforcement systems in place to provide incentives to businesses, and where consumers can be educated on how to make sustainable choices and are provided with affordable opportunities to do so.

In a nutshell, circular fashion is no longer a need-based trend due to environmental concerns, but rather a transformation opportunity. It transforms the interaction between business, society, and the environment by developing alternative models of growth that are regenerative as opposed to extractive. The process of circularity is complicated and demanding on resources, yet its outcomes of environmental sustainability, long-term profitability, and social accountability make it a necessity in the future of fashion. This chapter shows that much remains to be done to increase the scaling of circular practices with effective cooperation among actors, continued funding in innovation, and a cultural change in the way that the fashion industry is developed and used. By adopting these routes, the fashion market will go a step closer towards a sustainable future in which economic prosperity goes hand in hand with a healthy environment and society.

## REFERENCES


Blau, L., & Kimmig, R. (2024). *What's in the Alphabet Soup?* Disentangling Top Management Team Roles Related to Digital Transformation.

Bocken, N. M., De Pauw, I., Bakker, C., & Van Der Grinten, B. (2016). Product design and business model strategies for a circular economy. *Journal of industrial and production engineering, 33*(5), 308-320. https://doi.org/DOI: 10.1080/21681015.2016.1172124

Cachon, G. P., & Swinney, R. (2011). The value of fast fashion: Quick response, enhanced design, and strategic consumer behavior. *Management Science, 57*(4), 778–795. DOI: 10.1287/mnsc.1100.1303

Chen, H. L., & Burns, L. D. (2006). Environmental analysis of textile products. *Clothing & Textiles Research Journal, 24*(3), 248–261. DOI: 10.1177/0887302X06293065

Fletcher, K. (2013). *Sustainable fashion and textiles: design journeys*. Routledge., DOI: 10.4324/9781849772778

Gardetti, M. A., & Torres, A. L. (Eds.). (2017). *Sustainability in fashion and textiles: values, design, production and consumption*. Routledge., DOI: 10.4324/9781351277600

Gwilt, A. (2020). *A practical guide to sustainable fashion*. Bloomsbury Publishing., DOI: 10.5040/9781350067059

Henninger, C. E., Alevizou, P. J., & Oates, C. J. (2016). What is sustainable fashion? *Journal of Fashion Marketing and Management, 20*(4), 400–416. DOI: 10.1108/JFMM-07-2015-0052

Karim, R., Roshid, M., & Waaje, A. (2024a). Circular economy and supply chain sustainability. In Mızrak, F. (Ed.), *Strategic Innovations for Dynamic Supply Chains* (pp. 1–30). IGI Global Publishing., DOI: 10.4018/979-8-3693-3575-8.ch001

Karim, R., Waaje, A., & Roshid, Md. M., & Yeamin, Md. B. (2024b). Turning the Waste Into Wealth: Progressing Toward Global Sustainability Through the Circular Economy in Waste Management. In R. Sucheran (Ed.), *Sustainable Waste Management in the Tourism and Hospitality Sectors* (pp. 507–552). IGI Global Scientific Publishing. https://doi.org/DOI: 10.4018/979-8-3693-6110-8.ch019

Khattab, T. A., Abdelrahman, M. S., & Rehan, M. (2020). Textile dyeing industry: Environmental impacts and remediation. *Environmental Science and Pollution Research International, 27*(4), 3803–3818. DOI: 10.1007/s11356-019-07137-z PMID: 31838699

Ki, C. W., Chong, S. M., & Ha-Brookshire, J. E. (2020). How fashion can achieve sustainable development through a circular economy and stakeholder engagement: A systematic literature review. *Corporate Social Responsibility and Environmental Management*, *27*(6), 2401–2424. DOI: 10.1002/csr.1970

Kozlowski, A., Searcy, C., & Bardecki, M. (2018). The reDesign canvas: Fashion design as a tool for sustainability. *Journal of Cleaner Production*, *183*, 194–207. DOI: 10.1016/j.jclepro.2018.02.014

Macarthur, F. E. (2017). *A new textiles economy: Redesingning fashion's future Ellen Macarthur Foundation.*

Morgan, L. R., & Birtwistle, G. (2009). An investigation of young fashion consumers' disposal habits. *International Journal of Consumer Studies*, *33*(2), 190–198. DOI: 10.1111/j.1470-6431.2009.00756.x

Niinimäki, K. (2013). *Sustainable fashion: New approaches.* Aalto University.

Niinimäki, K., Peters, G., Dahlbo, H., Perry, P., Rissanen, T., & Gwilt, A. (2020). The environmental price of fast fashion. *Nature Reviews. Earth & Environment*, *1*(4), 189–200. DOI: 10.1038/s43017-020-0039-9

Pal, R., & Gander, J. (2018). Modelling environmental value: An examination of sustainable business models within the fashion industry. *Journal of Cleaner Production*, *184*, 251–263. DOI: 10.1038/s43017-020-0039-9

Pedersen, E. R. G., & Andersen, K. R. (2015). Sustainability innovators and anchor draggers: A global expert study on sustainable fashion. *Journal of Fashion Marketing and Management*, *19*(3), 315–327. DOI: 10.1108/JFMM-08-2014-0059

Pérez-Bou, S., Galán, M. V., & Rodríguez, P. E. (2021). 174 Fashion Stores as potential educators of conscious consumers. 2 case studies: H&M Group and Inditex. https://doi.org/DOI: 10.3217/978-3-85125-842-4-18

Pomponi, F., & Moncaster, A. (2017). Circular economy for the built environment: A research framework. *Journal of Cleaner Production*, *143*, 710–718. DOI: 10.17863/CAM.7204

Roshid, Md. M., Yeamin, Md. B., Waaje, A., Karim, R., & Kaswan, M. S. (2025). Understanding the Nexus of Industry 5.0 and Circular Economy for Enriched Lean and Environment Sustainability. In José Carlos Vieira de Sá, Shubham Sharma, Sunil Kumar, & Manjunath B.R. (Eds.), *Transformative Lean Six Sigma Techniques for the Quality 5.0 Paradigm* (pp. 349–380). IGI Global Scientific Publishing. https://doi.org/DOI: 10.4018/979-8-3373-0943-9.ch013

Sipari, J. (2022). Sustainability in luxury fashion: case: Stella McCartney. https://doi.org/DOI: 10.54691/8hp4yh68

Syniuk, A. (2021). Sustainable apparel marketing and consumption: An analysis of Patagonia's Worn Wear video marketing campaign.

Valerie, M., Elvina, J., Cristie, L., Litania, L., Tjoandra, M., & Widjojo, H. (2022). Key Success Factors in the Fast Fashion Business During a Pandemic. *Ultima Management: Jurnal Ilmu Manajemen*, *14*(2), 183–195. DOI: 10.31937/manajemen.v14i2.2680

Waaje, A., Karim, R., Roshid, M. M., Yeamin, B., & Meem, T. N. (2025). Integration of industry 4.0 with personalized medicine: A shift of the paradigm in healthcare. In Chemingui, H., & Lamloumi, M. (Eds.), *AI-Driven Personalized Healthcare Solutions* (pp. 27–69). IGI Global., DOI: 10.4018/979-8-3693-7858-8.ch002

WRAP. U. (2017). Valuing our clothes: The cost of UK fashion. *WRAP, London*.

# Chapter 11
# Circular Economy for Sustainability in the Global South:
## Transforming Industry, Consumption, and Climate Action in Bangladesh

**Md. Mehedi Hasan**
https://orcid.org/0000-0003-3824-6631
*Human Resource Management Discipline, Khulna University, Bangladesh*

**S. M. Shafeeul Islam**
*Human Resource Management Discipline, Khulna University, Bangladesh*

**Md. Solaiman Chowdhury**
https://orcid.org/0000-0002-7991-7270
*Department of Management Studies, University of Rajshahi, Bangladesh*

## ABSTRACT

*This chapter explores how circular economy (CE) principles can drive sustainable development in Bangladesh by aligning with Sustainable Development Goals (SDGs) 9, 12, and 13. The traditional linear economic model of "take, make, dispose" has led to rapid resource depletion, environmental pollution, and unsustainable industrialization, exacerbating the challenges faced by developing economies in the Global South. As one of the world's most climate-vulnerable countries, Bangladesh requires innovative strategies to ensure industrial growth, responsible consumption, and climate resilience. This chapter examines the role of CE in fostering sustainable industrial innovation (SDG 9) through eco-friendly production processes, industrial*

DOI: 10.4018/979-8-3373-2454-8.ch011

*symbiosis, and circular business models that enhance resource efficiency. It further highlights how CE supports responsible consumption and production (SDG 12) by promoting sustainable waste management and circular supply chains in industries such as textiles, plastics, and packaging.*

## 1. INTRODUCTION

To make sure that growth includes everyone, we have stayed ahead of the curve when it comes to sustainable development goals. The SDGs, which stand for "Transforming our World: the 2030 Agenda for Sustainable Development," have 17 goals and 169 objectives that aim to address a range of sustainable development issues. Bangladesh is at a key point in its quest for sustainable development (Uddin & Uddin, 2018). The country is full of vibrant culture and determined people. The journey is full of ups and downs that test the country's strength and spirit. The SDG agenda is a plan to help the planet, people, and the world as a whole by promoting peace, freedom, and an end to poverty. This is why all of the goals have been met by dealing with both problems and hopes for each country. Climate change has brought Bangladesh unwanted guests like rising sea levels, unpredictable rain, and extreme heat. Workers in garment factories, which are very important to our economy, are currently dealing with oppressive heat that affects their health and productivity. From 2020 to 2024, the number of dangerously hot days in Bangladesh rose by 42% compared to 2005 to 2009. This made factory floors feel more like cauldrons (Nabi et al., 2023). This is a big problem for long-term growth in the global south.

Goal 9 has 5 targets:

a. Building quality, reliable, sustainable, and resilient infrastructure,
b. Promoting inclusive and sustainable industrialisation,
c. Expanding small businesses with financial services,
d. Improving scientific research and
e. Technological capabilities.

Both the public and private sectors would need to make significant investments in industries and infrastructure, as well as work with international partners. Our farmers, the unsung heroes who feed the nation, are vulnerable to unforeseen floods. In 2024, catastrophic floods obliterated around 1.1 million metric tons of rice, transforming expansive fields into soggy disaster zones. This loss not only strained our food supply but also forced the country to increase rice imports, a heavy blow to our agricultural autonomy (Rahman et al., 2025). Projects are needed to improve services in the traditional country boat sector and enforce safety regulations to reduce accidents. Therefore, modernising every communication system by the government alone would be a challenging endeavour. Achieving this goal is troubling for a country sickened by geographical disadvantages. (The Financial Express, 2016).

Our ever-growing city, Dhaka, is like a teenager who outgrows its clothes. Rapid urbanization has led to too many people, traffic jams, and damage to the environment. The city's infrastructure is not keeping up with its quickly growing population, which is causing a lack of housing and overworked public services. There are fewer green spaces and more concrete jungles, which makes people want to get away from the noise and pollution (Akash et al., 2018). Goal 12 has 8 targets:

a) Implementing a 10-year plan for sustainable production
b) Sustainable consumption,
c) Managing natural resources sustainably,
d) Reducing food waste by at least 50% at the consumer level,
e) Cutting food losses in production,
f) Ensuring safe chemical
g) waste management and
h) Encouraging companies to adopt sustainable practices.

Goal 12's maximum aims are unlikely to have a significant impact on public and national policy because they do not directly address that subject (The Financial Express, 2017). Keeping in account the overpopulation and overconsumption, achieving this goal seems the most challenging.

The ministries of agriculture, jute and textiles, industries, food, planning, environment, and forests are key to achieving these goals. Goal 13 lays out some pretty clear targets, like boosting our resilience and ability to adapt to climate-related threats and natural disasters. It also emphasises the need to weave climate change considerations into national policies, plans, and strategies. Plus, it's all about raising awareness, enhancing education, and building institutional capacity to tackle climate change adaptation, mitigation, and impact reduction. The adverse global climate change is responsible for creating exigent environmental risks impacting both human and natural systems (Sustainable Development Solutions Network, 2022). With an ever-increasing youth bulge, Bangladesh has to take advantage of this opportunity. Currently, with a youth unemployment rate at 10.6%, and 29.8% of youth being neither in school, working, or in training, Bangladesh has a high rate of youth unemployment and underemployment (Cho, et al., 2019).

This is akin to a cricket team with a lot of energised players but no matches to play. We need to create opportunities that align with their interests and aspirations. Still, the adverse climate impedes the flourishing of many sectors that can create potential job sectors. Natural systems encompass a variety of environments, including heavy rainfall areas, atolls, coral reefs, glaciers, polar and alpine ecosystems, boreal and tropical forests, mangroves, and the last remnants of native grasslands, as well as prairie wetlands. Human encroachment is making Bangladesh's environment worse, which makes climate change worse and its effects worse (Klapper et al., 2016; Roshid et al., 2025). Water is at the heart of the dispute between Bangladesh and India over the Teesta River.

The installation of upstream barrages has greatly reduced the amount of water that flows into Bangladesh, which has hurt the farming and the lives of people in the Teesta Basin. It's like sharing a cup of tea and then finding out that the pot is empty before you've finished it (Mirchandani, 2016). Frequent natural disasters wreak havoc on crops, livestock, and infrastructure, throwing a wrench in the economic activities of rural areas, especially along the coast. There is a direct nexus between pauperism and intensifying climate change. So, massive efforts are needed, especially from government bodies like the Environment Ministry, the Planning Ministry, and the Disaster Management Department (SDG, 2015). The interconnectedness of the SDGS within a system is well acknowledged. We are to attain three goals, not just one. Some goals and targets have a more intense interaction than others. The reinforcement occurs between each other through certain targets. Our chapter here will cover how to overcome the adversities faced by this region and achieve the goals 09, 12 and 13.

This chapter provides a detailed overview of the circular economy concepts and then aligns them with the SDGs. It will also contextualise the global south and Bangladesh into the discussion.

## 2. THE LIMITATIONS OF THE LINEAR ECONOMY

Modern societal civilization is at a crossroads. Escalating environmental crises —including climate change, ecosystem destruction and collapse, resource depletion, pollution, and loss of biodiversity are highlighting just how unsustainable contemporary economic systems are. By the same token, however, there is a perpetual need to pursue economic growth and meet the growth demands of an increasing global population. The linear economic model, characterised as "take, make, dispose," is central to most global issues. For decades, this economic model has driven rampant industrial growth and consumerism. Take a natural source of raw material, make a product to satisfy our immediate need, and dispose of (usually improperly) our waste back into the environment. In this cycle, little thought is given to the long-term consequences on our planet or the well-being of communities (Elisha, 2020;

Roshid et al., 2025). The model views resources as limitless and disposability as fine. It ignores the emotional and environmental costs that accumulate over time.

While this system has historically supported industrial revolutions and economic growth, it is now seen as both ecologically and economically unsustainable. Global resource extraction has tripled in the past fifty years. Waste generation is expected to rise by 70% by 2050 if trends stay the same. Additionally, the rise of three billion more middle-class consumers by 2030 will further strain already stressed resources (Accenture, 2016).

A linear economy creates value by maximizing production and sales volumes. However, this model fails to account for the environmental externalities or the finite nature of Earth's resources (Bocken et al., 2016). For example, in 2012, only 40% of municipal waste in the EU was recycled. The rest was mainly sent to landfills or incinerated (Goyal, et al., 2018). In the same way, the fast fashion industry, as a linear model, produces around 92 million tonnes of textile waste every year and produces 10% of the world's carbon emissions (Sariatli, 2017).

## 3. KEY DRAWBACKS OF THE LINEAR ECONOMIC MODEL

### a. Resource Depletion

The linear economy's unending extraction of natural resources is hastening not only the depletion of critical ecosystems and raw materials but also worsening the state of our environment. The world's forests are disappearing, freshwater supplies are diminishing, and vital minerals and metals that we need to create technology are running out. The Amazon rainforest, which sequesters 90 billion tonnes of carbon, is on the verge of ecological collapse due to deforestation and the conversion of land for agriculture and other uses - if it collapses, we will reach a tipping point for this ecosystem and the region may then switch from being a net carbon sink to a carbon source and exacerbate global climate change (Lovejoy & Nobre, 2018; Jackson, 2009).

### b. Waste Generation and Pollution

In the linear model, waste is not a byproduct—it is a design flaw. Abandoned products then become pollutants and contribute to the degradation of the environment, as well as present threats to public health. Globally, an estimated 2 billion tonnes of municipal solid waste is generated each year, with at least 33% not being managed in an environmentally safe manner. A considerable amount of waste is sourced from plastics, which are entering marine ecosystems at safe, but alarming rates. Each year, there are 11 million tons of plastic that end up in the ocean.

## c. Economic Instability

Fluctuations in global commodity prices highlight the fragile economic nature embedded in linear models. The global chip shortage in 2021 wreaked havoc on production for the automobile and electronics industries because of supply chain disruptions and resource dependency (Shih, 2020), which similarly relates to the vulnerabilities linked with raw material dependency with women's supply (Bouton et al., 2016).

## d. Inefficient Material Use

The linear economy promotes a throwaway culture. Products are designed to only last a short time and are often not designed for reparability. With wastage of energy and labour, valuable materials go as such to incineration, while they could have been reused or recycled (Jordens, 2016). This increases production costs because companies have to continuously source new raw materials.

## e. Geopolitical Dependence and Input Risks

A dependency on essential raw materials, such as rare earth elements, can lead to geopolitical risks associated with the supply chain and procurement of materials. For example, China has over 60% of the supply of rare earth materials in the world. In the past several years, it has restrained exports of these materials, leading to higher costs and limited access for producers relying on these materials (Daigle & DeCarlo, 2021).

## f. Discouragement of innovation

The linear model does not incentivize innovation. Because it focuses on throughput and disposability, it makes people less likely to invest in a design that lasts, can be repaired, or can be recycled. As a result, economic creativity suffers, and circular business models, which depend on product-as-a-service, remanufacturing, or closed-loop systems, are pushed to the edges (Kirchherr et al., 2017).

## 4. THE URGENCY FOR TRANSITION: TOWARD A CIRCULAR ECONOMY

In view of the linear model's limitations and restrictions, moving towards a circular economy (CE) is not an option; it is a necessity. CE is a systemic approach that changes the growth narrative to one of I'm doubting only benefits, for society as a whole. CE is a system that requires that waste is no longer viable and that pollution is designed out, products and materials are kept in use and natural systems are regenerated (Macarthur, 2019).

Practical examples indicate how effective it is. The Netherlands looks to be completely circular by 2050. In addition, the CE initiatives added around €7.3 billion to its GDP and 54,000 jobs between 2010 and 2020. Sweden's "repair bonus", designed to get consumers to repair their goods (in particular, durable and electronic goods), supports local repair shops and aims to ameliorate electronic waste. In the garment sector of Bangladesh, which is traditionally linear, post-consumer textile recycling is now being introduced. Primarily through brands like H&M and Reverse Resources, post-consumer textile recycling aims to enhance resource efficiency by indigenizing the recycling process into a circular supply chain.

## 5. THE NECESSITY OF USING THE CONCEPTS OF THE CIRCULAR ECONOMY (CE) TO ATTAIN SUSTAINABILITY

The policy objectives of today's economies place a high premium on addressing environmental and social issues. Currently, most policies support the Circular Economy and Sustainability concerns through specific strategies and practices. Every single piece of plastic tossed aside, every resource squandered, and every harmful emission is just another ugly mark on the beautiful planet we're supposed to cherish (Nikolaou, Jones & Stefanakis, 2021). The air is polluted, the oceans are full of waste, and the future looks bleak for the next generations. However, the concept of sustainable development implies that the daily operations of economic and social actors should simultaneously promote economic growth, nature conservation, and social equity (Svensson et al., 2018; Stefanakis & Nikolaou, 2024). A portion of the current body of literature sees these two ideas (Circular Economy & Sustainability) as interchangeable, and occasionally as distinct (Pieroni et al., 2019).

We must see the circular economy (CE) as a moral duty to future generations, not just an option. CE encourages regeneration, reuse, and renewal. It breaks away from the wasteful linear model (Macarthur, 2013). This approach fosters an economy where products last longer, materials are reused, and waste is minimized. The shortage of natural resources significantly impacts the economy and production.

Large amounts of natural resources, which are dwindling, are required for the overproduction we see in modern economies. This overproduction aims to meet the growing demands of a rapidly increasing population (García-Barragán et al., 2019; Waaje et al., 2024). Many scholars believe that shifting to CE is essential for true sustainability. According to Geissdoerfer et al. (2017), the Circular Economy (CE) has benefits for both environment and economy—it embraces and creates systems that are efficient but at the same time resilient. This is now referred to by the familiar idea of CE, which drives economic actors to move towards circular thinking from a linear one. The need for action is clear; without it, the Earth will lose its ability to support life as we know it.

Now, let's explore how CE specifically contributes to sustainability:

## a. Efficiency of Resources and Reduction of Waste

CE aims to reduce waste by creating goods that endure longer, recovering resources, and encouraging reuse. This strategically supports sustainable development through resource efficiency and decreases ecological impact (Alamerew & Brissaud, 2019).

## b. Environmental Conservation

Circular economy (CE) benefits the environment, since it encourages responsible and efficient use of natural resources, and it has an environmental sustainability component that may reduce waste and enhance reuse (Nikolaou, Jones, and Stefanakis, 2021).

## c. Integration With the Sustainable Development Goals (SDGS)

The principles of the circular economy are tightly linked to several Sustainable Development Goals, such as affordable and clean energy (SDG 7), decent work and economic growth (SDG 8), responsible consumption and production (SDG 12), and climate action (SDG 13). By aligning business models with the SDGS, CE helps drive sustainability for both companies and the world (Dantas et al., 2021).

## d. Social Aspect

The CE is an important tool for achieving overall sustainability. It mainly focuses on environmental and economic issues (Stefanakis & Nikolaou, 2024). Its practices can also positively affect society by promoting fair employment and social justice.

## e. Encourage Technological and Business Model Innovation

CE inspires new ideas in technology and business plans that encourage sustainable practices. It sees itself as an important force for long-term sustainability by encouraging people to adapt to changing social and environmental needs.

We need to stop being greedy and wasteful and start being sustainable. This call for action needs to be heard by all, including individuals, businesses and governments, now. In short, using CE is vital to achieving sustainability because it assists in the overarching aims of sustainable development by being more resource-efficient, minimizing waste, and integrating environmental, social and economic considerations.

# 6. DEFINITION AND FUNDAMENTALS OF THE CIRCULAR ECONOMY

Since commodity producers are today facing significant constraints due to a lack of natural resources and unparalleled ecological degradation, modern economic realities necessitate a fundamentally different approach to the execution of economic activity. To achieve sustainable development that includes economic success, environmental safety, and improved quality of life, a new development paradigm is necessary. This isn't merely a model; it's a whole movement, a way of thinking, and maybe even a fresh start for a planet that's been ditched by sustainability for ages (Tambovceva & Titko, 2020). The circular economy is a system where nothing goes to waste. Instead of throwing things away, we reuse, repair, and recycle them to keep resources in use for as long as possible. The circular economy can also bring significant economic, social, and environmental benefits on a global scale, becoming a lodestar for new economic development (Lacy et al., 2014).

The circular economy envisions a shift from the existing paradigm of production and consumption, which is based on "end-of-life" ideas, to a new one. Promoting the circular economy aligns with several of the United Nations' 17 Sustainable Development Goals (SDGS), making it a key driver of sustainable development (Lacy et al., 2014). Forget the old take-make-dispose routine. Circular Economy (CE) is all about the make-use-return movement, where products are crafted to be repaired, reused, or remade. According to Kirchherr et al. (2017), the circular economy is a system that avoids waste by focusing on reducing, reusing, recycling, and recovering materials during production, distribution, and use. The circular economy provides the direction of material resources for reuse within the existing production cycle, which leads to several repercussions, both positive and negative. It was discovered that even if there are drawbacks, the advantages of the circular economy's growth are far greater (D'Amato et al., 2019).

One might choose the term that is most apparent to them. Being in the know about the principles and practices of the circular economy is crucial. One of the easiest ways to understand the idea of a circular economy comes from the quarterly journal *Government Europe*:

*The circular economy is a regenerative system in which resource input, waste, emissions, and energy leakage are minimized by slowing, closing, and narrowing energy and material loops; this can be achieved through long-lasting design, maintenance, repair, reuse, re-manufacturing, refurbishing, recycling, and re-purposing. (Deselnicu et al., 2018)*

The CE tenets are far more aspirational:

I. Design out waste and pollution.
II. Make sure to keep products and materials in circulation.
III. Help nature heal and grow again.

Over the past few years, the circular economy has significantly raised interest and continues to gain traction. The 2014 report *"Towards the Circular Economy: Accelerating the Scale-Up across Global Supply Chains,"* produced by the World Economic Forum in collaboration with the Ellen MacArthur Foundation and McKinsey & Company, identified five key tenets of the circular economy that guide its implementation across industries (World Economic Forum, 2014).

a) **Create waste-free designs:** Eliminating waste in the design process can lead to enhancements in both environmental impact and operational efficiency. The goal is to cut down waste by making products that can be taken apart and fixed, so they fit into natural and technical reuse cycles.
b) **Build resilience through heterogeneity:** A diversified system that is based on attributes like adaptability, flexibility, and versatility is less susceptible to shocks from the outside world.
c) **Rely on renewable energy sources:** Since production processes usually consume a hefty amount of energy, the ultimate goal for manufacturing systems should be to harness renewable resources.
d) **Look at the big picture:** On the analogy of the living systems, the circular economy is set to achieve optimisation of ecosystems instead of components of ecosystems. How various components of systems affect one another, hence it becomes important to understand.
e) **Waste is food:** The nourishing products and materials, be they technological or biological, have to strut their stuff through various uses in the economy towards gracefully returning their nutrients to nature.

From the Netherlands, Fairphone is said to be a smartphone where parts can be replaced one at a time, reducing e-waste and saving you from broken screens. IKEA is interested in leasing furniture to curb furniture waste. Granola is being made in Japan from leftover beer grains. Over the past two decades, there has been a new interest in business models among entrepreneurs, business strategists, and management researchers. In today's fiercely competitive global market, modeling in a manner that disrupts your own business model is the single most recognized step for maintaining the edge and the advantage. Due to the ever-changing market, which introduces new technologies, competitors, rules, and customer needs, there can be little question as to how sustainable any business model is (with new technologies, competitors, rules, and customer needs), which can quickly make old business models outdated or less profitable (Fonseca & Michie, 2024). A circular business model is "the rationale of how an organisation creates, delivers, and captures value with and within closed material circuits."

The Ellen MacArthur Foundation, which developed the concept of the circular economy, released a study outlining the four fundamental components of a circular economy:

## 6.1 Circular Economy Design

To boost product reuse, recycling, and cascading, businesses must develop essential skills in circular design. The design of circular products is important for making them last longer and recycling materials, as new design strategies focus on durability, easy maintenance, repair, and upgrading. For using circular designs in an economically successful way, it is necessary to pay attention to and operate in the following areas:

I. Material selection,
II. Standardized components,
III. Design for easy end-of-life sorting and recycling, i.e. easier separation or reusage of products and materials, and design-for-manufacturing,
IV. Taking into account the possible beneficial applications of by-products and waste.

## 6.2 Novel Business Strategies

To transition to a circular economy, business models must be inventive and either replace or create new possibilities. The companies hold the key role here. If business models and ideas make money, other companies will adopt them, helping these models spread more.

## 6.3 Reverse Iterations

It is essential to develop and acquire new skills to implement reverse cycles, cascades, and the ultimate return of materials to soil or the industrial production system. These abilities apply to delivery chain logistics, sorting, warehousing, risk management, power generation, and molecular biology/polymer chemistry. Every one of these systems needs to be budget-friendly and kind to the planet.

## 6.4 Factors and Conditions That Support the System

This block mostly focuses on education and legislation. Market mechanisms must be in charge of achieving greater resource productivity, encouraging broader material reuse, and establishing this habit. To achieve their ideas, they require backing from lawmakers, educational institutions, and opinion leaders. Here are a few examples to consider:

a) Working together.
b) Rethinking motivations.
c) Supplying an appropriate set of global environmental regulations.
d) Leading by example and driving up the scale quickly.
e) Funding availability.

Scholars like Geissdoerfer et al. (2017) believe that CE provides a strong framework for achieving sustainable growth while staying economically competitive. So, when the earth sighs under mountains of waste, the circular economy wraps its arms around her and promises, "I'll take better care of you this time." Additionally, she may believe us.

# 7. MONITORING CIRCULAR ECONOMY PROGRESS: INDICATORS, FRAMEWORKS, AND NATIONAL INTEGRATION

The circular economy is not just an attractive concept; it is a pragmatic and measurable way to move forward. However, to push any change with meaning, we need to know how far we have come, how far we have to go, and what we are already doing. Without strong measures, even the best intentions on CE will likely result in tokenistic and sometimes inaccurate CE programs. Monitoring and evaluation (M&E) will help to make adjustments quickly, maintain accountability, and help clarify what is actually working (Kirchherr et al., 2017). In situations like Bangladesh, where limited resources and institutional capabilities, combined with ecological fragility, exist, monitoring will help ensure that circularity becomes a real thing that is sought to connect with national objectives, including employment, sustainability, and inclusive development. Meanwhile, connectivity in the progress being made in the Circular Economy (CE) should be developed in line with both global and national Sustainable Development Goals (SDG), SDG9 (Industry and Innovation), SDG 12 (Responsible Consumption and Production), and SDG13 (Climate Action) (Parchomenko et al., 2019).

There are now a lot of monitoring systems around the world that keep an eye on CE transitions. These could be easily changed to fit the needs of Bangladesh and the Global South.

**Circularity Gap Report (Circle Economy, 2021):** This shows that in 2023, the world economy reached 7.2% circularity, meaning that a lot of work still remains to be done. In Bangladesh, the city can focus on issues relating to textile waste, plastic recycling, and electronic waste management (França, 2022).
**European Union CE Monitoring Framework:** Keeps track of ten important measures of competitiveness, such as secondary raw materials, waste management, output, and consumption. Governments in the Global South can use this framework to set goals and make their operations more effective (Llorente-Gonzâlez & Vence, 2019).
**OECD Circular Economy Indicator Framework:** Focuses on indicators around material flows, eco-innovation, and environmental effects, of particular value for countries that are still developing the underpinnings of a CE data system (Skvarciany et al., 2021).

Although worldwide tools offer models to implement, they must be contextually relevant practices and meaningful key performance indicators that match the local issues and current available capabilities. The following indicators may best exemplify Bangladesh's physical transition to CE:

**Material Circularity Indicator (MCI):** The percentage of resources reused in domestic production cycles (Rocchi, 2021).
**Municipal Solid Waste Recycling Rate (%):** Particularly in cities such as Dhaka, Chattogram, and Khulna.
**EPR Coverage Expansion:** % of sectors (textile, plastic, electronics) subject to the Extended Producer Responsibility regulation (Biller & McPeak, 2021).
**Eco-Designed Product Share:** The percentage of new products that were designed with modularity or disassembly in mind.
**Number of CE-Aligned Startups:** This is especially true for businesses that fix things, reuse things, or use eco-friendly packaging (Mishra et al., 2024).

- **Creating Green Jobs:** The number of new jobs in CE consulting, recycling, and circular logistics.
- **Carbon Emission Reduction Attributable to CE (%):** Can be estimated for certain industries (e.g., textiles, energy, agriculture) (Wang et al., 2024).

These indicators could be used to establish a national "CE Scoreboard" that is updated annually to assess trends, positives, negatives, and avenues for action.

## 8. SDG 9 (INDUSTRY, INNOVATION, AND INFRASTRUCTURE) AND CE'S RELATIONSHIP

Systemic innovation processes contain strategically deep inventions with a range of service innovations and new organisational frameworks. Systemic innovation processes are needed to facilitate the transition to the circular economy (CE) (De Jesus et al., 2018). CE requires sustainable business models (SBMs) that priorities the stakeholders' interests, encompassing both social and environmental aspects (Bocken et al., 2014). At the same time, the world struggles with the weight of traditional industrial practices like take, make, waste. CE rides in on a white horse like a gallant lead waiting patiently for its cue. Instead of the wasteful act of discarding, CE gives things a second chance and reuses them to keep them in use. Keeping things in use requires innovations that will break the conventional, mundane structure of a linear economy. This innovative act requires a new industry to grow to get this new infrastructure happening (Bag & Pretorius, 2022). If CE is the passionate lead actor, the same ambitious accomplice is SDG 9, which aims to build resilient infrastructure, promote inclusive and sustainable industrialization, and foster innovation.

According to several studies, new technological techniques in production and consumption patterns are necessary for sustainable growth and an efficient transition to CE (Androniceanu et al., 2021; Ahmed et al., 2022). Identifying SDG penetration requires both micro- and macro-level relationships. The core of their relationship revolves around transformation. SDG 9 calls for fair and lasting industry and strong infrastructure that helps people and the economy. CE shows how to do that by reusing resources and keeping them in the loop (Lee et al., 2016). Here, we can understand that the circular economy is the guide to a more innovative and industrialized economy, realizing SDG 09. Scholars argue that CE "redefines growth by focusing on positive society-wide benefits," offering a paradigm shift in how we value progress (MacArthur, 2013). In order to convert conventional economies into CE, technological advancements and processes are essential.

Using new technology and making and using things in a way that doesn't harm the environment should help many aspects of sustainability. This includes making eco-friendly supply chains, sustainable business models (SBM), sustainable circular production systems, and new product designs, among other things. Take Amsterdam, for example. The city has committed to becoming fully circular by 2050, leveraging CE principles in construction, transportation, and manufacturing to spearhead this transformation (Geerken et al., 2019). It is clearly understood that CE principles are the basis of such massive metamorphosis of innovation. Amsterdam's infrastructure is constructed on top of previous incarnations through the use of modular design, digital innovation, and recycled building materials. Now let's zoom into Bangladesh, where the Ready-Made Garment industry is a bit of pros and cons. Since CE principles came into play, brands like H&M and local pioneers have entered textile recycling hubs to minimize waste and maximize industrial efficiency (Muthu, 2020).

These are ways in which Circular Economy (CE) and SDG 9 infrastructure, innovation, and industry intersect:

a) **CE drives systemic innovation that aligns with SDG 9:** We need to completely change how we do things to move to a Circular Economy. This means coming up with new ways to provide services and new ways to organize ourselves. This directly supports SDG 9 by focusing on industries and new ideas that are good for the environment (De Jesus et al., 2018).
b) **CE relies on sustainable business models (SBMs):** CE demands SBMs that support the stakeholders' interests from the environmental and social contexts. These methods can support equitable growth, which is one of the main objectives of SDG 9 (Bocken et al., 2014).
c) **Current indicators might hinder CE and stifle innovation:** Recycling rate indicators can mislead or discourage innovation. Reforming these tools is necessary to support both CE practices and SDG 9 objectives, ensuring their efficacy (Di Maio & Rem, 2015).
d) **CE Needs Platforms for Multi-Level Collaboration:** Effective CE adoption relies on meso-level platforms that link company (micro) initiatives to national strategies (macro). This reflects the demand of SDG 9 for a comprehensive and inclusive approach to industrial growth (Meath et al., 2022; Trevisan et al., 2023).
e) **Policy and technology affect CE progress:** National economic structures, sectoral differences, and trade dynamics shape the success of CE and its contribution to industrial innovation (Androniceanu et al., 2021; Ahmed et al., 2022).
f) **CE Encourages the Development of Sustainable Infrastructure:** CE supports SDG 9's focus on building infrastructure that lasts and is strong by encouraging people to use fewer resources and plan for the long term (Antikainen & Valkokari, 2016).

CE is more than just a way to do math. It's a heartfelt message to the planet and its people. It says, "You are not expendable." When your present form ends, your value continues." SDG 9 responds, "Let's build something better together." When the industry starts to be concerned, innovation turns into something that cares, and infrastructure evolves from just being concrete to a solid expression of compassion.

## 9. THE PART CE PLAYS IN SDG 12 (RESPONSIBLE PRODUCTION AND CONSUMPTION)

A key part of SDG 12 is the circular economy. Closed-loop production has been recognised by researchers and advocates for decades. However, the concept of the circular economy has gained popularity recently (Verga & Khan, 2022). The circular economy focuses on eco-efficiency. This means saving money by using resources wisely. Adopting a circular economy and eco-efficiency helps us achieve several other SDGs, like SDG 6 (clean water and sanitation), SDG 7 (affordable and clean energy), SDG 8 (decent work and economic growth), and SDG 15 (life on land) (Schroeder et al., 2019). The Circular Economy (CE) is essential for SDG 12, which focuses on responsible production and consumption. By planning for waste and reusing resources, CE moves toward sustainability while the traditional "take-make-dispose" model leads to landfills. SDG 12 calls for a major change in how products and services are made and used. CE supports this shift by promoting closed-loop systems, efficient resource use, and sustainable product life cycles (Geissdoerfer et al., 2017).

Goal 12 encourages responsible consumption and production. A circular economy supports this goal by promoting smart resource use, reducing waste, and advocating eco-friendly consumption habits. This approach also helps us address other goals, such as clean water (SDG 6) and fighting climate change (SDG 13). Industries can effectively tackle global issues like water shortages, environmental damage, and greenhouse gas emissions by designing products for longevity and reusability. Family businesses, which are often steeped in their communities and customs, are ideal for implementing CE principles. Patagonia was not actually a family business, but they have inspired many family-run small clothing companies to adopt repair, reuse, and recycling practices. Their 'Worn Wear' initiative reduces the demand for new, virgin materials, and can serve as a model for textile families in developing societies (Bocken et al., 2016). Another example is Interface, a carpet company that uses old tiles and natural materials to provide contemporary, sustainable floors.

Transitioning to a circular economy is a great way to achieve several SDGs simultaneously. Not only does it offer a holistic, systematic way to tackle many global issues at once, such as resource scarcity and depletion, ecological degradation, and economic stagnation/unemployment, but CE is also appealing because it works in practice. Bangladeshi family-owned light engineering firms are crushing it with just one step, recycling metal scraps into shiny new tools- it's a small action that has a large impact in reducing industrial waste and environmental footprint. Using the Circular Economy helps the environment and makes businesses stronger and more competitive (Kirchherr et al., 2017).

## 10. THE ROLE OF CE IN SDG 13 (CLIMATE ACTION)

The limited empirical literature presents contradictory evidence regarding the relationship between CO2 emissions and indicators of the circular economy. Some studies suggest a correlation between circular economy practices and reduced CO2 emissions, whereas others find no supporting evidence for this assertion. Studies show that using advanced recycling methods and new waste management methods in a circular economy could cut carbon emissions by a large amount compared to traditional linear systems of production and consumption (Pao & Chen, 2022; Schwarz et al., 2021; Triguero et al., 2022). Climate change is more than just a scientific problem; it's a human problem that is destroying our forests, flooding our cities, and taking away our seasons. The Circular Economy is not just a trendy term; it's a movement that is changing the way we make, use, and enjoy life. SDG 13 wants quick action against climate change. The Circular Economy helps by lowering emissions and building stronger, smarter societies (Geissdoerfer et al., 2017).

The circular economy makes it possible to find long-term solutions to the climate crisis, especially through systemic change. It may not be clear right away how the two are related, but a circular economy can help meet a number of SDG 13 goals, such as Climate Action (Pao & Chen, 2022). CE's main goal is to cut down on emissions, and it does this not with flashy technology or groundbreaking new ideas, but with a very simple idea: efficiency. The Circular Economy alleviates the extraction of new materials and use of energy-intensive production processes associated with CO2 emissions through reuse, remanufacturing, and recycling and provides materials, including food, with a second life (Kirchherr et al., 2017). The making of things and food produces approximately 45% of emissions globally. The Circular Economy helps fix this problem (Macarthur, 2019). The Circular Economy's first goal is to fix the problem of how unsustainable the current Linear Economy is when it comes to making things. So it seems to be more closely related to SDG 12: Responsible Consumption. The climate benefits from a circular economy. Japan nailed the circular economy game back in the Edo Period, and it's still impressive (Hondroyiannis et al., 2024).

Utilizing circular economy practices reduces carbon emissions by optimising the use of material inputs and minimizing waste. This reduced need to extract raw materials and the associated emissions from their manufacture and end of life helps mitigate climate change. Additionally, repairing or refitting rather than replacing conserves energy and reduces CO2 emissions (Romero-Perdomo et al., 2022). Consider the Netherlands, where cities such as Amsterdam have implemented construction policies based on the circular economy, targeting a 50% reduction in raw material consumption by 2030. Not only does this cut down on emissions, but it also sparks some seriously creative business ideas—think companies turning buildings into chic, modular, recyclable services (Stahel, 2016). Finland is another good example. Its CE roadmap calls for increasing repair culture and prolonging product lifespans, which is a financially and environmentally sound approach.

Yet, the Circular Economy at present runs into limitations in addressing climate issues. It does not address climate injustices faced by those more vulnerable countries that currently experience climate impacts. Developed nations, hence, must accelerate the course of conversion towards a Circular Economy to lessen the burden on developing countries. Hence, they not only want to prevent climate change but are equally important in confronting what shall follow its aftermath. The common people haven't embarked on that part of the discussion yet, despite this stark threat posed by our quickly depleting carbon budget. The Circular Economy should stretch beyond the realms of a medium towards an SDG, but instead be the points at which better economies are worth a thought about human beings and the environment (Velenturf & Purnell, 2021). The link between CE and CO2 tags can sometimes be somewhat complicated. Speaking in broad terms, the researchers have exhibited something like the roller-coaster effect, with findings swinging back and forth depending on some factors such as the recycling rates or how much people adhered to the use of circular materials (Sikder et al., 2024). Piecing together all these opposing views is enough to at least conclude that, first and foremost, the Circular Economy incites carbon reduction by better resource use, intelligent consumption, and dependence on less fossil fuels.

## 11. CE'S CONTRIBUTION TO BANGLADESH'S LONG-TERM SUSTAINABILITY

The CE model has been proven to have the potential to act as a remedy to the serious problems of firms that create significant amounts of waste and pollution. The use of CE has made the firms economically and environmentally sustainable (Yin et al., 2023). The CE is an ever-unfolding topic that has also gained the attention of other industries in different nations, even though it was originally European in nature. Bangladesh has transformed from an economy that was failing to becoming a major industry in the global textile market, which is very inspiring. But as the looms rumble and the exports rise, our world quietly mourns. The ecological cost of the apparel and textile business is beyond debate, from rivers full of the latest fashion items to landfills overflowing with leftover fabric scraps. That is where the circular economy (CE) enters the picture, not as a term, but as a lifeline to sustainability and, quite frankly, sanity (Saha et al., 2022). Bangladesh textile industry improves environment, society, and economy with better results through the principle of circular economy (Tumpa et al., 2019).

Reduction of pollution, energy, and waste with the help of sustainable sourcing, reduction of waste, and optimisation of energy is undertaken by Bangladesh through the principle of the circular economy. It saves money and improves function by utilising saved energy and generating less waste by virtue of CE (Kazancoglu et al., 2020). Socially and ecologically responsible products increase brand value and consumer loyalty. CE increases the long-term growth of Bangladesh's textile industry by increasing environmental, social, and financial aspects, overcoming sustainability challenges, and creating a sense of competitiveness in a sustainable future (Habib et al., 2021). Circular economy naturally focuses on waste avoidance, ensuring products and materials are continuously in use, and restoring natural ecosystems. Contrary to the old 'take-make-dispose' way of thinking, the Circular Economy promotes 'reduce-reuse-recycle' or, as I affectionately term it, the three R's that can save us. Fashion, which comprises over 84% of Bangladeshi exports, has no choice in Bangladesh; CE must be used (Saha et al., 2022). Bangladesh's CE adoption faces the biggest challenges in the form of a lack of senior management leadership and commitment, a lack of experience and resources, a lack of enforcement of environmental laws, and a lack of supporting government policies (Rashid et al., 2024). For example, mountains of leftover fabric, defective zippers, and rejected labels find their way to the trash.

The Circular Economy's ideas can help give new life to things that have been thrown away. Companies like Desh Garments are already looking into ways to turn textile waste into insulation products. At the same time, local business owners are making stylish bags out of old denim. It's not just smart; it's also cool (Aka, 2022). By employing longevity design, closed-loop recycling, and waste reduction, the industry can have a reduced environmental impact and contribute to the country's economic growth. Bangladesh, which is based on resources, will gain a lot from the ideas of CE (Elf et al., 2022). Through principles such as longevity design, closed-loop recycling, and waste minimization, the industry can reduce its environmental impact while generating revenue (Mishra et al., 2023). To make sure that things last a long time, it is important to use sustainable practices like circular economy principles (Khan et al., 2021). Using circular methods in Bangladesh's garment industry can fix social and environmental problems and create new jobs and money. Bangladesh would benefit more from a circular economy system that uses the 3Rs (Reduce, Reuse, and Recycling) as a way to put the circular economy into practice (Yong, 2007). One of the most important parts of Bangladesh's circular economy policy is recycling.

People in Bangladesh have started to value recycling more in recent years, and Facebook's buying and selling of used items is helping to change the idea that using used items is bad (Arman & Mark-Herbert, 2021). Many factories and businesses also want to use fewer resources, which means they make less waste. But we don't recycle very much of the trash we make. There are 0.12 million poor people who work directly or indirectly in rubbish recycling in Dhaka, the capital city. Bangladesh's 522 cities and towns produce and throw away 13,332 tonnes of trash every day, or 4.86 million tonnes a year. This is bad for the environment (Chowdhury et al., 2014). CE also helps the economy stay stable. Companies reduce their dependence on unreliable global supply chains and new materials by reusing materials and making products last longer. These materials will change like the weather in Dhaka (Huang et al., 2021). In a world where resources are dwindling and the climate is shifting, we need this kind of stability. The overall efficiency of garbage collection in Bangladesh's major cities was only 55%, which is not enough to build a city that is good for the environment. The government of Bangladesh has been using the 3R garbage collection strategy since 2010 to try to deal with the country's constant production of trash.

In 2005, a recycling training center was launched with the UNDP (United Nations Development Programme), SEMP (Sustainable Environment Management Programme), and MoEF (Ministry of Environment and Forests) to train various professionals in waste recycling and energy saving (Jerin et al., 2022). Countries such as Italy and Taiwan are pioneering with Circular Economy (CE), capitalising on human capital, taking up digital technologies, and embracing green policies (Ranta, Aarikka-Stenroos & Mäkinen, 2018; Huang et al., 2021). Bangladesh possesses youth abundance, imagination, and vitality, but enormous possibilities will be lost if financial and technological issues still stand in the way (Saha et al., 2022).

## 12. CIRCULAR ECONOMY PRACTICES IN THE GLOBAL SOUTH

Circular economy (CE) practices in the Global South are emerging as important structural responses to environmental challenges, resource depletion, and socio-economic disparities. While traditional linear economic models focus on extract-produce-dispose processes, CE aims to design out waste, retain products and materials in use, and regenerate natural systems (MacArthur, 2013). CE practices in the Global South, especially in countries like Bangladesh, India, Kenya, and Brazil, occur not only for environmental improvement, but also for economic necessity and social innovation as well (Schroeder et al., 2019). For example, informal recycling networks are established across South Asia and sub-Saharan Africa, which engage in recovering, reclaiming, and repurposing waste materials at social and community levels, without any formal recognition or policy improvement. In Bangladesh, circular economy practices have been emerging in the textile and leather industries with the establishment of various pilots based on industrial symbiosis and waste-to-use concepts in order to mitigate pollution and resource use. Similarly, Kenya's green enterprises in plastic recycling and India's use of reverse logistics for improper disposal of electronics and packaging are examples of grassroots-level circular innovation and community resilience. Often, these circular economy practices do not occur under formal policy or institutional support, which may limit their potential benefit. Current integration of CE in national development agendas remains limited because of barriers such as low regulatory enforcement, limited access to technology, and insufficient stakeholder awareness. Nonetheless, CE offers tremendous potential in the Global South by creating green jobs, promoting sustainable urbanization, and enhancing climate resilience. With appropriate investment, capacity-building, and policy alignment, CE can be scaled up to support inclusive growth and sustainable development, contributing directly to the achievement of SDGs 9 (industry and innovation), 12 (responsible consumption), and 13 (climate action) in resource-constrained settings.

## 13. CHALLENGES AND POLICY RECOMMENDATIONS

The implementation of circular economy (CE) practices is particularly challenging in Bangladesh, as with many other Global South countries, and the practices remain disconnected from national development plans (Ahmed et al., 2022). The lack of a coherent regulatory framework and the enforcement of a regulatory framework that assiduously promotes circularity is a considerable obstacle. With respect to laws governing the environment, there are many existing frameworks, such as the Bangladesh Environment Conservation Act 1995, focused primarily on pollution control, but nothing is related to circularity, including waste valorization, product life-cycle management, or industrial symbiosis. Additionally, institutional fragmentation and limited inter-ministerial coordination undermine the creation of holistic CE policies. The other pressing barrier is financial and technological constraints. Many small- and medium-sized enterprises (SMEs), comprising a considerable share of the industrial base of Bangladesh, do not have access to green finance and investment incentives, nor affordable technology, to transition to a resource-efficient model.

Moreover, gaps in knowledge and awareness exist among all stakeholders, such as policymakers, industrial actors, and consumers. There is limited understanding of CE benefits, such as cost savings through resource recovery or new job creation in remanufacturing and recycling sectors. Educational institutions and training centers have yet to integrate CE concepts into mainstream curricula, leaving a skills gap that prevents implementation at scale. Cultural and behavioral inertia also inhibits transition processes, especially within urban areas with linear habits of consumption, and eliminates barriers with underdeveloped recycling infrastructure, as well as the formalization of the sector itself.

There are some multifaceted policy measures to overcome these constraints. First, establishing a national circular economy roadmap, like those in the EU or China, would provide a framework for a distinct and purposeful direction, including specific sectors, milestones, action points, and performance indicators, and aligning with Goals 9, 12, and 13 of the SDGs. Also, priority actions on circular economy as part of existing national plans, such as the Perspective Plan 2041 and the Eighth Five-Year Plan, to ensured that they fit within the national development landscape. Second, we need modalities of green finance (such as reduced taxes, subsidized loans, public-private partnerships, etc.) to help firms, especially small and medium enterprises (SMEs), switch to cleaner technologies and circular practices for sustainability, in addition to other forms of finance. Development partners are paying for building institutional capacity through technical colleges or other forms of hands-on training programs. These efforts can help address significant knowledge and skill gaps for all personnel. Campaigns to develop public awareness about consumption behaviour and producer responsibility are also very important in shifting perceptions. In addition, establishing extended producer responsibility (EPR) frameworks will legally obligate industries to take back and recycle their products. This will provide accountability and stimulate new ideas in product design. Lastly, digital applications and environmental monitoring systems that support data-driven decision-making can promote transparency, enable tracking of progress, and help regulators respond in the most appropriate manner. Bangladesh can accelerate progress toward a resilient, low-carbon, and inclusive circular economy model by prioritizing these strategies. Not only will this model assist Bangladesh in addressing its environmental and industrial challenges, but it will also help the country make tangible progress toward the SDGs.

## 14. CONCLUSION

The circular economy is more than a catchy phrase that appears in glamorous CSR brochures and climate conventions. At its heart, it is a massive shift in thinking about philosophy, economics, and the environment - a bold new way of contemplating how we produce, consume, and value resources. In Bangladesh, one of the countries striking a balance between rapid industrial growth and fragile nature, this should be a necessity, not just an option. The circular economy sits very well with SDGs 9 (Industry, Innovation and Infrastructure), 12 (Responsible Consumption and Production) and 13 (Climate Action), as we have demonstrated throughout this chapter. We've made it clear that these aren't just random goals on some global scoreboard; they're fundamental human challenges. They affect the air we breathe, the rivers we fish in, and even the growing waste under our kitchen sinks.

The take, make, dispose model of the linear economy has performed poorly. It brought us this far, but it's obvious that it's not going to take us much further. Researchers like Kirchherr et al. (2017) have warned that the biggest obstacle to a circular future isn't technology, it's the way we think. And in Bangladesh, a place where innovation often flourishes in the narrowest byways of resource constraint, perhaps this mindset shift is our greatest opportunity yet. The application of CE to Bangladesh's key sectors, textiles, agriculture, and plastics, reveals we've got quite a goldmine to mine ahead of us. Nabi et al. (2023) report that circular practices in the RMG sector will dramatically reduce emissions and increase job opportunities. Similarly, in the rural context, reprocessing agricultural waste into compost or biofuel can turn something useless into something precious. If executed effectively, as noted by Geissdoerfer et al. (2017), the Circular Economy (CE) not only generates economic value but also enhances social equity and promotes environmental regeneration, an uncommon trifecta in contemporary development discussions.

However, every good thing has its challenges. The circular transition in Bangladesh is hampered by legal loopholes, ineffective inter-agency collaboration, and a lack of awareness, sometimes verging on creative denial. While it sounds ridiculous to ask factories to do waste recycling when we don't give them the legal, financial and resources to do so, it is not going to work. For Bangladesh to move forward, it must act in multiple sectors, invest in clean technology, and integrate circular economy concepts into a variety of policies, including education and industry assistance. Above all, we should work toward a culture shift in which used does not equal useless and sustainability is a reality in homes and workplaces, not just a seminar topic.

In summary, the circular economy provides Bangladesh with a means not only to accomplish SDG 9, 12, and 13, but also to rethink the entirety of development itself. We need to stop using the planet like it is all-you-can-use and start sharing it responsibly more because it will never be replenished.

## REFERENCES


Accenture MVO. (2016). From rhetoric to reality. The circular economy index of Dutch businesses. Retrieved from https://www.ncbi.nlm.nih.gov/pubme

Ahmed, R. R., Streimikiene, D., Soomro, R. H., & Streimikis, J. (2022). Digital transformation and industry 4.0 initiatives for market competitiveness: Business integration management model in the healthcare industry. *Journal of Competitiveness*, *14*(4), 6–24.

Aka, D. C. (2022). Analyzing The Barriers to Adoption of Circular Economy Models by SMEs in Textile Industry Using Fuzzy SWARA Method. *Sosyal Araştırmalar ve Davranış Bilimleri*, *8*(17), 36–55.

Akash, M., Akter, J., Tamanna, T., & Kabir, M. R. (2018, February). The urbanisation and environmental challenges in Dhaka city. 7th International RAIS Conference on Social Sciences.

Alamerew, Y. A., & Brissaud, D. (2019). Circular economy assessment tool for end of life product recovery strategies. *Journal of Remanufacturing*, *9*(3), 169–185.

Androniceanu, A., Kinnunen, J., & Georgescu, I. (2021). Circular economy as a strategic option to promote sustainable economic growth and effective human development. Journal of International Studies, 14(1).

Antikainen, M., & Valkokari, K. (2016). A framework for sustainable circular business model innovation. *Technology Innovation Management Review*, *6*(7).

Arman, S. M., & Mark-Herbert, C. (2021). Re-commerce to ensure circular economy from consumer perspective. *Sustainability*, *13*(18), 10242.

Bag, S., & Pretorius, J. H. C. (2022). Relationships between industry 4.0, sustainable manufacturing and circular economy: Proposal of a research framework. *The International Journal of Organizational Analysis*, *30*(4), 864–898.

Biller, J. R., & McPeak, J. E. (2021). EPR everywhere. *Applied Magnetic Resonance*, *52*(8), 1113–1139.

Bocken, N. M., De Pauw, I., Bakker, C., & Van Der Grinten, B. (2016). Product design and business model strategies for a circular economy. Journal of Industrial and Production Engineering, 33(5), 308-320.

Bocken, N. M., Short, S. W., Rana, P., & Evans, S. (2014). A literature and practice review to develop sustainable business model archetypes. *Journal of Cleaner Production*, *65*, 42–56.

Bouton, S., Hannon, E., Rogers, M., Swartz, S., Johnson, R., Gold, A., & Staples, M. (2016). The circular economy: Moving from theory to practice. McKinsey Centre for Business and Environment.

Cho, S., Sultana, R., & Kwon, S. (2019). Social enterprise and sustainable development in Bangladesh and Korea: Opportunities and challenges. *Asian Social Work and Policy Review*, *13*(2), 189–198.

Chowdhury, A. H., Haque, M. R. U., & Hossain, T. (2014, May). Developing 3Rs (reduce, reuse & recycle) strategy for waste management for urban areas in Bangladesh: A socioeconomic and climate mitigation option. In *2014 3rd International Conference on the Developments in Renewable Energy Technology (ICDRET)* (pp. 1-6). IEEE.

D'Amato, D., Korhonen, J., & Toppinen, A. (2019). Circular, green, and bio economy: How do companies in land-use intensive sectors align with sustainability concepts? *Ecological Economics*, *158*, 116–133.

Daigle, B., & DeCarlo, S. (2021). *Rare earths and the US Electronics sector: supply chain developments and trends*. Office of Industries, US International Trade Commission.

Dantas, T. E. T., de-Souza, E. D., Destro, I. R., Hammes, G., Rodriguez, C. M. T., & Soares, S. R. (2021). How the combination of Circular Economy and Industry 4.0 can contribute towards achieving the Sustainable Development Goals. *Sustainable Production and Consumption*, *26*, 213–227.

De Jesus, A., Antunes, P., Santos, R., & Mendonça, S. (2018). Eco-innovation in the transition to a circular economy: An analytical literature review. *Journal of Cleaner Production*, *172*, 2999–3018.

Deselnicu, D. C., Militaru, G. H. E. O. R. G. H. E., Deselnicu, V., Zăinescu, G., & Albu, L. (2018, October). Towards a circular economy–a zero waste programme for Europe. In *Proceedings of the 7th International Conference on Advanced Materials and Systems* (Vol. 2018, pp. 18-20).

Di Maio, F., & Rem, P. C. (2015). A robust indicator for promoting circular economy through recycling. *Journal of Environmental Protection*, *6*(10), 1095–1104.

Elf, P., Werner, A., & Black, S. (2022). Advancing the circular economy through dynamic capabilities and extended customer engagement: Insights from small sustainable fashion enterprises in the UK. *Business Strategy and the Environment*, *31*(6), 2682–2699.

Elisha, O. D. (2020). Moving beyond take-make-dispose to take-make-use for sustainable economy. *Int. J. Sci. Res. Educ*, *13*(3), 497–516.

Fonseca, L., & Michie, R. (2024). Circular Economy Policies in Europe: Assessing Regional Policy Integration. Academic Press.

França, R., Nylén, E. J., Jokinen, A., & Jokinen, P. (2022). Filling the social gap in the circular economy: How can the solidarity economy contribute to urban circularity? In *Social and cultural aspects of the circular economy* (pp. 27–44). Routledge.

García-Barragán, J. F., Eyckmans, J., & Rousseau, S. (2019). Defining and measuring the circular economy: A mathematical approach. *Ecological Economics*, *157*, 369–372.

Geerken, T., Schmidt, J., Boonen, K., Christis, M., & Merciai, S. (2019). Assessment of the potential of a circular economy in open economies–Case of Belgium. *Journal of Cleaner Production*, *227*, 683–699.

Geissdoerfer, M., Savaget, P., Bocken, N. M., & Hultink, E. J. (2017). The Circular Economy–A new sustainability paradigm? *Journal of Cleaner Production*, *143*, 757–768.

Goyal, S., Esposito, M., & Kapoor, A. (2018). Circular economy business models in developing economies: Lessons from India on reduce, recycle, and reuse paradigms. *Thunderbird International Business Review*, *60*(5), 729–740.

Habib, M. A., Bao, Y., Nabi, N., Dulal, M., Asha, A. A., & Islam, M. (2021). Impact of strategic orientations on the implementation of green supply chain management practices and sustainable firm performance. *Sustainability*, *13*(1), 340.

Hondroyiannis, G., Sardianou, E., Nikou, V., Evangelinos, K., & Nikolaou, I. (2024). Circular economy and macroeconomic performance: Evidence across 28 European countries. *Ecological Economics*, *215*, 108002.

Huang, Y. F., Azevedo, S. G., Lin, T. J., Cheng, C. S., & Lin, C. T. (2021). Exploring the decisive barriers to achieve circular economy: Strategies for the textile innovation in Taiwan. *Sustainable Production and Consumption*, *27*, 1406–1423.

Jackson, T. (2009). *Prosperity without growth: Economics for a finite planet.* Routledge.

Jerin, D. T., Sara, H. H., Radia, M. A., Hema, P. S., Hasan, S., Urme, S. A., & Quayyum, Z. (2022). An overview of progress towards implementation of solid waste management policies in Dhaka, Bangladesh. *Heliyon*, *8*(2).

Jordens, J. W. (2016). Building a Collaborative Advantage within a Circular Economy: Inter-Organisational Resources and Capabilities of a Circular Value Chain (Master's thesis).

Kazancoglu, Y., Sagnak, M., Kayikci, Y., & Kumar Mangla, S. (2020). Operational excellence in a green supply chain for environmental management: A case study. *Business Strategy and the Environment*, *29*(3), 1532–1547.

Khan, S. A. R., Razzaq, A., Yu, Z., & Miller, S. (2021). Retracted: Industry 4.0 and circular economy practices: A new era business strategies for environmental sustainability. *Business Strategy and the Environment*, *30*(8), 4001–4014.

Kirchherr, J., Reike, D., & Hekkert, M. (2017). Resources, Conservation & Recycling Conceptualising the circular economy: An analysis of 114 definitions. Academic Press.

Kirchherr, J., Reike, D., & Hekkert, M. (2017). Conceptualizing the circular economy: An analysis of 114 definitions. *Resources, Conservation and Recycling*, *127*, 221–232.

Klapper, L., El-Zoghbi, M., & Hess, J. (2016). Achieving the sustainable development goals. The role of financial inclusion. http://www. ccgap. Org

Lacy, P., Keeble, J., McNamara, R., Rutqvist, J., Haglund, T., Cui, M., & Buddemeier, P. (2014). *Circular advantage: innovative business models and technologies to create value in a world without limits to growth.* Accenture.

Lee, B. X., Kjaerulf, F., Turner, S., Cohen, L., Donnelly, P. D., Muggah, R., & Gilligan, J. (2016). Transforming our world: Implementing the 2030 agenda through sustainable development goal indicators. *Journal of Public Health Policy*, *37*, 13–31.

Llorente-González, L. J., & Vence, X. (2019). Decoupling or 'decaffing'? The underlying conceptualization of circular economy in the European Union monitoring framework. *Sustainability*, *11*(18), 4898.

Lovejoy, T. E., & Nobre, C. (2018). Amazon tipping point. *Science Advances*, *4*(2), eaat2340.

MacArthur, E. (2013). *Towards the circular economy, economic and business rationale for an accelerated transition.* Ellen MacArthur Foundation.

Macarthur, E. (2019). How the circular economy tackles climate change. *Ellen MacArthur Found*, *1*, 1–71.

Meath, C., Karlovšek, J., Navarrete, C., Eales, M., & Hastings, P. (2022). Co-designing a multi-level platform for industry level transition to circular economy principles: A case study of the infrastructure CoLab. *Journal of Cleaner Production*, *347*, 131080.

Mentink, B. A. S. (2014). Circular business model innovation: a process framework and a tool for business model innovation in a circular economy. Delft University of Technology, 167.

Mirchandani, M. (2016). *The Teesta water dispute: Geopolitics, myth and economics.* Observer Research Foundation Issue Brief.

Mishra, A., Dutta, P., Jayasankar, S., Jain, P., & Mathiyazhagan, K. (2023). A review of reverse logistics and closed-loop supply chains in the perspective of circular economy. *Benchmarking*, *30*(3), 975–1020.

Mishra, R., Singh, R. K., & Rana, N. P. (2024). Digital orientation, digital eco-innovation and circular economy in the context of sustainable development goals. *Business Strategy and the Environment*, *33*(4), 2752–2770.

Muthu, S. S. (2020). Assessing the environmental impact of textiles and the clothing supply chain. Woodhead Publishing.

Nabi, M. H., Hasan, M., Chowdhury, A. T., Naz, F., & Hossain, M. (2023). The impact of climate change on the lives and livelihoods of readymade garment (RMG) workers: An exploratory study in selected readymade garment factories in Bangladesh. *BMC Public Health*, *23*(1), 2292.

Nikolaou, I. E., Jones, N., & Stefanakis, A. (2021). Circular economy and sustainability: The past, the present and the future directions. *Circular Economy and Sustainability*, *1*, 1–20.

Pao, H. T., & Chen, C. C. (2022). The dynamic interaction between circular economy and the environment: Evidence on EU countries. *Waste Management & Research*, *40*(7), 969–979.

Parchomenko, A., Nelen, D., Gillabel, J., & Rechberger, H. (2019). Measuring the circular economy-A Multiple Correspondence Analysis of 63 metrics. *Journal of Cleaner Production*, *210*, 200–216.

Pieroni, M. P., McAloone, T. C., & Pigosso, D. C. (2019). Business model innovation for circular economy and sustainability: A review of approaches. *Journal of Cleaner Production*, *215*, 198–216.

Prieto-Sandoval, V., Jaca, C., Santos, J., Baumgartner, R. J., & Ormazabal, M. (2019). Key strategies, resources, and capabilities for implementing circular economy in industrial small and medium enterprises. *Corporate Social Responsibility and Environmental Management*, *26*(6), 1473–1484.

Rahman, M. C., Sarkar, M. A. R., Kabir, M. J., Rahaman, M. S., Biswas, J. C., Aziz, M. A., & McKenzie, A. M. (2025). Vulnerability and adaptation in flood-prone ecosystems of Bangladesh: A case study of rice farming households. *Discover Sustainability*, *6*(1), 55.

Ranta, V., Aarikka-Stenroos, L., & Mäkinen, S. J. (2018). Creating value in the circular economy: A structured multiple-case analysis of business models. *Journal of Cleaner Production*, *201*, 988–1000.

Rashid, K. H. O., Al Aziz, R., Karmaker, C. L., Bari, A. M., & Raihan, A. (2024). Evaluating the challenges to circular economy implementation in the apparel accessories industry: Implications for sustainable development. Green Technologies and Sustainability, 100140.

Rocchi, L., Paolotti, L., Cortina, C., Fagioli, F. F., & Boggia, A. (2021). Measuring circularity: An application of modified Material Circularity Indicator to agricultural systems. *Agricultural and Food Economics*, *9*(1), 9.

Romero-Perdomo, F., Carvajalino-Umaña, J. D., Moreno-Gallego, J. L., Ardila, N., & González-Curbelo, M. Á. (2022). Research trends on climate change and circular economy from a knowledge mapping perspective. *Sustainability*, *14*(1), 521.

Roshid, Md. M., Yeamin, Md. B., Waaje, A., Karim, R., & Kaswan, M. S. (2025). Understanding the Nexus of Industry 5.0 and Circular Economy for Enriched Lean and Environment Sustainability. In José Carlos Vieira de Sá, Shubham Sharma, Sunil Kumar, & Manjunath B.R. (Eds.), *Transformative Lean Six Sigma Techniques for the Quality 5.0 Paradigm* (pp. 349–380). IGI Global Scientific Publishing. DOI: 10.4018/979-8-3373-0943-9.ch013

Roshid, Md. M., Waaje, A., Karim, R., Irtisham, M. R., & Kaswan, M. S. (2025a). Industry 5.0 and Organizational Sustainability: Integrating Cutting-Edge Technologies and Human Ingenuity. In Mahender Singh Kaswan, Rajeev Rathi, & Jose Arturo Garza-Reyes (Eds.), *Industry 5.0: Concepts and Strategies for Digital Transformation* (pp. 50–61). CRC Press. DOI: 10.1201/9781003535188-7

Saha, K., Dey, P. K., & Papagiannaki, E. (2022). Implementing circular economy in the textile and clothing industry. In *Supply chain sustainability in small and medium sized enterprises* (pp. 239–276). Routledge.

Sariatli, F. (2017). Linear economy versus circular economy: A comparative and analyzer study for optimization of economy for sustainability. *Visegrad Journal on Bioeconomy and Sustainable Development*, *6*(1), 31–34.

Schroeder, P., Anggraeni, K., & Weber, U. (2019). The relevance of circular economy practices to the sustainable development goals. *Journal of Industrial Ecology, 23*(1), 77–95.

Schwarz, A. E., Ligthart, T. N., Bizarro, D. G., De Wild, P., Vreugdenhil, B., & van Harmelen, T. (2021). Plastic recycling in a circular economy; determining environmental performance through an LCA matrix model approach. Waste Management, 121, 331-342.

SDG. (2015). Sustainable Development Goals: action towards 2030. UN.

Shih, W. (2020). Is it time to rethink globalized supply chains? *MIT Sloan Management Review, 61*(4), 1–3.

Sikder, M., Wang, C., Rahman, M. M., Yeboah, F. K., Alola, A. A., & Wood, J. (2024). Green logistics and circular economy in alleviating CO2 emissions: Does waste generation and GDP growth matter in EU countries? *Journal of Cleaner Production, 449*, 141708.

Skvarciany, V., Lapinskaitė, I., & Volskytė, G. (2021). Circular economy as assistance for sustainable development in OECD countries. Academic Press.

Stahel, W. R. (2016). Circular economy: A new relationship with our goods and materials. *Nature*.

Stefanakis, A. I., & Nikolaou, I. E. (Eds.). (2024). *Circular Economy and Sustainable Development: A Necessary Nexus for a Sustainable Future*. Springer Nature.

Sustainable Development Solutions Network. (2022). *Getting Started with the Sustainable Development Goals: A Guide for Stakeholders*. Sustainable Development Solutions Network.

Svensson, G., Ferro, C., Høgevold, N., Padin, C., Varela, J. C. S., & Sarstedt, M. (2018). Framing the triple bottom line approach: Direct and mediation effects between economic, social and environmental elements. *Journal of Cleaner Production, 197*, 972–991.

Tambovceva, T., & Titko, J. (2020). Introduction to circular economy. Ekonomikas un kulturas augstskola.

(The) Financial Express. (2016). Eradicating poverty through economic growth. The Financial Express. http://www.thefinancialexpress-bd.com

(The) Financial Express. (2017). SDG: Drawing up implementation roadmap. The Financial Express. http://www.thefinancialexpress-bd.com

Trevisan, A. H., Lobo, A., Guzzo, D., de Vasconcelos Gomes, L. A., & Mascarenhas, J. (2023). Barriers to employing digital technologies for a circular economy: A multi-level perspective. *Journal of Environmental Management, 332*, 117437.

Triguero, Á., Cuerva, M. C., & Sáez-Martínez, F. J. (2022). Closing the loop through eco-innovation by European firms: Circular economy for sustainable development. *Business Strategy and the Environment, 31*(5), 2337–2350.

Tumpa, T. J., Ali, S. M., Rahman, M. H., Paul, S. K., Chowdhury, P., & Khan, S. A. R. (2019). Barriers to green supply chain management: An emerging economy context. *Journal of Cleaner Production, 236*, 117617.

Uddin, M. J., & Uddin, M. N. (2018). Overview of the challenges to achieve sustainable development goals in Bangladesh. *The International Journal of Social Sciences and Humanities Invention, 5*(2), 4453–4460.

Velenturf, A. P., & Purnell, P. (2021). Principles for a sustainable circular economy. *Sustainable Production and Consumption, 27*, 1437–1457.

Verga, G. C., & Khan, A. Z. (2022). An introduction to the circular economy. Transitioning to a circular economy changing business models and business ecosystems, 15.

Waaje, A., Karim, R., & Roshid, Md. M., & Meem, T. N. (2024). Minimizing Waste in the Hospitality Industry: The Opportunities and Challenges of Lean Tourism. In R. Sucheran (Ed.), *Sustainable Waste Management in the Tourism and Hospitality Sectors* (pp. 125–166). IGI Global Scientific Publishing. DOI: 10.4018/979-8-3693-6110-8.ch006

Wang, N., Bai, Y., Guo, Z., Fan, Y., & Meng, F. (2024). Synergies between the circular economy and carbon emission reduction. *The Science of the Total Environment, 951*, 175603.

World Economic Forum. (2014). *Towards the Circular Economy: Accelerating the Scale-Up across Global Supply Chains.* Geneva: World Economic Forum, in collaboration with the Ellen MacArthur Foundation and McKinsey & Company.

Yin, S., Jia, F., Chen, L., & Wang, Q. (2023). Circular economy practices and sustainable performance: A meta-analysis. *Resources, Conservation and Recycling, 190*, 106838.

Yong, R. (2007). The circular economy in China. *Journal of Material Cycles and Waste Management, 9*, 121–129.

Chapter 12

# Digitalization and the Circular Economy:

## An Empirical Assessment of Material Footprint Across Developing and Developed Countries

**Tanvi Sanotra**
https://orcid.org/0009-0005-5419-7878
*Shri Mata Vaishno Devi University, India*

**Meenakshi Gupta**
https://orcid.org/0000-0002-7866-4961
*Shri Mata Vaishno Devi University, India*

## ABSTRACT

*This study shows the relationship between digitalization and circular economy. Material footprint is taken here as the approx. of circular economy. The material footprint helps to quantify the total raw material used to understand a country's consumption demand. It uses the panel data from 2000 to 2022 across 20 developed and developing countries. Material footprint is taken here as the independent variables including several proxies for digitalization, such as ICT service export, ICT goods export, individual using the internet, mobile subscription, telephone usage. Additionally, it includes GDP, FDI, and trade openness as controlled variables. The study used pooled OLS, fixed effect, random effect, and Hausman test. The findings from these results show the relationship between the material footprint and digitalization and integration of circular economy initiatives. The results also show that*

DOI: 10.4018/979-8-3373-2454-8.ch012

*developed countries have positive relationship with environmental sustainability whereas developing countries don't have positive relationship with environmental sustainability.*

## 1. INTRODUCTION

The importance of achieving environmental sustainability has grown in recent years as a result of increasing environmental degradation, biodiversity loss, and climate instability. Rapid industrialization and urban expansion, while driving global economic growth, have relied largely on fossil fuels and resource-intensive processes, resulting in a dramatic increase in greenhouse gas emissions and material overconsumption (Karim et al., 2025a). Traditional linear production processes, founded on the "take-make-dispose" paradigm, have accelerated resource depletion and contributed considerably to climate change. According to the UNEP 2023 report, the linear economy model accounts for around 80% of biodiversity loss and 90% of water stress globally. In this context, digitalization emerges as a disruptive force. Artificial intelligence (AI), the Internet of Things (IoT), and data analytics are being acknowledged for their ability to improve energy efficiency, streamline production, optimize resource use, and support environmental monitoring (UNIDO, 2022). These instruments also play an important role in enabling the circular economy, which is based on the concepts of reducing, reusing, repairing, and recycling resources. Unlike the linear economy, the circular economy aims to remove waste while also regenerating natural systems, maintaining long-term ecological equilibrium. Furthermore, the material footprint—an indicator of the total raw materials extracted to meet consumption demands—has emerged as an important sustainability metric. According to reports from the International Resource Panel (IRP) and the United Nations Environment Programme (UNEP), global material extraction has tripled since 1970, with over 100 billion metric tons predicted by 2022, leading to serious environmental consequences such as increased emissions and biodiversity loss. This increased demand has fueled the call to integrate digital innovation with circular practices in order to reduce material footprints and achieve Sustainable Development Goals.

Despite these improvements, a significant problem remains: while digitalization holds potential for long-term transformation, its adoption is unequal, particularly in developing nations where technological infrastructure and legislative backing are typically weak. Furthermore, while digital tools can improve circular practices and lessen environmental harm, their widespread use raises worries about expanding material demand, particularly for rare earth elements used in smart gadgets and data centers. This dual role of digitalization results in a complex relationship with environmental repercussions. The fundamental issue addressed in this paper is the influence of digitization in influencing material footprint and, consequently, how this relationship affects the adoption of circular economy methods. The study specifically investigates whether digital technologies contribute to decreased or increased material consumption, as well as whether this impact aids or hinders the shift from a linear to a circular economic model.

While industrialized countries like South Korea and Sweden are making rapid progress in digital circular integration, emerging markets like India and Brazil continue to encounter hurdles in digital infrastructure, waste management, and environmental regulation. As a result, recognizing the situations in which digitalization might positively contribute to sustainability is critical.

Furthermore, little research has been conducted to investigate the dual effect of digitalization on material use, which has the potential to both reduce and increase environmental impact. Many studies mention the circular economy concept but fail to empirically examine its interaction with digital tools in creating sustainability results. Liu et al. (2025) and Kurniawan et al. (2022) conducted research on the importance of digital technologies in enabling resource efficiency and low-carbon development, but comprehensive data from industrialized and developing countries remain sparse. Some experts have proposed policy changes, public-private partnerships, and green funding systems, but their practical implementation remains fragmented. There is also an urgent need for more integrative empirical studies that link digitization, material footprint, and circular economy adoption—particularly those that emphasize regional differences and suggest scalable solutions.

The purpose of this study is to empirically investigate the relationship between digitalization and material footprint, as well as how it influences the adoption of circular economy practices in selected developed and developing economies. The goal is to determine whether digital technologies considerably reduce material use and ease the transition from linear to circular economic systems. Furthermore, the study aims to discover differences between economies in order to highlight structural and policy-level factors influencing these processes. Unlike prior studies that used conventional indicators such as income or carbon emissions, this study takes a broader sustainability approach by concentrating on the material footprint. Existing studies have used a variety of methodologies, including panel cointegration models, dynamic ARDL, and structural equation modeling, but they frequently lack robustness across several model specifications. To address these limitations, this study used Pooled Ordinary Least Squares (POLS), the Hausman test, and both fixed and random effects models. These econometric techniques enable the testing of the robustness and consistency of results across multiple estimate approaches, while also accounting for cross-country variation. Using these technologies reveals how digitization affects material footprint, which supports long-term sustainability and the implementation of circular economy principles.

## 2. LITERATURE REVIEW

This chapter analyzes the evolution of academic and policy discourse on the circular economy (CE), digitization, material footprint (MF), and sustainability. It uses a combination of theoretical foundations and empirical research to discover gaps in the literature and establish the need for the current study. Theoretical talks about the circular economy began with a critique of the standard linear economic model. Kopnina (2019) divided CE thinking into three primary schools of thought: cradle-to-cradle, performance economy, and industrial ecology. These schools expanded the concept of sustainable manufacturing and consumption by utilizing longer product life cycles, closed-loop processes, and system-level integration. Earlier philosophers, such as Pearce and Turner (1989), argued for reuse, recycling, and resource recovery as a means of decoupling economic activity from environmental degradation. Hawken et al. (2013) and Korhonen et al. (2018) elaborated on these ideas, defining CE as a tool of promoting natural capitalism and behavior-driven change among producers and consumers.

On the empirical side, early implementations of CE frameworks appeared in the late 2010s. Blomsma et al. (2019) introduced the Circular Strategies Scanner, which assists organizations in developing practical strategies for CE adoption. Demestichas and Daskalakis (2020) used a readiness index to evaluate the preparedness of two Indian towns, Indore and Sagar, for Industry 4.0-based waste management solutions. Their findings indicated considerable geographical variations, highlighting the uneven rate of digital change in developing nations.

By 2021, scholars increasingly turned attention to material footprints and environmental stress. Sahoo et al. (2021) used cross-sectional augmented Dickey-Fuller (CSADF) and panel cointegration tests to show that urbanization and economic growth increased MF in BRICS countries, while human capital and trade openness had mitigating effects. Razzaq et al. (2021) employed method of moments quantile regression (MMQR) for 11 material-consuming nations and found that green innovation lowers MF, but infrastructure development raises it.

In 2022, critical perspectives on CE gained popularity. Bocken et al. (2022) evaluated 150 company cases and concluded that current CE models mainly focus on incremental change, advocating for a sufficiency-based CE strategy supported by policy and governance reform. Bellini and Bang (2022), citing interviews in Norway's AEC sector, highlighted data interoperability and knowledge shortages as major impediments to CE implementation. Meanwhile, Wang et al. (2023) investigated top emitter countries using augmented mean group estimators and discovered that CE occasionally increased emissions due to inadequate waste management techniques. Di et al. (2023) investigated 30 Chinese provinces and discovered that green initiatives resulted in a negative correlation between CE and emissions. Karlilar and Pata (2023) investigated the Environmental Kuznets Curve (EKC) in OECD countries and found that green taxes and innovation lower MF over time.

The year 2024 saw an increase in CE research using varied approaches. Chakraborty et al. (2024) showed that inadequate business procedures impede CE adoption in Australian SMEs, notwithstanding their interest in sustainability. Bhawna et al. (2024) emphasized how digitization improves supply chains using CE, albeit with restricted data breadth. Selvaraj et al. (2024) highlighted India's CE-aligned policies, such as increased producer responsibility. Meshram (2024) advocated for the 5R framework and green practices to lower the material footprint (MF). Tiwari et al. (2024) and Sikder et al. (2024) found that CE adoption cuts emissions significantly in high-income and EU nations, respectively.

Campoli et al. (2025) analyzed G20 nations on SDG 6 using DEA and the Malmquist Index, indicating significant CE adoption differences. Liu et al. found a correlation between improved digital infrastructure and financing and lower MF in 87 developing countries. Campoli, Jessica Suarez, et al. (2025) analyze the work done by G20 nations to achieve SDG 6 (clean water and sanitation) using a circular economy lens. The Data Envelopment Analysis Slack-Based quantify (SEA-SBM) is used to quantify efficiency, while the Malmquist Productivity Index (MPI) is used to track productivity. The findings indicate that there is a discrepancy between industrialized and poor countries. The UK, Germany, and Australia are ranking high among the developed countries, and South Africa, Saudi Arabia, and India show high efficiency among emerging nations. They also highlight the necessity to invest in resource-efficient technologies. Overall, the study proposes implementing sustainable water management practices that align with the circular economy paradigm. Public-private collaborations, financial aid, technology transfer, and training initiatives are required to expedite SDG 6 development.

Despite expansion, gaps remain. Most studies focus on emissions, but few investigate how digitization affects MF and CE adoption, particularly in cross-country contexts. To address this, the current study looks at the relationship between digitalization, material footprint, and the circular economy in both developed and developing countries.

## 3. DATA AND METHODS

### 3.1 Theoretical Justification

The current study attempts to properly justify each element before proceeding to data description. Under the environmentally sustainable paradigm, this study selects material footprint as an explanatory variable and uses it as a proxy for the circular economy. The material footprint estimates the total raw materials extracted to meet consumption demands and is a comprehensive indication of environmental pressure and resource intensity (Wiedmann et al., 2015). As the global economic landscape evolves toward sustainability, monitoring MF becomes critical for assessing a country's success in shifting from linear to circular models (Sanye-Mengual et al. 2019).

To investigate the relationship between material footprint and digitalization, the study employs multiple digitalization proxies, including ICT service exports, ICT goods exports, internet usage, mobile subscriptions, and telephone use. These variables were chosen because digital technology can improve resource efficiency, alleviate environmental stress, and promote circular business models. According to recent research, digitalization plays an important role in enabling circular economy initiatives by increasing sustainable resource use, closed-loop systems, and process efficiencies. Digital technologies such as IoT, AI, cloud computing, and blockchain improve traceability, waste minimization, and lifecycle management. Bhawna et al. (2024) contend that digitization directly promotes CE-oriented supply chain management by enabling real-time monitoring, predictive analytics, and inter-organizational collaboration. These digital innovations improve resource circularity while reducing material intensity. Similarly, Bellini and Bang (2022) emphasize that data management is a critical facilitator for CE, particularly in industries such as construction and manufacturing. Their findings from the Norwegian AEC industry show that, while the potential is high, limitations such as poor interoperability and a lack of data infrastructure might impede CE implementation. Thus, successful digitization necessitates strong governance and infrastructure support. Blomsma et al. (2019) create a circular strategy framework for manufacturers and argue that digitalization promotes CE-oriented innovation by enabling lifecycle thinking, product-as-a-service models, and modular design. Bocken et al. (2022) continue on to say that organizations moving toward "sufficiency-based" CE, which prioritizes low resource consumption and long-term product value, rely largely on digital solutions to handle complexity and scalability. Among its indicators, ICT service exports indicate a transition toward knowledge-based, resource-efficient economies. These services improve energy efficiency and carbon reduction, particularly in emerging countries (Gholami et al., 2010), while also encouraging innovation and efficient resource use (Niebel, 2018; Roahid et al., 2025). Furthermore, ICT goods exports demonstrate a country's technological advancement and assist CE by encouraging automation, real-time monitoring, and reduced material dependency. Kurnia et al. (2015) and Ali et al. (2018) both underline how ICT goods trade promotes cleaner production and environmental improvements. Internet use increases digital infrastructure by providing access to knowledge, fostering second-hand marketplaces, and increasing awareness about the environment (Karim et al., 2025b).Vu (2011) and Zhu et al. (2020) found that internet penetration is associated with higher economic efficiency and green production. Mobile subscriptions promote digital access and play an important role in CE adoption, particularly in developing countries. They enable trash tracking, smart recycling, and behavioral change (Marques & Ferreira, 2020; Jha & Prasad, 2020), thereby bridging the gap between technology and sustainability practices. Even telephone use, while more traditional, helps to digital inclusion. Telephony helps to disseminate knowledge and promote environmental planning in developing countries (Qureshi, 2009; Czernich et al., 2011), complementing larger digital efforts. In order to account for broader economic factors, GDP, FDI, and trade openness are included. The relationship between GDP and the environment follows the Environmental Kuznets Curve, which shows that pollution rises with income but falls at a certain point (Grossman & Krueger, 1995; Shahbaz et al., 2013). FDI can provide green innovations, but it also risks shifting pollution-intensive sectors to nations with lax regulations (Zarsky, 1999; Cole & Elliott, 2005). Similarly, trade openness can improve sustainability through technology transfer (Antweiler et al., 2001), but it may also exacerbate environmental stress if not properly regulated (Dean, 2002). This analysis uses panel data from 20 countries—ten developed and ten developing—spanning the years 2000 to 2022. A complete table of variable descriptions is included at the end to define and explain each variable utilized in the analysis.

## 3.2 Description of the Variable and Its Sources

*Table 1. Description of the variable and its sources*

| Variable (dependent) | Definition | source |
|---|---|---|
| Material footprint | Material footprint tracks the total volume of the natural resources needed for the production of goods and services used in an economy, encompassing both local extraction and imported materials. In developed countries, people's consumption levels are high for fossil fuels and metal ores, but on the other hand, developing countries have low consumption levels as compared to developed countries. | resource panel |

Independent variable: the study contains eight explanatory variables related to digitalization and economic growth.

*Table 2.*

| | | |
|---|---|---|
| ICT SERVICES EXPORT | It includes the export of digital services like cloud computing, digital platforms, IT assistance, and software-as-a-service (SaaS). Developed countries have great digital infrastructure and skilled workforce, which makes them lead the ICT services export, whereas developing countries have limited digital infrastructure, but they are gradually catching up with time. | world-development-indicators |
| ICT GOODS EXPORT | ICT goods export represent the global trade of hardware-based products such as semiconductors, computers, telecommunications equipment, and consumer electronics. Developed countries have industrial investment and high investment when it comes to ICT goods production like AI-powered hardware, telecom equipment, etc. however, on the other hand, developing countries have low investment support related to ICT goods production. | world-development-indicators |
| MOBILE SUBSCRIPTION | Mobile subscription tracks the total number of mobile connections, which facilitates individuals and businesses with wireless mobile networks. It is key indicator of digitalization. In contrast, developed countries have higher adoption of smartphones, which leads to an increase in the usage of mobile subscriptions, but developing countries have a low adoption rate of smartphones, which leads to a decrease in the usage of mobile subscriptions. | world-development-indicators |
| FIXED TELEPHONE SUBSCRIPTION | Fixed telephone subscription refers to the total number of landline connections that provide communication services. Decline in the telephone usage in developed countries lead to the adoption of fiber-optic and VoIP services, whereas developing countries are more in favor of mobile networks, as they are more practical and cost effective as compared to fixed telephone subscriptions. | world-development-indicators |
| GROSS DOMESTIC PRODUCTION | GDP refers to the total value of all the final goods and services produced within the country at a particular period of time. Developed countries typically have high GDP per capita, it shows that strong and advance industrial sector but on the other side, developing countries have low GDP per capita. Economic growth brings opportunity but challenges also. | world-development-indicators |
| FOREIGN DIRECT INVESTMENT | FDI refers to the investment of money by a foreign entity in the business of another country by owning some share of that business. Developed countries are the main source for FDI and they often invest in the developing countries. For the growth of the country FDI is needed by all the countries. | our world in data |
| INDIVIDUAL USING THE INTERNET | It is an important estimator of digitalization. It also refers how many people in the country that have access to the internet, which will be used for online services. People in developed countries have high speed internet and have high digital literacy as compared to developing countries but is it improving due to government initiatives. | world-development-indicators |
| TRADE OPENNESS | Trade openness tells us about engagement in international trade by reducing tariffs and quotas. Developed countries often export goods related to financial services, technology, and pharmaceuticals, whereas developing countries typically trade goods related to agriculture, raw materials, and textiles. | world-development-indicators |

Source: Author's own estimation

### 3.3 Country Selection and Case Study

The developed countries in this study were chosen for their leadership in digital innovation, advanced circular economy (CE) initiatives, and access to high-quality longitudinal data that enable robust empirical analysis. Norway, Denmark, Sweden, and Finland have consistently ranked at the top of the Digital Economy and Society Index (DESI) and the Global Innovation Index, underscoring their superior digital infrastructure, investment in research and development, and long-standing commitment to sustainable practices (OECD, 2020). These Nordic countries have historically prioritized environmental sustainability, making them early adopters of CE frameworks and technological solutions aimed at minimizing resource consumption. For example, Finland was one of the first countries globally to implement a national CE strategy, developed by Sitra, which encompasses digital waste tracking systems, real-time resource optimization tools, and innovative product-as-a-service models designed to extend product lifecycles and reduce waste (Sitra, 2019). In Western Europe, France, Germany, and Italy stand out for their progressive CE legislative frameworks. France's AGEC Law (Anti-Gaspillage pour une Économie Circulaire, 2020) seeks to drastically reduce waste generation, promote eco-design, and enhance consumer transparency through digital labeling and repairability scores. Germany's KrWG Circular Economy Act emphasizes extended producer responsibility and lifecycle resource efficiency while actively incorporating Industry 4.0 tools such as AI and IoT in waste classification and recycling logistics. Italy, though sometimes overlooked, has also adopted several digital CE initiatives in manufacturing and packaging sectors.

In East Asia, Japan and South Korea offer technologically advanced models of CE adoption. Their strategies integrate digital tools into green growth and innovation policies, focusing on smart waste management infrastructure, extended producer responsibility (EPR) systems, and data-driven smart manufacturing. Government-led programs incentivize businesses to develop closed-loop production processes using sensors, AI, and predictive analytics (UNIDO, 2021). Moreover, Switzerland, known for its strict environmental laws and high innovation capacity, encourages circularity through policy instruments, fiscal incentives, and public-private collaborations, particularly in the construction and chemical industries.

The high standard of living, advanced industrial structures, and higher per capita consumption in these developed countries contribute to larger material footprints, which in turn provide both the urgency and the opportunity to implement impactful CE measures supported by digital technologies. This intersection of affluence, innovation, and environmental concern makes these nations ideal candidates for examining the digital-CE nexus in empirical research. Their policies and practices not only serve as benchmarks for global CE advancement but also demonstrate how digitalization can act as a catalyst for systemic environmental transformation. The developing countries chosen include large emerging markets, economically diverse regions, and countries with significant material consumption and environmental footprints. These nations—India, China, Brazil, South Africa, Russia, Singapore, Mexico, Thailand, Pakistan, and the Philippines—represent a broad spectrum of regional contexts and development stages. Despite considerable disparities in digital maturity, governance quality, and infrastructure capacity, these countries are actively engaging in sustainability dialogues and implementing reforms aligned with circular economy (CE) principles. Their commitment to CE is evident, although its scale and effectiveness vary significantly across contexts.

India, China, and Brazil, all members of the BRICS bloc, are rapidly industrializing economies that face a dual challenge: minimizing environmental degradation while supporting inclusive economic development. India's Digital India Initiative and E-Waste Management Rules (2016) aim to increase digital access and improve regulatory oversight of electronic waste, though informal waste handling remains a significant concern. China's Circular Economy Promotion Law (2008) marked one of the earliest institutionalized CE policies in the Global South and has since evolved to include smart city initiatives and digital waste-sorting infrastructure. Brazil's National Solid Waste Policy (NSWP) emphasizes integrated waste management and has initiated the use of digital tracking systems in urban waste management programs. Despite their different governance styles, all three countries illustrate how CE policies are increasingly embedded in broader digital and industrial development agendas (Razzaq et al., 2021; Liu et al., 2025). South Africa represents the African continent and provides a unique example of CE application in a resource-intensive economy. Its newly introduced Circular Economy Roadmap targets key sectors such as plastics, construction, and packaging, although implementation remains limited by institutional weaknesses and socio-economic disparities. Russia, classified as an upper-middle-income country, faces regulatory and data transparency challenges, but select urban regions have piloted smart environmental solutions, including digital monitoring of waste and emissions. Meanwhile, smaller and more digitally advanced economies like Singapore and Mexico offer different insights. Singapore's Zero Waste Masterplan (2019) and Smart Nation initiative reflect an integrated approach to CE, leveraging high-tech solutions for waste tracking, behavioral nudging, and resource efficiency. Mexico, though facing policy enforcement issues, has made progress in aligning its industrial and environmental strategies, particularly through digital innovations in packaging and waste logistics. Other developing countries such as Thailand, Pakistan, and the Philippines are exploring CE models through pilot projects and city-led initiatives. These nations often rely on international support and NGO partnerships to implement smart waste management systems, though progress is slowed by regulatory fragmentation, budgetary constraints, and limited technical expertise.

In this context, emerging literature underscores the importance of tailoring CE strategies to local conditions in developing countries. Approaches that focus on lean tourism, community-based resource recovery, and personalized waste management systems show particular promise (Waaje & Roshid, 2025; Karim et al., 2024). These insights reinforce that, despite varied capacities, countries at different stages of development can adopt effective and innovative CE frameworks by leveraging context-sensitive digital tools, local knowledge systems, and participatory governance structures.

### 3.3.1 Case Study

#### *Germany: Circular Infrastructure and Industry 4.0*

Germany's KrWG Circular Economy Act and the German Resource Efficiency Programme (ProgRess) place a premium on integrating environmental protection with digitally enhanced industrial operations. German companies, particularly those in the industrial, automotive, and chemical sectors, are rapidly relying on Industry 4.0 technologies such as IoT, AI, robotics, and blockchain to enable resource-efficient production, real-time waste tracking, and closed-loop supply chains (Waaje et al., 2025). These technologies advance not only environmental goals, but also cost efficiency and competitiveness. Germany also pushes digital twin simulations and predictive analytics in product design to increase product life and minimize material intensity, in line with the EU's CE standards and climate obligations. According to Roshid et al. (2025), these technologies support Industry 5.0 principles—where human-centric innovation meets machine intelligence—providing platforms for circularity through precision manufacturing and personalized product cycles. Germany's focus on sustainability through digital upscaling aligns with this advanced paradigm.

#### *Denmark: Energy Loops and Smart Waste Grids*

Denmark is world-renowned for its highly efficient waste-to-energy (WtE) systems and smart waste grid infrastructure, making it a pioneer in combining circular economy (CE) concepts with digital innovation. The Amager Bakke (Copenhill) plant in Copenhagen is a prime example of CE success, converting household and industrial waste into electricity and district heating for thousands of homes while also serving as a ski slope and public park, demonstrating how sustainability and urban design can coexist. Behind this functionality lies a sophisticated digital monitoring system that uses RFID technology, IoT sensors, and GIS mapping to track waste flow from households to sorting and recovery centers in real time. These tools ensure efficient waste categorization, minimize landfill use, and optimize energy recovery processes. In addition, Denmark has installed smart bins with fill-level sensors and predictive analytics to advise municipal collection schedules, lowering emissions and operational costs. The country's CE strategy is bolstered by citizen engagement platforms and smartphone apps that promote proper waste sorting behavior, demonstrating a model in which digital tools strengthen both infrastructure and public participation. Denmark's scenario is consistent with the concept of personalized CE infrastructure, as mentioned in research on AI-enhanced planning and smart systems (Waaje et al., 2025). As a result, the system is socially integrated, adaptive, and efficient.

### *Japan: Waste-to-Wealth and Eco-Town Models*

Japan has a long history of waste minimization, resource productivity, and industrial symbiosis. Its pioneering Eco-Town effort, which began in the late 1990s, is still functioning today thanks to integration with smart city frameworks and digital waste flow monitoring. In locations such as Kitakyushu, industrial byproducts from one facility are reused as raw materials for another, resulting in a network of interconnected eco-industrial systems. These transactions are tracked and optimized with IoT-enabled systems, real-time data analytics, and blockchain technology to assure traceability, compliance, and optimal resource efficiency (Waaje et al., 2024). Japan's emphasis on technical innovation reinforces its vision of a low-carbon, circular economy. Japan also used AI-powered predictive analytics to improve collection, recycling, and resale operations. The case exemplifies Yeamin's (2024) strategy, which focuses on waste-to-wealth transitions in which waste is transformed into a raw resource via smart mapping and digital valorization.

### *India: Digital Access and Informal Sector Challenges*

India stands out from the Global South in that it combines increasing urbanization, a big informal sector, rising consumerism, and expanding digital infrastructure. Several smart towns have used the Digital India Initiative to implement sensor-based waste bins, GIS monitoring for garbage trucks, and online public dashboards to improve municipal solid waste management. While India's E-Waste Management Rules (2016) include Extended Producer Responsibility (EPR) and require formal e-waste collection channels, enforcement is limited due to the informal recycling sector's dominance, a lack of standardized procedures, and insufficient waste segregation infrastructure (Jha and Prasad, 2020).Nonetheless, India has witnessed promising digital start-ups like as Recykal and Karo Sambhav, which provide mobile-based systems for waste collection, reverse logistics, and recycling tracking. These advances are consistent with Roshid et al. (2024), who suggest that transforming waste into wealth necessitates not only top-down governmental reform, but also grassroots digital innovation involving local communities, small enterprises, and informal workers in inclusive circular systems.

### *Brazil: Lean Tourism and Sustainable Material Use*

In Brazil, material footprint concerns are prevalent in industries like as hospitality and tourism. The National Solid Waste Policy, introduced in 2010, laid the groundwork for waste reduction and recycling, but implementation has been patchy. Recent research indicates that incorporating digital circularity into tourism—via lean inventory systems, sensor-based resource tracking, and predictive waste analytics—can dramatically reduce environmental pressure (Waaje et al., 2025). Brazil's experience contributes to the growing acceptance of sector-specific CE measures, particularly in resource-intensive industries. Karim et al. (2024) emphasize that digital CE models tailored to industry contexts (e.g., tourism, hospitality, construction) boost both efficiency and local sustainability outcomes.

### *Pakistan and Thailand: Digital CE at Early Stages*

Countries such as Pakistan and Thailand are testing digital technologies in CE through restricted pilot initiatives. In Pakistan, the government has collaborated with private sector companies to establish recycling apps, digital trash marketplaces, and school-based awareness initiatives. Although these initiatives are dispersed, they represent the early adoption of what Waaje et al. (2025) refer to as "adaptive digital circularity," in which CE concepts are gradually entrenched inside current economic structures through the use of mobile platforms and low-cost technologies.

## 3.4 Model Specification

The current research examines the role of digitalization on environment sustainability and how the concept of circular economy can be adopted in the context of developed and developing countries. Here material footprint is taken as dependent variables, and it is the proxy of circular economy. ICT goods export, ICT services export, fixed telephone usage, individual using the internet, and mobile subscriptions is used as the measure of digitalization whereas FDI, GDP and trade openness are taken as controlled variables for representing the macroeconomic factors.

The pooled ordinary least squares (POLS) method of estimation is used here so that we can analyze the relationship between material footprint and digitalization. Additionally, a Hausman test is used to choose whether a fixed effect model will be used or random effect model will be used. It will also help us to understand how unobserved heterogeneity has an impact on results.

The methodology for this is below for both developed and developing countries:

$$MFit = \alpha + \beta 1ICT\_SEit + \beta 2ICT\_GEit + \beta 3MSit + \beta 4FTSit + \beta 5GDPit$$

$$+ \beta 6FDIit + \beta 7IUIit + \beta 8TOit + \varepsilon it$$

As we have used logarithmic form to estimate the effect of various components in MF so have to convert the equations according to the usage of the logarithmic form.

## Model

This study explains how digital and economic elements affect material footprint to investigate their impact on adoption of circular economy and environmental sustainability. The Material footprint represents the quantity of raw materials used to sustain the economy. It is taken here as the proxy of the circular economy because it represents the amount of raw material used for meeting the consumption. ICT goods export, ICT services export, fixed telephone usage, individual using the internet, and mobile subscriptions are used as the proxy of digitalization. Furthermore, FDI, GDP and trade openness are taken as controlled variables. The comparison between the developed and developing countries is being conducted to understand whether digitalization reduces material footprint or increases the material footprint. Through this we can understand whether the concept of circular economy can be adopted or not.

Mathematically it can be written as:

$$lnMFit = \alpha + \beta 1ICT\_SEit + \beta 2ICT\_GEit + \beta 3MSit + \beta 4FTSit + \beta 5lnGDP + \beta 6FDIit + \beta 7IUIit + \beta 8TOit + \varepsilon it$$

In terms of developed and developing countries this model is estimated through a fixed effect model.

$$lnMFit = \alpha + \beta 1ICT\_SEit + \beta 2IC\_GEit + \beta 3MSit + \beta 4FTSit + \beta 5lnGDP$$

$$+ \beta 6FDIit + \beta 7IUIit + \beta 8TOit + \varepsilon it$$

**MF** = Material Footprint (proxy for circular economy);
**ICT SE** = ICT Services Export (% of total services export);
**ICT GE** = ICT Goods Export (% of total goods export);
**IUI** = Individuals Using the Internet (% of population);
**MS** = Mobile Cellular Subscriptions (per 100 people);
**FTS** = Fixed Telephone Subscriptions (per 100 people);
**GDP** = Gross Domestic Product;

**FDI** = Foreign Direct Investment;
**TO** = Trade Openness (% of GDP);
**ln** = Natural logarithm;
**αi** = Intercept for each cross-section;
**β1–β8** = Coefficients of independent variables;
**εit** = Error term for country *i* at time *t*.

### 3.5 Estimation Strategy

According to the statements in the literature, most existing research on the circular economy (CE), material footprint (MF), and digitalization are either single-country focused or confined to emissions, leaving a lack in cross-country analysis, particularly when comparing developed and developing countries. To address this, the current study employs a structured estimating technique on panel data from 20 countries (10 developed and 10 developing) spanning the years 2000-2022. First, unit root tests (PP- Fisher Chi-square, ADF-Fisher and Im-Pesaran-Shin) are used to ensure stationarity. This is critical for avoiding misleading results, especially as macroeconomic variables frequently exhibit trends and differ across development levels (Baltagi, 2008). Next, the Pooled Ordinary Least Squares (POLS) model is used as a baseline. However, POLS assumes homogeneity and ignores country-specific characteristics, which is problematic in multi-country studies. An F-test can assist determine whether heterogeneity is significant. The Hausman test is used to determine whether to use Fixed Effects (FE) or Random Effects (RE). If the test indicates a correlation between individual effects and regressors ($p < 0.05$), the FE model is preferred to capture structural variations such as regulations or digital capacity. Otherwise, RE is utilized to make more efficient cross-country predictions. This layered method guarantees that model selection is consistent with developmental variation, increasing the robustness of the study's findings.

## 4. RESULTS

This subsection presents essential pre-estimation tests to validate the panel data structure and guide model selection. The analysis begins with stationarity tests to confirm the stability of variables over time, followed by Pooled OLS estimation as a baseline. To account for country-level heterogeneity across developed and developing economies, the Hausman test is applied to determine whether Fixed Effects or Random Effects models are more appropriate for long-run estimation.

*Table 3. Unit root analysis of developing countries*

| | Im, Pesaran, and shin W-state | | ADF- Fisher Chi-square | | PP- Fisher Chi-square | |
|---|---|---|---|---|---|---|
| **Variables** | **Level** | **First Difference** | **Level** | **First Difference** | **Level** | **First Difference** |
| **Ln MF** | 0.0150 | | 0.0166 | | 0.0004 | |
| **FDI** | 0.0000 | | 0.0000 | | 0.0000 | |
| **ICT GE** | 0.0090 | | 0.0137 | | 0.0018 | |
| **ICT SE** | 0.8109 | 0.0000 | 0.0199 | | 0.0001 | |
| **Fixed telephone usage** | 0.9643 | 0.0000 | 0.5394 | 0.0000 | 0.5766 | 0.0000 |
| **Individual using internet** | 1.0000 | 0.0033 | 0.9994 | 0.0002 | 0.9952 | 0.0000 |
| **Mobile subscription** | 0.0468 | | 0.0823 | 0.0000 | 0.0927 | 0.0000 |
| **Ln GDP** | 0.2011 | 0.0000 | 0.2936 | 0.0000 | 0.5031 | 0.0000 |
| **Trade openness** | 0.5522 | 0.0000 | 0.4196 | 0.0000 | 0.2981 | 0.0000 |

Source: Author's own estimation

The Im-Pesaran-Shin W-stat test is a panel unit root test developed by Im, Pesaran, and Shin (2003), that determines the stability of variables over a panel dataset. It helps to account for heterogeneity in the dynamics and error variance across cross-sections. In this study, we use this model to understand whether variables like MF, GDP, FDI, etc. are stationary or not. Here variables like MF, FDI, and ICT GE are already stationary at level, whereas ICT SE, fixed telephone usage, GDP, and individuals using the internet are not stationary at level; it indicates a high p-value. These variables are stationary at first difference. These findings justify using panel econometric models that accommodate non-stationary data, such as fixed/random effects with different variables.

The ADF-Fisher chi-square test is a panel unit root test based on Maddala and Wu's (1999) method that examines the stationarity of variables in a panel dataset. It combines individual ADF tests for each cross-section and aggregates them into a single statistic using Fisher's approach.

This test is suitable for panels with unbalanced and diverse data. The results show that variables like FDI, ICT SE, ICT GE, and MF are stationary at level, whereas variables like GDP, TO, FTS, internet usage, and MS are stationary at first difference.

The Phillips-Perron Fisher chi-square test is a well-known panel unit root test that evaluates the stationarity of variables across multiple cross-sections. Compared to the ADF test, the PP test is non-parametric and corrects for serial correlation and heteroskedasticity in the error terms without adding lagged difference terms (Phillips & Perron, 1988). The results show that variables like MF, FDI, ICT GE, and ICT

SE are stationary at level, whereas variables like fixed telephone usage, MS, GDP, and TO are non-stationary at level and stationary at first difference.

*Table 4. Unit root test for developed countries*

| | Im, Pesaran, and shin W-state | | ADF- Fisher Chi-square | | PP- Fisher Chi-square | |
|---|---|---|---|---|---|---|
| **Variables** | **Level** | **First Difference** | **Level** | **First Difference** | **Level** | **First Difference** |
| **Ln MF** | 0.0129 | | 0.0353 | | 0.0000 | |
| **FDI** | 0.0000 | | 0.0000 | | 0.0000 | |
| **ICT GE** | 0.0000 | | 0.0002 | | 0.0000 | |
| **ICT SE** | 0.8666 | 0.0000 | 0.4931 | 0.0000 | 0.4508 | 0.0000 |
| **Fixed telephone usage** | 0.9358 | 0.0000 | 0.5492 | 0.0000 | 0.9136 | 0.0000 |
| **Individual using internet** | 0.0000 | | 0.0000 | | 0.0000 | 0.0000 |
| **Mobile subscription** | 0.0028 | | 0.0000 | | 0.0000 | 0.0000 |
| **Ln GDP** | 0.0000 | | 0.0001 | | 0.0000 | |
| **Trade openness** | 0.9560 | 0.0000 | 0.8994 | 0.0000 | 0.9990 | 0.0000 |

Source: author's own estimation

The Im-Pesaran-Shin W-stat test (Im, Pesaran, & Shin, 2003) shows that variables like FDI, ICT GE, internet usage, and MS are stationary at level, whereas ICT SE, TO, and FTS are stationary at first difference. The ADF-Fisher chi-square is a panel unit root test (Maddala & Wu, 1999) that shows FDI, ICT GE, internet usage, MS, and GDP are stationary at level, whereas ICT SE, FTS, and TO are stationary at first difference. The Phillips-Perron-Fisher chi-square test (Phillips & Perron, 1988) shows that variables like MF, FDI, ICT GE, internet usage, MS, and GDP are stationary at level, whereas ICT SE, FTS, and TO are stationary at first difference.

The comparison of the unit root test shows that developed countries tend to have more stable and predictable variable behavior while developing countries show more volatile trends.

*Table 5. POLS model of developing countries*

| Variable | coefficient | Std. Error | T-statistic | Probability |
|---|---|---|---|---|
| **FDI** | -0.0378 | 0.0092 | -4.1012 | 0.0001 |
| **Fixed telephone subscription** | 0.0003 | 0.0047 | 0.0667 | 0.9468 |
| **ICT GE** | 0.0151 | 0.0021 | 6.9962 | 0.0000 |

continued on following page

*Table 5. Continued*

| Variable | coefficient | Std. Error | T-statistic | Probability |
|---|---|---|---|---|
| **ICT SE** | -0.0012 | 0.0028 | -0.4526 | 0.6512 |
| **hIndividual using internet** | -0.0115 | 0.0020 | -5.5580 | 0.0000 |
| **MS** | -0.0020 | 0.0010 | -1.9451 | 0.0530 |
| **Trade openness** | 0.0038 | 0.0007 | 4.8508 | 0.0000 |
| **Ln GDP** | 1.3220 | 0.0368 | 35.8949 | 0.0000 |
| **constant** | -17.4385 | 0.9501 | -18.3535 | 0.0000 |
| **R-squared** | 0.9338 | **F-statistic** | 389.9304 | |
| **Adjusted R-squared** | 0.9314 | **prob(F-statistic)** | 0.0000 | |

Source: Author's own estimation

The pooled OLS model serves as a baseline to assess the link between material footprint (MF)—a proxy for circular economy—and selected digital and economic indicators. FDI shows a significant negative impact on MF, supporting earlier findings that green technology inflows via FDI reduce resource use (Cole & Elliott, 2005; Zarsky, 1999). FTS has an insignificant positive impact, implying minimal influence on sustainability (Qureshi, 2009).

ICT GE has a significant positive effect, indicating that hardware trade increases material demand (Kurnia et al., 2015; Ali et al., 2018). In contrast, ICT SE shows an insignificant negative effect, aligning with views that service-based ICT promotes dematerialization (Gholami et al., 2010; Niebel, 2018). Internet usage and mobile subscriptions both negatively impact MF, supporting digitalization's role in reducing physical resource needs (Zhu et al., 2020; Marques & Ferreira, 2020). Trade openness and GDP show significant positive relationships with MF, indicating that economic and trade expansion increase material use, consistent with the Environmental Kuznets Curve hypothesis (Grossman & Krueger, 1995; Dean, 2002). The model is statistically strong with $R^2 = 0.9338$ and a significant F-statistic (389.93, $p = 0.0000$).

*Table 6. POLS model of developed countries*

| Variable | coefficient | Std. Error | T-statistic | Probability |
|---|---|---|---|---|
| **FDI** | 5.43E | 0.0037 | 0.0146 | 0.9884 |
| **Fixed telephone subscription** | 0.0026 | 0.0013 | 1.9803 | 0.0489 |
| **ICT GE** | 0.0471 | 0.0023 | 0.8973 | 0.3705 |
| **ICT SE** | 0.0024 | 0.0027 | 0.8973 | 0.3705 |
| **Individual using internet** | -0.0020 | 0.0012 | -1.6674 | 0.0968 |

continued on following page

*Table 6. Continued*

| Variable | coefficient | Std. Error | T-statistic | Probability |
|---|---|---|---|---|
| **MS** | -0.0010 | 0.0009 | -1.0389 | 0.3000 |
| **Trade openness** | -0.0025 | 0.0010 | -2.4290 | 0.0159 |
| **Ln GDP** | 0.9914 | 0.0234 | 42.2523 | 0.0000 |
| **constant** | -9.5083 | 0.6396 | -14.8657 | 0.0000 |

*Table 7.*

| R-squared | 0.9594 | F-statistic | 653.1563 |
|---|---|---|---|
| **Adjusted R-squared** | 0.9578 | **prob(F-statistic)** | 0.0000 |

Source: Author's own estimation

The POLS results for developed countries show that FDI is positive but insignificant, suggesting limited influence on material footprint (MF), which aligns with Cole and Elliott (2005), who found that strong regulations in developed nations reduce FDI's environmental impact. FTS has a positive and significant effect on MF, indicating legacy infrastructure may still contribute to environmental pressure (Qureshi, 2009). ICT GE and ICT SE are positively but insignificantly associated with MF, meaning digital exports alone don't reduce resource use without proper infrastructure (Kurnia et al., 2015). Internet usage and mobile subscriptions have negative but insignificant effects, indicating some potential for dematerialization, as supported by Zhu et al. (2020) and Marques and Ferreira (2020). TO has a negative and significant effect on MF, reflecting trade's role in technology diffusion (Antweiler et al., 2001). GDP is positively and significantly linked with MF, consistent with the Environmental Kuznets Curve (Grossman & Krueger, 1995). The high $R^2$ (0.9594) and significant F-statistic (653.16) confirm the model's robustness. These findings suggest that developed countries benefit more from digitalization due to better infrastructure and regulation (Bhawna et al., 2024; Liu et al., 2025).

Overall it shows that in developed countries digitalization contributed more effectively to sustainability as compare to developing countries where it is more complex due to infrastructure gaps, different stages of development and regulatory differences.

*Table 8. Hausman test of developing countries*

| Test type | Hausman test | | | |
|---|---|---|---|---|
| Chi-square statistics | 302.220743 | | | |
| Degree of freedom | 8 | | | |
| P-value | 0.0000 | | | |
| Preferred model | Fixed effect is preferred ($p < 0.05$) | | | |
| **Variable** | **Fixed** | **Random** | **Var. (diff.)** | **Probability** |
| **FDI** | -0.0058 | -0.0183 | 0.0000 | 0.0000 |
| **FTS** | -0.0023 | 0.0036 | 0.0000 | 0.0436 |
| **ICT GE** | 0.0079 | 0.0110 | 0.0000 | 0.0890 |
| **ICT SE** | 0.0097 | 0.0020 | 0.0000 | 0.0004 |
| **Individual using internet** | -0.0055 | -0.0047 | 0.0000 | 0.0925 |
| **Mobile subscription** | -0.0004 | -0.0046 | 0.0000 | 0.0000 |
| **Trade openness** | 0.0026 | 0.0017 | 0.0000 | 0.1499 |
| **Ln GDP** | 0.71466 | 1.0995 | 0.0010 | 0.0000 |

Source: Author's own estimation

The Hausman test is employed to decide between fixed and random effects models by testing whether unique errors are correlated with regressors. In this table, the chi-square statistic is 302.22 with a p-value of 0.0000, which is less than the 5% significance level. Thus, the null hypothesis is rejected, and the fixed effects model is preferred over random effects. This indicates that in the context of developing countries, fixed effects better capture unobserved heterogeneity across countries. This result aligns with the guidance provided by Hausman (1978), which recommends fixed effects when regressors correlate with individual effects. Moreover, Baltagi (2008) emphasizes that fixed effects are useful in analyzing within-entity variation, which is crucial when countries differ in policy, infrastructure, and economic structure.

*Table 9. Hausman test of developed countries*

| Test type | Hausman test | | | |
|---|---|---|---|---|
| Chi-square statistics | 126.768038 | | | |
| Degree of freedom | 8 | | | |
| P-value | 0.0000 | | | |
| Preferred model | Fixed effect is preferred ($p < 0.05$) | | | |
| **Variable** | **Fixed** | **Random** | **Var. (diff.)** | **Probability** |
| **FDI** | -0.0002 | 0.0000 | 0.0000 | 0.0000 |
| **FTS** | -0.0019 | -0.0010 | 0.0000 | 0.0000 |
| **ICT GE** | -0.0011 | -0.0013 | 0.0000 | 0.7968 |
| **ICT SE** | -0.0020 | -0.0041 | 0.0000 | 0.0012 |
| **Individual using internet** | -0.0054 | -0.0071 | 0.0000 | 0.0000 |
| **Mobile subscription** | 0.0010 | 0.0004 | 0.0000 | 0.0000 |
| **Trade openness** | 0.0006 | -0.0015 | 0.0000 | 0.0000 |
| **Ln GDP** | 0.2650 | 0.5239 | 0.0007 | 0.0000 |

Source: Author's own estimation

The Hausman test is used to choose between fixed and random effects models to determine the more appropriate estimation strategy in panel data analysis. In this case, the chi-square value is 126.77 with a p-value of 0.0000, which is below the 5% significance threshold. This leads to the rejection of the null hypothesis, indicating that the fixed effects model is more suitable than the random effects model. The result suggests that unobserved heterogeneity across countries is correlated with the explanatory variables. Therefore, the fixed effects model better captures the true relationships in developed countries. As Hausman (1978) originally proposed, and as supported by Baltagi (2008), when such correlation exists, the fixed effects estimator yields consistent and unbiased results. The Hausman test thus confirms that fixed effects are appropriate for analyzing the association between material footprint, digitalization, and economic variables.

*Table 10. Result of fixed effect of developing countries*

| Variable | coefficient | Std. Error | T-statistic | Probability |
|---|---|---|---|---|
| **FDI** | -0.0058 | 0.0043 | -1.3640 | 0.1740 |
| **FTS** | -0.0023 | 0.0041 | -0.5558 | 0.5789 |
| **ICT GE** | 0.0079 | 0.0023 | 3.3324 | 0.0012 |
| **ICT SE** | 0.0097 | 0.0029 | 3.2822 | 0.0012 |

continued on following page

*Table 10. Continued*

| Variable | coefficient | Std. Error | T-statistic | Probability |
|---|---|---|---|---|
| **Individual using internet** | -0.0055 | 0.0009 | -5.5933 | 0.0000 |
| **Mobile subscription** | -0.0004 | 0.0006 | -0.6598 | 0.5101 |
| **Trade openness** | 0.0026 | 0.0007 | 3.3888 | 0.0008 |
| **Ln GDP** | 0.7146 | 0.0412 | 17.31802 | 0.0000 |
| **C** | -1.2763 | 1.0864 | -1.1747 | 0.2414 |
| **R-squared** | 0.9909 | **F-statistic** | 1370.78 | |
| **Adjusted R-squared** | 0.9902 | **prob(F-statistic)** | 0.0000 | |

Source: Author's own estimation

The fixed effects model results largely reflect existing literature. The positive and significant effect of ICT goods exports (ICT GE) on material footprint (MF) confirms the findings of Kurnia et al. (2015) and Ali et al. (2018), who argue that ICT hardware contributes to resource consumption despite enabling digital operations. Similarly, the positive effect of ICT service exports (ICT SE) aligns with Bellini and Bang (2022), who show that without robust infrastructure, digital services may still demand material inputs. The negative and significant effect of internet usage supports Zhu et al. (2020) and Vu (2011), who highlight its role in boosting efficiency and reducing material consumption. The insignificant effect of FDI and fixed telephone subscriptions (FTS) aligns with Zarsky (1999) and Qureshi (2009), suggesting these factors only indirectly influence sustainability. The positive effect of GDP and trade openness (TO) is in line with the Environmental Kuznets Curve (Grossman & Krueger, 1995), indicating rising material demand with economic growth and global trade exposure.

Here the $R^2$ (0.9909) is significantly high, showing that over 99% of MF variation is explained. The model's **F-statistic (1370.78, p = 0.0000)** confirms it is highly statistically significant.

*Table 11. Result of fixed effect of developed countries*

| Variable | coefficient | Std. Error | T-statistic | Probability |
|---|---|---|---|---|
| **FDI** | -0.0002 | 0.0014 | -0.1467 | 0.8834 |
| **FTS** | -0.0019 | 0.0007 | -2.6942 | 0.0076 |
| **ICT GE** | -0.0011 | 0.0027 | -0.4014 | 0.6885 |
| **ICT SE** | -0.0020 | 0.0023 | -0.9042 | 0.3669 |
| **Individual using internet** | -0.0054 | 0.0008 | -6.3445 | 0.0000 |
| **Mobile subscription** | 0.0010 | 0.0005 | 1.7800 | 0.0765 |

continued on following page

*Table 11. Continued*

| Variable | coefficient | Std. Error | T-statistic | Probability |
|---|---|---|---|---|
| **Trade openness** | 0.0006 | 0.0010 | 0.6554 | 0.5129 |
| **Ln GDP** | 0.2650 | 0.0468 | 5.6558 | 0.0000 |
| **C** | | | | |

*Table 12.*

| R-squared | 0.9944 | F-statistic | 2238.466 |
|---|---|---|---|
| **Adjusted R-squared** | 0.9940 | **prob(-F-statistic)** | 0.0000 |

Source: Author's own estimation

According to the Hausman test, a fixed-effects model is appropriate for developed countries. The results show that FDI and FTS have negative and significant effects on material footprint (MF), supporting literature that highlights their role in promoting cleaner technologies and improved infrastructure (Zarsky, 1999; Czernich et al., 2011). Internet usage also reduces MF, consistent with Vu (2011) and Zhu et al. (2020). However, ICT GE and ICT SE have negative but insignificant impacts, suggesting that developed nations may have already optimized digital efficiency. Interestingly, mobile subscriptions show a positive and significant effect, indicating rising digital use may still increase resource consumption (Marques & Ferreira, 2020). Trade openness and GDP positively influence MF, reaffirming that economic growth and trade still drive material demand.

These results align with the literature review, which noted that developed countries benefit more from digitalization due to stronger infrastructure and policies (Bhawna et al., 2024; Campoli et al., 2025). In contrast, developing countries face greater challenges in translating digital growth into sustainability.

## 5. POLICY RECOMMENDATIONS

Several policy lessons emerge from global experiences that developing countries can adapt to strengthen their circular economy (CE) transitions.

### 5.1 Develop Material Efficiency Indicators

Japan has developed a Material Flow Accounts system (MFS), which tracks the flow of materials through its economy. This system helps to monitor the consumption, production, and recycling rate of the materials. Developing countries can work in collaboration with international organizations like UNEP, which can help in adopting these material flow monitoring systems. This will help to promote waste minimization and responsible resource extraction.

## 5.2 France: AGEC Law (Anti-Waste for a Circular Economy)

In 2020, France introduced this law to prevent waste generation and accelerate the concept of a circular economy. This policy promotes reducing the usage of single-use plastic, recycling and repairing the devices, and transparency across the supply chain system (Bellini & Bang, 2022). The AGEC law is further strengthened by government support for Industry 5.0 paradigms, which integrate human creativity with automation and personalized data-driven CE solutions (Roshid et al., 2025). Developing countries can take inspiration from this model and prioritize the creation or strengthening of circular economy laws that include digital provisions—such as e-waste tracking systems, repairability scores, and product traceability mechanisms. These legal instruments should be supported by a digital ecosystem of government dashboards and interoperable databases for monitoring compliance.

## 5.3 South Korea's Circular Economy Policy

Korea's policy on smart green new deals and circular economy is focused on achieving net-zero emissions and promoting digital innovation. Its main focus is to invest in green infrastructure, low-carbon sectors, urban digital twins, and circular economy techniques in both construction and manufacturing. Developing countries can work on software to create local digital twins by using IoT devices.

## 5.4 Finland: Digital Waste Trading and Public-Private Platforms

Finland's national vision for CE includes the Materiaalitori platform, a digital marketplace for waste exchange and industrial byproducts. This platform, which uses real-time data and geospatial mapping, aids in resource conservation by facilitating industrial symbiosis, a core CE principle. The country's emphasis on AI-powered garbage sorting and smart tracking devices promotes low-waste cities and effective recycling systems. The study shows Finland's approach as a paradigm for data interoperability and public-private collaboration in CE (Karim et al., 2024). The technology integrates environmental data and logistics, allowing businesses and communities to make automated decisions. The success of such models in nations with high digital maturity demonstrates the value of integrated digital-CE ecosystems. Developing countries should encourage the development of digital marketplaces for secondary materials that connect businesses, waste generators, and recyclers. Governments can partner with tech companies and local authorities to launch pilot platforms similar to Finland's model, adapted for informal sector integration and mobile accessibility.

## 6. CONCLUSION

This study investigates the relationship between digitalization and material footprint (MF) in both developed and developing countries. For the methodological framework, we employed panel data techniques including stationarity tests, Pooled Ordinary Least Squares (POLS), the Hausman test, and both fixed and random effects models to ensure robust and consistent estimations. In developed countries, variables like individuals using the internet have both had a negative impact on material footprint, as increasing digital connectivity helps to reduce material consumption. Mobile subscription, on the other side, had a positive impact; it means higher material consumption, which leads to higher MF. The findings for the developed countries show that internet usage has a negative impact on MF, whereas ICT services and ICT goods exports have a positive impact on the MF, reflecting that in developing countries digital infrastructure expansion is still material intensive. The circular economy-related policies have already been adopted by the developed countries. For example, the 5R concept includes reduce, reuse, recycle, refuse, and repurpose, which helps to extend the life cycle of resources for a longer period of time, resulting in lower MF and more sustainable consumption habits. In contrast, developing countries face several challenges such as lack of education, insufficient investment, and poor infrastructure; all this creates issues in circular economy adoption. This study highlights that digitalization paired with circular economy-related policies helps to reduce the environmental impact. In the end, this study highlights that environmental sustainability is determined by how countries deploy and manage resources. Implementation of the methods is helpful and necessary to achieve SDGs and net-zero emissions. In essence, achieving sustainability depends not only on advances in technology but also on collaborative, inclusive, and conscious policy efforts.

The novelty of this paper lies in its threefold contribution. First, it provides a systematic comparison between developed and developing countries, illustrating how the effects of digitalization on material footprints differ by stage of economic development. Unlike most studies that treat countries homogeneously, this research reveals structural divergences that shape sustainability outcomes. Second, it integrates the concept of circular economy with material footprint and digitalization indicators, creating a novel framework that connects environmental pressure with digital transformation metrics. Third, the paper goes beyond theoretical arguments by empirically establishing the relationship between digitalization and material footprint, offering robust evidence that adds depth to policy and academic debates. Together, these contributions reinforce the importance of context-specific CE strategies, where digitalization acts not merely as a driver of growth but as a catalyst for long-term sustainability.

## 7. LIMITATIONS

The study faces several limitations that must be acknowledged. First, there is a lack of symmetrical data quality, as developing countries often lack current and consistent records, reducing the validity of cross-country comparisons, while developed countries benefit from more extensive and reliable datasets. Second, countries are at different phases of circular economy adoption: developed economies have advanced significantly in terms of regulations and technologies, whereas many developing nations remain in the early stages, making direct comparisons difficult. Third, gaps in policy implementation also exist; weak enforcement structures in developing countries undermine the efficiency of circular economy initiatives, whereas developed countries perform better due to strong governance, effective regulatory systems, and financial support. In addition, socioeconomic differences create distinct consumption patterns—developed countries, with higher income levels, generally have greater material consumption compared to developing ones. Finally, there is a lack of region-specific circular economy patterns; while developed countries often rely on high-tech and policy-driven initiatives, developing nations tend to depend on informal systems such as local repair practices, manual recycling, and community-based resource recovery. Because the policies adopted in developing countries are often not tailored to regional contexts, their implementation remains challenging.

## REFERENCES


Ali, M., Liu, H., & Su, J.-W. (2018). ICT and environmental sustainability: A global perspective. *Environmental Science and Pollution Research International, 25*(6), 5231–5242. DOI: 10.1007/s11356-017-0699-z PMID: 28528500

Amager Resource Center. (n.d.). From waste to energy – Amager Bakke. Retrieved October 15, 2025, from https://a-r-c.dk/amager-bakke/from-waste-to-energy/

Antweiler, W., Copeland, B. R., & Taylor, M. S. (2001). Is free trade good for the environment? *The American Economic Review, 91*(4), 877–908. DOI: 10.1257/aer.91.4.877

Baltagi, B. H. (2008). *Econometric analysis of panel data* (4th ed.). John Wiley & Sons.

Bellini, A., & Bang, S. (2022). Barriers for data management as an enabler of circular economy: An exploratory study of the Norwegian AEC-industry. *IOP Conference Series. Earth and Environmental Science, 1122*(1), 012047. DOI: 10.1088/1755-1315/1122/1/012047

Bhawna, K. S., Singh, R. K., & Kazancoglu, Y. (2024). Bridging the gap: A systematic analysis of circular economy, supply chain management, and digitization for sustainability and resilience. *Operations Management Research.* DOI: 10.1007/s12063-024-00490-4

Blomsma, F., Brennan, G., & Tennant, M. (2019). Developing a circular strategies framework for manufacturing companies to support circular economy-oriented innovation. *Journal of Cleaner Production, 241*, 118271. DOI: 10.1016/j.jclepro.2019.118271

Bocken, N. M. P., Espinosa, N., & Kemp, R. (2022). The sufficiency-based circular economy—An analysis of 150 companies. *Frontiers in Sustainability, 3*, 899289. DOI: 10.3389/frsus.2022.899289

Brazil. (2010). Law No. 12.305/2010 – National Solid Waste Policy. https://www.wiego.org/wp-content/uploads/2019/09/Pereira-Brazilian-Waste-Policy.pdf

Brazilian Government. (2010). Law No. 12.305/2010 – National Policy on Solid Waste. Retrieved October 15, 2025, from https://braziliannr.com/brazilian-environmental-legislation/law-no-12305-brazilian-national-policy-solid-waste/

Campoli, J. S. (2025). Progress of G20 nations on the 6th sustainable development goal under the circular economy perspective. *Journal of the Knowledge Economy*. Advance online publication. DOI: 10.1007/s13132-024-02475-x

Chakraborty, A., Mehta, V., & Khan, R. (2024, August). Empirical evidence on circular economy adoption in Australian small and medium enterprises. *Journal of Cleaner Production*, *467*, 142958. https://doi.org/10.1016/j.jclepro.2024.142958

Cole, M. A., & Elliott, R. J. R. (2005). FDI and the capital intensity of "dirty" sectors: A missing piece of the pollution haven puzzle. *Review of Development Economics*, *9*(4), 530–548. DOI: 10.1111/j.1467-9361.2005.00291.x

Czernich, N., Falck, O., Kretschmer, T., & Woessmann, L. (2011). Broadband infrastructure and economic growth. *Economic Journal (London)*, *121*(552), 505–532. DOI: 10.1111/j.1468-0297.2011.02420.x

Dean, J. M. (2002). Does trade liberalization harm the environment? A new test. *The Canadian Journal of Economics. Revue Canadienne d'Economique*, *35*(4), 819–842. DOI: 10.1111/0008-4085.00156

Demestichas, K., & Daskalakis, E. (2020). Information and communication technology solutions for the circular economy. *Sustainability*, *12*(18), 7272. DOI: 10.3390/su12187272

Di, K., Chen, W., Zhang, X., Shi, Q., Cai, Q., Li, D., Liu, C., & Di, Z. (2023). Regional unevenness and synergy of carbon emission reduction in China's green low-carbon circular economy. *Journal of Cleaner Production*, *420*, 138436.https://doi.org/10.1016/j.jclepro.2023.138436

France. (2020). Law No. 2020-105 of 10 February 2020 on combating waste and the circular economy [Loi n° 2020-105 du 10 février 2020 relative à la lutte contre le gaspillage et à l'économie circulaire]. Legifrance. https://www.legifrance.gouv.fr/jorf/id/JORFTEXT000041553759

Germany. (2012). Circular Economy Act (KrWG) – Act to promote circular economy and ensure environmentally compatible waste management [Kreislaufwirtschaftsgesetz (KrWG) – Gesetz zur Förderung der Kreislaufwirtschaft und Sicherung der umweltverträglichen Bewirtschaftung von Abfällen]. Gesetze im Internet. https://www.gesetze-im-internet.de/krwg/

Gholami, R., Sulaiman, A. B., Ramayah, T., & Molla, A. (2010). Senior managers' perception on green information systems (IS) adoption and environmental performance: Results from a field survey. *Information & Management*, *50*(7), 431–438. DOI: 10.1016/j.im.2013.01.007

Grossman, G. M., & Krueger, A. B. (1995). Economic growth and the environment. *The Quarterly Journal of Economics, 110*(2), 353–377. DOI: 10.2307/2118443

Hausman, J. A. (1978). Specification tests in econometrics. *Econometrica, 46*(6), 1251–1271. DOI: 10.2307/1913827

Hawken, P., Lovins, A. B., & Lovins, L. H. (1999). Natural capitalism: Creating the next industrial revolution. Little, Brown and Company. https://doi.org/10.1007/978-1-349-62753-0

Im, K. S., Pesaran, M. H., & Shin, Y. (2003). Testing for unit roots in heterogeneous panels. *Journal of Econometrics, 115*(1), 53–74. DOI: 10.1016/S0304-4076(03)00092-7

International Energy Agency (IEA). (2021, July 16). Korean New Deal: Green New Deal, Digital New Deal, and Stronger Safety Net. Retrieved October 15, 2025, from https://www.iea.org/policies/11514-korean-new-deal-digital-new-deal-green-new-deal-and-stronger-safety-net

Jha, M. K., & Prasad, S. (2020). Role of mobile technology in sustainable consumption in developing countries. *Journal of Cleaner Production, 258*, 120585. DOI: 10.1016/j.jclepro.2020.120585

Karim, R., Bhowmik, R. C., Faruq, M. O., & Roshid, Md. M., Islam, S., & Ramachandaran, S. D. (2025). Harnessing growth for ecological sustainability through exploring the drivers of load capacity factor in G7 nations. *Discover Sustainability, 6*(1), 821. DOI: 10.1007/s43621-025-01757-x

Karim, R., Roshid, M., & Waaje, A. (2024). Circular economy and supply chain sustainability. In Mızrak, F. (Ed.), *Strategic Innovations for Dynamic Supply Chains* (pp. 1–30). IGI Global Publishing. DOI: 10.4018/979-8-3693-3575-8.ch001

Karim, R., Waaje, A., & Roshid, Md. M. (2025). Digital Circular Economy for Attaining Sustainable Development Goals. In *Effects of Digitalization and Circular Economy on Sustainable Policy and Climate Change Prevention* (pp. 211–240). IGI Global. DOI: 10.4018/979-8-3693-9909-5.ch008

Karlilar, S., & Pata, U. K. (2023). Determinants of Material Footprint in OECD Countries: The Role of Green Innovation and Environmental Taxes. *Natural Resources Forum*, 49(1), 100–15. DOI: 10.1111/1477-8947.12379

Kitakyushu City. (n.d.). Kitakyushu Eco-Town Center. Kitakyushu City. Retrieved October 15, 2025, from https://sdgs-kitakyushu.iges.jp/facility/kitakyushu-eco-town-center/

Kopnina, H. (2019). Green-washing or best practice? Using circular economy principles to teach sustainability. Journal of Cleaner Production, 219, 106–114. https://doi.org/10.1016/j.jclepro.2019.06.241

Korhonen, J., Honkasalo, A., & Seppälä, J. (2018). Circular economy: The concept and its limitations. Ecological Economics, 143, 37–46. https://doi.org/10.1016/j.ecolecon.2017.06.041

Kurnia, S., Karnali, R. J., & Rahim, M. M. (2015). A qualitative study of business-to-business electronic commerce adoption within the Indonesian grocery industry: A multi-level perspective. *Electronic Commerce Research and Applications*, *14*(6), 512–524. DOI: 10.1016/j.elerap.2015.09.004

Kurniawan. (2022). Transformation of Solid Waste Management in China: Moving towards Sustainability through Digitalization-Based Circular Economy. *Sustainability*, 14(4), 2374. .DOI: 10.3390/su14042374

Liu, N. (2025). Pathway to sustainable consumption and production: The role of climate-related development finance, digitalization, and financial development on material productivity and energy intensity of developing countries. *Energy & Environment.* Advance online publication. DOI: 10.1177/0958305X251315401

Liu, N., Borojo, D. G., Yushi, J., & Liu, X. (2025). Pathway to sustainable consumption and production: The role of climate-related development finance, digitalization, and financial development on material productivity and energy intensity of developing countries. Energy & Environment. https://doi.org/10.1177/0958305X251315401

Maddala, G. S., & Wu, S. (1999). A comparative study of unit root tests with panel data and a new simple test. *Oxford Bulletin of Economics and Statistics*, *61*(S1), 631–652. https://doi.org/10.1111/1468-0084.0610s1631

Marques, A. C., & Ferreira, A. (2020). Digital transformation and environmental sustainability in developing countries: The role of mobile technologies. *Environmental Science and Pollution Research International*, *27*, 21867–21877. DOI: 10.1007/s11356-020-08478-5

Meshram, K. (2024). The Circular Economy, 5R Framework, and Green Organic Practices: Pillars of Sustainable Development and Zero-Waste Living. *Discover Environment*, 2(1), 147. .DOI: 10.1007/s44274-024-00177-4

Ministry of Environment. Forest and Climate Change. (2016). E-Waste (Management) Rules, 2016. Government of India. https://www.wbpcb.gov.in/writereaddata/files/E-WASTE%20M%20Rules%202016.pdf

Ministry of Sustainability and the Environment. (2019). Zero Waste Masterplan. Singapore Government. https://www.mse.gov.sg/resources/zero-waste-masterplan.pdf

Motiva Oy. (2025, August 21). Materiaalitori – the platform for wastes and side streams. Retrieved October 15, 2025, from https://www.motiva.fi/en/solutions/material_efficiency/materiaalitori

National People's Congress of China. (2008). Circular Economy Promotion Law of the People's Republic of China. https://www.greenpolicyplatform.org/sites/default/files/downloads/policy-database/CHINA%29%20Circular%20Economy%20Promotion%20Law%20%282008%29.pdf

Niebel, T. (2018). ICT and economic growth–Comparing developing, emerging and developed countries. *World Development*, *104*, 197–211. DOI: 10.1016/j.worlddev.2017.11.007

OECD. (2020). *OECD Digital Economy Outlook 2020*. OECD Publishing. DOI: 10.1787/bb167041-

Pearce, D. W., & Turner, R. K. (1989). Economics of natural resources and the environment. Johns Hopkins University Press. https://doi.org/10.4324/9781315070212

Pesaran, M. H. (2007). A simple panel unit root test in the presence of cross-section dependence. *Journal of Applied Econometrics*, *22*(2), 265–312. DOI: 10.1002/jae.951

Phillips, P. C. B., & Perron, P. (1988). Testing for a unit root in time series regression. *Biometrika*, *75*(2), 335–346. DOI: 10.1093/biomet/75.2.335

Qureshi, S. (2009). Social and economic perspectives on the role of information and communication technology for development. *Information Technology for Development*, *15*(1), 1–3. DOI: 10.1002/itdj.20106

Razzaq, A.. (2021). Investigating the asymmetric linkages between infrastructure development, green innovation, and consumption-based material footprint: Novel empirical estimations from highly resource-consuming economies. *Resources Policy*, *74*, 102302. DOI: 10.1016/j.resourpol.2021.102302

Razzaq, A., Ajaz, T., Li, J. C., Irfan, M., & Suksatan, W. (2021). Investigating the asymmetric linkages between infrastructure development, green innovation, and consumption-based material footprint: Novel empirical estimations from highly resource-consuming economies. *Resources Policy*, *74*, 102302. https://doi.org/10.1016/j.resourpol.2021.102302

Recykal. (n.d.). Recykal: Transforming waste management with technology. Retrieved October 15, 2025, from https://www.recykal.com/

Roshid, Md. M., Waaje, A., Karim, R., Irtisham, M. R., & Kaswan, M. S. (2025). Industry 5.0 and Organizational Sustainability: Integrating Cutting-Edge Technologies and Human Ingenuity. In Mahender Singh Kaswan, Rajeev Rathi, & Jose Arturo Garza-Reyes (Eds.), *Industry 5.0: Concepts and Strategies for Digital Transformation* (pp. 50–61). CRC Press. DOI: 10.1201/9781003535188-7

Roshid, Md. M., Yeamin, Md. B., Waaje, A., Karim, R., & Kaswan, M. S. (2025). Understanding the Nexus of Industry 5.0 and Circular Economy for Enriched Lean and Environment Sustainability. In José Carlos Vieira de Sá, Shubham Sharma, Sunil Kumar, & Manjunath B.R. (Eds.), *Transformative Lean Six Sigma Techniques for the Quality 5.0 Paradigm* (pp. 349–380). IGI Global Scientific Publishing. DOI: 10.4018/979-8-3373-0943-9.ch013

Sahoo. (2021). Determinants of Material Footprint in BRICS Countries: An Empirical Analysis. *Environmental Science and Pollution Research*, 28(28), 37689–704. .DOI: 10.1007/s11356-021-13309-7

Sanye-Mengual, E., Ardente, F., & Beylot, A. (2019). Measuring circularity: From the material footprint to the circular economy in the EU. *Journal of Cleaner Production*, *229*, 361–374. DOI: 10.1016/j.jclepro.2019.05.035

Selvaraj, V. K., Subramanian, J., & Sai Dutt, S. (2024). Integrating circular economy principles in achieving net zero: Toward sustainable futures — India's perspective. *Materials Circular Economy*, *6*, 53. https://doi.org/10.1007/s42824-024-00144-7

Shahbaz, M., Solarin, S. A., & Sbia, R. (2013). Does energy intensity contribute to $CO_2$ emissions? A trivariate analysis in selected African countries. *Ecological Indicators*, *50*, 215–224. DOI: 10.1016/j.ecolind.2014.11.017

Sikder, M., Wang, C., Rahman, M. M., Yeboah, F. K., Alola, A. A., & Wood, J. (2024). Green logistics and circular economy in alleviating $CO_2$ emissions: Does waste generation and GDP growth matter in EU countries? *Journal of Cleaner Production*, *449*, 141708. https://doi.org/10.1016/j.jclepro.2024.141708

Sitra. (2019). *Leading the cycle – Finnish road map to a circular economy 2016–2025.* https://www.sitra.fi/en/projects/circular-economy/

Smart Nation and Digital Government Office. (2024). Smart Nation 2.0: Transforming Singapore for the future. Singapore Government. https://file.go.gov.sg/smartnation2-report.pdf

Tiwari, S., Mohammed, K. S., Mentel, G., Majewski, S., & Shahzadi, I. (2024). Role of circular economy, energy transition, environmental policy stringency, and supply chain pressure on $CO_2$ emissions in emerging economies. Geoscience Frontiers, 15, 101682. https://doi.org/10.1016/j.gsf.2023.101682

UNIDO. (2021). *The Future of Industrialization in a Post-Pandemic World*. United Nations Industrial Development Organization.

United Nations Conference on Trade and Development. (2024, July 10). Chapter II: Digitalization trends and the material footprint. https://unctad.org/system/files/official-document/der2024_ch02_en.pdf

Vu, K. M. (2011). ICT as a source of economic growth in the information age: Empirical evidence from the 1996–2005 period. *Telecommunications Policy*, *35*(4), 357–372. DOI: 10.1016/j.telpol.2011.02.008

Waaje, A., Karim, R., Roshid, Md. M., & Meem, T. N. (2024). Minimizing Waste in the Hospitality Industry: The Opportunities and Challenges of Lean Tourism. In R. Sucheran (Ed.), *Sustainable Waste Management in the Tourism and Hospitality Sectors* (pp. 125–166). IGI Global Scientific Publishing. DOI: 10.4018/979-8-3693-6110-8.ch006

Waaje, A., Karim, R., Roshid, M. M., Yeamin, B., & Meem, T. N. (2025). Integration of industry 4.0 with personalized medicine: A shift of the paradigm in healthcare. In Chemingui, H., & Lamloumi, M. (Eds.), *AI-Driven Personalized Healthcare Solutions* (pp. 27–69). IGI Global. DOI: 10.4018/979-8-3693-7858-8.ch002

Wang, M., Hossain, M. R., Si Mohammed, K., Cifuentes-Faura, J., & Cai, X. (2023). Heterogeneous effects of circular economy, green energy and globalization on $CO_2$ emissions: Policy-based analysis for sustainable development. *Renewable Energy*, *211*, 789–801. https://doi.org/10.1016/j.renene.2023.05.033

Wiedmann, T. O., Schandl, H., Lenzen, M., Moran, D., Suh, S., West, J., & Kanemoto, K. (2015). The material footprint of nations. *Proceedings of the National Academy of Sciences of the United States of America*, *112*(20), 6271–6276. DOI: 10.1073/pnas.1220362110 PMID: 24003158

Yeamin, M. (2024). Waste-to-wealth transitions through smart mapping and digital valorization. Academic Press.

Zarsky, L. (1999). Havens, halos and spaghetti: Untangling the evidence about FDI and the environment. In *Foreign Direct Investment and the Environment* (pp. 47–74). OECD Publishing. DOI: 10.1787/9789264173639-

Zhu, B., Wang, P., Zhang, X., & Chevallier, J. (2020). Internet development and environmental sustainability: Evidence from a panel of 86 countries. *Environmental Science and Pollution Research International*, *27*(32), 39840–39851. DOI: 10.1007/s11356-020-09938-y

# Chapter 13
# Converging Strategic Innovations and Cybernetic Systems:
## A Cyber-Physical Framework for Circular Economy Transition

**Patcha Bhujanga Rao**
https://orcid.org/0000-0003-4736-8497
*Jain University, Bangalore, India*

**Preethi Inampudi**
https://orcid.org/0009-0006-1339-0417
*Surana College, India*

## ABSTRACT

*This research proposes a cyber-physical system that integrates cybernetic systems thinking with strategy concepts to facilitate an easy transition to a circular economy. This strategy accounts for how complex it would be for data and resources to flow through circular systems. It leverages concepts from cybernetics, flexible systems, and Industry 5.0 to create systems resilient and responsive in real time. Legislative inertia and un-integrated digital systems may at times make it difficult for existing linear-to-circular shifts to occur. The framework indicates how the economic, environmental, and technical spheres are interconnected. In order to assist in solving these issues, sensor-driven intelligence, feedback mechanisms, and decision automation are all combined to streamline resource utilization and make industry society more robust, transparent, and people-oriented. Based on the find-*

DOI: 10.4018/979-8-3373-2454-8.ch013

*ings, cyber-physical coordination according to cybernetic back-loops and circular reasoning could accelerate the transition towards sustainability in all industries by a significant margin.*

## 1. INTRODUCTION

There is growing demand for a fundamental shift from linear models of production towards more sustainable ones, as natural systems decline at an accelerating rate and global flows of resources become more unstable. One of those that has attracted attention is the circular economy (CE), where economic growth is decoupled from material consumption by extending product lifespans, reusing them, and designing them so they can regenerate. Circularity seems to be promising in theory, but becomes challenging to execute massively due to technical, infrastructural and organisational obstacles. The efficacy and speed of this is usually interrupted by the lack of ecosystem, legacy industry practices and suitable governance structure.

These advances have created new opportunities for cybernetic thinking and cyber-physical systems (CPS). Much research exists on CPS designs in smart production, autonomous transport, and intelligent infrastructure. However, limited research has examined how CPS can enable a circular economy. But bringing together real-time data gathering, distributed control, and systemic feedback loops in CE strategies can create the scaffolding needed for dynamic resource optimization and resilience-based design (Waaje et al., 2024). Cybernetics, developing from control theory and systems logic, has much in common with the thinking that makes circularity work.

These models rely on self-regulation, adaptability, and closed-loop mechanisms. Integrating the imperative of strategic innovation with cybernetic system design enables more harmonious and responsible framework to implement the circular economy. This approach exceeds individual technology solutions and deal with the complexities of the socio-technical ecosystem as a whole. This document presents an innovative cyber-physical system to facilitate circular economy change.

Industry 5.0 represents a paradigm shift in industrial practices, emphasizing the collaboration between humans and machines, with a focus on sustainability and personalization. This is not just about technological automation; it is an understanding at the systems level of how digital, physical, and human domains are all intertwined. It also reflects a shift in how businesses approach strategy thinking: they are shifting from individual sustainability initiatives to integrating information into production and usage processes. It should be stressed that this process does not purport to address all the issues that arise in the course of the CE shift. Rather, it seeks to establish a theory and practice foundation upon which to construct more sophisticated, data-informed, and responsive strategies. By emphasizing the relationship between strategic innovation and cybernetic design, we aim to illustrate new paths for sustainable change that are both technologically feasible and consistent with long-term systemic resilience.

## 2. REVIEW OF THE CONCEPT

Cybernetics has not yet become commonly employed in circular economy literature, though it is beginning to gain traction. Static models have been found in earlier research to be poor at demonstrating just how complex CE systems are (Geissdoerfer et al., 2017). Nica et al. (2023) applied economic cybernetics to demonstrate how feedback-based governance can be utilized for modeling adaptive responses in economic, environmental, and technological domains.

Industry 5.0 literature (e.g., Breque et al., 2021; Rejeb et al., 2025) affirms shifting towards people-centered systems and being totally connected with technology even further. These results are timely in linking purpose of strategy with operating facts amidst CE transformation. However, more empirical studies are required to ensure that these models are applicable in several industries and environments.

### 2.1. Rethinking Economic Paradigms: From Linear to Circular

According to Górecki (2019), for decades the central premise of economics has been that things must be extracted, made, used, and discarded. It served as an effective paradigm for short -term economic development, but has led to resource depletion, environmental destruction, and excessive waste (Geissdoerfer et al., 2017). CE, however, is a sustainable approach that has a closed-loop system where the material and energy remain within the system for as long as they can (Ellen MacArthur Foundation, 2013). CE is not a recycling plan; it is a significant shift in the way things are designed, produced, and consumed such that they are sustainable in the long term. Academics have long advocated that CE must view things from the systems-level perspective, across industries and from individual sustainability initiatives (Kirchherr, Reike, & Hekkert, 2017). Genuine implementation - particularly at scale - continues to be patchy, in part because of institutional resistance, technology deficits, and sclerotic digital infrastructure (Murray, Skene, & Haynes, 2017).

### 2.2. The Role of Strategic Innovation in Enabling Circularity

When strategic innovation aligns with sustainable objectives, it can be a strong driver of CE transformation. It involves intentionally transforming business models, value chains, and government systems so that social and ecological issues can go hand-in-hand with economic development (Boons & Lüdeke-Freund, 2013). Such transformation can be observed in novel service-based models such as product-as-a-service and extensive producer responsibility (Lacy & Rutqvist, 2015).

Bocken et al. (2014) state that innovation for sustainability must consider technical and business model perspectives. In this manner, circularity can be embedded in an organization's strategy rather than being tacked on as an add-on. Some firms have made some progress in this direction, but these are frequently not as robust as they might be because they lack systemic feedback and adaptive learning processes.

## 2.3. Cybernetics: A Way to Control and Feedback Systems

Cybernetics is the study of how systems are controlled and communicate. Cybernetics has much to share with us about how to apply CE. Cybernetics stems from the work of Wiener (1948) and deals with feedback loops, dynamic balance, and self-controlling systems. These concepts are close to circularity, which similarly attempts to minimize chaos through reuse and re-setting back and forth.

Applied economic cybernetics has recently demonstrated that it can be applied with several factors on the model complex, dynamic system that interact applied to complex, dynamic systems that interact in non-linear patterns. (Nica et al. 2023). The use of the autoregressive distributed lag (ARDL) model has allowed researchers to observe short and long -term determinants such as CE variables such as GDP per capita, waste production and labor productivity.

Although cybernetics has been applied to robots, biology, and control theory for decades, it is relatively new to be applied to CE research. Nevertheless, this intersection point has tremendous potential for designing intelligent, adaptable governmental systems that can respond to changing economic and environmental signals.

## 2.4. Cyber-Physical Systems: Bridging the Digital and Material Worlds

Through their sensitized motors, data networks and real-time analytics, Cyber-Physical Systems (CPS) bring the virtual and physical more vividly to life (Lee, Bagheri, & Kao, 2015; Roshid et al., 2025a). In the context of CE (Reuter, 2016), CPS can be the foundation for smart tracking of repair needs, avoiding waste, and anticipating repair requirements.

Angelopoulos et al. (2023) established how CPS can enable citizens to make data-informed decisions in circular smart cities through crowdsensing, IoT, and AI-optimization. These systems allow real-time monitoring and facilitate civic engagement through the addition of feedback loops for citizens.

It should be noted, however, that although CPS systems are theoretically feasible, they prove to be difficult to implement due to sociotechnical problems such as privacy, interoperability issues, and institutions unprepared to handle complex, multi-agent data environments. (Liu et al., 2022).

## 2.5. The convergence of Industry 5.0 and human-centric technology

Patyal et al. (2021), with the progress in CPS, the shift towards Industry 5.0(Roshid et al., 2025b) is a shift in full automation (as in industry 4.0) and a change in mindset towards those systems in which human and machine cooperate to exchange information. Privatization, flexibility and ecological balance industries are important in Industry 5.0 (Breque et al., 2021).

Following Rejeb et al. (2025), Industry 5.0 technologies such as human-in-the-loop AI, tracking with blockchain, and flexible manufacturing are aligned with CE principles since they facilitate localised production systems that do not waste resources.

In a surprising turn, their review also emphasizes that for various groups to ensure and prepare innovation plans simultaneously, to ensure that technology is used for good ends and not only for profit. In this way, CE transformation can be enhanced by embracing the industry 5.0 as both a technical advancement and a philosophical movement that puts humans at the forefront of digital transformation processes.

## 2.6. Cyber-Physical–Cybernetic Structures for Circular Transitions

With the literatures under discussion, a convergent framework is created that integrates cyber-physical systems with strategic innovation and relies on the self-regulating principle of cybernetics. The proposed cyber-physical systems-cybernetic framework translates endeavors to circular economy not as discrete technical procedures but as time-evolving processes operated by feedback, coordination, and strategy planning.

This perspective also causes you to consider data differently - no longer as waste but as a resource that can be repurposed. According to Zocco et al. (2024), in a circular system, data flows need to be controlled with the same strategic intent as material flows. This will produce a "circularity of information" which increases the flexibility of the system.

However, care is warranted. Like with any systems approach, integration complexity can produce some surprising effects such as too much information, not being able to make a decision, or system vulnerability if done incorrectly. Therefore, implementation must occur in phases, consider the circumstance, and be guided by continuous thinking and evaluation.

*Figure 1. Cyber-Physical Systems-Cybernetic Framework for Circular Economy Implementation*

**Figure 1 : Cyber-Physical-Cybernetic Framework for Circular Economy Implementation**

**Source:** Researcher's Compilation through smart draw

*Source: Researcher's compilation using SmartDraw.*

## 2.7. Comparing Conventional and cyber-physical-Systems-Cybernetic Circular Economy Models

Re-designing material flow and implementing commercial models such as recycling, reuse, and product-service systems are the main goals of traditional circular economy (CE) approaches. Although these strategies enhance the efficiency of resources, they often depend on delayed response mechanisms and static planning. On the other hand, cyber-physical systems-cybernetic models combine adaptive decision-making, algorithmic response, and real-time sensing. Decentralized governance, future-oriented maintenance, and dynamic resource flow optimization are made possible by this combination. The model provides a stronger, adaptable, and scalable method of applying circularity by integrating feedback loops in both digital and material processes.

## 2.8. Final Thoughts and Plans for the Future

Strategic innovation, cybernetics, and cyber-physical systems can collectively create a smart, adaptable, and dynamic circular economy. This convergence of concepts addresses implementation obstructions by providing a feedback-controlled framework, participatory intelligence, and real-time accountability.

This integration may be further reinforced in the future by examining specific applications in industries, urban development and agricultural land. In order to guarantee that cyber-physical systems as circular systems not only promote stability but also promote equity and inclusion, automation, sensing and social effects of algorithm governance should be carefully examined.

Overall, this cross-interdisciplinary perspective is expected to be beneficial because the circular economy transitions from theory to practice. It can facilitate transitions in obscure, interconnected conditions.

# 3. REVIEW OF THE CONCEPTUAL LITERATURE

Circular economy (CE) has evolved from an aspirational sustainability paradigm to a strategic imperative for addressing pressing global challenges, including lack of resources, excessive waste generation and environmental degradation (Ellen MacArthur Foundation, 2013; Geissdoerfer et al., 2017). Nevertheless, despite its increasing prominence in scholarly and policy discourses, CE's implementation remains fragmented, especially in references where technical infrastructure, institutional coordination and innovation trajectories have not been concurrently upgraded (Kirchherr, Reike, & Hekkert, 2017; Murray, Skene, & Haynes, 2017). This review advances the argument that meaningful CE transitions require the development of a comprehensive structure that systematically integrates the dynamics of cybernetic systems, cyber-physical infrastructure, and innovation, while system theory, transition studies, and strategic rules are informed by insight.

## 3.1. From Linear to Circular: Rethinking the Way Economics Works

The transition from linear to circular economic systems is not simply a technical change in how things are organized, but it is also a fundamental change in how people think about energy, materials, and value, Górecki (2019).

Increasingly, more individuals are stating that linear systems-gathering, making, using, and disposing - are not ecologically sustainable (Geissdoerfer et al., 2017). Whereas the circular systems attempt to maintain the value of resources through healing and regenerative loops (Ellen MacArthur Foundation, 2013), they are underpinned by concepts such as biomimicry (Benyus, 2002) and cradle-to-cradle design (McDonough & Braungart, 2002). But CE is not only recycling; it requires modifications to design, production, inventory, use, repair, reuse, refurbish, recycle, and remake. That implies that government models and innovation strategies must be more than incremental because it has to enable profound changes (Stahel, 1982; Loorbach, Frantzeskaki, & Avelino, 2017).

### 3.2. Strategic Innovation as a Means to Accelerate the Transition to Circular Economy

To create sustainable futures, strategic innovation involves changing value propositions, working models, and institutional rules intentionally (Boons & Lüdeke-Freund, 2013). When applied to CE, this involves altering the manner by which firms produce goods, engage with customers, and govern supply chains. One can observe such changes in real life in concepts such as product-service systems (PSS), extended producer responsibility (EPR), and sharing platforms (Bocken et al., 2014; Lacy & Rutqvist, 2015).

However, many companies continue to view sustainability as an add-on, and not as a strategic objective. Even the most well-intentioned innovation initiatives do not always succeed as they might because they are short on system-wide feedback, adaptive learning, and interdisciplinary collaboration (De Jesus & Mendonça, 2018). Applying cybernetic thinking to strategic innovation efforts might be the solution for this situation

### 3.3. Cybernetic Design for Feedback, Control, and Adaptation

Wiener (1948) formally established cybernetics in 1948 as the study of how systems communicate and regulate themselves through feedback loops and self-regulation. The objectives of the circular economy (CE) reflect these principles, highlighting adaptability, closed-loop design, and structural resilience.

The viable system model (VSM) Beer (1979) further extended these concepts within the organisational governance, proposing that the long -term existence of any system depends on the ability to see, interpret and react to both inner and external signs. In the more recent work, Nica, Chiri , and Delcea (2023) in their study demonstrated how economic cybernetics can support decision making by following non-economic interactions of socio-economic and biological factors. In this context, the autoregressive distributed lag (ARDL) models are applied to analyze temporary relations between CE indicators such as energy efficiency, GDP growth and waste generation.

Ultimately, the potentiality of cybernetics lies in its ability to close the loop between data generation and actionable intelligence, which enables real -time, adaptive control in circular systems (Chiri ă & Georgescu, 2024).

## 3.4. Cyber-Physical Systems: The Digital Backbone of Circularity

Cyber-Physical Systems (CPS) are the enabling infrastructure (Liu et al., 2022) for CE that allows data to be gathered, transmitted, and processed in real-time through material flows (Lee, Bagheri, & Kao, 2015). CPS with sensors, motors, and networked analytics, as noted by Angelopoulos et al. (2023), can monitor usage, predict failure, and optimize performance of the product, thereby lengthening the product's lifespan and reducing waste.

CPS are also crucial in producing "intelligent goods" that can speak of how they are utilized and what needs to be repaired. This enables closed-loop processes and reverse supply chains (Rajala et al., 2018). Despite this, their implementation is not a problem free. CPS inclusion in CE references is still being interrupted by concerns such as data sharing, privacy issues and organisational inertia (Garrido et al., 2024; Rejeb et al., 2025).

## 3.5. Industry 5.0: Human-Centric and Ethical Digitalization

Following the technically focused argument of the industry 4.0, the industry focuses 5.0 people and emphasizes resilience, personalization and ethical (Breque, De Nul, and Petridis, 2021). This perspective serves to support CE objectives through the encouragement of production systems that include for all, decentralized and flexible.

Rejeb et al. (2025) evokes that human-in-loop machines such as learning, digital twins and blockchain-based tracking can make CE models more transparent and participation. In practice, it means that purely passing beyond the efficiency-driven systems and actively involving individuals in the design and monitoring of processes. The co-design of innovations with stakeholders further reflects democratic values, especially those who are bound by the long-term discovery of adaptability and stability in governance.

## 3.6. Systems Transitions and Governance: Multi-Level Perspective and Strategic Niches

Loorbach et al. (2017) To see how these organisational and technological shifts fit into a broader systemic frame, it helps to draw on models from sustainability transitions research. The Multi-Level Perspective (MLP) argues that sociotechnical change unfolds through the interaction innovations, incumbent regimes, and larger landscape forces (Geels, 2002). In parallel, Strategic Niche Management (SNM) demonstrates how protected niches can foster emerging concepts-such as early computer control systems (Kemp, Schot, & Hoogma, 1998)-until they mature enough to be applied on a wider scale.

Both approaches contribute to cybernetics by offering temporal and structural accounts of systemic transformation. In other words, they describe how innovations emerge first, attain validity, and eventually the way the major paradigms are understood are replaced (Weber & Rohracher, 2012). What is in these principles is the emphasis on learning processes, the creation of network agents and the presence of institutional support-all of them are necessary to maintain long-term changes.

### 3.7. Reflexive and Adaptive Governance in Circular Systems

Not only must smart CE systems be reactive, meaning they can assess and modify their own rules and feedback mechanisms as circumstances shift (Meadowcroft, 2009). To accomplish this, government frameworks must emphasize openness, stakeholders, and making the same decisions repeatedly. Economic cybernetics provides us with the essential instruments we must have for this type of government, and online platforms enable us to implement these concepts on a mass level.

Data flows are to be given the same attention as material flows; that is data circularity (Zocco at al., 2024) and it is extremely crucial in this context. Ensuring data is usable, secure, and accessible allows circle systems to function as truly versatile and inclusive.

## 4. CASE ILLUSTRATIONS AND HYPOTHETICAL SCENARIOS

To inspect the practical implications of this interaction, one may investigate a manufacturer who produces products and applies a product-as-a-service strategy. Using integrated CPS technology, the company may track real -time use, worn and returned because of built-in CPS technologies. Cybernetic systems analyze this information, facilitates adaptive maintenance programs, optimize inventory management, and products forecast a lifetime. The introduction of a service-oriented model is not only a commercial innovation, but also on intelligent response systems that help maintain value over time.

A comparable trend is seen at a town or city level working with waste can apply the sensor network in waste receptacles and pickup vehicles to monitor the waste being generated. This data, when artificial response is integrated with loops, can provide dynamic passage design, quick policy modifications and increased citizen engagement. As a result, waste management transforms in an active and responsible system from a reactive service.

## 4.1. Philips Lighting - Service-Based Lighting Systems

Philips has reversed its business model from selling lighting equipment to offering "lighting as a service." This is possible due to the convergence of cyber-physical systems. Philips collaborated with Schiphol Airport to implement the smart light system with sensors on how much energy is consumed, how often space is used, and how efficiently the lights function is used. Cybernetic feedback loops enable the changes in real -time, maintenance to be projected and minimize energy consumption. This new concept not only expands the length of time, the length of the products remains useful, but it also applies a performance-based approach to ensure customers' desires and achieve environmental objectives, Panarotto et al. (2017).

## 4.2. Loop Industries - Molecular Recycling Innovation

Loop industries provides a useful illustration of a cyber-physical circular system employing advanced depolymerization technologies to recycle pet plastic at the molecular level. This makes this approach specific. The real -time process is the integration of the process control: depending on the characteristics of the material input, the system automatically adjusts temperature and pressure such as variables. In doing so, it applies continuous monitoring and cybernetic principles of continuous monitoring and adaptive response to customize the recycling results. The gathered information is considered to continue to make the system more efficient and improve the percentage of material return. The new closed-loop system demonstrates how intelligent investments in CPS can turn recycling more efficient than normal, Romero et al. (2018).

## 4.3. Renault - Circular Manufacturing at Flins Refactory

French Renault Flins Refactory is an excellent case of a circular ecosystem on a manufacturing level. The factory employs CPS technologies across all parts repair, battery recovery, and remanufacturing. By infixing monitors and feedback-systems into production lines, the movement of materials and the condition of its machines are constantly monitored by Renault. Reconfiguring value chains and transportation systems to favor backward flows, with the support of real-time cybernetic coordination, is a strategic innovation in this instance.

### 4.4. Amsterdam City – Circular Smart City Strategies:

Amsterdam's circular strategies employ both strategic urban planning and cyber-physical systems closed-loop feedback-systems. The city employed IoT sensors to collect data on the employment of resources, amount of waste yielded, and utilization of energy. Cybernetic analysis assists in creating adaptive policies enabling such actions as adjusting the prices for garbage collection and initiatives to green buildings. The circular plan places a significant focus on being human-centric and open aligning with Industry 5.0 values within a systems-level government system.

### 4.5. Bundled Laundry Services – IoT-Product-as-a-Service Model

The Dutch firm Bundles leases washing machines equipped with IoT sensors. The software monitors how much detergent is consumed in real-time to optimize storage, energy consumption, and repair planning. Customers are encouraged to establish lasting habits through feedback dashboards. This builds a user ecosystem with plenty of input. This model integrates strategy innovation, CPS infrastructure, and robot behaviour feedback loops to enable the business to function and benefit the environment.

## 5. RESEARCH METHODOLOGY

### 5.1. Research Design

The aim of this measurable, empirical research is to examine how significant, helpful, and challenging individuals believe it is to incorporate strategic innovation, cybernetic systems, and cyber-physical systems technologies into circular economy (CE) initiatives. The key objective is to gather thoughts from individuals who are engaged in sustainability-driven transformation, for instance, practitioners, legislators, and technologists. An organized poll instrument was devised to obtain standard information from numerous respondents. Statistical analysis and results that could be used in other contexts were possible.

### 5.2. Population and Sampling

Individuals this study is targeting are employees who deal with sustainable operations, digital transformation, industrial robots, and the circular economy.

Respondents were selected from four basic groups:

1. Manufacturing and industrial engineering
2. Environmental and sustainability consulting
3. Public policy and smart government
4. Managing technology and new ideas

One strategy employed was a technique known as "purposeful sampling" to ensure that the individuals who participated had the appropriate subject knowledge. 450 questionnaires were posted via professional networks, business groups, and university mailing lists. 317 valid responses were returned, which is equivalent to 70.4% of individuals approached.

## 5.3. Instrument Development

A substantial body of relevant literature was reviewed before developing the survey instrument (Rejeb et al., 2025; Angelopoulos et al., 2023; Zocco et al., 2024). The instrument was organized into five sections:

1. Demographics and Organisational Profile
2. Perceptions of Circular Economy Maturity
3. Adoption of Cyber-Physical Systems and Digital Enablers
4. Strategic Innovation and Organisational Readiness
5. Perceived Value and Challenges of Cybernetic Feedback Integration

Each item was measured using a five-point Likert scale, where 1 indicated "Strongly Disagree" and 5 indicated "Strongly Agree." Content and face validity were ensured through expert review, which included three academics and two industry professionals.

## 5.4. Pilot testing and reliability

A pilot test was conducted with thirty participants to evaluate the clarity, utility and stability of the survey tool. Depending on their response, minor adjustments were made to the wording and structure of several items. To assess internal stability, Cronbach's alpha was calculated for all multi-item constructs, each of which demonstrated strong reliability ($\alpha = 0.78$).

## 5.5. The data collection process

A digital poll tool (Google Forms/Qualtrics) was employed to survey the data over four weeks. Individuals who participated were assured of privacy and anonymity, and they provided consent prior to participating. Accompanying the tool was a brief overview of the aims of the study and some clarification of key concepts, including cybernetics, cyber-physical systems, and CE shift.

## 5.6. Data Analysis Methods

The data collected were uploaded to SPSS 28.0 and SmartPLS 4.0 to be analysed. The following methods were employed:

There are statistical test types:

a) Descriptive (means, frequencies, and standard deviations);
b) Exploratory (EFA) to ensure that the fundamental concepts are right;
c) Confirmatory (CFA) via structural equation modelling (SEM); and
d) Regression analysis will examine the relationships between CPS adoption, strategic Innovation, and CE preparedness.
e) Moderation analysis will observe how organisational size and digital maturity influence the outcome of integrating cybernetic systems.

Statistical significance was at $p < 0.05$. Missing values were deleted singly, and the data were tested for normality, multicollinearity, and heteroscedasticity to ensure that they could be properly interpreted.

## 5.7. Considerations regarding ethics

The study adhered to the school's ethical guidelines. Individuals who participated were informed that they could stop at any time and without penalty. No data that could be used to identify someone was kept, and the data was used for study purposes only.

## 5.8. A Reason for Choosing a Method

In order to bridge the theoretical disciplines and practical cases in CE implementation, an empirical approach was selected incorporating respondents. There is an increasing number of studies that wish to understand how digitalisation, circularity, and adaptive systems interact (Nica et al., 2023; Ghobakhloo et al., 2023). The approach we selected enables us to comprehend how technological convergence influences transitions to sustainability from the stakeholders' perspective.

## 5.9. The aims of the research

Grounded in the theoretical framework and research design, the empirical study was guided by the following aims:

1. To assess how strategic innovation practices shape organisational readiness for circular economy (CE) transition.
2. To examine the role of cyber-physical system (CPS) adoption in enabling real-time responsiveness and closed-loop feedback within CE operations.
3. To evaluate the mediating function of cybernetic feedback mechanisms in strengthening CE effectiveness and adaptive capacity.
4. To analyze the moderating influence of organisational digital maturity on the relationship between CPS adoption and CE performance.

## 5.10. Development of Hypotheses

The following hypotheses were developed for literature and previously mentioned theoretical constructions, empirical tests:

H1: Strategic innovation is positively associated with organisational readiness to circular economy (CE) transition.

H2: Adopting cyber-physical systems (CPS) greatly increases the effectiveness of CE implementation.

H3: Cybernetic feedback mechanisms mediate relationships between CPS and CE effectiveness.

H4: Organisational digital maturity moderates relationships between CPS adoption and CE effectiveness, such as relationship is strong in digital mature organizations.

## 6. EMPIRICAL ANALYSIS

### 6.1. Descriptive Statistics - demographic factors and key organisational data

*Table 1. Descriptive Statistics with the demographic factors and key organisational data*

| Demographic/Organisational Factor | Category | Percentage / Mean |
|---|---|---|
| **Total valid responses** | - | 317 |
| **Gender** | Male | 58% |
| | Female | 42% |
| **Organisational Role** | Middle/Senior Managers | 45% |
| | Technical Roles | 28% |
| | Policy/Strategy Positions | 27% |
| **Mean Organisational Experience (years)** | - | 9.6 |
| **Organizations with Circular Initiatives** | - | 62% |
| **Organizations with CPS Integration** | - | 41% |
| **Organizations with Cybernetic Feedback** | - | 27% |

**Source:** Research Compilation

The sample data set comprised of 317 complete, valid responses from employees in a broad variety of sectors, including industry, technology, policy, and ecology. The gender distribution was fairly even with 58 percent claiming to be male and 42 percent claiming to be female.

45% of respondents held middle or senior management roles, 28% occupied technical roles, and 27% worked in policy/strategy positions..6 years of organisational work experience on average, indicating that the respondents were relatively experienced and enjoyed a great deal of professional interaction. This level of knowledge renders the opinions provided more credible, particularly in the case of complex and systematic organisational projects.

Interestingly, 62 percent of the organizations present had already initiated circular economy (CE) projects. This indicates that the number of people interested in green and clean ways of going about things is growing. But while only 41% indicated they were actively investing in integrating cyber-physical systems (CPS), and even fewer at 27% indicated they had installed formal cybernetic feedback-systems. These figures indicate that there is a large implementation gap between strategic intent and system feedback integration. This indicates that the organisation might need some assistance in increasing and developing its capabilities.

## 6.2. Exploratory Factor Analysis (EFA) – Factor Loadings

*Table 2. Exploratory Factor Analysis (EFA) – Factor Loadings*

| Construct | Item | Loading |
|---|---|---|
| **Strategic Innovation** | SI1 | 0.82 |
| | SI2 | 0.79 |
| | SI3 | 0.84 |
| **CPS Adoption** | CPS1 | 0.80 |
| | CPS2 | 0.76 |
| | CPS3 | 0.81 |
| **Cybernetic Feedback** | CF1 | 0.85 |
| | CF2 | 0.83 |
| **CE Readiness** | CE1 | 0.78 |
| | CE2 | 0.82 |
| **Digital Maturity** | DM1 | 0.88 |
| | DM2 | 0.86 |

**Note:** *All factor loadings > 0.65 (acceptable), extracted using Principal Component Analysis with Varimax Rotation.*

**Source:** Research Compilation

The underlying structure of the measurement instrument was analyzed using exploratory factor analysis (EFA) with principal component extraction and varimax rotation. These five factors, each having its origin from a different theory, were found: Strategic Innovation, CPS Adoption, Cybernetic Feedback, CE Readiness, and Digital Maturity.

With values greater than the commonly accepted mean of 0.65, all of the items loaded highly, thus establishing the convergent validity of the constructs. Individual questions measuring Strategic Innovation (SI1–SI3) scored between 0.79 and 0.84, and thus represent this construct well. Similarly, items in CPS Adoption (CPS1–CPS3) loaded high, and these scores ranged from 0.76 to 0.81.

Two of the items, CF1 = 0.85 and CF2 = 0.83, reflected the meaning of Cybernetic Feedback. Both of these items had high loadings, which means that individuals always understood this dimension. Moreover, for CE Readiness, item loadings were well above the cut-off (CE1 = 0.78; CE2 = 0.82), which means that the underlying mental theme is separable and meaningful.

Unexpectedly, however, the construct Digital Maturity had the strongest factor loadings of all of them. DM1 and DM2 loaded at 0.88 and 0.86, respectively. These results could be interpreted as meaning that the people who answered actually knew what digital capability and infrastructure were, or that this dimension is more critical when it comes to executing CPS and CE.

Overall, the EFA results heavily support the dimensionality of the measurement model, thus confirmatory analysis needs to be conducted. The existence of this robust factor structure supports the usefulness of the instrument in monitoring the intricate interrelationship between strategic, technical, and flexible skills in models for shifting to a circular economy.

## 6.3. Confirmatory Factor Analysis (CFA) – Reliability and Validity

*Table 3. Confirmatory Factor Analysis (CFA) – Reliability and Validity*

| Construct | CR (Composite Reliability) | AVE (Average Variance Extracted) |
|---|---|---|
| Strategic Innovation | 0.84 | 0.63 |
| CPS Adoption | 0.81 | 0.59 |
| Cybernetic Feedback | 0.85 | 0.67 |
| CE Readiness | 0.80 | 0.61 |
| Digital Maturity | 0.86 | 0.68 |

**Thresholds:** CR > 0.70, AVE > 0.50 confirm construct reliability and convergent validity.
**Source:** Research Compilation

It was concluded that the measurement model was even superior after a confirmatory factor analysis (CFA) had been conducted in order to ensure both construct reliability and convergent validity. These findings supply strong factual evidence of how reliable and coherent the measure tool is.

All the constructs had composite reliability (CR) values well above the recommended cut-off of 0.70, indicating a high level of internal consistency of the indicators. E.g., the CR of strategic innovation was 0.84, while CPS adoption and cybernetic feedback were 0.81 and 0.85 respectively. Both CE Readiness and Digital Maturity also met the criteria with CR values of 0.80 and 0.86,. indicating that the constructs are reliable.

For convergent validity, the average variance extracted (AVE) for all constructs was greater than the minimum of 0.50. This indicates that a large percentage of the variance is explained by the underlying factors. For CPS Adoption, the AVE value was 0.59, and for Digital Maturity, it was 0.68. Notably, Strategic Innovation and

CE Readiness received AVEs of 0.63 and 0.61, respectively. Cybernetic Feedback, on the other hand, had a strong convergent validity with an AVE of 0.67. When examined collectively, these CR and AVE values indicate that the constructs are psychometrically well-defined, and thus they support theory consistency and measurement validity. The findings indicate that the latent variables are well-specified and that the accompanying signs do an adequate job of representing the intended mental dimensions. Due to this, the measurement model is believed to be valid enough to be employed for additional structural studies.

## 6.4. Hypothesis Testing – Path Coefficients

The structural model was evaluated using SmartPLS, with attention given to the strength, significance, and direction of the hypothesised relationships among constructs. Table 4 reports the standardised path coefficients (β), together with their corresponding *t*-values and *p*-values, and provides a summary of hypothesis testing outcomes.

*Table 4. Hypothesis Testing – Path Coefficients*

| Hypothesis | Path / Relationship | β (Beta) | t-value | p-value | Result |
|---|---|---|---|---|---|
| H1 | Strategic Innovation → CE Readiness | 0.52 | 9.67 | < 0.001 | Supported – Strong Positive Effect |
| H2 | CPS Adoption → CE Effectiveness | 0.41 | 7.89 | < 0.001 | Supported – Significant Positive Effect |

**Source:** Research Compilation

Results provide strong support for hypotheses. Strategic innovation practices showed a strong and statistically significant impact on the readiness of the circular economy (CE) ($\beta = 0.52$, $t = 9.67$, $p < 0.001$), confirming that the organisational transformation towards circulatory is operated closely with further and new strategies. In a similar vein, the passage from CPS adoption to CE effectiveness was positive and statistically important ($\beta = 0.41$, $T = 7.89$, $P < 0.001$), underlines the importance of cyber-physical systems in providing digital infrastructure required for circular operations. Together, these conclusions validate the proposed theoretical models and suggest that both strategic orientation and digital integration serve as important enablers to carry forward CE effectiveness.

## 6.5. Mediation and Moderation Results

To extend the assessment of the structural model, mediation and moderation analyses were carried out in SmartPLS. The mediation analysis tested whether cybernetic feedback mechanisms act as an intervening variable in the relationship between CPS adoption and circular economy (CE) outcomes. In parallel, the moderation analysis examined whether the strength of this relationship varies depending on the level of organisational digital maturity. The results are summarised in Table 5.

*Table 5. Mediation and Moderation Results*

| Effect Type | Path | β (Beta) | p-value | Interpretation |
|---|---|---|---|---|
| Mediation (H3) | CPS → Cybernetic Feedback → CE effectiveness | 0.18 | < 0.01 | Partial mediation; adaptive feedback enhances impact |
| Moderation (H4) | CPS × Digital Maturity → CE effectiveness | 0.27 | < 0.05 | Significant moderation; digital maturity strengthens relationship |

**Source:** Research Compilation

These results indicate that cybernetic feedback functions as a partial mediator in the pathway from CPS adoption to improved CE effectiveness ($\beta = 0.18, p < 0.01$). This suggests that while the direct influence of CPS on CE is substantial, the introduction of feedback mechanisms significantly amplifies this effect - underscoring the value of adaptive control systems in reinforcing circular performance.

In parallel, the moderating role of digital maturity ($\beta = 0.27, p < 0.05$) highlights a critical contextual condition: organizations with more advanced digital infrastructures and capabilities derive greater benefit from CPS implementation. Put differently, digital maturity does not merely support CPS adoption - it actively magnifies its contribution to circular economy effectiveness by facilitating smoother integration of feedback loops and data-driven decision-making processes.

Taken together, these findings provide empirical validation for the theoretical proposition that successful CE transitions are contingent not only on technological integration but also on systemic adaptability and contextual readiness. The inclusion of feedback-rich mechanisms and mature digital environments thus emerges as a dual imperative for organizations seeking to operationalize circular principles at scale.

## 7. DISCUSSION

The findings of this study highlight the value of integrating strategic innovation, cybernetic systems, and cyber-physical infrastructure as reinforcing pillars in advancing circular economy (CE) practices. By combining conceptual analysis with empirical evidence, several insights emerge that deepen our understanding of how these elements interact.

First, the results clearly support the view that strategic innovation functions as a central enabler of CE readiness. Organisations that embed sustainability into their core value proposition, while also experimenting with forward-looking business models and governance reforms, display a stronger capacity to move beyond linear production logics. The observed effect ($\beta = 0.52$, $p < 0.001$) confirms that CE adoption cannot be reduced to a matter of technology or regulation alone - it is also a strategic orientation.

Second, the study validates the role of cyber-physical systems (CPS) as digital backbones for CE effectiveness. Their positive effect ($\beta = 0.41$, $p < 0.001$) shows how CPS enables real-time monitoring, predictive maintenance, and intelligent lifecycle management, effectively bridging the digital–material divide. Yet, the evidence also suggests that CPS implementation on its own is insufficient. The contribution of cybernetic feedback mechanisms as a mediating factor emerges as particularly important.

The mediation analysis confirms partial mediation between CPS adoption and CE effectiveness ($\beta = 0.18$, $p < 0.01$). This highlights the importance of feedback-rich systems that allow organisations to adapt continuously to environmental and operational changes. In this sense, cybernetics moves beyond its role as a conceptual framework and operates as a practical governance tool - one that enables data to function not just as a record, but as a driver of ongoing adaptation and resilience.

Finally, digital maturity is shown to moderate the CPS–CE relationship ($\beta = 0.27$, $p < 0.05$). Organisations with higher levels of digital readiness are better positioned to integrate CPS and cybernetic feedback in a scalable and coherent way. This finding underlines the point that technological investment alone does not deliver transformation. It must be matched with institutional capabilities and digital competences that allow potential to be converted into performance.

## 7.1. Policy Inertia and Organisational Resistance

Despite the technical ability of cyber-physical systems and cybernetic systems, their adoption often confronts resistance within organizations and policy environments. The regulatory structure may be slow to develop, resulting in a mismatch between technical capabilities and compliance requirements. Cultural inertia and risk averting can further disrupt experimentation with new models. To overcome these obstacles, organization regulators can establish sandboxes that allow controlled tests of innovations. Public-private involvement can also enable collaborative policy adaptation and create shared belief in emerging practices. Training and change management programs are necessary to align organisational culture with dynamic governance paradigms.

## 7.2. Stepwise Framework for Systemic Integration

The transition to a cyber-physical systems-cybernetic circular economy can follow a structured pathway, Pagoropoulos et al. (2017):

Step 1: Assess digital and organisational readiness using standardized maturity models.
Step 2: Identify priority processes for CPS and feedback integration (e.g., maintenance, supply chains).
Step 3: Develop pilot projects in low-risk operational areas to validate data flows and control loops.
Step 4: Scale up through modular architectures and cross-functional governance structures.
Step 5: Institutionalize adaptive learning and continuous improvement via embedded feedback analytics.

## 7.3. Critical Reflections on Feasibility and Digital Optimism

Although this chapter highlights the predictive potential of cyber-physical and cybernetic systems in driving circular economy (CE) transitions, it is equally important to recognize that scalability is not guaranteed. Fragmented data infrastructures can create ecosystem or vendor lock-in, while inconsistent interoperability may lead to uneven implementation and inflated expectations. Overreliance on automation also carries risks, as system overloads or cyberattacks can compromise resilience.

Examples from smart city projects illustrate how data complexity can generate unintended vulnerabilities, reinforcing the need for caution. To mitigate these risks, organisations should adopt modular system architectures, ensure continuous human oversight, and prioritize transparency and accountability in automated response systems. Such safeguards can help balance innovation with reliability, ensuring that technological integration such as maintenance and supply chains.

# 8. IMPLICATIONS

The implications of these findings extend across both practice and theory. On the practical side, policymakers and organisational leaders must recognize that circular economy (CE) transition requires more than piecemeal technological upgrades. What is needed is a systems-level approach that combines strategic foresight, cybernetic design, and human-centered innovation. Digital tools should not stand in isolation but must be embedded within broader feedback structures that promote learning, responsiveness, and coordination across sectors.

For scholars and theorists, the study adds to the ongoing conversation on economic cybernetics and systems thinking by grounding these perspectives in empirical evidence. It shows that cybernetics - once regarded largely as a theoretical abstraction -can be applied in managing the complexity of socio-environmental transitions. In this sense, the circular economy provides not only a field of application but also a proving ground for the practical value of cybernetic principles.

## 8.1. Ethical and Societal Considerations

The implementation of cyber-physical systems and cybernetic systems in a circular economy also raises important moral and social inquiries. Algorithm decision-making may introduce biases that affect marginalized groups, particularly if training data lacks multifariousness or transparency. Continuous collection of data and processing can challenge privacy and agency, especially in consumer-support applications. Additionally, there is a risk that the benefits of digitally increased circular systems can be concentrated among organizations in advanced economies, which can increase global digital division. To ensure just and responsible changes, it is necessary to include confidentiality-by-design principles, establish clear governance standards, and engage stakeholders in the development of algorithms and data infrastructure. Public policy can play an important role by being compulsory and supporting capacity in less digitally mature areas.

While this chapter focuses primarily on advanced industrial contexts, it is essential to identify that many economies rely on informal practices and have restricted access to digital infrastructure. In such an environment, enforcing cyber-physical systems-cybernetic models can be challenging due to infrastructure deficiency, skill intervals, and regulatory fragmentation. However, hybrid approaches that combine low-technical solutions (e.g., manual tracking) with modular digital tools can provide aggressive passage for circularity. Future research should check that the informal sector-such as waste picking and repair economies that can be integrated into feedback-driven circular systems without displacing or increasing inequality.

## 8.2. Data Interoperability and Privacy Solutions

To enable seamless data exchange across heterogeneous systems such as IoT networks and blockchain platforms, organisations should adopt interoperability frameworks like ISO/IEC 21823 alongside open APIs that support data portability.

Standardised protocols, including MQTT and OPC UA, further facilitate secure and real-time data flows between devices and platforms. At the same time, embedding privacy-by-design principles and employing encrypted ledger technologies can help safeguard sensitive information while could be maintaining transparency and traceability across value chains (Karim et al., 2025).

## 9. LIMITATIONS

As with any research endeavor, this study has certain limitations. The survey-based design, though useful for capturing perceptions and organisational trends, is inherently subject to potential response biases and the risk of sectoral overrepresentation. In addition, the cross-sectional nature of the analysis restricts causal inference and prevents observation of how dynamics evolve over time.

Another limitation concerns the generalizability of the findings across regions and industries. Although the sample was relatively diverse, levels of digital maturity and circular economy (CE) integration are likely to vary considerably across different contexts. Future research should therefore investigate sector-specific and region-specific dynamics, as well as adopt longitudinal approaches, to further refine and validate the framework.

*Table 6. Challenges and Solutions for cyber-physical systems-Cybernetic CE*

| Challenge | Suggested Solution or System Component |
|---|---|
| Data fragmentation | Adoption of common interoperability standards (e.g., ISO/IEC 21823) |
| Technology readiness gaps | Capacity-building programs and pilot implementation projects |
| Organisational resistance | Change management frameworks and participatory co-design |
| Cybersecurity risks | Privacy-by-design architectures and continuous monitoring |
| Regulatory inertia | Regulatory sandboxes and adaptive policy experimentation |
| Digital divide | Inclusive infrastructure investment and skills development |

Source: Researcher's Compilation

## 10. SUGGESTIONS FOR FUTURE RESEARCH

Future research could benefit from longitudinal case studies that trace the evolution of cyber-physical and cybernetic systems over time, especially in contexts shaped by shifting regulatory environments. Such studies would provide richer insight into learning curves, adaptation pathways, and institutional change. Qualitative explorations could explore how organisational culture, leadership, and employee engagement influence the implementation of feedback-based systems.

Beyond this, there is scope to extend the framework into sectors such as agriculture, healthcare, and education fields where CPS and CE principles are beginning to take hold but remain comparatively underexplored. Finally, ethical and societal considerations, including questions of data privacy, digital equity, and algorithmic accountability, merit closer attention as CE systems grow increasingly dependent on real-time data processing and autonomous decision-making.

## 10.1. Sectoral and Regional Applicability

This framework model has broad potential relevance beyond manufacturing. cyber-physical systems in the healthcare industry can monitor the use of medical equipment, plan maintenance, and enhance patient outcomes. Sensor networks can improve nutrient and water cycles in agriculture, promoting circular resource management. Open-source platforms and modular IoT devices can facilitate gradual adoption in areas with low levels of digital maturity could explore how organisational culture, leadership, and employee engagement influence. To comprehend the particular facilitators and limitations of these sectoral and regional extensions, targeted research is necessary. For example, integrating IoT-based soil monitoring in agriculture can create dynamic feedback on nutrient cycles (Karim et al., 2024), while CPS-enabled tracking in hospitals can optimize inventory and waste flows in real-time (Waaje et al., 2025).

# 11. CONCLUSION

This study has proposed and empirically validated a novel framework for advancing circular economy (CE) transitions through the convergence of strategic innovation, cybernetics, and cyber-physical systems. The findings demonstrate that such transitions cannot be reduced to technical upgrades alone; rather, they are embedded within broader systemic, organisational, and cognitive architectures.

By by clarifying both the mediating role of cybernetic feedback and the moderating influence of digital maturity, the study offers a more nuanced account of how digital infrastructures and strategic foresight intersect to enable sustainable transformation. The insights presented here aim to inform both scholarly debate and policy design, supporting the development of circular systems that are not only efficient but also adaptive, inclusive, and resilient in the face of growing complexity.

## REFERENCES


Angelopoulos, C. M., Katos, V., & Kostoulas, T.. (2023). IDEAL-CITIES - A trustworthy and sustainable framework for circular smart cities. In *Proc. 15th Annual Int. Conf. on Distributed Computing in Sensor Systems (DCOSS 2019)* (pp. 443–450). https://doi.org/DOI: 10.1109/DCOSS.2019.00089

Beer, S. (1979). The viable system model. *Viable Systems Model, Wiley, Chicester.*

Benyus, J. M. (2002). *Biomimicry: Innovation inspired by nature*. Harper Perennial.

Bocken, N. M. P., Short, S. W., Rana, P., & Evans, S. (2014). A literature and practice review to develop sustainable business model archetypes. *Journal of Cleaner Production*, *65*, 42–56. DOI: 10.1016/j.jclepro.2013.11.039

Boons, F., & Lüdeke-Freund, F. (2013). Business models for sustainable innovation: State-of-the-art and steps towards a research agenda. *Journal of Cleaner Production*, *45*, 9–19. DOI: 10.1016/j.jclepro.2012.07.007

Breque, M., De Nul, L., & Petridis, A. (2021). Industry 5.0: Towards a sustainable, human-centric and resilient European industry. Publications Office of the European Union. https://op.europa.eu/en/publication-detail/-/publication/468a892a-5097-11eb-b59f-01aa75ed71a1

Chiri ă, N., & Georgescu, I. (2024). Cybernetics analysis of the circular economy from Romania. In Ivan, I., Ro ca, I. D., & Pîrjan, A. D. (Eds.), *Proceedings of the 22nd International Conference on Informatics in Economy* (pp. 360–373). Springer. https://doi.org/DOI: 10.1007/978-981-99-6529-8_27

De Jesus, A., & Mendonça, S. (2018). Lost in transition? Drivers and barriers in the eco-innovation road to the circular economy. *Ecological Economics*, *145*, 75–89.

Ellen MacArthur Foundation. (2013). Towards the circular economy (Vols. 1–3). Ellen MacArthur Foundation. https://ellenmacarthurfoundation.org/towards-the-circular-economy

Garrido, S., Muniz, J.Jr, & Ribeiro, V. B. (2024). Operations management, sustainability & Industry 5.0: A critical analysis and future agenda. *Cleaner Logistics and Supply Chain*, *10*, 100141. DOI: 10.1016/j.clscn.2024.100141

Geels, F. W. (2002). Technological transitions as evolutionary reconfiguration processes: A multi-level perspective and a case-study. *Research Policy*, *31*(8–9), 1257–1274. DOI: 10.1016/S0048-7333(02)00062-8

Geissdoerfer, M., Savaget, P., Bocken, N. M. P., & Hultink, E. J. (2017). The circular economy - a new sustainability paradigm? *Journal of Cleaner Production, 143*, 757–768. DOI: 10.1016/j.jclepro.2016.12.048

Ghobakhloo, M., Iranmanesh, M., Foroughi, B., Tseng, M. L., Nikbin, D., & Khanfar, A. A. (2023). Industry 4.0 digital transformation and opportunities for supply chain resilience: A comprehensive review and a strategic roadmap. *Production Planning and Control*. Advance online publication. DOI: 10.1080/09537287.2023.2252376

Górecki, J. (2019). Circular economy maturity in construction companies. *IOP Conference Series. Materials Science and Engineering, 471*(11), 112090. DOI: 10.1088/1757-899X/471/11/112090

Karim, R., Roshid, M., & Waaje, A. (2024). Circular economy and supply chain sustainability. In Mızrak, F. (Ed.), *Strategic Innovations for Dynamic Supply Chains* (pp. 1–30). IGI Global Publishing., DOI: 10.4018/979-8-3693-3575-8.ch001

Karim, R., Waaje, A., & Roshid, Md. M. (2025). Digital Circular Economy for Attaining Sustainable Development Goals. In D. C. Yıldırım, S. Yıldırım, & V. Kandpal (Eds.), Effects of Digitalization and Circular Economy on Sustainable Policy and Climate Change Prevention (pp. 211–240). IGI Global. https://doi.org/ DOI: 10.4018/979-8-3693-9909-5.ch008

Kemp, R., Schot, J., & Hoogma, R. (1998). Regime shifts to sustainability through processes of niche formation: The approach of strategic niche management. *Technology Analysis and Strategic Management, 10*(2), 175–198.

Kirchherr, J., Reike, D., & Hekkert, M. (2017). Conceptualizing the circular economy: An analysis of 114 definitions. *Resources, Conservation and Recycling, 127*, 221–232.

Lacy, P., & Rutqvist, J. (2015). *Waste to Wealth – The Circular Economy Advantage*. Palgrave Macmillan., DOI: 10.1057/9781137530707

Lee, J., Bagheri, B., & Kao, H. A. (2015). A cyber-physical systems architecture for Industry 4.0-based manufacturing systems. *Manufacturing Letters, 3*, 18–23. DOI: 10.1016/j.mfglet.2014.12.001

Liu, Y., Ma, X., Zhang, X., Zhu, Q., & Li, Y. (2022). A framework of digital technologies for the circular economy: Digital functions and mechanisms. *Business Strategy and the Environment, 31*(6), 2445–2460. DOI: 10.1002/bse.3015

Loorbach, D., Frantzeskaki, N., & Avelino, F. (2017). Sustainability transitions research: Transforming science and practice for societal change. *Annual Review of Environment and Resources*, *42*, 599–626. DOI: 10.1146/annurev-environ-102014-021340

McDonough, W., & Braungart, M. (2002). Design for the triple top line: New tools for sustainable commerce. *Corporate Environmental Strategy*, *9*(3), 251–258.

Meadowcroft, J. (2009). What about the politics? Sustainable development, transition management, and long-term energy transitions. *Policy Sciences*, *42*(4), 323–340. DOI: 10.1007/s11077-009-9097-z

Murray, A., Skene, K., & Haynes, K. (2017). The circular economy: An interdisciplinary exploration of the concept and application in a global context. *Journal of Business Ethics*, *140*(3), 369–380. DOI: 10.1007/s10551-015-2693-2

Nica, I., Chiri ă, N., & Delcea, C. (2023). Towards a sustainable future: Economic cybernetics in analyzing Romania's circular economy. *Sustainability*, *15*(19), 14433.

Pagoropoulos, A., Pigosso, D. C. A., & McAloone, T. C. (2017). The emergent role of digital technologies in the circular economy: A review. *Procedia CIRP*, *64*, 19–24. DOI: 10.1016/j.procir.2017.02.047

Panarotto, M., Wall, J., & Larsson, T. (2017). Simulation-driven design for assessing strategic decisions in the conceptual design of circular PSS business models. *Procedia CIRP*, *64*, 25–30. DOI: 10.1016/j.procir.2017.03.026

Patyal, V. S., Sharma, S. K., & Dhir, S. (2021). Mapping the links between Industry 4.0, circular economy and sustainability: A systematic literature review. *Journal of Enterprise Information Management*, *34*(3), 850–876. DOI: 10.1108/JEIM-05-2021-0197

Rajala, R., Hakanen, E., Mattila, J., Seppälä, T., & Westerlund, M. (2018). How do intelligent goods shape closed-loop systems? *California Management Review*, *60*(3), 20–44. DOI: 10.1177/0008125618759685

Rejeb, A., Rejeb, K., Keogh, J. G., & Süle, E. (2025). (in press). When Industry 5.0 meets the circular economy: A systematic literature review. *Circular Economy and Sustainability*.

Reuter, M. A. (2016). Digitalizing the circular economy: Circular economy engineering defined by the metallurgical internet of things. *Metallurgical and Materials Transactions. B, Process Metallurgy and Materials Processing Science*, *47*(6), 3194–3220. DOI: 10.1007/s11663-016-0735-5

Romero, D., Stahre, J., Wuest, T., Noran, O., Bernus, P., Fast-Berglund, Å., & Gorecky, D. (2018). Digital lean cyber-physical production systems: The emergence of digital lean manufacturing and the significance of digital waste. In Nääs, I. (Eds.), *IFIP Advances in Production Management Systems* (pp. 67–75). Springer., DOI: 10.1007/978-3-319-99704-9_2

Roshid, Md. M., Karim, R., Waaje, A., Islam, S., & Bhowmik, R. C. (2025). Mapping the carbon finance research: A bibliometric exploration of innovation in sustainable finance. *Development and Sustainability in Economics and Finance*, *5*, 100048. DOI: 10.1016/j.dsef.2025.100048

Roshid, Md. M., Waaje, A., Karim, R., Irtisham, M. R., & Kaswan, M. S. (2025). Industry 5.0 and organizational sustainability: Integrating cutting-edge technologies and human ingenuity. In M. S. Kaswan, R. Rathi, & J. A. Garza-Reyes (Eds.), Industry 5.0: Concepts and Strategies for Digital Transformation (pp. 50–61). CRC Press. https://doi.org/DOI: 10.1201/9781003535188-7

Stahel, W. R. (1982). *The product life factor. An Inquiry into the Nature of Sustainable Societies. The Role of the Private Sector*. Houston Area Research Centre.

Waaje, A., Karim, R., & Roshid, Md. M., & Meem, T. N. (2024). Minimizing waste in the hospitality industry: The opportunities and challenges of lean tourism. In R. Sucheran (Ed.), Sustainable Waste Management in the Tourism and Hospitality Sectors (pp. 125–166). IGI Global. https://doi.org/DOI: 10.4018/979-8-3693-6110-8.ch006

Waaje, A., Karim, R., Roshid, M. M., Yeamin, B., & Meem, T. N. (2025). Integration of industry 4.0 with personalized medicine: A shift of the paradigm in healthcare. In Chemingui, H., & Lamloumi, M. (Eds.), *AI-Driven Personalized Healthcare Solutions* (pp. 27–69). IGI Global., DOI: 10.4018/979-8-3693-7858-8.ch002

Weber, K. M., & Rohracher, H. (2012). Legitimizing research, technology and innovation policies for transformative change: Combining insights from innovation systems and multi-level perspective in a comprehensive "failures" framework. *Research Policy*, *41*(6), 1037–1047. DOI: 10.1016/j.respol.2011.10.015

Wiener, N. (1948). *Cybernetics: Or control and communication in the animal and the machine*. MIT Press.

Zocco, F., Haddad, W. M., Corti, A., & Malvezzi, M. (2024). *A unification between deep-learning vision, compartmental dynamical thermodynamics, and robotic manipulation for a circular economy*. IEEE., DOI: 10.1109/ACCESS.2024.3493755

# Chapter 14
# Eco-Innovation as a Driver of the Circular Economy: Multi-Criteria Approach

**Ivana Janjić Papakosmidis**
http://orcid.org/0000-0003-3142-8467
*Innovation Center, University of Nis, Serbia*

**Miljana Talić**
https://orcid.org/0000-0003-1017-8788
*Innovation Center, University of Nis, Serbia*

**Žarko Jovan Rađenović**
https://orcid.org/0000-0002-3679-7870
*Innovation Center, University of Nis, Serbia*

## ABSTRACT

*Eco-innovation is the critical element that ensures a shift to the circular economy and plays a significant role in the protection of the environment, building new environmentally friendly consumption behaviour and increasing the efficiency of production. The purpose of this research is to rank EU27 countries based on a selection of Eco-Innovation Index sub-indicators and identify the alternative that best reflects the interconnection between eco-innovation and the circular economy. This paper explores the application of multi-criteria decision-making (MCDM) methods to evaluate EEI sub-indicators. In the MCDM model, the sub-indicators weight coefficients were calculated by applying the CRITIC method. The optimal alternative is considered to be the one that demonstrated significant progress in*

DOI: 10.4018/979-8-3373-2454-8.ch014

*applying eco-innovation principles during the analyzed period from 2020 to 2024. The originality of this research is fundamentally linked to the evaluation of the eco-innovation performances of EU member states using a portfolio of MCDM methods.*

## 1. INTRODUCTION

The circular economy is an innovative economy that fosters research and innovative activities in all sectors of the economy (European Parliament, 2023). The circular economy, as an initiative aimed at achieving sustainability, deals with innovations, especially those related to the environment, primarily to reduce resource exploitation (Ghormare et al., 2024).

In the knowledge-based economy, improving the eco-innovation capability requires R&D efforts. R&D activities play a central role in driving innovation at the level of companies and national economies (Rađenović et al, 2023, p. 35). The implementation of R&D-driven innovation activities propels development and the adoption of eco-innovation technologies. For boosting eco-innovation performance, rationally planning and allocating R&D activities has become essential. Although capabilities associated with a higher R&D investment are not always addressed to eco-innovation, R&D activities and the knowledge creation they generate are necessary for adopting CE practices (Khan et al., 2020).

The concept of eco-innovation is widely accepted in the European Union (EU), as indicated by numerous initiatives aimed at promoting sustainable development, developing environmental awareness, and eco-innovation. A special composite index is used to assess the level of development of eco-innovation in member states. The Eco-Innovation Index (EEI) provides essential insights into how EU member states are incorporating eco-innovations into their economies. EEI examines countries based on their commitment to environmentally sustainable innovation, highlighting factors such as resource efficiency, waste reduction, green technologies, the development of eco-friendly products and services, and government environmental and energy R&D appropriations and outlays.

This chapter aims to fill the gaps identified in previous research in the field of eco-innovation and circular economy, providing a clearer insight into the interrelationship between the aforementioned concepts. Due to the growing importance of eco-innovation in the transition to a circular economy, the paper seeks to address the challenges related to the quantification and ranking of eco-innovation performance, as well as to explain the role of eco-innovation within the broader concept of a circular economy. The paper contributes to the understanding of how specific eco-innovation indicators can influence the implementation of circular practices in the EU27 countries, which provides a basis for future policies and strategies for sustainable development.

In order to evaluate EEI sub-indicators, this research applies the multi-criteria decision-making (MCDM) methods. By utilizing the PROMETHEE methods through the Visual PROMETHEE Academic software, decision-makers, researchers, and practitioners can effectively assess eco-innovation performances across various economic entities, fostering informed decisions toward a sustainable future. The methodology incorporates the CoCoSo (Combined Compromise Solution) and ARAS (Application of Additive Ratio Assessment) MCDM methods, comparing their ranking outcomes with those of the previously mentioned approach. Different approaches to examining weight coefficients for selected criteria in the form of sub-indicators are crucial in addressing multi-criteria challenges. A key aspect of the research's originality centers around establishing weight coefficients through CRITIC analysis, alleviating concerns regarding subjective author biases in prioritizing the analyzed sub-indicators. Furthermore, correlating different ranking methods paves the way for original insights into the current conditions within the EU's eco-innovation landscape.

The paper is organized as follows. The paper begins with an overview of the background and literature review on the key concepts addressed by the research. After that, the authors present a conceptual framework related to the circular economy and eco-innovations, with a special reference to the role of digital transformation in the development of the circular economy and a conceptual framework for measuring eco-innovation. In the following chapters, the applied methodological framework is explained, and solutions and recommendations based on the results of the analysis are offered. At the end of the paper, directions for future research, implications for policymakers, and final remarks are presented.

## 2. BACKGROUND

Due to the increasing problem of environmental degradation (Milanović et al., 2022), but also due to the inability to balance economic growth, consumption of limited resources, and the use of renewable energy sources, new approaches such as sustainable development and circular economy are gaining importance (Talić et al., 2023; Durán-Romero et al., 2020). The goal of the circular economy was to address the pressing global ecological challenges brought by the linear economy (Stahel, 2019). A feature of the linear economy paradigm is that waste resulting from the production process is released into the environment, assuming boundlessness and easy availability of material resources. Due to economic and technological development, the linear economy model using the "take-make-dispose" approach is no longer able to meet all the needs. The problems that arise in a linear economy model are reflected in wasteful value extraction, waste landfills, the problem of trash, a worsening environmental catastrophe, a bias against sustainable development programs, and loss of competitive advantage (Sharma et al., 2021). The essence of the circular economy is expressed in the definition given by the Ellen MacArthur Foundation (2025), which defines the circular economy as "a system where materials never become waste and nature is regenerated". The European Parliament (2023) defines CE as "a model of production and consumption, which involves sharing, leasing, reusing, repairing, refurbishing and recycling existing materials and products as long as possible" to extend the life cycle of products. The circular economy is based on three principles: fostering system effectiveness, optimising resource yields, and enhancing natural capital (Rashid & Malik, 2023).

The transition from a linear economy to a circular economy has many obstacles related to finances, society, politics, jurisdiction, development of environmentally friendly technologies and innovations (Stankevičienė & Nikanorova, 2020). This transition requires significant changes in existing production and consumption systems, which highlights the key role of innovation, especially eco-innovation, as a necessary factor in a sustainable economy (De Jesus et al., 2021; Sehnem et al., 2022). Eco-innovations are new solutions that lead to the avoidance or reduction of harmful impacts on the environment, in accordance with the characteristics and objectives of environmental changes (Pichlak & Szromek, 2022). The role of eco-innovations is to enable the transition to a circular economy model that implies numerous changes at different levels.

## 3. LITERATURE REVIEW

The topic of eco-innovation and circular economy is very popular in academic and research circles, which brings different views of these concepts, their connections, and different ways of quantifying their effects.

The study by Juracka et al. (2024) deals with the topic of eco-innovation and its measurement to provide a comparative analysis of the performance of EU countries in the field of eco-innovation. This study provides valuable insight into the field of eco-innovation in the EU. It uses the Eco-Innovation Index as a basis, which is an instrument for monitoring eco-innovation at the level of EU countries. Measuring the performance of countries in the field of eco-innovation is a process that should be constantly monitored and adapted at the level of each country (Juracka et al., 2024). Measuring and assessing eco-innovations at the country level can be a limitation, as recognized in this research. Therefore, conducting further research in the field of eco-innovation measurement is crucial to overcome the obstacles that an inadequate measurement approach can cause.

Another paper by Constantin et al. (2023) examines the issue of eco-innovation and digital inclusion by exploring how the efficiency of resource use in EU countries can be increased. The paper applies cluster analysis, and the results show that EU countries with a more successful economy achieve better performance in terms of eco-innovation and digital inclusion.

According to Kobryń and Prystrom (2017), the Eco-Innovation Scoreboard can be an essential instrument for considering the level of eco-innovation of countries, as well as comparing their performance with other countries' achievements. In its paper, they analyzed the level of environmental innovations in the European Union countries, and for this purpose, the multi-criteria approach was used. They defined clusters, in which countries are classified according to their performance in eco-innovation, allowing for the identification of leading countries, as well as those behind them.

A certain number of research studies focus on segments of the Eco-innovation index as a measure of the eco-innovation performance of EU countries, taking into account individual sub-indicators that this composite index includes. For example, Pakulska and Rutkowska (2024) use resource efficiency as a sub-indicator of the Eco-innovation index to examine EU diversity in the performance achieved in this area. The results of the research showed that there is a considerable gap between individual EU countries regarding innovation.

Stojanović et al. (2022) published a study where the multi-criteria approach was used in the context of innovation. According to the authors (Stojanović et al., 2022), a country must invest in innovative activities if it wants to achieve competitiveness in the global market. This research paper focused on the innovative competitiveness of the Western Balkan countries.

Based on the given literature review, several issues can be identified from which the significance of this chapter arises. First, even though not so new, the concept of eco-innovation is often the subject of different and sometimes controversial definitions, and does not have a universally accepted definition. Since it is based on ecological technologies, eco-innovation is characterized by significantly greater variability compared to traditional innovations (Tomala & Urbaniec, 2021, p. 73). Despite different interpretations, applying the concept of eco-innovation in practice can be a challenge. In academic literature, the term eco-innovation is often equated with terms such as green, ecological, and sustainable innovation, which are frequently considered synonymous.

Furthermore, empirical studies often have doubts about quantifying the effects of eco-innovations, as well as measuring success in implementing circular practices. Therefore, it is necessary to address the issues of measuring performance in the circular economy, especially long-term effects, and to offer more approaches to this problem.

## 4. CONCEPTUAL FRAMEWORK

### 4.1 Circular economy

Circularity in the economy is increasing in importance as a response to the problems of improper waste disposal and the increasing accumulation of garbage (Karim et al., 2024a). Issues such as pollution, climate change, and depletion of natural resources continue to represent key global challenges that must be addressed immediately systemically and comprehensively (Karim et al., 2024b). Taking into account the limited nature of natural resources, CE emphasizes the need to create a resource-efficient and economically efficient system (Bril et al, 2023; Roshid et al., 2025a). In addition to being seen as one of the key initiatives for the transition to a sustainable economy (European Commission, 2020), the circular economy has become a strategic factor for business and industry. Industry should have more responsibility towards the environment (Holly et al., 2023; Hák et al., 2016) and strive for greater resource independence.

Circular economy is an approach that has emerged to represent a different way of human activity in relation to nature (Geissdoerfer et al., 2017; Latif et al., 2025). This approach proposes to update the traditional linear model with one that relies on the principles of reduction, reuse, and recycling (Prieto-Sandoval et al., 2018). The fundamental characteristic of the circular economy is the desire to reduce the burden on nature by keeping resources in use for as long as possible. Essentially, the goal is to create a closed-loop system that ensures a product is not discarded after use, minimizing waste (EESC, 2016). There are three basic principles underlying CE: "1) eliminate waste and pollution; 2) circulate products and materials; 3) regenerate nature" (EMF, 2025). Maximizing a product's use throughout its life cycle and reusing it to generate new value is the aim of the CE (Geissdoerfer et al., 2017; Roshid et al., 2025b).

The combination of innovative ecosystem elements and the circular economy can have numerous advantages and lead to the achievement of sustainable development goals, which is why sustainable innovations that lead to both ecological and economic growth are encouraged (Alka et al, 2024; Ghormare et al., 2024). The circular economy encourages research and innovation activities in all economic sectors (European Parliament, 2023).

Understanding the role of eco-innovations in the new, circular economy is very important for promoting sustainable development and introducing systemic changes at all levels of a country's development. Eco-innovations are of great importance in modern conditions due to the promotion of optimal use of resources and efficiency in various sectors (Rađenović et al., 2024).

## 4.2 Eco-innovation

To encourage the transition to a sustainable economy and to solve significant environmental and social challenges, multi-level efforts are necessary. The digital circular economy (DCE) is increasingly recognized as a key driver of sustainable development, supporting the transition from a linear to a resource-efficient and renewable circular economic model (Karim et al., 2025). Through the application of innovative technologies and solutions, the efficiency of resource distribution can be increased and resource disparities can be reduced (Rađenović et al., 2025).

Despite numerous economic challenges and the increasing consequences of climate change, there is a growing need to introduce sustainable practices in all sectors of society and the economy (Wajee et al., 2024; Wajee et al., 2025). Eco-innovation encompasses the creation, development, and implementation of new products, processes, and services that lead to significant improvements in social sustainability and reductions in environmental impact while maintaining economic viability (Peyravi et al., 2023). Initially, the concept of eco-innovation was focused on maintaining a balance between commercial value and environmental responsibility, while more recent studies highlight eco-innovation as a systematic enabler of green transitions and circular economic models (Hojnik et al., 2024). According to some researchers (Carrillo-Hermosilla et al., 2010), eco-innovation is a sustainable tool that contributes to environmental and economic benefits, improving environmental performance. Kemp et al. (2019) state that eco-innovation represents a response to environmental problems, but at the same time, it can also be a natural reaction to high resource prices. An innovation is considered ecological if it leads to significant progress in promoting sustainable development, improving environmental resilience to the impact of business activities, and achieving more efficient and responsible use of natural resources (European Commission, 2012). Eco-innovation combines innovation policy with environmental policy, including the principles of sustainable development (Androniceanu & Sabie, 2022). At the same time, it represents a segment of eco-entrepreneurship, as it allows companies to generate significant revenue by solving environmental problems (Hajdukiewicz & Pera, 2023, p. 147). According to the EU definition, eco-innovation is: “any innovation that reduces the use of natural resources and decreases the release of harmful substances across the whole life-cycle” (EIO, 2012).

To successfully respond to global social challenges and achieve a sustainable competitive advantage, R&D and innovation must form the basis for the creation of modern technical solutions (Janjić et al., 2022; Krstić et al., 2021; Radjenović et al., 2025). In globalized and contemporary conditions, for improving competitiveness and fostering sustainable development at local, national, and regional levels, eco-innovation represents a vital strategy. A sustainable innovation system depends heavily on absorptive capacity, which is primarily driven by extensive R&D activities (Miao et al., 2021). Variations in eco-innovation performance across countries are often linked to differences in R&D input levels. Therefore, R&D is fundamental for enabling sustainable innovations and meeting climate change objectives (Sarpong et al., 2023).

Improving eco-innovation performance requires an integrated approach that links public support, business capacity, and strategic investments in R&D. Increased environmental awareness at the national and global level has highlighted the importance of establishing effective and binding environmental protection policies (Peyravi et al., 2023). Command-type legal regulations can oblige enterprises to implement measures to protect the environment and reduce its degradation, promoting them to adopt environmentally friendly materials and technologies, thereby improving eco-innovation performance and environmental efficiency (Zhiwei et al., 2025). Enhancing eco-innovation performance can be achieved by implementing an eco-organizational management system and adopting eco-friendly operating processes/products (Phey et al., 2021). R&D activities can reduce the emission of pollutants and contribute to the efficient utilisation of natural resources, which in turn can enhance green innovation performance (Du et al., 2019).

R&D investments are crucial for the economic and social development of each country (Biurrun, 2022). Public support for R&D plays an essential role in national economic policies and represents one of the key determinants of sustainable economic growth (Hameed et al., 2018) and enhancing eco-innovation performance. In the CE agenda, the government is a major actor and, through public policies in the form of financial support and regulations, can help mitigate the impact of economic activities on the environment and positively influence the CE (Arthur et al., 2023). The member states' governments of the EU are increasingly investing in public R&D in the energy and environmental sectors to advance the circular economy and achieve its goals (Garrido-Prada et al., 2021). On the other hand, R&D activities have been considered as a key driver of the policy transition toward a more resource-efficient and circular economy.

## 4.3 Digital Transformation's Role in the Circular Economy

The twenty-first century has witnessed two parallel revolutions that are fundamentally reshaping economic systems, business models, and environmental management approaches. The first is the digital transformation revolution, characterized by the rapid advancement and deployment of Industry 4.0 technologies that are creating unprecedented capabilities for data collection, analysis, and automated decision-making. The second is the circular economy revolution, driven by growing recognition of the unsustainability of linear "take-make-dispose" economic models and the urgent need for regenerative approaches that minimize waste and maximize resource utilization (Kirchher et al., 2017). The concept of digital transformation has evolved significantly since its initial emergence in business and management literature, where it was primarily understood as the use of digital technologies to improve business processes and customer experiences (Rosa et al., 2020).

In the circular economy context, digital transformation takes on additional dimensions related to environmental impact, resource efficiency, and systemic change that extend beyond individual organizational boundaries. Contemporary understanding of digital transformation in circular economy contexts recognizes several key characteristics that distinguish it from traditional approaches to technology adoption. One of them involves the integration of multiple digital technologies into coherent systems that can address complex, multi-dimensional challenges related to resource flows, waste management, and sustainable production. In addition, it also requires new forms of collaboration and data sharing across organizational and sectoral boundaries, challenging traditional approaches to competitive advantage and intellectual property protection. Digital transformation approach in circular economy contexts often involves the creation of new business models and value propositions that fundamentally alter the relationship between producers, consumers, and other stakeholders. These new models frequently rely on continuous data flows and real-time optimization rather than discrete transactions and static product offerings. Last but not least, it requires new approaches to governance and regulation that can address the complex interactions between digital technologies, environmental outcomes, and social impacts.

Industry 4.0, often referred to as the Fourth Industrial Revolution, represents a fundamental transformation in manufacturing and industrial processes characterized by the integration of cyber-physical systems, Internet of Things (IoT) technologies, blockchain, cloud computing, and artificial intelligence (Lopes de Sousa Jabbour et al., 2018). While Industry 4.0 was initially conceptualized primarily in terms of manufacturing efficiency and competitiveness, its implications for circular economy implementation have become increasingly apparent as the technologies have matured and been deployed at scale. The integration of Industry 4.0 and circular economy principles is also driving the development of new business models that challenge traditional approaches to manufacturing and industrial organization. Product-as-a-service models, enabled by IoT monitoring and predictive maintenance, allow manufacturers to retain ownership of products while providing functionality to customers. This model creates incentives for designing products for durability and repairability while providing manufacturers with detailed data about product performance and usage patterns. The Internet of Things represents perhaps the most fundamental technological enabler of digital-circular economy integration, providing the sensory infrastructure necessary for real-time monitoring, tracking, and optimization of resource flows throughout economic systems. IoT technologies create unprecedented visibility into previously opaque processes, enabling the identification of inefficiencies, waste streams, and optimization opportunities that were previously invisible or economically unfeasible to address. However, achieving this seamless integration is a major technical hurdle. One of the most significant challenges is interoperability, the ability of different digital systems, devices, and applications to connect, communicate, and exchange data in a coordinated manner. In a typical circular ecosystem, data must be shared among a diverse array of stakeholders, including manufacturers, logistics providers, consumers, and recyclers, who often use proprietary systems with different data formats, protocols, and architectures.

## 5. METHODOLOGICAL FRAMEWORK

In the context of the Eco-Innovation Index (EEI) and its sub-indicators, a ratio scale of measurement is used, complemented by normalization relative to the EU average. This permits comparisons between sub-indicators with different measurement units and consistent application of mathematical operations (Do, 2024). Many indicators, such as GHG emissions productivity (emissions per unit of GDP), ecological patents, academic publications, or the number of ISO 14001 certificates, possess equal intervals and a meaningful zero point (i.e., zero represents the absence of a measure), thereby meeting the criteria for a ratio scale (Stevens, 1946). Consequently, the ratio scale enables addition, subtraction, averaging, and computation of ratios, as outlined in Stevens's typology of measurement scales. Sub-indicators are collected from diverse sources and presented as absolute values. Each sub-indicator is normalized using the "distance-to-reference" method, setting the EU average at 100, where countries above the EU average score above 100, and those below score below 100. Following normalization, the unweighted average of all sub-indicators is calculated to yield the overall eco-innovation index for each country. In the EEI framework, the ratio scale supports quantitative interpretation of differences and relationships across countries or periods. As a composite index, EEI employs standardization (normalization) of sub-indicator values, since sub-indicators are expressed in different units and cannot be directly compared without conversion to a dimensionless scale. However, multiple normalization techniques have been chosen based on the nature of the data and the MCDM method. For EEI sub-indicators (which are based on a ratio scale), the min-max normalization is applied to establish a dimensionless scale. Given the heterogeneity of measurement units among sub-indicators measuring various phenomena, these must be rendered dimensionless before applying MCDM analysis. In the Eco-innovation methodology, normalization is performed per indicator, not across the entire dataset. Therefore, in the MCDM methods used in this study, normalization should be repeated to ensure compatibility among all criteria. To compare sub-indicators and subsequently use them as criteria for ranking countries, a dimension-less scale is created via min-max transformation (linear scaling), ensuring all inputs lie within the same range after scaling. This dimensionless scale allows fair comparison of indicators expressed in different units and prevents skewing caused by indicators with large numeric ranges. It also enables equal weighting, if weights are given as percentages summing to 1, while preserved dimensionlessness ensures proper application of weights.

The Multi-Criteria Decision-Making (MCDM) process is a structured approach for evaluating and ranking a set of alternatives based on multiple, often conflicting, criteria. When applied to the assessment of sustainability performance, such as the ranking of European countries or cities, the process becomes especially valuable for integrating diverse environmental, economic, and social indicators into a single decision-support framework. Methods such as PROMETHEE, CoCoSo, and ARAS follow the general logic of MCDM but differ in their mathematical underpinnings and sensitivity to criteria relationships. The outcomes of the research greatly rely on the chosen indicators, and utilizing other indicators may result in varying conclusions ( Rađenović & Boshkov, 2023). The applied methodology combines triangularity between the CoCoSo (Combined Compromise Solution) and ARAS (Application of Additive Ratio Assessment) MCDM methods, comparing their ranking outcomes with the aforementioned PROMETHEE method. The empirical ranking of sustainability and eco-innovation performance among EU countries was conducted using three Multi-Criteria Decision-Making (MCDM) methods: PROMETHEE, CoCoSo, and ARAS. Each technique brings distinct methodological advantages and constraints, rendering them suitable for varying decision-making contexts. (Table 1):

*Table 1. Applied MCDM comparison*

| Method | Strengths | Weaknesses |
|---|---|---|
| **CoCoSo** | + Combines additive and multiplicative models<br>+ Transparent scoring system | – Sensitive to outliers<br>– Can produce unstable rankings under inconsistent weighting |
| **PROMETHEE** | + Captures decision-maker's risk attitudes via fuzzy preference functions<br>+ Handles both gain and loss contexts (asymmetry) | – Requires careful selection of preference functions<br>– Computationally more intensive |
| **ARAS** | + Straightforward implementation<br>+ Intuitive comparison with ideal alternative | – Assumes linear utility functions<br>– May oversimplify multidimensional trade-offs |

Source: Authors' observation after literature review

Consequently, this research aims to rank EU27 countries based on a selection of EEI sub-indicators and identify the alternative that best exemplifies the synergy between eco-innovation and circular economy. The optimal alternative is defined as the one demonstrating significant progress in adopting eco-innovation principles during the analyzed five-year period from 2020 to 2024 (Figure 1).

*Figure 1. Descriptive statistics for the analyzed EEI sub-indicators*

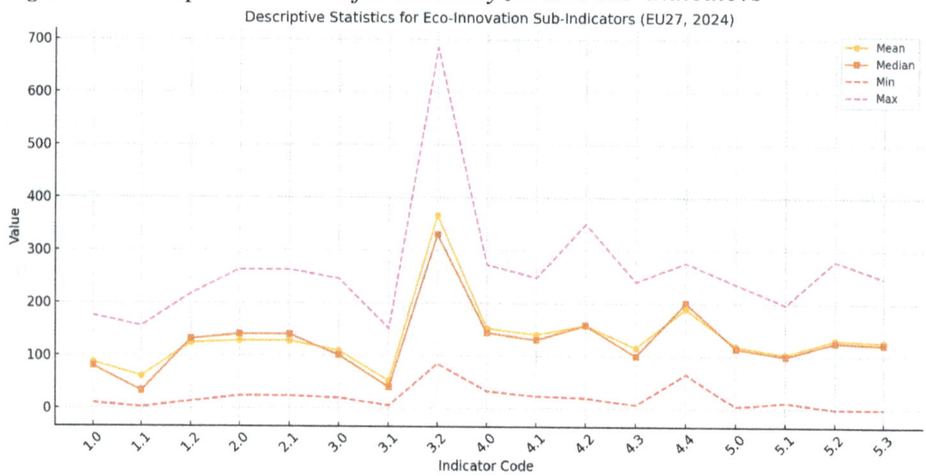


*Note*: The numbers on the x-axis represent the indicators whose names are given in Table 2.*
*Source: Authors' calculation*

Weight coefficients for these sub-indicators within the MCDM models were established using the CRITIC method. The CRITIC (CRiteria Importance Through Intercriteria Correlation) method provides an objective way to assign weights to indicators (Table 2) based on their variability and conflict among each other (Diakoulaki et al., 1995). This method is especially well-suited for assessing complex sustainability and eco-innovation systems. Unlike subjective weighting schemes such as AHP or expert-based scoring, CRITIC leverages statistical properties, standard deviation, and correlation to derive objective weights that capture both the informational richness and the discriminatory power of each criterion. In the context of eco-innovation, where numerous indicators span environmental, economic, technological, and social domains, CRITIC ensures that no single dimension is over- or under-emphasized without empirical justification. It systematically assigns higher weights to indicators that not only exhibit high variability across countries but also provide unique, non-redundant insights relative to others. For example, eco-innovation-related media coverage received the highest weight in our analysis, suggesting that public discourse and perception around innovation is both volatile and uniquely informative. Conversely, well-established metrics like eco-patents or academic publications, while important, display lower variance and higher correlations, leading to lower CRITIC weights. Applying CRITIC in combination with Multi-Criteria Decision-Making (MCDM) methods such as ARAS, CoCoSo, and PROMETHEE allows for more balanced and data-driven composite scores (Rađenović & Rajić, 2025). This is particularly beneficial for cross-country comparisons where political, institutional, and economic contexts differ substantially. The use of CRITIC minimizes the risk of bias and enhances the transparency and reproducibility of rankings. Ultimately, CRITIC supports the strategic alignment of policy tools with empirically dominant drivers of eco-innovation performance.

*Table 2. CRITIC method results for weighted coefficients*

| **1. Eco-Innovation Inputs** | |
|---|---|
| **1.0**: Government environmental and energy R&D appropriations and outlays (% of GDP) | 0.065 |
| **1.1**: Total R&D personnel and researchers (% of total employment) | 0.075 |
| **1.2**: Green early-stage investments (% of GDP) | 0.063 |
| **2. Eco-Innovation Activities** | |
| **2.0**: Implementation of sustainable products among SMEs (% of SMEs) | 0.062 |
| **2.1**: Number of ISO 14001 certificates (per million population) | 0.062 |
| **3. Eco-Innovation Outputs** | |
| **3.0**: Eco-innovation-related patents (per million population) | 0.042 |
| **3.1**: Eco-innovation-related academic publications (per million population) | 0.042 |
| **3.2**: Eco-innovation-related media coverage (per million population) | 0.076 |
| **4. Resource Efficiency Outcomes** | |
| **4.0**: Material productivity (GDP/Domestic Material Consumption) | 0.042 |

continued on following page

*Table 2. Continued*

| 1. Eco-Innovation Inputs | |
|---|---|
| **4.1**: Water productivity (GDP/Total freshwater abstraction) | 0.071 |
| **4.2**: Energy productivity (GDP/Gross available energy) | 0.069 |
| **4.3**: GHG emission productivity (GDP/Total GHG emissions) | 0.044 |
| **4.4**: Waste generation per capita (kg per capita) | 0.072 |
| **5. Socio-Economic Outcomes** | |
| **5.0**: Employment in eco-industries (% of total employment) | 0.049 |
| **5.1**: Turnover in eco-industries (% of total turnover) | 0.051 |
| **5.2**: Exports of eco-innovation products (% of total exports) | 0.063 |
| **5.3**: Number of green jobs created (per million population) | 0.054 |
| **Total value of weighting coefficients:** | 1.000 |

Source: Authors' calculation

## 6. SOLUTIONS AND RECOMMENDATIONS

Circular economy and eco-innovation indicators can be powerful instruments that provide valuable information on the progress of the initiatives and policies related to sustainable development. Denmark, Sweden, Luxembourg, France, and Germany top the eco-innovation rankings. Denmark leads in both ARAS and CoCoSo due to its systemic focus on circular economy, SME green innovation adoption, and water/energy productivity (Table 3). Ranked as the first one across both ARAS and CoCoSo, though surprisingly ranked 20th in Φ, this divergence suggests strong consistency in weighted scores but less distinctiveness in directional indicator space. Denmark's eco-innovation strategy is embedded in its Circular Economy Action Plan, combining R&D investments, green job creation, and export-led environmental tech solutions. Its leadership in resource efficiency (water, energy) and high adoption of green technologies among SMEs is well-documented. Sweden exhibits strength in green technology deployment and clean mobility policies. Strong performance across all MCDM rankings, Sweden's CoCoSo and ARAS scores reflect a balanced eco-innovation ecosystem. A robust high-tech research ecosystem, strong public-private collaboration in clean technology, and ambitious national targets for climate neutrality and sustainable mobility underpin Sweden's policy framework. Luxembourg's standout performance stems from digital and green finance synergies. The second place in ARAS and the third place in CoCoSo, but a lower Φ rank suggests a specialized profile, possibly excelling in specific niche indicators such as green finance or circularity, while being more neutral in others.

France balances industrial decarbonization and innovation scaling, while Germany continues leveraging industrial transformation, albeit with slower indicator diversity. France holds a consistent position across all methods, suggesting broad but not radical leadership in eco-innovation. Its strategy includes industrial decarbonization, circular economy initiatives, and the Greentech Innovation program, targeting green startups and scaling. On the other hand, Germany's score reflects strong industrial eco-efficiency and innovation capacity, but its lower Φ rank may indicate less diversity in indicator performance or a heavy reliance on traditional sectors that are adapting gradually. Namely, policies such as the Resource Efficiency Programme (ProgRess) and funding for industrial transformation are key levers.

*Table 3. Ranking comparison through the different applied MCDM methods according to EEI sub-indicators values and their weights.*

| Country | ARAS Score | ARAS Rank | CoCoSo Score | CoCoSo Rank | Phi Score (PC1) | Phi Rank |
|---|---|---|---|---|---|---|
| **Denmark** | 0.73 | 1 | 1.00 | 1 | -0.06 | 20 |
| **Sweden** | 0.70 | 3 | 0.73 | 2 | -0.04 | 17 |
| **Luxembourg** | 0.71 | 2 | 0.54 | 3 | -0.05 | 19 |
| **France** | 0.6 | 4 | 0.41 | 4 | -0.07 | 22 |
| **Germany** | 0.59 | 5 | 0.41 | 5 | -0.06 | 21 |
| **Netherlands** | 0.56 | 6 | 0.39 | 6 | -0.05 | 18 |
| **Czechia** | 0.56 | 7 | 0.38 | 7 | 0.01 | 10 |
| **Slovenia** | 0.55 | 8 | 0.38 | 8 | -0.03 | 16 |
| **Spain** | 0.54 | 9 | 0.37 | 9 | -0.01 | 14 |
| **Estonia** | 0.52 | 10 | 0.35 | 10 | 0.04 | 3 |
| **Lithuania** | 0.50 | 11 | 0.34 | 11 | 0.03 | 8 |
| **Latvia** | 0.49 | 12 | 0.34 | 12 | -0.0 | 13 |
| **Belgium** | 0.46 | 13 | 0.31 | 13 | -0.03 | 15 |
| **Slovakia** | 0.43 | 14 | 0.29 | 14 | 0.03 | 7 |
| **Croatia** | 0.42 | 15 | 0.29 | 15 | 0.03 | 6 |
| **Cyprus** | 0.42 | 16 | 0.29 | 16 | 0.04 | 4 |
| **Malta** | 0.42 | 17 | 0.29 | 17 | 0.0 | 12 |
| **Greece** | 0.39 | 18 | 0.27 | 18 | 0.01 | 11 |
| **Romania** | 0.34 | 19 | 0.23 | 19 | 0.05 | 2 |
| **Hungary** | 0.3 | 20 | 0.2 | 20 | 0.03 | 5 |
| **Poland** | 0.28 | 21 | 0.19 | 21 | 0.03 | 9 |
| **Bulgaria** | 0.27 | 22 | 0.19 | 22 | 0.07 | 1 |

Source: Authors' calculation

The Phi coefficient (PC1) reflects each country's relative performance along the sustainability preference axis. Higher values indicate more substantial alignment with eco-innovation drivers. Bulgaria, Romania, and Estonia rank highest in 2024, while several Western economies are positioned lower due to indicator variance or normalization effects. The disparities in performance are primarily the result of different starting positions of countries' development (Rađenović et al., 2020). The GAIA chart displays countries in a two-dimensional space based on their performance profiles across 17 eco-innovation sub-indicators. Red vectors represent the influence of each criterion on the two principal components. Proximity between countries reflects similarity, while distance from the origin indicates strength or uniqueness in performance across indicators (Figure 2).

*Figure 2. GAIA matrix quadrants with EEI sub-indicators' influence on the EU27 ranking*

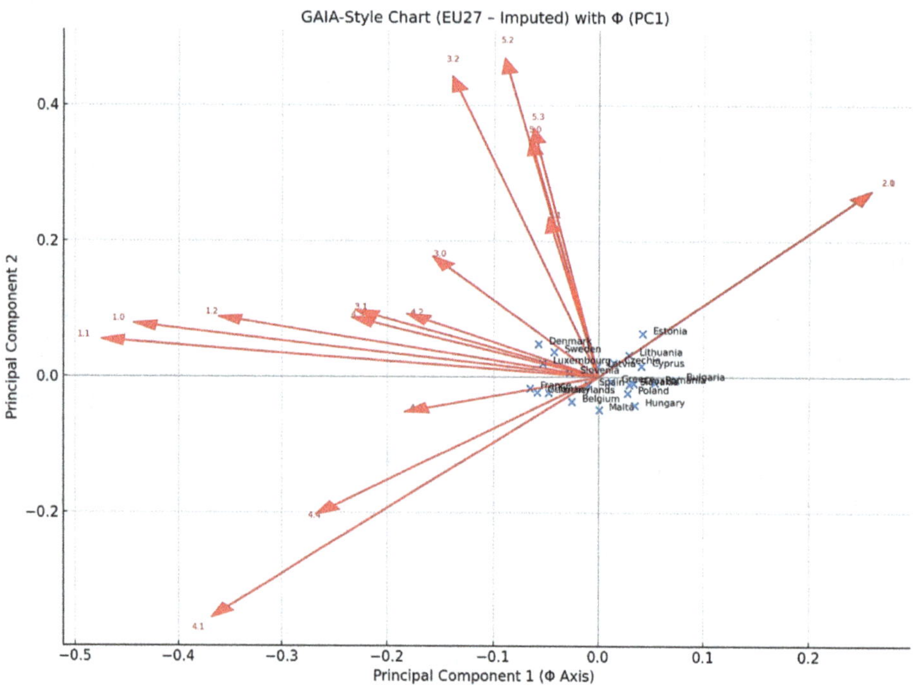


*Source: Authors' calculation*

The PROMETHEE network (Figure 3) chart illustrates the simulated pairwise preference flows between the top 10 countries based on Φ (PC1 from PCA). Arrow thickness and annotated values indicate the strength of dominance from one country over another. This enables strategic comparisons and benchmarking. Bulgaria appears to occupy a central, dominant position in the network, indicated by numerous outgoing arrows to other countries. This suggests that Bulgaria has a strong net positive flow ($\Phi^{+}$) relative to most other EU member states in the ranking. Bulgaria's dominance could be attributed to strong relative performance on specific criteria with high CRITIC weights, such as waste generation per capita, water productivity, or ISO 14001 certification density, especially if these indicators exhibit both high variance and low correlation with others.

*Figure 3. PROMETHEE Network for the analyzed EU27 based on the eco-innovation performance Phi flow*

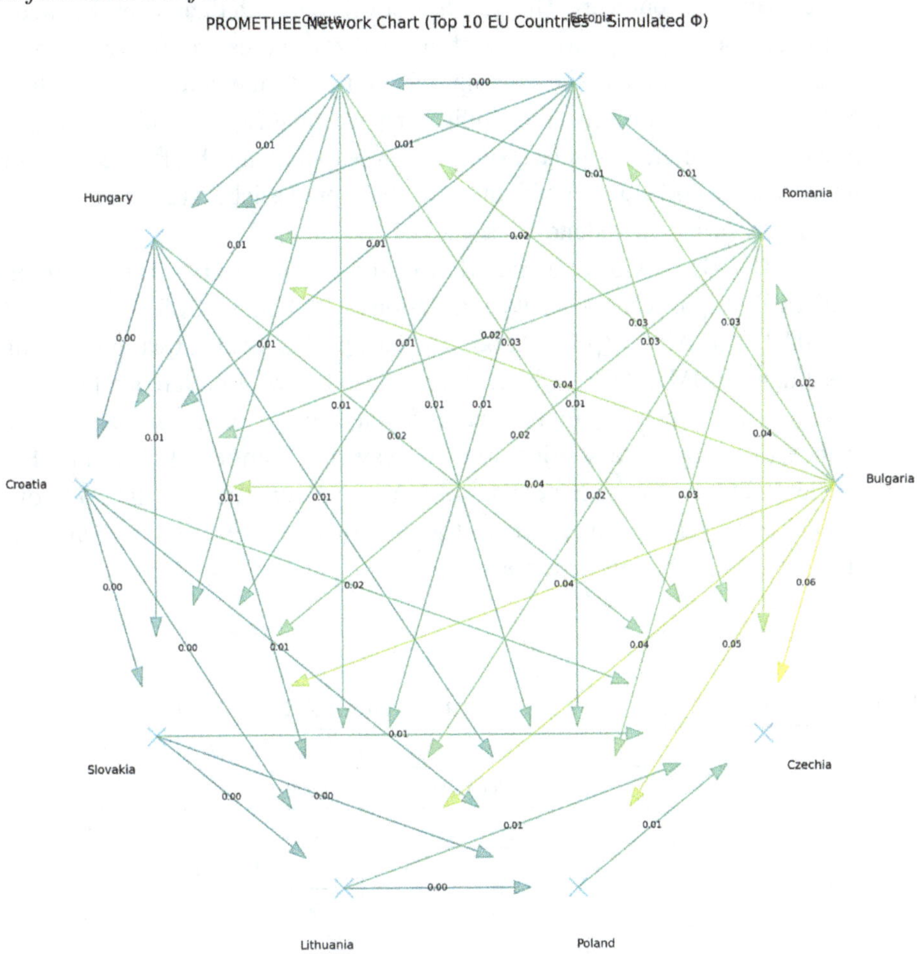


*Source: Authors' calculation*

The Spearman's rank correlation (Table 4) coefficients provide insights into the convergent and divergent tendencies among the three Multi-Criteria Decision-Making (MCDM) methods applied: CoCoSo, ARAS, and PROMETHEE. Each method reflects a distinct conceptual and computational logic for aggregating multiple eco-innovation indicators, and the observed correlations reveal how closely their resulting rankings align:

- CoCoSo vs. ARAS: A strong positive correlation is observed between CoCoSo and ARAS rankings ($\rho = 0.832$, $p < 0.01$), suggesting a high level of concordance in how these two additive-utility-based models evaluate the performance of countries. This result is consistent with their shared structural logic, where aggregation through normalized scores and proximity to an ideal solution dominate the ranking behavior. Both methods appear to favor countries with uniformly high indicator performance, particularly in areas like resource efficiency and eco-innovation outputs. The observed correlation illustrates the collective contribution of the input variables to the variance in the output variable.(Rađenović et al., 2020).
- PROMETHEE vs. CoCoSo and ARAS: In contrast, PROMETHEE exhibits statistically significant negative correlations with both CoCoSo ($\rho = -0.806$, $p < 0.01$) and ARAS ($\rho = -0.563$, $p < 0.01$). This inverse relationship suggests that PROMETHEE, which incorporates pairwise preference flows and gain-loss asymmetries, highlights different aspects of performance dominance that aggregate scoring models may not capture. PROMETHEE is particularly sensitive to outlier performances and criteria-specific trade-offs, which likely leads to rank reversals for countries that score exceptionally well in a few criteria but underperform in others.

*Table 4. Spearman's rank correlation for the applied MCDM methods*

| **MCDM** | | | **CoCoSo** | **ARAS** | **PROMETHEE** |
|---|---|---|---|---|---|
| **Spearman's rho** | **CoCoSo** | Correlation Coefficient | 1,000 | ,832** | -,806** |
| | | Sig. (2-tailed) | . | ,000 | ,000 |
| | **ARAS** | Correlation Coefficient | ,832** | 1,000 | -,563** |
| | | Sig. (2-tailed) | ,000 | . | ,006 |
| | **PROMETHEE** | Correlation Coefficient | -,806** | -,563** | 1,000 |
| | | Sig. (2-tailed) | ,000 | ,006 | . |

**. Correlation is significant at the 0.01 level (2-tailed).

Source: Authors' calculation

## 7. FUTURE RESEARCH DIRECTIONS

Several avenues for future research can enhance the analytical depth, policy relevance, and methodological robustness of such assessments. Future research could be focused on expanding the scope of indicators used to rank EU countries. Future studies can further enhance the understanding and evaluation of eco-innovation by linking eco-innovation indicators to the Sustainable Development Goals (SDGs), thereby strengthening the policy relevance and international comparability of the results. To capture a more comprehensive picture of CE dynamics and to understand different development paths towards circular models, future research could apply similar analysis to other regions outside the European Union, such as the Western Balkans, North Africa, and emerging markets. It is highly recommended to perform a cross-sectoral analysis of industry-specific patterns not visible in this macro-level research.

Expanding beyond the current application of ARAS, CoCoSo, and PROMETHEE, future studies could test hybrid or ensemble MCDM models, for example, combining PROMETHEE with AHP or ELECTRE for more context-sensitive ranking, or integrating entropy and CRITIC for comparative weighting validation. Further methodological refinement is needed through sensitivity and robustness testing across multiple MCDM algorithms, normalization techniques, and weight perturbations. This would ensure that rankings are stable and resilient under different modeling assumptions. While this study focused on EU-level country comparisons, future research could explore regional-level rankings (NUTS2 regions) or sector-specific eco-innovation performance (bioeconomy, energy tech, or sustainable transport). This would increase granularity and inform more targeted policy design. Additionally, incorporating stakeholder-derived preference functions using participatory methods like AHP, Delphi, or conjoint analysis could align the evaluation more closely with policy priorities and citizen values.

## 8. IMPLICATIONS FOR POLICYMAKERS

Although eco-innovation is a key factor in establishing a circular model, the success of this process could depend on a few policy instrument gaps - overcoming the lack of specific financial instruments for SMEs, insufficient digital infrastructure for CE monitoring, the absence of synergy between sectoral policies, and a lack of support for international cooperation.

To support companies and SMEs investing in eco-innovations, policymakers should primarily develop and expand the application of economic instruments that support enterprises in developing and adopting eco-innovative solutions. The introduction of policies, such as green innovation vouchers, green subsidies, tax reliefs, and favourable green credits, can significantly reduce financial barriers and encourage circular innovation in the economy.

Decision makers should identify existing technological gaps and direct efforts towards the development of interoperable digital platforms at the national level, which would create the conditions for a reliable, efficient application of circular economy principles and support CE value chains and real-time data collection. To ensure synergy between the circular economy and digitalisation, it is essential that decision-makers from different sectors cooperate, as well as with other key stakeholders (business, civil society). Policymakers should create programs to develop export-oriented innovation centres and innovation hubs that focus on circular solutions.

Based on the results of multi-criteria methods for eco-innovation ranking, decision makers at the national and regional level should encourage strategic cooperation between countries with developed and balanced eco-innovation ecosystems (like Denmark and Sweden) and countries that excel in specific segments of the circular economy – such as digital technologies (e.g., Estonia), green financial mechanisms (e.g., Luxembourg), or efficient waste management and certification (e.g., Bulgaria).

## 9. CONCLUSION

The transition to a circular economy is undoubtedly a complex and challenging activity for all actors involved in this process - from national authorities to economic entities and consumers. In the conditions in which modern society finds itself today, the implementation of mechanisms specific to the circular economy is no longer just an option, but an obligation, to ensure long-term sustainable development and a future with fewer environmental problems. Due to the macroeconomic impacts of the transition to a new, circular economic model, it is becoming increasingly popular among business and government leaders. The implementation of circular economy practices creates a long-term sustainable system, which seeks to provide a favorable framework for the development of all economic activities with radically reduced adverse effects. For the transition to sustainability to be successful, synergistic action is needed from all actors in society.

These findings highlight an important methodological consideration: method choice substantially influences outcome interpretation. CoCoSo and ARAS provide a more consensus-oriented ranking. In contrast, PROMETHEE introduces a differentiated perspective, especially valuable when decision-makers are interested in identifying relative performance gaps and risk-adjusted rankings. Given the observed divergence, especially between PROMETHEE and the other two methods, it is advisable in future research to use triangulation or ensemble approaches to synthesize findings. Another truly significant aspect is that methods such as PROMETHEE, VIKOR, CRITIC, TOPSIS, and others are utilized. Doing so can enhance the robustness and interpretability of eco-innovation rankings, particularly when informing high-stakes policy decisions. It can improve further analysis in this area of research related to composite indices creation using MCDM methods (Vranjanac et al. 2023, p.82). PROMETHEE's sensitivity to asymmetric dominance and pairwise comparison favors countries that consistently outperform others, even if not uniformly excellent across all indicators. Bulgaria may be benefiting from this effect if it exceeds thresholds in specific strategic dimensions that PROMETHEE amplifies through its preference modeling. Bulgaria's positioning signals that even countries traditionally perceived as lower-performing in aggregate MCDM models like ARAS or CoCoSo can emerge as leaders under pairwise, risk-aware assessments such as PROMETHEE. Both Denmark and Finland exhibited strong mid-to-high rankings across MCDM methods, but didn't dominate any specific index. Their consistent performance in resource efficiency and innovation infrastructure suggests opportunities for global leadership if targeted actions are taken to promote cross-border partnerships in green technologies and digital circular systems. Furthermore, scale up eco-startup ecosystems via green venture capital and export-oriented innovation hubs, encouraging circular procurement models across public sectors to amplify domestic demand for sustainable solutions. This case reinforces the need for method triangulation because what is visible as dominance in PROMETHEE may not hold in additive models. The choice of MCDM method, therefore, has a direct impact on strategic insights derived for policy formulation.

## ACKNOWLEDGMENT

The paper is a part of research financed by the Ministry of Science, Technological Development, and Innovation of the Republic of Serbia, as agreed in decision no. 451-03-136/2025-03/200371.

## REFERENCES

Alka, T. A., Raman, R., & Suresh, M. (2024). Research trends in innovation ecosystem and circular economy. *Discover Sustainability*, *5*(1), 323. DOI: 10.1007/s43621-024-00535-5

Androniceanu, A., & Sabie, O. M. (2022). Overview of green energy as a real strategic option for sustainable development. *Energies*, *15*(22), 8573. DOI: 10.3390/en15228573

Arthur, E. A., Gyamfi, S., Gerstlberger, W., Stejskal, J., & Prokop, V. (2023). Towards Circular Economy: Unveiling Heterogeneous Effects of Government Policy Stringency, Environmentally Related Innovation, and Human Capital within OECD Countries. *Sustainability*, *15*(6), 4959. DOI: 10.3390/su15064959

Biurrun, A. (2022). New evidence toward solving the puzzle of innovation and inequality. The role of institutions. *Economics of Innovation and New Technology*, *31*(8), 729–750. DOI: 10.1080/10438599.2020

Bril, H., Kell, G., Rasche, A., . . .. (2023). *Sustainability, Technology, and Finance Rethinking How Markets Integrate ESG*. Routledge. https://www.routledge.com/Sustainability-Technology-and-Finance-Rethinking-How-Markets-Integrate-ESG/Bril-Kell-Rasche/p/book/9781032200545?srsltid=AfmBOopGnYcjquS0_WPY9j5khy4n9yPtxoSXjACMEEwWbJy1a86E041X

Carrillo-Hermosilla, J., del Río, P., & Könnölä, T. (2010). Diversity of eco-innovations: Reflections from selected case studies. *Journal of Cleaner Production*, *18*(10-11), 1073–1083. DOI: 10.1016/j.jclepro.2010.02.014

Constantin, A.-M., Voica, O.-M., Silvestru, C.-I., Icociu, V.-C., Silvestru, R.-C., & Grecu, A. (2023). The Efficiency of Resource Utilization in EU-28 Countries through Eco-Innovation and Digital Inclusion. *Sustainability*, *15*(24), 16924. DOI: 10.3390/su152416924

De Jesus, A., Lammi, M., Domenech, T., Vanhuyse, F., & Mendonça, S. (2021). Eco-Innovation Diversity in a Circular Economy: Towards Circular Innovation Studies. *Sustainability, 13*(19), 10974. DOI: 10.3390/su131910974

Diakoulaki, D., Mavrotas, G., & Papayannakis, L. (1995). Determining objective weights in multiple criteria problems: The CRITIC method. *Computers & Operations Research, 22*(7), 763–770. DOI: 10.1016/0305-0548(94)00059-H

Do, D. T. (2024). Assessing the impact of criterion weights on the ranking of the top ten universities in Vietnam. *Engineering, Technology &. Applied Scientific Research, 14*(4), 14899–14903. DOI: 10.48084/etasr.7607

Du, J. L., Liu, Y., & Diao, W. X. (2019). Assessing regional differences in green innovation efficiency of industrial enterprises in China. *International Journal of Environmental Research and Public Health, 16*(6), 940. DOI: 10.3390/ijerph16060940 PMID: 30875986

Durán-Romero, G., López, A. M., Beliaeva, T., Ferasso, M., Garonne, C., & Jones, P. (2020). Bridging the gap between circular economy and climate change mitigation policies through eco-innovations and Quintuple Helix Model. *Technological Forecasting and Social Change, 160*(120246), 1–13. DOI: 10.1016/j.techfore.2020.120246

EESC. (2016). *The Circular Economy: Beneficial for All. The Threefold Challenge: Climate Change, Competitiveness & Dependence on Raw Materials*. European Economic and Social Committee., DOI: 10.2864/442377

EIO. (2012). *Closing the eco-innovation gap: An economic opportunity for business*. Annual Report 2011. Brussels. Eco-Innovation Observatory. European Commission.

Ellen MacArthur Foundation (EMF). (2025). *What is a circular economy?* https://www.ellenmacarthurfoundation.org/topics/circular-economy-introduction/overview

European Commission. (2012). *Eco-innovation: The key to Europe's future competitiveness* (EUR 25271 EN). Publications Office of the European Union. https://doi.org/DOI: 10.2777/31522

European Commission. (2020). *Circular economy action plan: for a cleaner and more competitive Europe*. Directorate-General for Communication, Publications Office of the European Union. https://data.europa.eu/doi/10.2779/05068

European Parliament. (2023). *Circular economy: definition, importance, and benefits*. https://www.europarl.europa.eu/topics/en/article/20151201STO05603/circular-economy-definition-importance-and-benefits

Garrido-Prada, P., Lenihan, H., Doran, J., Rammer, C., & Perez-Alaniz, M. (2021). Driving the circular economy through public environmental and energy R&D: Evidence from SMEs in the European Union. *Ecological Economics, 182*, 106884. DOI: 10.1016/j.ecolecon.2020.106884

Geissdoerfer, M., Savaget, P., Bocken, N. M., & Hultink, E. J. (2017). The Circular Economy–A new sustainability paradigm? *Journal of Cleaner Production, 143*, 757–768. DOI: 10.1016/j.jclepro.2016.12.048

Ghormare, R., Fatima, S., Grover, P., Phutela, N., Kandpal, V., & Santibanez Gonzalez, E. (2024). Exploring the paradigm shift towards sustainability: A systematic literature review on circular economy and eco-innovation. *AIMS Environmental Science, 11*, 940–959. DOI: 10.3934/environsci.2024047

Hajdukiewicz, A., & Pera, B. (2023). Eco-innovation in the European Union: Challenge for catching-up economies. *Entrepreneurial Business and Economics Review, 11*(1), 145–164. DOI: 10.15678/EBER.2023.110108

Hák, T., Janoušková, S., & Moldan, B. (2016). Sustainable development goals: A need for relevant indicators. *Ecological Indicators, 60*, 565–573. DOI: 10.1016/j.ecolind.2015.08.003

Hameed, T., von Staden, P., & Kwon, K. (2018). Sustainable Economic Growth and the Adaptability of a National System of Innovation: A Socio-Cognitive Explanation for South Korea's Mired Technology Transfer and Commercialization Process. *Sustainability, 10*, 1397. DOI: 10.3390/su10051397

Hojnik, J., & Ruzzier, M., Konečnik, Ruzzier, M., Sučić, B., Soltwisch, B., & Rus, M. (2024). Review of EU projects with a focus on environmental quality: Innovation, eco-innovation, and circular-economy elements. *International Journal of Innovation Studies, 8*(1), 1–12. DOI: 10.1016/j.ijis.2023.10.001

Holly F., Kolar G., Berger M., Fink S., Ogonowski P., & Schlund S. (2023). Challenges on the way to a circular economy from the perspective of the Austrian manufacturing industry. *Frontiers in Sustainability, 4*, 1243374. DOI=DOI: 10.3389/frsus.2023.1243374

Janjić, I., Krstić, B., & Milanović, S. (2022). The Impact of R&D Activity on the Business Performance of High-Technology Companies. *Facta Universitatis. Series. Economics and Organization, 19*(4), 253–271. DOI: 10.22190/fueo220707018j

Juracka, D., Nagy, M., Zvarikova, K., & Popescu, G. (2024). Eco-Innovation: Analysis of EU Member States Performance in the EU Eco-Innovation Index 2013–2022. *Systems, 12*(11), 452. DOI: 10.3390/systems12110452

Karim, R., Roshid, M., & Waaje, A. (2024b). Circular economy and supply chain sustainability. In Mizrak, F. (Ed.), *Strategic Innovations for Dynamic Supply Chains* (1st ed., pp. 1–30). IGI Global Publishing., DOI: 10.4018/979-8-3693-3575-8.ch001

Karim, R., Waaje, A., & Roshid, Md. M., & Yeamin, Md. B. (2024a). *Turning the Waste Into Wealth: Progressing Toward Global Sustainability Through the Circular Economy in Waste Management.* In R. Sucheran (Ed.), Sustainable Waste Management in the Tourism and Hospitality Sectors (pp. 507–552). IGI Global Scientific Publishing. https://doi.org/DOI: 10.4018/979-8-3693-6110-8.ch019

Karim, R., Waaje, A., & Roshid, Md. M. (2025). *Digital Circular Economy for Attaining Sustainable Development Goals.* In Durmus Çagri Yildirim, Seda Yildirim, & Vinay Kandpal (Eds.), Effects of Digitalization and Circular Economy on Sustainable Policy and Climate Change Prevention (pp. 211–240). IGI Global. https://doi.org/DOI: 10.4018/979-8-3693-9909-5.ch008

Kemp, R., Arundel, A., Rammer, C., Miedzinski, M., Tapia, C., Barbieri, N., Tűrkeli, S., Bassi, A. M., Mazzanti, M., Chapman, D., Diaz López, F., & McDowall, W. (2019). *Maastricht Manual on Measuring Eco-Innovation for a Green Economy.* Innovation for Sustainable Development Network. Maastricht, The Netherlands. https://www.inno4sd.net/uploads/originals/1/inno4sd-pub-mgd-02-2019-fnl-maastrich-manualecoinnovation-isbn.pdf

Khan, O., Daddi, T., & Iraldo, F. (2020). Microfoundations of dynamic capabilities: Insights from circular economy business cases. *Business Strategy and the Environment, 29*(3), 1479–1493. DOI: 10.1002/bse.2447

Kirchherr, J., Reike, D., & Hekkert, M. (2017). Conceptualizing the circular economy: An analysis of 114 definitions. *Resources, Conservation and Recycling, 127*, 221–232. DOI: 10.1016/j.resconrec.2017.09.005

Kobryń, Andrzej & Prystrom, Joanna. (2017). Multi-Criteria Evaluation of the Eco-Innovation Level in the European Union Countries. *Problemy Ekorozwoju*. 12. 15-26. https://www.researchgate.net/publication/316853533

Krstić, B., Janjić, I., Jovanović, M., & Milanović, S. (2021). The Significance of R&D Activities for Managing Intellectual Capital of Enterprises. *BH Ekonomski Forum, 14*(1), 117–132. https://doi.org/DOI: 10.5937/bhekofor2101117K

Latif, M. N., Ivanović Đukić, M., Talić, M., & Stefanović, S. (2025). FINANCIAL EFFECTS OF THE CIRCULAR ECONOMY: HOW CIRCULAR ECONOMY PRACTICES BOOST PERFORMANCE OF SMALL AND MEDIUM ENTERPRISES IN SELECTED EUROPEAN SOUTHEAST COUNTRIES. *Economics of sustainable development, 9*(1), 75-85. DOI: 10.5937/ESD2501075N

Lopes de Sousa Jabbour, A. B., Jabbour, C. J. C., Godinho Filho, M., & Roubaud, D. (2018). Industry 4.0 and the circular economy: A proposed research agenda and original roadmap for sustainable operations. *Annals of Operations Research, 270*(1–2), 273–286. DOI: 10.1007/s10479-018-2772-8

Miao, M., Saide, S., Ratna, S., & Muflih, M. (2021). Business Continuity Innovation in Disruption Time: Sociotechnical Systems, Business Analytics, Virtual Business, and Mediating Role of Knowledge Absorptive Capacity. *IEEE Transactions on Engineering Management, 71*, 1–12. DOI: 10.1109/TEM.2021.3086306

Milanović, S., Talić, M., & Rađenović, Ž. (2022). Social entrepreneurial intention factors of tourism management students. In V. Bevanda & S. Štetić (Eds.) 6th International Thematic Monograph: *Modern Management Tools and Economy of Tourism Sector in Present Era.* UDEKOM: Belgrade, 567–583. https://doi.org/DOI: 10.31410/tmt.2021-2022.567

Pakulska, J., & Rutkowska, M. (2024). Resource efficiency indicators in measuring eco-innovation in EU countries. *Economics and Environment, 87*(4), 446, 1-14. https://doi.org/DOI: 10.34659/eis.2023.87.4.446

Peyravi, B., Peleckis, K., & Jakubavičius, A. (2023). Eco-Innovation Performance of Lithuania in the Context of European Environmental Policy: Eco-Innovation Indicators and Efficiency. *Sustainability, 15*(4), 3139. DOI: 10.3390/su15043139

Phey, C. C., Cheah, J., & Amran, A. (2021). The moderating effect of market turbulence in the Malaysian technology industry. *Journal of Cleaner Production, 283*, 124556. DOI: 10.1016/j.jclepro.2020.124556

Pichlak, M., & Szromek, A. R. (2022). Linking eco-innovation and circular economy: A conceptual approach. *Journal of Open Innovation, 8*(3), 1–14. DOI: 10.3390/joitmc8030121

Prieto-Sandoval, V., Jaca, C., & Ormazabal, M. (2018). Towards a consensus on the circular economy. *Journal of Cleaner Production, 179*, 605–615.

Rađenović, T., Krstić, B., Janjić, B., & Jovanović-Vujatović, M. (2023). The Effects of R&D Performance on the Profitability of Highly Innovative Companies. *Strategic Management, 28*(3), 34–45. DOI: 10. 5937 /straman2200034r

Rađenović, Ž., & Boshkov, T. (2023). SMART CITIES SUSTAINABILITY ACHIEVEMENTS RANKING: PCA WEIGHTING. *SCIENCE International Journal, 2*(4), 141–146. DOI: 10.35120/sciencej0204141r

Rađenović, Ž., Janjić, I., & Talić, M. (2024). Targeting eco-innovation performance among EU countries: Cluster outcome. *SCIENCE. International Journal (Toronto, Ont.)*, *3*(1), 189–194. DOI: 10.35120/sciencej0301189r

Rađenović, Ž., Janjić, I., Talić, M., Đokić, M., Rađenović, T., Boskov, T., & Krstić, B. (2025). Digital mapping of business performance indicators of agricultural holdings in Serbia. *Ekonomika Poljoprivrede*, *72*(1), 225–240. DOI: 10.59267/ekoPolj2501225R

Rađenović, Ž., Krstić, B., & Marković, M. (2020). SMART FARMING IN AGRICULTURAL INDUSTRY: MOBILE TECHNOLOGY PERSPECTIVE. *Ekonomika Poljoprivrede*, *67*(3), 925–938. DOI: 10.5937/ekoPolj2003925R

Rađenović, Ž. J., & Rajić, M. N. (2025). Two-step cluster analysis for energy performance indicators comparison. *Thermal Science*, (00), 56–56. DOI: 10.2298/TSCI250205056R

Radjenović, Ž., Talić, M., & Janjić, I. (2025). Data Analysis on Greening SMEs through Resource Efficiency: K-Means Algorithm. Trivedi, S., Balusamy, B., Gernal, L.M., & Al-Khasawneh, M.A. (Eds.) *Green Engineering for Optimizing Firm Performance: AI and Automation for Sustainable Technologies* (1st ed.). In Chapman and Hall/CRC. https://doi.org/DOI: 10.1201/9781003626527

Rashid, S., & Malik, S. H. (2023). Transition from a Linear to a Circular Economy. In Bandh, S. A., Mall, F. A., & Hoang, A. T. (Eds.), *Renewable Energy in Circular Economy. Circular Economy and Sustainability*. Springer., DOI: 10.1007/978-3-031-42220-1_1

Rosa, P., Sassanelli, C., & Terzi, S. (2020). Towards circular business models: A systematic literature review on classification frameworks and archetypes. *Journal of Cleaner Production*, *236*, 117696. DOI: 10.1016/j.jclepro.2019.117696

Roshid, Md. M., Waaje, A., Karim, R., Irtisham, M. R., & Kaswan, M. S. (2025a). Industry 5.0 and Organizational Sustainability: Integrating Cutting-Edge Technologies and Human Ingenuity. In Mahender Singh Kaswan, Rajeev Rathi, & Jose Arturo Garza-Reyes (Eds.), *Industry 5.0: Concepts and Strategies for Digital Transformation* (pp. 50–61). CRC Press. https://doi.org/DOI: 10.1201/9781003535188-7

Roshid, Md. M., Yeamin, Md. B., Waaje, A., Karim, R., & Kaswan, M. S. (2025b). Understanding the Nexus of Industry 5.0 and Circular Economy for Enriched Lean and Environment Sustainability. In José Carlos Vieira de Sá, Shubham Sharma, Sunil Kumar, & Manjunath B.R. (Eds.), *Transformative Lean Six Sigma Techniques for the Quality 5.0 Paradigm* (pp. 349–380). IGI Global Scientific Publishing.

Sarpong, D., Boakye, D., Ofosu, G., & Botchie, D. (2023). The three pointers of research and development (R&D) for growth-boosting sustainable innovation. *Technovation*, *122*, 102681. DOI: 10.1016/j.technovation.2022.102581

Sehnem, S., de Queiroz, A. A. F. S. L., Pereira, S. C. F., dos Santos Correia, G., & Kuzma, E. (2022). Circular economy and innovation: A look from the perspective of organizational capabilities. *Business Strategy and the Environment*, *31*(1), 236–250. DOI: 10.1002/bse.2884

Sharma, N. K., Govindan, K., Lai, K. K., Chen, W. K., & Kumar, V. (2021). The transition from linear economy to circular economy for sustainability among SMEs: A study on prospects, impediments, and prerequisites. *Business Strategy and the Environment*, *30*(4), 1803–1822. DOI: 10.1002/bse.2717

Stahel, W. R. (2019). *The circular economy: A user's guide*. Routledge-Taylor and Francis Group., DOI: 10.4324/9780429259203

Stankevičienė, J., & Nikanorova, M. (2020). ECO-INNOVATION AS A PILLAR FOR SUSTAINABLE DEVELOPMENT OF CIRCULAR ECONOMY. *Business: Theory and Practice*, *21*(2), 531–544. DOI: 10.3846/btp.2020.12963

Stevens, S. S. (1946). On the theory of scales of measurement. *Science*, *103*(2684), 677–680. DOI: 10.1126/science.103.2684.677 PMID: 17750512

Stojanović, I., Puška, A., & Selaković, M. (2022). A multi-criteria approach to the comparative analysis of the global innovation index on the example of the Western Balkan countries. *Economics Innovative and Economics Research Journal*, *10*(2). Advance online publication. DOI: 10.2478/eoik-2022-0019

Talić, M., Stefanović, S., & Milenković, A. (2023). Eco-innovations in small and medium-sized enterprises: analysis of the regulatory framework in the Republic of Serbia. In *Proceedings of 54th International Scientific Conference: Smart and sustainable economy: Trends and perspectives* (pp. 211-219). University of Niš, Faculty of Economics.

Tomala, J., & Urbaniec, M. (2021). Eco-innovation development in selected European and Asian countries: A comparative analysis. *Ekonomia i Środowisko, 79*(4), 74–88. https://doi.org/DOI: 10.34659/2021/4/28Yadda+2

Vranjanac, Ž., Rađenović, Ž., Rađenović, T., & Živković, S. (2023). Modeling circular economy innovation and performance indicators in European Union countries. *Environmental Science and Pollution Research International*, *30*(34), 81573–81584. DOI: 10.1007/s11356-023-26431-5 PMID: 36918486

Waaje, A., Karim, R., & Roshid, Md. M., & Meem, T. N. (2024). *Minimizing Waste in the Hospitality Industry: The Opportunities and Challenges of Lean Tourism.* In R. Sucheran (Ed.), Sustainable Waste Management in the Tourism and Hospitality Sectors (pp. 125–166). IGI Global Scientific Publishing. https://doi.org/DOI: 10.4018/979-8-3693-6110-8.ch006

Waaje, A., Karim, R., Roshid, M. M., Yeamin, B., & Meem, T. N. (2025). Integration of industry 4.0 with personalized medicine: A shift of the paradigm in healthcare. In Chemingui, H., & Lamloumi, M. (Eds.), *AI-Driven Personalized Healthcare Solutions* (pp. 27–69). IGI Global., DOI: 10.4018/979-8-3693-7858-8.ch002

Zhiwei, L., Feng, L., & Pan, G. (2025). A meta-analysis of the factors influencing corporate eco-innovation. *Cleaner Waste Systems, 11*, 100260. DOI: 10.1016/j.clwas.2025.100260

## KEY TERMS AND DEFINITIONS

**Circular Economy:** A circular economy is an economic model aimed at reducing environmental impact by minimizing the extraction of natural resources and the generation of waste. It emphasizes maintaining the value of products, materials, and resources in the economy for as long as possible through practices such as repair, recycling, refurbishment, and redesign.

**Eco-Innovation:** Eco-Innovation represents new solutions that enable the transition to a circular economy by reducing the use of natural resources and the emission of harmful substances throughout the product life cycle, in line with the goals of environmental change and sustainable development.

**Research and Development:** Research and Development (R&D) is a process that encompasses systematic and creative activities, varying by characteristics, aimed at expanding the body of knowledge and discovering new applications of existing knowledge.

**European Union:** The European Union (EU) is a political and economic union composed of 27 member states, predominantly located in Europe, which operate through common institutions to promote integration, economic cooperation, and shared regulatory frameworks.

**CoCoSo:** A relatively new CoCoSo multi-criteria decision aid approach (Combined Compromise Solution) that integrates the exponentially weighted product model (MEP) with simple additive weighting (SAW).

**ARAS:** The ARAS method operates on the principle that the utility degree of each alternative is directly proportional to the sum of its normalized and weighted performance values across all criteria.

**PROMETHEE:** The Preference Ranking Organization Method for Enrichment of Evaluations is an outranking method for a finite set of alternative actions to be ranked and selected among criteria, which are often conflicting.

**Spearman's correlation:** Spearman's rank correlation coefficient or Spearman's ρ, measures the rank correlation (statistical dependence between the rankings of two variables). It assesses how well the relationship between two variables can be described using a monotonic function.

# Chapter 15
# Effect of Strategic Innovation on the Success of the Circular Economy With the Mediating Role of Competitive Advantage in Ethiopian Large-Scale Manufacturing Firms

**Tafese Niguse**
https://orcid.org/0009-0008-6536-6392
*Bule Hora University, Ethiopia*

**Shashi Kant**
https://orcid.org/0000-0003-4722-5736
*Bule Hora University, Ethiopia*

**Aynetu Terefe Alem**
https://orcid.org/0009-0003-3814-9548
*Bule Hora University, Ethiopia*

**Kenesa Kebede**
https://orcid.org/0009-0004-1860-2752
*Bule Hora University, Ethiopia*

**Sofi Badi Mohammed**
*Bule Hora University, Ethiopia*

**Zerihun Kinde Alemu**
https://orcid.org/0000-0002-8546-2882
*Ethiopian Civil Service University, Ethiopia*

## ABSTRACT

*This study considers the circular economy, competitive advantage, and the success of strategic innovation in Ethiopian large-scale manufacturing firms. A quantitative*

DOI: 10.4018/979-8-3373-2454-8.ch015

*research design was used to analyze the correlations between strategic innovation, alongside competitive advantage, and the performance success of the circular economy. A total of 384 participants were selected for the study to achieve adequate statistical power for dependable analysis results. The researcher used simple random sampling to select participants of the study. Exploratory factor analysis and confirmatory factor analysis were applied to confirm the reliability and validity of the variables utilized. A structural equation model was carried out to examine correlation between variables and mediation using SPSS V.26 and AMOS V.26 software. The outcome of the study highlighted that strategic innovation has a major impact on the success of the circular economy, both directly and indirectly through competitive advantage. The results underline how important it is for companies to develop an innovative culture in order to improve their competitive posture and the efficiency of sustainable development. In addition to providing managers with ideas on navigating the subtleties of sustainable commercial practice, the results contribute to theoretical knowledge on the way in which these constructions interact. This study emphasizes the value of a complete strategy, which includes stability, competitive strategy and innovation.*

## INTRODUCTION

The critical global role of strategic innovation in circular economy practices exceeds all possible assessments while countries fight against environmental decline and diminishing resource availability. The circular economy is developing as a transformational model for business sustainability, as it helps companies maintain continuous resource use while reducing waste and establishing regenerative practices. Strategic innovation brings both environmental necessity and economic benefits, as it enables companies to enhance their business performance while ensuring sustainable operations (Karim et al., 2025a). Strategic innovation functions as an essential organizational mechanism to support change in market conditions and promote operational efficiency while responding to consumer needs for eco-friendly products and services (Bekata & Kero, 2025).

Strategic innovation developed through management theory began with traditional innovation theory, which includes extensive strategic thinking that supports business targets and market trends. Business organizations must restructure their product offerings and operational structures by integrating sustainability principles according to this approach (Roshid et al., 2025; Karim et al., 2025b). Strategic innovation research demonstrates multiple aspects that show that creative approaches between organizations create better competitive results. Research demonstrates that businesses which adopt strategic innovation gain better capability to conquer circular economy complexities while beating their market competitors (Arfaoui et al., 2022).

Large manufacturing firms in developing countries like Ethiopia face multiple practical hurdles when implementing these concepts to achieve competitive advantages through strategic innovation decisions. Lack of proper infrastructure, together with restricted funding opportunities and scarce skilled personnel, prevents firms from implementing strategic innovation strategies. Firms experience difficulties implementing circular economy practices because the cultural and regulatory environments may not provide sufficient support for change (Demessie & Shukla, 2024).

Analysis of contemporary business practices shows that researchers currently lack sufficient data to understand strategic innovation effects on circular economy success for Ethiopian manufacturing companies. Few studies investigate innovation or the circular economy as individual topics, yet no research combines them with competitive advantage as a mediation variable. The study seeks to address these existing knowledge gaps through examination of how strategic innovation and competitive advantage affect the success of circular economy operations in Ethiopian manufacturing firms (Kwarteng et al., 2022).

Future researchers and policymakers will find practical guidance in this chapter through frameworks and strategies that help overcome identified challenges so that the findings can develop sustainable manufacturing sector policies. The study examines the impact of strategic innovation on circular economy outcomes for large manufacturing companies in Ethiopia, while also evaluating competitive advantage as a crucial link between these two variables. The study investigates this question to generate insights that will help theoretical progress and practical approaches, which combine to create a more sustainable and competitive manufacturing sector in Ethiopia.

## BACKGROUND OF THE STUDY

A circular economy introduces profound operational changes to conducting business, especially within the manufacturing industry. The model works to reduce the link between economic expansion and resource depletion so supply chains attain sustainable and resilient operation. Sustainable development in Ethiopia depends critically on large-scale manufacturing industries, so the nation needs to adopt circular economy principles. Various obstacles stand in the way of execution strategies meant to lead this transformation (Shebeshe & Sharma, 2024).

Ethiopian large-scale manufacturing companies may improve their competitive edge by taking a number of practical initiatives to embrace circular economy techniques. First, they should invest in research and development to devise new methods and processes that reduce waste. Second, collaborating with suppliers and consumers helps optimize resource utilization and promotes environmentally friendly behaviors throughout the supply chain. Third, companies should use circular design concepts to make products last longer and make recycling easier. Furthermore, giving staff training and development will create a competent workforce that can drive innovation. Finally, companies may show their commitment to the circular economy and keep track of their progress by setting KPIs to quantify their sustainability performance (Arfaoui et al., 2022).

The textile industry of Ethiopia stands as an ideal demonstration of the potential of the circular economy because its sector shows promising signs for its implementation. Although the government provides backing to the industry, its manufacturers still face difficulties due to inefficiencies and waste control issues. An Ethiopian research on textile companies showed that multiple businesses understand circular economy advantages yet face technical shortages and public waste recycling restrictions. Low uptake of strategic innovation occurs due to the lack of a standard regulatory framework, which makes sustainable practice implementation more challenging (Shebeshe & Sharma, 2024).

Post-harvest losses in the food processing sector create substantial problems because they occur frequently. Principal studies showed that food manufacturing operations need advanced technologies, which enable them to create useful products from various waste streams. An Ethiopian dairy processing company struggled when trying to establish a waste-to-value project, despite inadequate access to relevant technology and specialized knowledge. The industrial challenge highlights how fundamental strategic innovation functions as a solution for practical adoption barriers of circular economy practices (Suchek et al., 2021).

The implementation of circular economic principles shows significant challenges in the Ethiopian construction industry. The attempt to recycle construction materials by some firms has met opposition because of traditional industry practices and low stakeholder awareness levels. A construction company initially expressed interest in sustainable building but failed to advance innovative recycling methods when workers lacked sufficient training and awareness about the practice's benefits (Sufa et al., 2023).

The investigated real-world examples demonstrate all the various hurdles which Ethiopian businesses encounter when executing strategic innovation methods under circular economy structures. The realization of progress faces resistance from insufficient training along with limited access to technology and inadequate infrastructure and missing regulatory support. The research investigates the problems facing this sector that block the success of a circular economy transition through strategic innovation implementation in Ethiopian large-scale manufacturing. Knowledge about these forces helps policymakers alongside industrial leaders to create sustainable manufacturing systems which are resilient (Triguero et al., 2022).

## STATEMENT OF THE PROBLEM

Many challenges hinder the effective implementation of strategic innovation aimed at achieving circular economy objectives, despite this framework gaining recognition for promoting sustainable development in large-scale Ethiopian manufacturing. The desired manufacturing environment will enable optimal resource management and waste reduction, while promoting economic development without compromising environmental sustainability. Currently, many organizations face major obstacles that prevent them from using innovative approaches and implementing circular practices (Kwarteng et al., 2022; Roshid et al., 2025).

Research on strategic innovation and the success of circular economies reveals conflicting data regarding their relationship in current studies. Research studies on strategic innovation reveal conflicting findings, indicating that these advantages do not apply uniformly across all market environments. The lack of clear results shows a theoretical gap since new programming needs to recognize industry-specific challenges alongside regional barriers, mainly affecting developing countries such as Ethiopia (Simane et al., 2024).

Insufficient research demonstrates how strategic innovation links with competitive advantage for enhancing circular economy performance. Academic research minimizes its attention on this fundamental connection, which prevents firms from understanding effective strategies to utilize their distinctive capabilities for sustainable practice implementation. The existing scrutiny of circular economy implementation in developed economies creates a contextual gap, which leads to limited empirical evidence about Ethiopian manufacturers encountering exclusive barriers and prospects in their adoption efforts. The scenario faces additional hurdles due to practical issues that challenge Ethiopian firms, including poor infrastructure conditions, restricted access to modern technology, and weak regulatory systems. Strategic innovation faces significant obstacles due to these factors, which simultaneously hinder the implementation of circular economy practices and discourage business innovation (Arfaoui et al., 2022).

This research intends to examine strategic innovation and its impact on the circular economy success of Ethiopian large-scale manufacturing firms, yet specifically emphasizes competitive advantage as the mediator. This research fills the identified gaps in contradictory evidence and theoretical and practical challenges by generating valuable findings that future researchers and policymakers can use for guidance. This research establishes a conceptual model as a tool to help firms handle circular economy challenges so they can adopt sustainable methods that fulfill development targets and environmental protection efforts. The research findings will establish themselves as essential knowledge for stakeholders who seek to advance innovation together with sustainability in Ethiopia's manufacturing marketplace (Simane et al., 2024).

## LITERATURE REVIEW

### Definition and Origin of Terms

**Strategic innovation:** Strategic innovations are closely tied to the development of new business strategies and models, which fundamentally alter the way businesses create and deliver value to their customers. This term combines the concepts of "strategies," which encompass long-term plans and actions aimed at achieving organizational objectives, with "innovation," which refers to the introduction of new ideas, products, or methods. The origin of this term can be traced to before the early 2000s, when scientists began to distinguish between simple innovation and broader strategic change. He emphasizes that innovation should focus not only on products and technologies but also on how businesses function and interact with their environment. Companies that excel in strategic innovation are often those that can anticipate market trends, adapt to changes, and create unique value propositions that set them apart from competitors (Sufa et al., 2023).

**Circular Economy:** The circular economy is an economic model aimed at minimizing waste and making the most of resources. This is in strong contrast to the traditional linear economy following the "simplex model". Instead, the circular economy expands its lifecycle and reduces environmental impact, as it seeks to create a system with closed circuits where resources are recycled, repaired, and processed. This concept gained power at the end of the 20th century, with early debates on sustainable development and resource effectiveness. It should be noted that the term is widely recognized thanks to the work of the Ellen MacArthur Foundation, created in 2010. The circular economy is based on systematic thinking, highlighting the relationship between economic activity and environmental systems, and seeking to develop a regenerative economy that benefits both people and the planet (Shebeshe & Sharma, 2024).

**Competitive Advantage:** refers to the attributes that allow the organization to go beyond their competitors. These attributes can be obtained from various sources, including the cost structure, products of products, brand reputation and customer service. The term was popularized by Michael Porter in the 1980s and described how companies could achieve competitive advantage through differentiation or cost-leadership strategies. Competitive advantages are often seen as dynamic concepts that companies must constantly evolve to maintain advantages in market development. Understanding how to use unique power can lead to lasting success and market leadership, and it thus plays a crucial role in an organization's strategic planning (Triguero et al., 2022).

## Related Theories

When studying the intersection of strategic innovations, competitive advantages and a circular economy, several theoretical frameworks give valuable information.

RBV theory: One of the outstanding theories is the perspective based on resources (RBV), which claims that the company's resources and capabilities are decisive determinants of its competitive advantage. According to this theory, unique resources include innovative opportunities, technological examination, and qualified human capital, which enable firms to differentiate themselves in the market. In the context of the circular economy, RBV emphasizes the importance of investing in sustainable resources and practices, which increase operational efficiency and contribute to long-term stability (Bekata & Kero, 2025).

**Dynamic capability theory:** Another corresponding theory is the dynamic capacity theory, which expands upon the Resource-Based View (RBV) by focusing on a company's ability to adapt, integrate, and reconfigure its resources in response to changing market conditions. This theory is particularly relevant for strategic innovation because it highlights the need for companies to develop opportunities that allow them to constantly introduce innovations and use new ones within the framework of the circular economy. It highlights how companies developing dynamic capabilities can better navigate the complexity of stable practices and meet consumer and regulatory development requirements (Demessie and Shukla, 2024).

**Theory of diffusion of innovation:** The theory of innovation propagation plays an important role in understanding how innovation is accepted and implemented in an organization. This theory examines the process by which new ideas and technologies propagate among users and how a variety of factors affect the effects of benefits, compatibility, and complexity. In the context of the circular economy, this theory helps to unravel the challenges companies face in adopting innovative practices and social dynamics that promote or hinder this transition (Shebeshe & Sharma, 2024).

While all these theories offer valuable insights, the resource-based view (RBV) serves as the primary theoretical lens for this study. RBV is especially relevant, as it enables the study of how Ethiopian production firms can utilize their unique resources and opportunities to achieve a competitive advantage when transitioning to a circular economy. This study focuses on concrete and intangible assets owned by companies and examines how to optimize these resources through strategic innovations to address practical problems related to implementing circular economy principles. This theoretical framework guides an analysis of interactions with strategic innovation, competitive advantage, and the success of circular economies, ultimately contributing to a deeper understanding of Ethiopia's stable production methods.

## Empirical Literature Review

Research on strategic innovation's influence on circular economy achievements has strongly increased recently because it exposes complex links between these concepts across different contexts. Firms need strategic innovation as their key transformation tool to adopt circular economy approaches because these innovations create essential business models and operational approaches which priorities resource-efficient sustainability (Demessie & Shukla, 2024).

Strategic innovation produces value for both the individual company and wider society. Companies holding circular economy principles by implementing strategic innovative approaches achieved better marketplace standing coupled with a growing customer base. Strategic innovation proves to be an outside-facing method that shapes how external actors see circular businesses while building vital social connections, which contribute to circular initiative success (Arfaoui et al., 2022).

Organizations drilling innovative resource management tools like recycling strategies and waste reduction reported necessary operational efficiency improvements along with cost savings in their research. The successful deployment of circular economy frameworks received support from these improvements, which demonstrated a beneficial cycle between strategy innovation and circular economy end results (Tolossa et al., 2024).

Strategic innovation in product design and supply chain management proved successful for companies in the automotive industry. Companies that applied sustainable design elements such as modularity alongside recyclability components could minimize waste production and extract maximum value from resources to build enduring business advantages. Core business process integration of strategic innovation leads to successful circular economy transition because of its vital role in shaping company success (Kwarteng et al., 2022).

The literature acknowledges that numerous barriers exist during the implementation of strategic innovation for circular economy frameworks. Businesses understand how strategic innovation delivers potential benefits, though they face problems with organizational resistance, along with inadequate funding and specialist talent deficits. The integration of circular economy principles faces obstacles that make it difficult to implement successfully, so it requires organizational dedication backed by ample resources (Shebeshe & Sharma, 2024).

Research studies also analyze competitive advantage as a facilitator that affects this connection. Organizations that connect innovative initiatives to their competitive strategies develop resource efficiencies and reduced waste output that drive economic results as well as environmental conservation (Sufa et al., 2023).

## H1: STRATEGIC INNOVATION HAS SIGNIFICANT EFFECT ON SUCCESS OF CIRCULAR ECONOMY

### Effect of Strategic Innovation on Competitive Advantage

Organizations from all sectors can use strategic innovation as their main tool to boost their competitive position. Through innovative business model transformation and new value proposition development, organizations can establish clear customer differentiators and better address market modifications while creating environmentally friendly business practices that appeal to customers (Simane et al., 2024).

Organizational competitive advantage strongly benefits from strategic innovation by developing specially tailored solutions that fulfill evolving consumer requirements. Innovative firms that implement new techniques and technologies produce sustainable products that surpass basic customer needs. Organizations that implement strategic innovation successfully obtain better market opportunities to claim increased market share among loyal customers (Bekata & Kero, 2025).

The adoption of strategic innovation regularly leads to enhancements in operational efficiency. Organizations can achieve lower prices and operation efficiency through process redesign and technology integration while increasing work output. Aligning strategic innovation into operational processes generates earnings efficiency that strengthens business profits and enables additional investment into innovation to establish a competitive advantage (Shebeshe & Sharma, 2024).

Competitive advantage strongly depends on strategic innovation because companies need to adjust to market changes effectively. Businesses that develop dynamic capabilities and their capacity to handle environmental transformations show stronger resistance against disruptions. Companies equipped with strong strategic innovation structures demonstrate rapid product or service development abilities when adapting to consumer needs and regulatory demands. Organizations need this speed in response systems because they operate within sectors where technology progresses quickly, as well as sectors that face changing sustainability standards (Demessie & Shukla, 2024).

Strategic innovation creates essential relationships and networks that are crucial for obtaining competitive business advantages. The active utilization of partners and suppliers and consumers during innovative projects enables firms to achieve access to fresh concepts together with new resources and modern technologies. Organizations that apply open innovation principles, according to Chesbrough (2003), use external knowledge to improve their strategic projects. Firms that engage in such collaborative collaborations accelerate their innovation procedures while simultaneously building market leadership by creating an expanded network of connections that share common goals (Bekata & Kero, 2025).

## H2: STRATEGIC INNOVATION HAS A SIGNIFICANT EFFECT ON COMPETITIVE ADVANTAGE

### Effect of Competitive Advantage on the Success of Circular Economy

Competitive advantage functions as a key determinant for organizations to achieve successful circular economy practice deployment. Organizations that utilize their exclusive assets and capabilities with available resources will successfully handle circular economy complexities, which improve sustainability together with long-term business survival. The implementation of the circular economy gains success through competitive advantage by providing sustainable product offerings that are distinct from industry competitors (Kwarteng et al., 2022). Companies with superior competitive advantages can dedicate resources towards developing creative products which combine structural strength with ease of recycling and sustainable resource practices. Companies which lead technological developments can develop new products that fulfil customer needs while reducing resource use and waste generation at the same time. Through product differentiation, organizations obtain leadership status in sustainability that generates environmentally aware customer acquisition along with increased brand loyalty (Suchek et al., 2021).

A competitive advantage helps businesses optimize their supply chains, as this represents a key element for the circular economy to succeed. Established strong interconnections between companies and suppliers, and stakeholders enable more efficient resource-recovery systems and logistics operations. organizations that maintain a competitive position obtain superior suppliers' terms for sustainable materials and develop recycling program through take-back systems. The capability enables them to bring their production cycle closed, which enables them to decrease waste generation while implementing circular resource management principles (Tolossa et al., 2024).

The competitive edge of organizations opens doors to better financial resources, enabling them to invest in circular economy research and development. Strategic financial structures achieved through leverage enable them to carry out investments in innovative waste reduction methods and resource-efficient technologies. Strong financial positions in companies lead them to embrace innovative sustainable practices, which generate better operational performance and deliver support for circular economy goals, according to research (Tolossa et al., 2024).

Organizations that achieve a competitive advantage will develop innovative cultures that sustain essential changes needed for circular business operations. Organizations that integrate sustainability into their fundamental strategy tend to motivate staff to creatively solve circular economy problems. The correct organizational alignment strengthens circular economy implementation and develops a staff dedicated to sustainability targets (Demessie & Shukla, 2024).

Competitive advantage relates to the success of a circular economy in a manner that may present complexities. Entirely short-term, financially driven firms discover challenges implementing sustainable initiatives since those sustainable practices demand prolonged return periods. Organizations must ensure their competitive strategies maintain their sustainability commitment, as failure to do so results in both negative image damage and consumer distrust (Shebeshe & Sharma, 2024).

## H3: COMPETITIVE ADVANTAGE HAS SIGNIFICANT EFFECT ON SUCCESS OF CIRCULAR ECONOMY

### Mediation of Competitive Advantage Among Strategic Innovation and Success of Circular Economy

organizations can implement successful circular economies through strategic innovation by establishing competitive advantages, which mediate this relationship. Strategic innovation provides businesses with the tools to create new business frameworks as well as distinct products and operational processes which support sustainability goals. The adoption of innovative approaches allows businesses to develop special market offers which attract environmentally responsible consumers. Distinctive differentiation between businesses enables competitive advantages by providing firms the opportunity to exploit their special resources, which distinguishes them from market competitors (Suchek et al., 2021).

Competitive advantage serves as a relationship enhancer because it enables successful execution of circular economy principles. Organizations that extract their unique assets, such as technological excellence, operational competencies and market standing, can produce substantial environmental outcomes from their strategic innovations. The competitive advantage of a firm that invents sustainable products enables it to pull in a dedicated customer segment while strengthening circular practices and enlarging its business reach (Tolossa et al., 2025).

A competitive position improves how organizations use their resources, which stands as a central requirement within a circular economy model. Strong competitive enterprises tend to allocate funds into efficient resource management technologies along with waste reduction programs. Closing the loop in production cycles depends on this ability, which helps organizations recover materials for reuse to sustain their industry leadership (Bekata & Kero, 2025).

Sustainable development, improved resource efficiency, and less waste are the hallmarks of the circular economy's success. There is a synergistic impact when competitive advantage mediates the link between strategic innovation and the success of the circular economy. Significant gains in sustainability indicators, including reduced carbon emissions and higher recycling rates, are more likely to be attained by businesses that successfully innovate while utilizing their competitive advantages. According to empirical research, businesses with significant competitive advantages are better equipped to adopt and apply circular economy principles, which enhances operational effectiveness and reduces costs (Shebeshe & Sharma, 2024).

## H4: COMPETITIVE ADVANTAGE MEDIATES THE RELATIONSHIP BETWEEN STRATEGIC INNOVATION AND THE SUCCESS OF THE CIRCULAR ECONOMY

## Conceptual Framework

Based on theoretical and empirical literature, the following conceptual framework was developed.

*Figure 1. Conceptual framework*

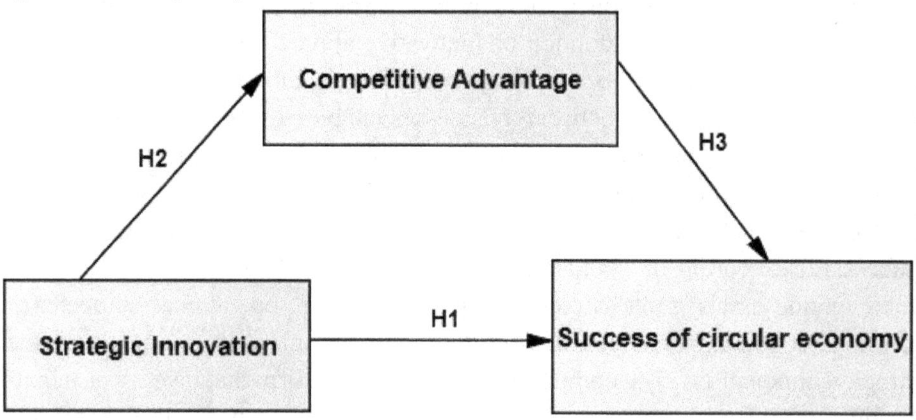


*Source: Authors*

## RESEARCH METHODOLOGY

A quantitative research design analyzed the correlations between strategic innovation alongside competitive advantage and the performance success of circular economics. A total of 384 participants were selected for the study to achieve adequate statistical power for dependable analysis results. Simple random sampling was used for data collection because it focused on individuals with experience and knowledge about sustainability practices in their organizations. The research strategy enabled researchers to obtain significant data, which produced valuable findings about their research queries.

The research data showed appropriateness for factor analysis according to a Kaiser-Meyer-Olkin (KMO) test that produced a value above 0.6. The sample showed appropriate characteristics for performing factor analysis according to the KMO results. The analysis employed exploratory factor analysis (EFA) to reveal the organizational structure among strategic innovation variables, competitive advantage variables, and circular economy success variables. A factor reduction of the data through EFA generated specific factors representing the main study constructs. The factors were refined using rotation methods to enhance the interpretability of the output.

The study adopted Structural Equation Modeling (SEM) using AMOS software to test the theoretical links between constructs after identifying the factors. Using SEM, researchers tested both direct and indirect relationships between strategic innovation and the success of the circular economy, with competitive advantage acting as a mediating factor. Model fit indices demonstrated excellent performance, thereby confirming all established theoretical relationships and facilitating a more comprehensive analysis of the underlying system processes.

## SCALE DEVELOPMENT

The research instrument was developed through a comprehensive literature review of strategic innovation, competitive advantages, and circular economies. The research adopted existing measurement items from previous investigations, which underwent necessary adjustments to match the present study framework. The research began by examining extensive literature to identify key components and variables linked to the constructs. The three dimensions of strategic innovation consisted of product, process, and business model innovation. Competitive advantage assessment depended on three elements: cost leadership, differentiation, and brand reputation.

Resource efficiency, waste reduction, and stakeholder involvement served as the measures to evaluate the success of the circular economy.

Researchers generated an initial set of items after the literature study to depict the fundamental aspects of the constructs at hand. The items underwent expert evaluation to establish both their validity as well as their applicability. The subject matter experts' feedback enabled developers to make necessary adjustments, resulting in a clear, concise item collection which corresponded to the constructs. Diverse measurement items used Likert-type scales that let respondents indicate their agreement position or behavior frequency for strategic innovation, competitive advantage, and circular economy elements.

The team incorporated the scale into the survey, which was given to the total 384 participants. The developed scale sought a complete understanding of construct relationships, therefore enabling extensive evaluation of gathered survey data. The extensive development process of the research instrument made it reliable and valid for future statistical analysis in the study.

## DATA ANALYSIS

### Data Adequacy Test

*Table 1. KMO and Bartlett's Test*

| **Kaiser-Meyer-Olkin Measure of Sampling Adequacy.** | | **.850** |
|---|---|---|
| Bartlett's Test of Sphericity | Approx. Chi-Square | 1419.742 |
| | Df | 86 |
| | Sig. | .003 |

Source: Authors

Table 1 shows vital data information that verifies the suitability of the analysis data. Strong sampling adequacy emerges from the Kaiser-Meyer-Olkin (KMO) Measure of Sampling Adequacy value of 0.850. The analysis accepts KMO values exceeding 0.6, while numbers approaching 1.0 indicate high variable correlation within the study framework. The KMO result of 0.850 demonstrates that the sample used in this research is appropriate for factor analysis because the variables demonstrate substantial common variance.

The sphericity test provides information about whether the variables in the correlation matrix show separation from an identity matrix, so uncorrelated variables exist. The Chi-Square test reached a value of 1419.742 while using 86 degrees of freedom and demonstrated a significance level of 0.003. A significant p-value of less than 0.05, along with this evidence, indicates that the variables show meaningful correlations. The combined research outcomes demonstrate that the analyzed data have sufficient characteristics to enable faith-based factor analysis, enabling efficient studies of underlying principles within strategic innovation alongside competitive advantage, as well as the circular economy success model.

*Table 2. Total Variance Explained*

| Component | Initial Eigenvalues | | | Extraction Sums of Squared Loadings | | | Rotation Sums of Squared Loadings | | |
|---|---|---|---|---|---|---|---|---|---|
| | Total | % of Variance | Cumulative % | Total | % of Variance | Cumulative % | Total | % of Variance | Cumulative % |
| 1 | 4.55 | 32.240 | 32.240 | 4.55 | 32.240 | 32.240 | 3.00 | 21.450 | 21.450 |
| 2 | 1.41 | 9.792 | 42.024 | 1.41 | 9.792 | 42.024 | 2.20 | 15.488 | 36.939 |
| 3 | 1.32 | 9.090 | 51.110 | 1.32 | 9.090 | 51.130 | 2.00 | 14.170 | 51.130 |
| 4 | .949 | 6.780 | 57.892 | | | | | | |
| 5 | .870 | 6.221 | 64.117 | | | | | | |
| 6 | .771 | 5.520 | 69.632 | | | | | | |

Extraction Method: Principal Component Analysis.
Source: Authors

The results of the factorial analysis are presented in Table 2, which also contains detailed information on the general dispersion explained by each component identified by the analysis of the main components (ACP). The first component explains an initial value of 4.55, representing 32.24% of the total dispersion, indicating a significant portion of the database's structure. When considered collectively, this factor accounts for 32.24% of the variation explained. With an eigenvalue of 1.41 for the second component, which accounts for 9.79% of the variance, the cumulative variance explained is 42.02%. Because its own value is 1.32, the third component also contributes, explaining 9.09% of the variance, typically 51.11%. Together, these three elements represent more than half of the variation, emphasizing their value to understand the relationships between structures. Despite having lower eigenvalues, these components nevertheless contribute to the overall variance. The eigenvalue of the fourth component is 0.949 (6.78%), that of the fifth component is 0.870 (6.22%), and that of the sixth component is 0.771 (5.52%). However, it is often considered a threshold for key factors between 4 and 6, as its value falls below 1.0. No additional explanatory power is provided.

Because the rotation sums of squared loadings further clarify the variance described by each component upon rotation, the factors may be easier to comprehend. 51.13% of the variance is explained by the first three rotated components alone, whereas the first component still accounts for the majority of the variation reported. This research demonstrates that the proposed components are effective in summarizing the data, supporting a better understanding of the relationships among strategic innovation, competitive advantage, and circular economy performance. Taking everything into account, the results indicate that the proposed constructs are well-represented and lend themselves to reliable analysis for further comprehension.

## Confirmatory Principal Component Examination

The principal component arrangement of a set of guided proxies is verified using a type of statistics known as confirmatory principal component analysis. Based on the underlying frame job and correlatives among observed proxies, investigations can evaluate hypotheses utilizing CFA. The inquiry employed CFA under Table 3 to evaluate the hypothesis that an association exists between the guided proxies and the latent concepts that support them.

*Table 3. Covariance*

| Covariance | | | Approximation | S.E. | C.R. | P | Hy. |
|---|---|---|---|---|---|---|---|
| **Competitive advantage** | **<-->** | **Strategic innovation** | **.245** | **.026** | **8.235** | *** | **H2** |
| Competitive advantage | <--> | Success of circular economy | .305 | .029 | 7.688 | *** | **H3** |
| Strategic innovation | <--> | Success of circular economy | .252 | .025 | 7.828 | *** | **H1** |

Source: Authors

The statistical measures that correlate to the covariances between the concepts of strategic innovation, competitive advantage, and the performance of the circular economy are shown in Table 3. The covariances show how strongly these concepts are related to one another, highlighting the interdependencies between them. This study finds that the association coefficient is 0.245, with a standard error of 0.026 (S.E.). A key critical report on the relationship (c.r.) shows an important positive link between these two structures. This link is substantive and supports Hypothesis 2 (H2). The present study maintains a positive correlation between strategic innovation and competitive advantage, as indicated by the asterisk (***) that indicates statistical implications.

In contrast, the results show that the competitive advantage has a value of 0.305, with a C.R. of 7.688 and an S.E. Furthermore, the statistical significance of this association confirms Hypothesis 3 (H3). This conclusion confirms that competitive advantage has a beneficial effect on the functioning of the circular economy. According to a strong correlation, companies with a competitive advantage are more likely to succeed in their efforts in the circular economy. There are 0.252 correlation between the success of a circular economy and strategic innovations, with C.R. 7.828 and S.E. 0.025. The statistical significance of this result supports Hypothesis 1 (H1), which asserts a positive correlation between strategic innovation and the success of the circular economy. Positive connections in this case suggest that firms engaged in strategic innovation have a higher likelihood of successfully implementing their circular economy strategies.

*Table 4. Validity Concern*

| | CR | AVE | MSV | MaxR (H) | SI | CA | SCE |
|---|---|---|---|---|---|---|---|
| SI | 0.729 | 0.640 | 0.216 | 0.849 | **0.651** | | |
| CA | 0.760 | 0.661 | 0.135 | 0.761 | 0.168 | **0.660** | |
| SCE | 0.789 | 0.600 | 0.270 | 0.791 | 0.384 | 0.190 | **0.681** |

Note: SI= Strategic innovation; CA= Competitive advantage; SCE= Success of circular economy
Source: Authors

The component validity metrics measured in the study – strategic innovation (SI), competitive advantage (CA), and circular economy (SCE) success – are comprehensively outlined in Table 4. Table 4 comprehensively outlines each metric for the reliability (CR), mean variance extraction (AVE), maximum variance (MSV), maximum variance (MAX), and maximum integrity (MAX) constructs. The Compound Reliability (CR) value indicates the internal consistency of the structure, for CR 0.729, CA CR 0.760, SCE A CR 0.789. All values exceed the recommended threshold of 0.7. This indicates that the structure is being measured reliably. This high reliability is necessary to ensure that the tools used in the research produce consistent results.

Compared to measurement error, the average variation is extracted (Ave), which evaluates the percentage of variation that captures the concept. The AVEs for SI, CA and SCE are 0.640, 0.661, and 0.600, respectively. Each value for the avenue above 0.5 cut suggests that the structure considers a significant part of the corresponding indicator change. This shows that the elements used to test each structure are actually captured by suspected concepts that demonstrate good convergence validity.

The largest shared variation between constructs is shown by the maximum shared variation (MSV), which is crucial for assessing discriminant validity. While CA's MSV is 0.135 with SI and 0.270 with SCE, SI's MSV is 0.216 with regard to CA and 0.270 with regard to SCE. The idea that the structures differ from each other is confirmed by the fact that these MSV values are lower than the corresponding AVA values, which confirms their discriminating reliability.

Finally, the evaluation of the structures of the greatest reliability is determined by the greatest reliability (maxr (h)). The structures retain a high degree of reliability even under the most extreme dependability conditions, as indicated by the values, which vary from 0.761 for CA to 0.849 for SI. This strengthens the measurement models' overall resilience.

## Mediating Role Examination

*Figure 2. Conceptual framework*

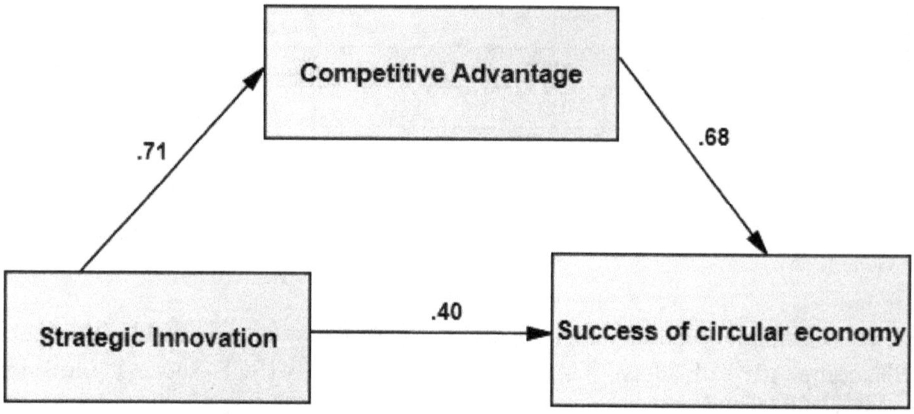


## Model Fit Indices

*Table 5. indices for Model Fit*

| Sig. | Chi-Sq | RMR | Fitness Goodness | Fitness Confirmatory | TLI | RMSEA |
|---|---|---|---|---|---|---|
| 0.004 | 1.960 | .042 | 0.936 | 0.918 | 0.917 | .035 |

Source: Authors

Several indices evaluating the suggested model's fit are shown in Table 5, offering important information about how well the model captures the underlying data structure. Chi-Square (Chi-Sq), Root Mean Square Residual (RMR), goodness-of-fit metrics, the significance level (Sig.), and other confirmatory fit indices like the Tucker-Lewis Index (TLI) and Root Mean Square Error of Approximation (RMSEA) are examples of key fit indices. The adjustments to the model are statistically significant, as evidenced by the reported level of significance (SIG.) of 0.004. In general, the zero hypotheses that the model claims to be mediocre differ significantly from the observed data, as evidenced by the lower values of P. This importance shows how well the models reflect the connections between designs.

1.960 is the Chi-Square (Chi-Sq) value. A low chi-square value usually indicates a satisfactory factory fit, but it must be read in conjunction with the p-value and the degrees of freedom. Although it is still necessary to consider other correspondence indicators for an in-depth evaluation, the relatively small chi-squared value in this case indicates that the model accurately represents the observed data. The average difference between the real and expected values is represented by the average square residue (RMR), which is reported to be 0.042. In general, an RMR value below 0.05 indicates that the model is suitable for the data.

According to a good indicator, the indicator is the fit index, which is to be checked at 0.918, and the total score is 0.936. The two numbers are higher than the universally recognized 0.90 breakpoint, suggesting that the model appropriately considers data changes and confirms the relevance of the model structure. Additionally, the Tucker-Lewis Index (TLI) reported of 0.917 is higher than the suggested cutoff of 0.90. This implies that, while taking into consideration the intricacy of the suggested structure, the model offers a better match than a baseline model. Finally and above all, the average quadratic error indicated in the approximation (RMSEA) is 0.035, well below the appropriate limit of 0.06. This low RMSEA value suggests that the model successfully reflects the underlying relationships with negligible error, indicating a close adjustment between the model and the data observed.

*Table 6. Regression Examination*

| **Relative** | | | **Approx.** | **S.E.** | **C.R.** | **P** | **Ass.** |
|---|---|---|---|---|---|---|---|
| **Competitive advantage** | **<---** | **Strategic innovation** | **.859** | **.138** | **7.578** | ******* | **H2** |
| Success of circular economy | <--- | Competitive advantage | .332 | .135 | 2.850 | *** | H3 |
| Success of circular economy | <--- | Strategic innovation | .960 | .195 | 7.961 | *** | H1 |

Source: Authors

The findings of the regression analysis, looking at the connections between competitive advantage, strategic innovation, and the success of the circular economy, are shown in Table 6. A thorough understanding of the interactions between these constructs is provided by the evaluation of each connection using coefficients, standard errors (S.E.), critical ratios (C.R.), and significant levels (P). The regression coefficient and standard error for the path from strategic innovation to competitive advantage are 0.859 and 0.138, respectively. This association has a highly significant critical ratio of 7.578 ($p < 0.001$). Hypothesis 2 (H2) argues that greater levels of strategic innovation lead to greater competitive advantages, supported by this strong coefficient, which indicates a significant positive association. The results underscore the importance of innovations in enabling businesses to enhance their market competitiveness.

At $p < 0.001$, the regression coefficient for the path from competitive advantage to the success of the circular economy is 0.332, with an S.E. of 0.135 and a C.R. of 2.850. According to this conclusion, competitive advantage has a beneficial impact on the performance of circular economy projects, which is consistent with Hypothesis 3 (H3). Businesses that gain a competitive edge are more likely to succeed in their circular economy initiatives, according to the coefficient, which is still significant even if it is smaller than that of the prior relationship.

Lastly, the regression coefficient of 0.960, S.E. of 0.195, and C.R. of 7.961 for the route from Strategic Innovation to the Success of the Circular Economy are statistically significant ($p < 0.001$). Strategic innovation is a major predictor of success in the circular economy, as seen by this high coefficient, which confirms Hypothesis 1 (H1). The high score supports the notion that creative approaches are essential to sustainability initiatives by indicating that innovation directly leads to improved results in circular economy practices.

*Table 7. Mediating Role Effect*

| | Influence | worth | Path Influence |
|---|---|---|---|
| Strategic innovation→ Success of circular economy | Direct Influence | .40 | Direct influence stated |
| Strategic innovation → Competitive advantage → Success of circular economy | Indirect Influence | .71*.68=.49 | Indirect Influence Ensued |
| | Whole influence | .89 | Partial mediation |

Source: Authors

Competitive advantage mediates the link between strategic innovation and the success of the circular economy, as Table 7 illustrates. The findings shed light on how these characteristics interact and affect overall results by demonstrating both direct and indirect implications.

Strategic innovation has a direct impact of 0.40 on the success of the circular economy. This figure illustrates how strategic innovation directly contributes to the success of circular economy projects. The direct influence highlights how crucial innovation is to reaching sustainability objectives, implying that businesses that emphasize creative approaches may directly increase their efficacy in the circular economy.

The indirect influence, on the other hand, is determined by taking into account the route that leads from strategic innovation to competitive advantage and ultimately to the success of the circular economy. An indirect effect of 0.49 is obtained by multiplying the coefficients in this indirect path: 0.71 (the influence of strategic innovation on competitive advantage) by 0.68 (the influence of competitive advantage on the success of the circular economy). According to this assessment, the indirect effect is significant, but it supplements strategic innovation's direct influence rather than taking its place.

When direct and indirect benefits are included, the overall impact of strategic innovation on the success of the circular economy comes to 0.89. With both direct impacts and the mediated approach through competitive advantage, this high cumulative value indicates that strategic innovation has a strong overall influence. The study concludes that there is a degree of mediation in the link. A significant amount of the impact is still direct, even while competitive advantage is essential for enabling the success of circular economy efforts through strategic innovation. This research highlights the importance of both direct and mediated paths in comprehending the role that strategic innovation plays in the circular economy.

## DISCUSSION

The conclusions of this study highlight how important strategic innovation is to creating a competitive advantage and ultimately stimulating the success of the circular economy. Innovative practices are important for businesses in the hopes of thriving in the sustainable commercial industry, as they demonstrated the strong direct effect of strategic innovation on the success of the circular economy. This highlights how important it is for businesses to concentrate and invest in innovation, particularly with regard to circular economy activities, as an important component of their planning.

Strategic innovation has a coefficient of 0.40, which means it has a direct effect on the performance of the circular economy. This shows how important it is for reaching sustainability objectives. The indirect impact, which is 0.49, comes from the route via competitive advantage (0.71) and its effect on the success of the circular economy (0.68). The total impact is 0.89 when you add together both the direct and indirect impacts. This shows that strategic innovation has a large overall effect. Even while competitive advantage is important, a lot of the effect is still direct. This shows how important both paths are for understanding this connection.

In addition, thanks to a competitive advantage, the study shows that strategic innovations have a serious indirect influence on the success of a circular economy. Depending on this intermediate effect, innovations have greater influence when they lead to a competitive advantage, even if they directly contribute to lasting results. Successful initiatives on the circular economy will most likely be implemented by companies that use innovation to separate themselves on the market. This study emphasizes the relationship between innovation and competitive positioning, which indicates that for businesses to succeed in their sustainable development initiatives, they must contribute to both.

The findings also indicate that a key component of the connection between strategic innovation and the success of the circular economy is competitive advantage. This supports the notion that gaining a competitive edge is a prerequisite for converting inventive endeavors into fruitful results rather than just a result of innovation. Companies that have successfully leveraged innovation to gain competitive advantages are better equipped to meet the challenges of the circular economy. This improves overall performance.

This study emphasizes the importance of understanding how these aspects interact. Organizations should use an integrated approach. This includes innovation, competitive strategies, and stability, based on strong bonds and essential routes. In doing so, companies can establish a synergy that supports wider social objectives, including environmental stability and controllability, in addition to obtaining an economic advantage.

## CONCLUSION

This study offers strong proof of the essential contributions that competitive advantage and strategic innovation bring to success in the circular economy. According to this study, strategic innovation improves competitive advantage and indirectly affects stability outcomes. This bilateral approach underscores the importance of innovation, enabling businesses to overcome obstacles associated with the circular economy. The results show that businesses need to make strategic innovation a key prioritization and fundamental component of their business planning. Companies can gain a competitive advantage and contribute to the development of innovative crops, which helps them acquire unique skills that distinguish them in the market. This competitive advantage plays an important role in the reality of the circular economy, indicating an important link between innovation and competitive strategy. The report also highlights the need for a fully sustainable development strategy. This emphasizes that companies include collaboration between innovation, competitive positioning, and environmental responsibility. Long-term success depends on understanding how these components interact, as companies increasingly operate in environments where stability is crucial.

## Managerial Implications

The conclusion of this study shows that companies seeking to improve their indicators within the framework of a circular economy have a certain number of key management results. First of all, as an important element of strategic planning, managers should identify strategic priorities for innovation. Leaders can not only meet consumers' needs but also promote the creation of new solutions that promote sustainability goals and develop an organizational culture that values innovation and experiments. By focusing on innovation, the company can distinguish itself from its competitors and develop unique and valuable offers that ultimately increase its competitive advantage. Managers should also understand how important a competitive advantage is to reduce the gap between strategic innovation and the success of a circular economy. This implies that the tactics that improve the competitive position of the company should correspond to innovation.

To ensure that they respond well to market demands and use sustainable technologies, managers need to regularly assess the competitive environment and change their innovation strategies accordingly. This ongoing evaluation will enable companies to not only adapt to evolving consumer preferences but also to identify new opportunities for sustainable growth. By nurturing a culture of innovation and agility, organizations can better position themselves to thrive in an ever-changing marketplace. Emphasizing collaboration and encouraging diverse perspectives within teams can further enhance this culture. Investing in employee training and development will also empower staff to contribute creatively, ultimately driving the organization towards long-term success. In addition to optimizing resources, this alignment establishes the company as a leader in sustainability. The research also highlights critical intermittent cooperation in business. Managers need to contribute to inter-sectoral collaborations, such as R&D, marketing, operations, and others, to effectively support the circular economy with creative ideas. The creation of complete solutions that address several facets of sustainability might result from this integrated approach, increasing the overall impact of the organization's actions. Such collaborative efforts not only foster innovation but also enhance the organization's reputation by attracting customers who prioritize sustainability. Ultimately, by embracing these strategies, companies can drive meaningful change while reaping long-term economic benefits.

## Practical Implications

This report has important practical results for businesses striving to launch effective circular economy projects. Companies must first fund training and development initiatives that educate their employees on the innovation potential. Organizations can develop their staff. This is not only about creating creative ideas but also about turning these ideas into useful applications that contribute to sustainability and provide employees with the training and education they need.

To promote the exchange of innovation and best practices, businesses should also think about forming alliances with external teams such as suppliers, customers, and experts. Working together will lead to the exchange of resources, skills, and information, improving the creation of creative solutions suitable for circular economy principles. These collaborations can also provide insightful information about consumer preferences and industry trends that can help direct strategic choices.

Businesses should also closely monitor and assess their performance in terms of sustainability and innovation criteria. Organizations can make data-driven choices, ensure accountability, and identify areas for improvement by establishing a framework for monitoring progress. This continuous assessment has an important role to play, as businesses can adapt to the changing environment of the circular economy, stay competitive, and use their creative potential effectively.

## Theoretical Implications

The findings of this study contribute significantly to the case of documents already available in the areas of stability, innovation and strategic management, particularly in relation to the circular economy. Showing a robust link between strategic innovation, competitive advantage and circular economy success, this study enriches our knowledge of how these aspects interact and influence organizational outcomes. This study maintains the premise that innovation is not only a lonely success engine but is closely linked to competitive positioning in the market. By highlighting the mediating function of competitive advantage, this achievement expands the theoretical framework surrounding innovation management and provides a more sophisticated perspective on how companies use ingenious capabilities for sustainability.

The study also emphasizes the value of a comprehensive approach to organizational performance research, recommending that future theories in this area consider interdependencies across different strategic components. Scholars may develop more comprehensive models that capture the complexity of today's corporate contexts by integrating ideas from competitive strategies, innovation theories, and sustainability frameworks. Such integrative models could lead to improved decision-making processes, enabling organizations to navigate the dynamic landscape of market demands and environmental challenges more effectively. Ultimately, this holistic view may foster a culture of continuous improvement and resilience, positioning firms to thrive in an increasingly competitive world. The study also highlights the need for further empirical research to investigate other factors that may influence the dynamics between these categories in various situations, such as the impact of corporate culture and external environments.

## Policy Implication

Policymakers should focus on encouraging strategic innovation to enhance the success of the circular economy in large-scale industrial companies in Ethiopia. This can be achieved by providing companies with incentives to engage in research and development, as well as fostering cooperation between businesses and universities. Also, establishing rules that promote resource efficiency and long-term practices can help you stay ahead of the competition. It is essential to have training programs that help individuals enhance their skills, as a trained workforce is necessary for implementing new ideas. Policymakers should also make it easier for businesses to access funding for new projects that align with the principles of the circular economy. This will enable businesses to transition to more sustainable practices, becoming more competitive in the market.

## Recommendations

The research's conclusions and implications suggest several recommendations for researchers and practitioners. First and foremost, businesses need to promote an innovative culture that rewards experimentation and risk. Initiatives like idea contests, innovation workshops, and cross-departmental projects that encourage innovative cooperation can help achieve this. By allowing people to contribute to the innovation process, businesses may boost their capacity to find and develop sustainable solutions. Furthermore, it is important that organizations often modify competitive environments and change their innovative plans accordingly. Organizations can better combine creative efforts with market requirements, conduct competitive analysis and market research, and provide valuable insights into customer preferences and future trends. In addition to improving economic advantage, this strategic alignment will make it easier to successfully adopt circular economy principles.

It is advised that scholars investigate how outside alliances and partnerships promote sustainability and innovation. Future research can study how partnerships with other parties, such as suppliers, customers and educational establishments, contribute to creative solutions within the framework of the circular economy. Longitudinal studies that control the influence of strategic innovations over time can also shed light on factors that contribute to sustainable stability/success projects.

## Future Directions

Future studies on competitive advantages, strategic innovations, and the viability of the circular economy should explore various intriguing paths to enhance our understanding of these complex connections. One of the most important areas is the study of how corporate culture contributes to creativity. For practitioners seeking to enhance their creative skills, analyzing how different cultural characteristics—such as risk tolerance, adaptability, and teamwork affect companies' ability to innovate sustainably may provide useful insights. Additionally, examining the role of leadership in fostering an environment that encourages experimentation and embraces failure could further illuminate the dynamics at play. By understanding these elements, organizations can better position themselves to leverage their unique cultural attributes in pursuit of sustainable innovation.

Another important direction is to investigate the external influences that affect the relationship between these conceptions. This entails examining how market conditions, technical developments, and regulatory frameworks impact organizations' innovation strategies and their ability to gain a competitive edge in the circular economy. Businesses can more effectively navigate challenges and adapt their strategies to shifting market demands by understanding how these external influences impact organizational behavior. The dynamics of partnerships and co-operation in stimulating innovations in the circular economy can be the subject of a future study, which will demonstrate how companies can leverage their connections with key stakeholders, including suppliers, consumers, and university institutions, to enhance their innovation capabilities. The long-term consequences of innovation, including competitive advantages and sustainability results, will also be identified in longitudinal studies that control for the influence of strategic innovation over time. Such studies can also shed light on whether this research connection can be applied to other situations and industries, providing a more complete understanding of how companies continue to achieve success in the circular economy.

## REFERENCES


Arfaoui, N., Le Bas, C., Vernier, M. F., & Vo, L. C. (2022). Innovation strategies and implementation of various circular economy practices: findings from an empirical study in France. *Journal of Innovation Economics & Management*, I141-34. .DOI: 10.3917/jie.pr1.0141

Bekata, A. T., & Kero, C. A. (2025). Modeling the significance of strategic orientation for innovation capabilities and enterprise performance: Evidence from Ethiopian SMEs. *Cogent Business & Management, 12*(1), 2440122. DOI: 10.1080/23311975.2024.244012

Demessie, G. T., & Shukla, A. (2024). Drivers and outcomes of sustainable marketing strategy in the African context: The role of competitive advantage and strategic proactivity as mediating and moderating variables. *Cogent Business & Management, 11*(1), 2348442. DOI: 10.1080/23311975.2024.2348442

Karim, R., Bhowmik, R. C., Faruq, M. O., & Roshid, Md. M., Islam, S., & Ramachandaran, S. D. (2025a). Harnessing growth for ecological sustainability through exploring the drivers of load capacity factor in G7 nations. *Discover Sustainability, 6*(1), 821. DOI: 10.1007/s43621-025-01757-x

Karim, R., Waaje, A., & Roshid, Md. M. (2025b). Digital Circular Economy for Attaining Sustainable Development Goals. In Durmuş Çağrı Yıldırım, Seda Yıldırım, & Vinay Kandpal (Eds.), *Effects of Digitalization and Circular Economy on Sustainable Policy and Climate Change Prevention* (pp. 211–240). IGI Global. DOI: 10.4018/979-8-3693-9909-5.ch008

Kwarteng, A., Simpson, S. N. Y., & Agyenim-Boateng, C. (2022). The effects of circular economy initiative implementation on business performance: The moderating role of organizational culture. *Social Responsibility Journal, 18*(7), 1311–1341. DOI: 10.1108/SRJ-01-2021-0045

Roshid, Md. M., Yeamin, Md. B., Waaje, A., Karim, R., & Kaswan, M. S. (2025a). Understanding the Nexus of Industry 5.0 and Circular Economy for Enriched Lean and Environment Sustainability. In José Carlos Vieira de Sá, Shubham Sharma, Sunil Kumar, & Manjunath B.R. (Eds.), *Transformative Lean Six Sigma Techniques for the Quality 5.0 Paradigm* (pp. 349–380). IGI Global Scientific Publishing. DOI: 10.4018/979-8-3373-0943-9.ch013

Roshid, Md. M., Waaje, A., Karim, R., Irtisham, M. R., & Kaswan, M. S. (2025b). Industry 5.0 and Organizational Sustainability: Integrating Cutting-Edge Technologies and Human Ingenuity. In Mahender Singh Kaswan, Rajeev Rathi, & Jose Arturo Garza-Reyes (Eds.), *Industry 5.0: Concepts and Strategies for Digital Transformation* (pp. 50–61). CRC Press. DOI: 10.1201/9781003535188-7

Shebeshe, E. N., & Sharma, D. (2024). Sustainable supply chain management and organizational performance: The mediating role of competitive advantage in Ethiopian manufacturing industry. *Future Business Journal, 10*(1), 47. DOI: 10.1186/s43093-024-00332-6

Simane, B., Malcolm, R., O'Meara, N., Oremo, F., Geleta, Y., & Ahmedin, A. (2024). Knowledge, attitudes, and practices on Circular Economy among senior managers of Ethiopian textiles and Agro-food Processing companies. *Circular Economy and Sustainability, 4*(4), 3093–3117. DOI: 10.1007/s43615-023-00342-6

Suchek, N., Fernandes, C. I., Kraus, S., Filser, M., & Sjögrén, H. (2021). Innovation and the circular economy: A systematic literature review. *Business Strategy and the Environment, 30*(8), 3686–3702. DOI: 10.1002/bse.2834

Sufa, W. G., Birbirsa, Z. A., & Mogose, T. E. (2023). Assessing the Mediating Effect of Competitive Advantages in the Relationship between Marketing Strategy and Firm Performance of the Ethiopian Textile Manufacturing Industry. Academic Press.

Tolossa, A. T., Singh, M., & Gautam, R. K. (2024). Linking business sustainable development with entrepreneurial marketing practices: Evidence from small and medium enterprises in Ethiopia. *Business Strategy & Development, 7*(3), e412. DOI: 10.1002/bsd2.412

Tolossa, A. T., Singh, M., & Gautam, R. K. (2025). Bridging Competitive Advantage and Sustainable Development: The Moderating Role of Value Creation for Stakeholders in Ethiopian SMEs. *Business Strategy & Development, 8*(1), e70066. DOI: 10.1002/bsd2.70066

Triguero, Á., Cuerva, M. C., & Sáez-Martínez, F. J. (2022). Closing the loop through eco-innovation by European firms: Circular economy for sustainable development. *Business Strategy and the Environment, 31*(5), 2337–2350. DOI: 10.1002/bse.3024

# Chapter 16
# Strategic Innovation and Systemic Transformation for a Circular Economy: Theories, Practices, and Global Implications

**Muhammad Usman Tariq**
https://orcid.org/0000-0002-7605-3040
*Abu Dhabi University, UAE & University College Cork, Ireland*

## ABSTRACT

*Strategic innovation for the circular economy changes traditional linear production models by prioritizing sustainability, resource efficiency, and long-term value-added. In this article, businesses and political decision-making-makers will explore ways to reduce waste, expand the lifecycle of product life services, and regenerate natural systems using circular design, digital technology, and innovative business models. Through case studies and theoretical framework conditions, we examine the role of strategic planning, stakeholder collaboration, and systematic changes in promoting intersectoral circularity. Critical enablers such as product as a service model, reverse logistics, and material innovation have been critically acclaimed. The study also addresses regulatory gaps, cultural resistance, and scalability issues. Ultimately, it strategically integrates circular principles into core organizational practices to support global sustainability goals.*

DOI: 10.4018/979-8-3373-2454-8.ch016

## INTRODUCTION

As the Global Footprint Network notes, in 2017, the global population was living on 1.7 Earths, so there is a gap between demand and planetary capacity. The linear model has its functioning principles: extraction of limited natural resources, mass production, consumption, and discarding products as waste. This model can be considered the main force of economic growth over several decades. However, this has come at the cost of extreme environmental degradation, depletion of natural resources, and increased climate change (Zorpas, 2024). The alternative idea circulating the circular economy involves a completely different view of economic processes, where the process is not connected to the use of resources and its impact on the environment. A circular economy aims to replace the concept of end-of-life with the concept of end-of-use by circulating products and materials for as long as possible, making products durable and repairable, and regenerating and reusing materials. The aim is to achieve a sustainable system where resources circulate in an endless chain, the amount of new materials used is kept to a minimum, and environmental harm is minimal. This change requires a rethinking of manufacturing and consumption processes and a revamping of the way things are done in industries, governments, and individual lives (Karim et al., 2024).

Experimentation is essential for promoting this transition. Innovation within the circular economy is not restricted to product development. Nevertheless, it includes the entire business model, value chain, and organizational culture. New business models, such as product-as-a-service, redefine what products do and how people use them, making it desirable that their products are longer-lived, maintained, and recycled. Companies have an incentive to keep products durable, repairable, and compatible with recycling; by retaining their ownership, they are interested in making products long-lasting, easy to repair, and recyclable (Nie et al., 2024). This shift in the incentive structure guarantees that economic value creation is in harmony with environmental stewardship. It is a strategic innovation in the circular economy that involves digital transformation. Sustainability plays a vital role in the development of a circular economy by ensuring that ecological and social aspects are factored into decision-making activities. These practices involve prioritizing the production of sustainable products, using renewable or reusable materials, using less energy during production, and ensuring fair and ethical labor practices. To achieve circular and sustainable practices, cooperation between industries is often required, particularly in industries with complex and global supply chains. Organizations must cooperate with their suppliers, regulators, consumers, and competitors to establish industry standards, share the best strategies, and invest collectively in innovative solutions. Therefore, it is essential to create efficient policy frameworks and offer institutional support. Simultaneously, educational institutions and civil society organizations can contribute by creating more awareness, skills, and positive actions.

The importance of the circular economy on a global level is underpinned by the fact that several of the United Nations Sustainable Development Goals (SDGs) are aligned with it. Sdg 9 emphasizes the role of good infrastructure and inclusive industrialization, which are essential for developing circular manufacturing processes and supply chains. The concepts of circular economy encourage a shift in industries by underlining the redesign of infrastructure to enable recycling, remanufacturing, and recovery of materials. SDG 12 goes hand-in-hand with circularity, where a minimal amount of waste should be produced, and there should be responsible management of natural resources, as well as the promotion of sustainable consumption patterns (Herrero et al., 2023). The circular economy framework addresses these goals through the reuse of products, waste elimination, and a shift towards a consumer attitude that values retained product value over a disposable nature. Finally, SDG 13 focuses on the urgency of climate change action. By decreasing resource extraction, lowering energy usage, and curbing waste and emissions, the circular economy can play a substantial role in combating climate change.

Despite its potential, implementing a circular economy involves numerous challenges and obstacles. One significant challenge is the absence of economic incentives to promote linear models. In numerous sectors, extracting raw materials and discarding products is often more cost-effective and practical than investing in recycling facilities, reverse logistics, or redesigning products. The financial disparity between circular and traditional practices discourages investment in circular initiatives, especially among small and medium enterprises that lack the necessary funds to pursue long-term innovative strategies. Another critical obstacle is the regulatory misalignment. Current policies and standards are primarily designed using linear models, which may impede the implementation of circular practices (Suphassomboon and Vassanadumrongdee, 2022). Waste classification regulations may prohibit the reuse of materials classified as waste, even if they are deemed safe and functional. Addressing these barriers necessitates regulatory changes, alignment of standards across borders, and implementation of incentives, such as subsidies for circular technologies or penalties for excessive waste production.

Cultural and behavioral resistance are barriers to adopting circular models. Consumers and businesses may hesitate to alter their habits and beliefs regarding ownership, convenience, and product value. The sharing economy is one such example, where one challenges the idea that when you own something, you feel important or safe. Likewise, people may not be drawn to circular product designs that are durable or easily reconfigured when they are expected to drive new product sales through frequent upgrades. Addressing these issues will require targeted awareness campaigns, educational programs, and success story sharing to emphasize the practicality and benefits of circular practices. User-centered design should be a priority for companies to ensure that circular products exceed customer expectations in terms of quality, appearance, and usability. There are important research questions regarding the size of the circular economy (Le, 2023). A key question is: What are the most effective strategic innovations to drive circular transitions? To answer this question, it is necessary to investigate how business model innovation, technological breakthroughs, and policy frameworks can facilitate circularity. Scientists should examine how other developments, such as digital passports for products, closed-loop manufacturing systems, and decentralized recycling systems, can influence the system to yield important changes. The most important question is how businesses can make their operations, supply chains, and products circular. It involves looking into ways of involving suppliers, performance analysis, material selection, and life-cycle evaluation. It also involves understanding the organizational changes required to develop circular thinking, such as learning new skill sets, changes in leadership approaches, and increases in cross-functional collaboration.

It is equally important to note the impediments and weaknesses faced by businesses in moving to circular economy models. This study focuses on determining industry-specific barriers, assessing the effectiveness of existing policy interventions, and analyzing the impact of consumer behavior on market trends. Finally, the issue of how technology can aid and boost the systems of a circular economy is worth considering. Further research is needed to explore how artificial intelligence, the Internet of Things, and blockchain can enhance resource efficiency, traceability, and transparency. For instance, AI can create circular products by simulating material properties and streamlining the production processes. Simultaneously, IoT devices can facilitate product-as-a-service models by tracking usage patterns and initiating maintenance actions. Blockchain technology can facilitate the verification of recycled materials and ensure adherence to sustainability standards. The research questions require the cooperation of various disciplines to be addressed. The circular economy is multidisciplinary, cutting across environmental science, industrial design, economics, information technology, and public policy (Abu-Bakar et al., 2024). Effective solutions will require interaction between academia, industry, and civil society, and the use of local and indigenous knowledge systems. Furthermore, international cooperation is essential for coordinating standards, sharing winning policies, and creating international markets for circular products and services. Innovation in this respect must include both technological developments and institutional practices. This must include adopting more sophisticated tools and systems and creating new governance systems, financial systems, and cultural stories that promote long-term thinking and collective responsibility. The circular economy represents a more sustainable, resilient, and inclusive future. Challenging established production and consumption patterns will compel societies to re-examine the ideas of value, progress, and prosperity. Nevertheless, its economic, societal, and environmental benefits are tremendous and far-reaching (Fischer et al., 2021).

## THEORETICAL FOUNDATIONS OF CIRCULAR ECONOMY AND INNOVATION

The principles of a circular economy offer an optimistic system for understanding how people and companies can decouple wealth generation from ecological destruction. Fundamentally, the circular economy is a call to stop thinking in terms of a linear system of take-make-dispose and start thinking in terms of a cyclic system of reduce-reuse-recycle. The principles within this paradigm shift are the 3R approach: reduce, reuse, and recycle. These three measures are not short-lived but are used to change the entire system of production and consumption. The focus on reduction is on the minimization of the volume of resources used and waste generated at the source and encouraging the efficient use of each of the materials and energy. It also focuses on prevention rather than treatment, and industries find themselves revisiting product design, process efficiencies, and patterns of consumer demand. Reuse targets the extension of the useful life of goods and parts through repair, refurbishment, and repurposing, reducing the consumption of new materials and energy. Recycling, although frequently discussed in public conversations, is regarded as a last resort in the hierarchy of recovering materials only when products are no longer usable in their current state. Figure 1 provides an overview of the listing of product-life extension, resource recovery, circular inputs, and information-sharing platforms.

*Figure 1. Circular Business Models and Examples (Source: Author)*

| Business Model | Core Idea | Example |
|---|---|---|
| **Product-life extension** | Design for durability, repair, upgradability | **Patagona** repair services |
| **Resource recovery** | Recover materials/ components from waste streams | **Renault** remanufacturing plant |
| **Circular inputs** | Use renewable/ recyclable/ biodegradabble inputs | **Unilever** blo-based packaging |
| **Sharing platforms** | Maximise asset use via sharing | **Airbnb** for accommdation |

Although the 3R framework laid the foundation for early sustainability practices, it is insufficient for systemic transformation. Recycling, in particular, often addresses symptoms rather than root causes and cannot alone offset rising consumption levels. Although the 3r model is a solid foundation, circular economy thinking has expanded to encompass more intricate and interconnected concepts. It encompasses regenerative design, closed-loop systems, cradle-to-cradle product development, and biomimicry (Herrador & Van, 2024). These advanced frameworks acknowledge that materials, products, and systems should be designed to harmonize with ecological processes rather than oppose them. Regenerative systems strive to reduce harm and restore and enhance natural and social resources. For instance, a regenerative agricultural system can improve soil fertility, biodiversity, and water retention, rather than merely recycling nutrients. By adopting a circular economy approach, the focus shifts from a purely technical solution to a philosophical perspective that challenges the traditional relationship between humans and the environment.

The most important aspect of circular thinking is the difference between linear and circular business models. Linear models presuppose unlimited resources and a lack of repercussions from waste disposal. Raw materials undergo a series of processes to become goods, which consumers use and then dispose of, usually without regard to the impact that ensues past the point of purchase. This model is not sustainable in the long run because it relies on the extraction of new materials and dumping into nature, which is already highly stressed (Opoku et al., 2022). The linear model is also vulnerable to economic and geopolitical shocks, resource scarcity, and regulatory pressure, making it a risky model in the modern globalized economy. Circular business models take the opposite direction, striving to eliminate waste and aiming to keep existing materials and resources in use for longer. Circularity has different forms, including prolonging the life of products, reclaiming resources, consuming circular resources, sharing platforms, and products as a service. The circular economy lends itself to various aspects of value extraction in each model. Product life extension strategies aim to prolong the life of products by designing features that foster durability, repair, and upgrading. Resource recovery models collect and reprocess materials in used products or waste streams and then reuse those materials in production processes. Circular inputs lack finite or toxic materials compared to renewable, recyclable, or biodegradable materials (Monciardini et al., 2024). Sharing platforms help make the best use of underutilized assets, as they allow such resources to be used by more than one person. Simultaneously, product-as-a-service arrangements change the focus of ownership to access, which leads manufacturers to focus on the durability and performance of their products rather than high levels of sales.

The advantages of adopting circular business models are numerous and diverse. Circularity can open up new sources of income, lower expenses, and enhance resilience by diversifying supply chains and decreasing reliance on raw materials. Environmentally, it helps conserve resources and reduces greenhouse gas emissions, pollution, and biodiversity loss. Socially, it can create new jobs in sectors such as repair, remanufacturing, and sustainable logistics, while promoting more inclusive and community-focused consumption habits. Circular business models necessitate systems thinking, a comprehensive approach that considers the interconnections among the design, production, usage, and post-use phases of the product lifecycle. Strategic innovation is crucial for facilitating and expediting the transition to a circular economy (Ding et al., 2023). It encompasses technological advancements and organizational, behavioral, and policy changes that facilitate systemic transformations. Design thinking is one of the most effective tools in this regard. Fostered by empathy, creativity, and iteration, design thinking empowers businesses and policymakers to reshape problems, envision diverse futures, and collaboratively develop solutions with users and stakeholders. In the realm of circularity, design thinking promotes collaboration among individuals from various disciplines to create products, services, and systems that are practical, visually pleasing, and sustainable. This challenges the traditional balance between performance and sustainability, leading to the development of innovative solutions such as modular smartphones, refillable packaging, and biodegradable textiles.

Digitalization is also a transformative force in circular innovation. Connected sensors can provide up-to-date information on product status, use, and location, which can be used to conduct predictive maintenance, optimize product utilization, and drive circular services. Machine learning allows the examination of large amounts of data, uncovering areas of waste, revealing opportunities to optimize processes, and anticipating demand, resulting in more precise and efficient production. Blockchain offers transparency and accountability, creating trust in recycled materials, responsible sourcing, and product provenance. These technologies facilitate circular approaches and create new value propositions and business models that are more sustainable. However, technology alone is insufficient without conducive policies and institutional frameworks. To achieve circular goals, it is important to enact strategic policies that align market incentives with circular goals, overcome structural barriers, and establish a level playing field for sustainable innovation. The policy instruments would include a variety of mechanisms, including the development of standards, the provision of financial incentives, guidelines on public purchases, and investment in infrastructure and skills development projects. For example, the extended producer responsibility laws ask manufacturers to take responsibility for the full life cycle of their products, which should internalize environmental externalities and encourage superior design. Through the adoption of eco-labeling and transparency, customers are empowered to make informed choices, which serves as an incentive for their support of circular products. By encouraging circular goods and services, related laws facilitating public procurement can help grow the market and stimulate innovation (Tariq, 2025).

Institutional change is equally important. This entails restructuring governance systems, stakeholder relationships, and value networks to facilitate circular outcomes. When public and private entities, industry groups, and sectors come together, they can combine their resources, share risks, and accelerate the learning process. Educational institutions and research organizations are crucial for building the knowledge base, skills, and thought leadership required for circular transitions. Grassroots movements and civil society organizations can bring about cultural transformation, fight for environmental justice, and ensure that institutions are held responsible. These actors work together to create an environment that encourages innovation, allows for the growth and expansion of circular practices, and helps establish them as standards within the industry. A crucial aspect of strategic innovation is assessing and conveying progress. Traditional financial metrics often overlook circular strategies such as resource efficiency, environmental restoration, and social inclusion, which fail to capture their true value (Gu et al., 2024). New indicators and reporting frameworks are needed to assess trade-offs and inform decision-making to accurately measure circular performance. Material flow analysis, life cycle assessment, circularity indicators, and sustainability reporting standards are some of the tools that can help businesses and policymakers monitor progress and provide guidance on how to act. The circular economy is not a silver bullet but rather an adaptable model that must be adapted to individual ways, resources, and objectives. In some locations, the focus can be on industrial symbiosis and the manufacturing efficiency of resources (Chau et al., 2024). In other cases, it could be related to the integration of the informal sector, waste valorization, or sharing initiatives at the community level. The interpretation and application of circularity are determined by cultural values, social norms, and historical legacies. For this reason, the process of circular transitions requires bottom-up and inclusive approaches that recognize local expertise, empower the disadvantaged, and promote social justice.

The circular economy is a highly interdisciplinary concept with connections to environmental science, economics, systems theory, industrial ecology, design, and innovation studies. They offer a panoramic view that can help us rethink our practices of creating value, what we mean by progress, and how we can restore human systems to work within the boundaries of the planet (Wilson et al., 2024). Strategic innovation, facilitated by design thinking, digital tools, and the right policies, can transform theory into practice, paving the way for change. In this era of urgent problems, such as climate change, biodiversity loss, and resource scarcity, the circular economy is not only a criticism of the existing process but also a constructive and optimistic guide to how such problems can be solved for a more sustainable, equitable, and resilient future.

## Circular Product Design and Innovation

The circular design and innovation of products are key to creating a sustainable economy, where resources are used more efficiently, waste is minimized, and value is constantly restored by reuse, repair, and renewal. The circular design also contrasts with the linear model of production, in which goods are produced with a limited life span and later discarded. Instead, the circular design aims to maximize the life cycle of materials and components so that they can be reintroduced into the system at the end of their use. The paradigm shift begins at the very beginning of the product life cycle: the design phase. Designers should now consider how the items will perform during use and how they could be disassembled, repaired, made more useful, or re-purposed. Consequently, the focus of eco-design and green product development is to develop durable, low-maintenance, and flexible products for long-term use. Durability is a significant factor in eco-design because products that last longer will not have to be replaced often, reducing the amount of resource extraction, production emissions, and disposed quantities (Williams et al., 2024).

Ensuring durability involves selecting durable materials and components, designing with mechanical strength in mind, and designing against wear. The possibility of a product being repaired or restored by the user or technician is also important. The complexity of integrated electronics and the enclosed components of many modern devices have made repair almost impossible, creating a culture of planned obsolescence. Circular design subverts the convention of linear design by prioritizing accessibility, encouraging user contributions to product longevity and supporting third-party repair economies.

Recyclability is a fundamental principle of eco-design. The success of recycling a product depends on its ability to be disassembled and its components separated into pure material streams. This necessitates the thoughtful consideration of how materials are joined, secured, or combined. Adhesives, for instance, frequently impede recycling efforts by contaminating material flows or making separation difficult. Designers should consider using single-material components, snap-fit assemblies, and standardized parts to simplify recycling. To promote sustainable practices, it is crucial to minimize or replace complex labels, coatings, and treatments with eco-friendly alternatives that do not hinder material recovery. Crucially, when designing for recyclability, the practicality of recycling in real-world scenarios must be considered, ensuring that the product design aligns with current or future recycling infrastructure. Material innovation is crucial in the process of creating circular products. Materials such as plastics and metals typically have significant environmental costs because of the extraction, processing, and disposal involved. Advancements in eco-friendly, reusable, and renewable materials are reshaping the potential for sustainable design (Trivediet al., 2024). Materials that can naturally break down, such as those made from corn starch, algae, or cellulose, provide a solution for reducing the impact of waste on the environment. When constructed correctly, these materials can decompose into harmless substances when subjected to industrial composting or natural environmental conditions.

Nevertheless, adopting this technology must be handled cautiously to prevent unintended outcomes, such as the contamination of recycling streams or premature deterioration (Khan et al., 2024). Closed-loop materials, such as aluminum, glass, and specific plastics, can be recycled repeatedly without a significant decline in quality. Their utilization promotes a system in which the consumption of raw materials is reduced and existing materials are continuously reused. Renewable materials, including bamboo, cork, wool, and bio-based polymers, offer promising alternative solutions. These resources are naturally replenished and can decrease reliance on fossil-based or limited resources. Their utilization frequently contributes to the growth of rural economies and decentralized production systems, bolstering resilience and social value (Seyi-Lande et al., 2024). Material innovation involves discovering substitutes for harmful substances and maximizing resource utilization throughout the entire lifecycle. Lightweighting, for instance, decreases the material content of products and enhances energy efficiency during transportation and usage. Researchers are investigating self-healing materials, responsive textiles, and smart composites to improve product durability and performance, aligning with circularity goals (Tariq, 2025). End-of-life strategies are where circular thinking completes the cycle. Unlike landfilling or incineration, circular end-of-life options prioritize the preservation of product and material values. The remanufacturing process entails disassembling a previously used item, thoroughly examining and cleaning its components, replacing or enhancing worn-out parts, and then reassembling it to match its original quality. This is important when handling complex, high-value items such as machinery, electronics, and auto parts (Karaca et al., 2024). Remanufactured products have the potential to yield enormous cost and resource savings over new production with quality and reliability. The trick behind remanufacturing is the vision during design: products must be easily dismantled, their components must be standardized, and they must have a traceability feature that guarantees their authenticity and performance data.

The idea of upcycling is to recycle used materials or products into new objects of equal or greater value. This may involve the creation of waste material into stylish accessories or the reutilization of old furniture for a different purpose. Upcycling reduces waste and maximizes the aesthetic and cultural value of products. It often involves the participation of local artisans, designers, and communities to create unique objects. Although it may not be as large-scale and consistent as industrial recycling, upcycling is essential for creating a culture of reuse and inspiring innovation. It encourages customers to value products that people regularly discard and become participants in the circular economy. PaaS models mark a radical departure from the traditional ownership-based consumption model. Instead of selling products, companies provide access to products and maintain ownership and responsibility for maintenance, upgrade cycles, and disposal. Instead of being careful about spending money on the product itself, customers are ready to pay the price of the perceived value of a product or service. This model encourages manufacturers to produce goods that will have a more prolonged life, easy repair, and repurpose because the cost and benefits of the product over time are the responsibility of the producer (Tutore et al., 2024). Examples of leasing models include renting electronics, subscribing to clothing services, or paying for appliance usage. PaaS promotes better resource utilization, as products can be shared among users or repurposed once used. It also strengthens customer relationships, as companies stay engaged in the product experience by providing maintenance, support, and regular updates.

Design considerations must prioritize flexibility, adaptability, and value retention to achieve success. Products should be designed to be easily upgraded, have a timeless appearance, and be durable enough to withstand repeated use. Incorporating digital technologies can improve the value proposition by enabling remote monitoring, optimizing usage, and predicting maintenance needs. Products are equipped with sensors that gather information on wear and performance, enabling service providers to be notified of potential problems before they lead to product failure. It enhances user satisfaction and optimizes the economic and environmental benefits derived from each unit. Engaging consumers is crucial for driving innovation in circular products. The effectiveness of long-lasting, fixable, and service-oriented products relies on consumers' openness to embracing new routines and standards. By implementing educational initiatives, clear communication, and intuitive design, we can encourage a shift in mindset from a disposable culture to one that values sustainability. By equipping consumers with the necessary information and resources to fix, enhance, or exchange products, we can empower them to take ownership and accountability of their purchases (Singh et al., 2024). Additionally, companies must ensure that circular products are easily accessible, visually appealing, and reasonably priced. Design should not prioritize sustainability over functionality or aesthetics. However, it should be noted that environmentally conscious decisions can also be practical and appealing. Incorporating circular principles into product development necessitates an organizational transformation. Collaboration between design, engineering, procurement, marketing, and sustainability teams is crucial to ensure that goals are aligned and that systemic solutions are implemented. Embedding lifecycle thinking into decision-making processes is crucial, and tools such as lifecycle assessment, material flow analysis, and sustainability scorecards can provide the necessary support. Incentive structures, performance metrics, and innovation frameworks must be adapted to encourage long-term value creation rather than focusing solely on short-term profitability. Within the broader economic environment, circular product innovation helps make economies more resilient, competitive, and environmentally friendly. Businesses can hedge against price fluctuations, political uncertainties, and challenges in government control by reducing their dependence on new resources and outward supply webs. Circular products incur lower total costs and produce new value flows as they are reused, refurbished, and provided as services. They can also fulfill the growing consumer interest in brands that have an ethical focus, are sustainable, and have a purpose behind them. The transition to circular product design does not come without challenges. Lack of technical capabilities, infrastructural deficits, consumer skepticism, and organizational reluctance may become barriers (Alka et al., 2024). However, the growing effectiveness of case studies, emerging enabling technologies, and the necessity to overcome environmental problems contribute to this movement.

## Circular Supply Chain and Manufacturing Processes

Adopting a circular economy will require a complete transformation of supply chains and production processes, with an emphasis on integrating sustainable practices and utilizing digital technologies. Sustainable supply chain management (SCM), which redefines the processes of procurement, logistics, production, and distribution by considering sustainability, is the basis of this change. CM exceeds traditional efficiency measures by considering environmental and social concerns along the value chain. It also requires positive working relationships with suppliers, open procurement processes, and a commitment to ethical and sustainable sourcing. The production processes are energy-efficient and have little impact on the environment (Chabowski et al., 2025). One way to do this is to emphasize not just the cost and quality of purchases, but also the impact that purchase will have on the environment and the possibility of end-of-life recovery, making sustainability a primary consideration instead of an add-on or afterthought at the end of the decision-making process. In circular supply chains, relationships with suppliers are even more vital because manufacturers must rely on an intricate network of partners to achieve their sustainability goals. Creating long-term partnerships based on trust and common benefits is important for setting the same objectives and collectively dealing with risks. To enhance communication and coordination, organizations can use collaborative planning, forecasting, and replenishment systems to ensure a smooth and sustainable flow of materials across the network (Strippoli et al., 2024). Such partnerships can also be used to develop circular innovations such as reusable packaging, modular product design, and shared transportation systems. Transparency is the key to these alliances. Blockchain and other technologies can be used to ensure traceability when all parties can ensure the sources of raw materials, ingredients, composition, and environmental effects, leading to transparency and reducing the chances of greenwashing (Sehnem et al., 2024). Logistical factors must be modified to fit circularity.

Traditionally, logistics involves transporting manufactured products to the place of consumption. In contrast, circular logistics should not ignore the reverse flow of products for maintenance, refurbishment, or recycling. The reverse logistics process requires setting up infrastructure, systems, and partnerships that ensure the efficient withdrawal and processing of used goods (Tariq, 2025). The rise of industrial production, especially within the mechanisms of Industry 4.0 and Industry 5.0, significantly supports the concept of circular manufacturing. Industry 4.0 is the integration of cyber-physical systems, data analytics, and the Internet of Things (IoT) in manufacturing. These technologies enable constant monitoring, proactive maintenance, and informed decision-making, which are critical for maximizing resource consumption and reducing waste. By collecting and analyzing the data on every production stage, manufacturers can identify areas of ineffectiveness, predict future equipment failure, and reduce downtime, thus maximizing the use of machines and resources. A digital twin is one of the most promising tools developed with the advent of Industry 4.0, which is a virtual representation of a physical asset, system, or process. Digital twins allow manufacturers to test various scenarios, analyze optimization, and forecast without disturbing real-time processes. This aspect encourages proper resource management, as it allows for proper planning and rapid response to changing conditions, including changes in material supply or energy prices. It emphasizes the necessity of collaborating with intelligent systems, the aim being the manufacture of more robust, responsive, and inclusive systems. In circular production, Industry 5.0 advances more efficient production methods and focuses on environmental sustainability and social equity (Phonthanukitithaworn et al., 2024).

Human employees are regarded as valuable inputs to innovation and ethical decision-making, not as cogs in a machine. This perspective focuses on the role of training, employee engagement, and cultural change in the transition to circular manufacturing. Ai can accumulate large volumes of operational information, helping to discover possibilities for material replacement, process improvement, or energy expenditure. Machinery and products using IoT keep track of how they are used, how they are worn, and how they perform at all times so that necessary decisions can be made on whether to maintain them, use them again, or recycle them. Blockchain ensures that the validity and integrity of supply chain data are accurate and reliable and that information can be shared securely and transparently among the various parties involved. Another major aspect of Industry 4.0 that enhances sustainability is predictive analytics, which allows companies to anticipate future events and take prompt action to manage their resources. It reduces overproduction, stock destruction, and energy wastage. Predictive models can also be used to determine maintenance cycles, extend the lives of equipment, and prevent failures that can cause material loss or environmental pollution. Predictive analytics is applicable in a circular economy to observe the product lifecycle, wear and tear of parts, and optimize reverse logistics processes (Brogi & Menichini, 2024). Through these tools, one can make informed decisions based on statistics and ensure that business prosperity does not conflict with environmental sustainability.

Efficient resource and waste management are the basic goals of circular manufacturing. To achieve these objectives, a holistic approach is required that involves selecting the right materials, designing and manufacturing products, and managing waste. It involves the selection of the right inputs and partnering with suppliers to ensure that the inputs meet sustainability standards. In certain instances, manufacturers can invent materials that they own or partner in joint ventures to develop environmentally friendly materials (Arranz et al., 2024). Product design is a potential leverage point. Durability, modularity, and disassembly should be prioritized to maximize the chances of reuse, repair, and recycling of a product. Modular design allows changes in or upgrades of individual components without disposing of the entire product. Conversely, disassemblable products aid in the easy recovery of materials at the end of their life cycle (Khan & Khurshid, 2024).

## Business Models for the Circular Economy

The shifting environment of business operations due to environmental concerns and resource constraints has brought about new innovative products, such as product-as-a-service (PaaS) and the sharing economy in general. These models are a way of leaving traditional ownership models and opening towards service-based models that do not contradict circular economy models, as they readjust the operations of companies towards the best end. In a pay-as-you-go model, a company does not sell products or services to consumers but gives them access to products or services through a subscription or pay-per-use basis. It not only transforms the mode through which we use products but also has a considerable influence on the way products are designed and managed through their lifecycle. Businesses in the pay-as-you-go system will continue to own the products they make, which is an incentive to create products that last so they can be reused or recycled. As the company is responsible for the product over the entire cycle of the product, there will be a strong incentive to increase its usability and reduce waste at the end of the product's life. The shift to service provisioning facilitates the relationship between businesses and clients, encouraging long-term associations and creating multi-year revenue streams (Tariq, 2025). Figure 2 provides an overview of sustainable intentions and how businesses can intervene.

*Figure 2. Intention–Action Gap in Consumer Behaviour (Source: Author)*

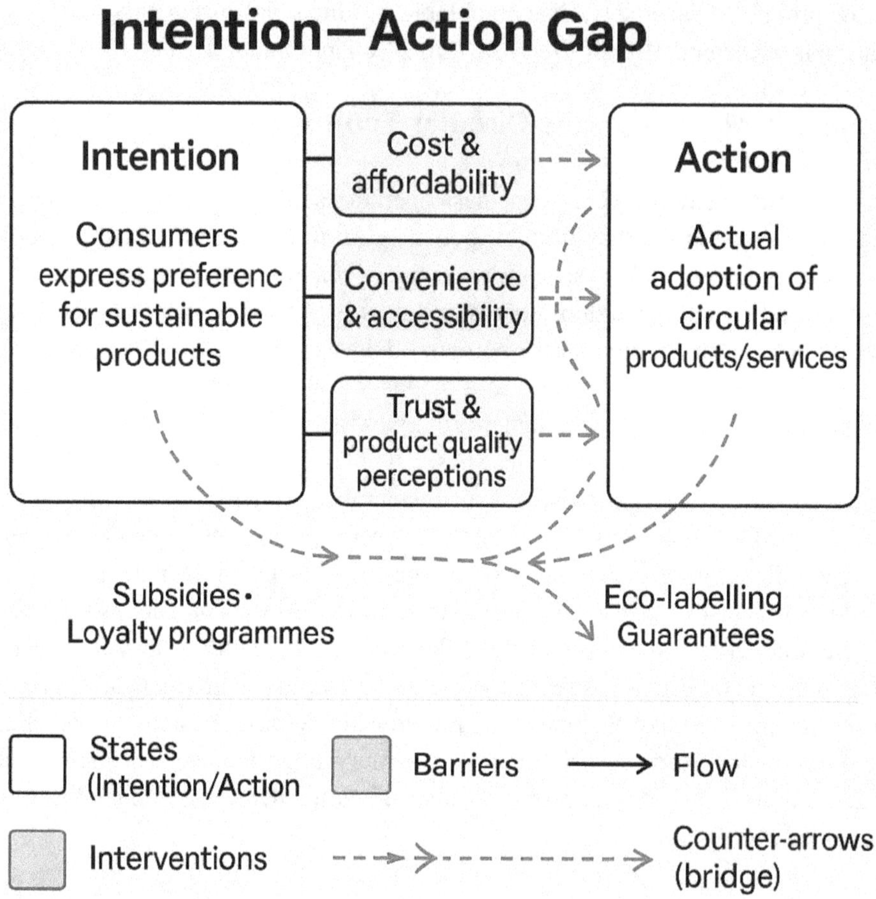


The transition to a pay-as-you-go system requires significant changes to both the processes and mentality of a company. Companies must develop new logistics, customer service, asset tracking, and product upkeep expertise. Moreover, priced strategies must consider long-term value delivery instead of short-term deals. Some companies may find it difficult to commit funds to long-term investments and asset management. Nevertheless, these barriers are usually outweighed by the lower costs and environmental responsibility. Moreover, with the growing consciousness of the environment, service-based models are becoming more attractive, particularly when they are convenient, flexible, and have a lower environmental impact. Reverse logistics and take-back programs are closely related to PaaS and service-based solutions and are important methods for closing the loop in a circular economy. The main idea behind these systems is to reintroduce used, broken, or obsolete objects into the production and consumption processes through reuse, repair, reconditioning, or recycling (Oghazi et al., 2024). Take-back programs help businesses reclaim valuable raw materials, reduce their dependency on material resources, and reduce the volume of waste that ends up in landfills. Such programs also provide useful information on the performance of the products and the way they are used by consumers, which can be used to make better product designs in the future and improve services.

To establish a feasible reverse logistics model, several necessary factors are required, including easy collection, transportation logistics, processing facilities, and incentives to motivate customer participation. For example, companies can establish specific drop-off points in retail stores or free pick-ups to encourage returns (Speranza, 2018). Once the products have been collected, they need to be sorted, analyzed, and sent to the appropriate channels, that is, to be repaired, to extract the components, or to be recycled. Technology is invaluable for monitoring these materials, compliance, and the efficiency of the reverse flow process. Large-scale take-back programs require businesses to devote resources to training, partnerships, and even lobbying policies to ease the implementation process. Effective reverse logistics programs are strongly dependent on consumer participation. By making consumers aware of the positive environmental benefits of returning used products and making the process easy and rewarding, businesses can effectively engage consumers. Designed rewards, rebates on subsequent purchases, or loyalty points may encourage consumers to keep products rather than dispose of them. In some cases, companies can use online platforms and tools to provide information, organize pick-ups, or track returns, all of which add to the user experience.

It is also important to build trust, as customers should have confidence that returned items are handled responsibly and that their attempts are doing some good. Knowledge of consumer behavior plays a critical role in any field of circular economy application, particularly in retail and marketing. The success of eco-friendly products and circular models depends on their design and operational viability, as well as consumer attitudes, willingness, and motivation to adopt new behaviors. Although sustainability is recognized by many as being significant, not much is done because of factors such as inertia, cost, or the lack of an appealing alternative. The problem recounted, the gap between what people say they want to do and what they actually do, is an old one (Yosep et al., 2021). The only way to bridge this gap is by having in-depth knowledge of consumer values and factors affecting their decision-making. Digital marketing platforms present good opportunities to personalize communications and communicate directly with customers. Social media initiatives, partnerships with micro-influencers, and interactive content can also increase awareness and the feeling of community around circular behaviors. Consumers can also be engaged in more effective ways to eliminate the gap between their intention and action towards sustainability, such as engaging them in dialogue, asking about their sustainable processes, and building new solutions to existing problems through feedback loops. Loyalty programs that reward sustainable habits can motivate good behavior and provide unending rewards. Organizations that can openly share their efforts to communicate their initiatives, whether in product lifecycles, environmental impact reductions, or progress towards circularity goals, build credibility and trust. In addition, educating consumers not only about the individual product benefits of circularity but also about the system benefits (e.g., fewer carbon emissions, less pollution) can move the discourse beyond the product and towards a lifestyle change. Retailers have a unique opportunity to play a part in the circular economy by applying these principles to their physical and digital estates. Retail stores can develop designated spaces for circular products, repair services, and customer returns or exchanges (Rabby et al., 2021). They can post product stories, add sustainability filters, and provide detailed descriptions of take-back policies or product certifications online. External partnerships for resale or refurbishment can increase the product life cycle and widen the customer base. The growing popularity of e-commerce, a system where used products find their way into the market via formal channels, is one example of how the circular economy can align with the growing preferences of consumers to be cost-efficient and environmentally responsible.

## POLICY, GOVERNANCE, AND INSTITUTIONAL FRAMEWORKS

Policy, governance, and institutional changes cannot be excluded from the factors that lead to the growth and development of the circular economy. On a fundamental level, the circular economy is a response to the linear economic system of resource extraction, utilization, and disposal. A transition towards a system that emphasizes reuse, repair, recycling, and regeneration requires a massive reorganization of the regulatory system and institutional framework. Policy is a crucial element in the transition process, providing the guidelines, drivers, and machinery necessary to transform the actions of both the general populace and business communities (Mishra et al., 2024). Different regulatory mechanisms are being introduced worldwide to promote circular economy practices and guide businesses and societies to refocus their economic activities on long-term sustainability. Globally established policy frameworks, such as the sustainable development goals (SDGs) of the United Nations, can provide a detailed guide to countries wishing to integrate the circular economy into their policies regarding the economy and the environment. SDG12 which is based on responsible consumption and production, perfectly fits the concept of a circular economy, which deals with less waste and maximum utilization of available resources. Some nations, such as the European Union, have established national roadmaps and strategies for a circular economy (Karim et al., 2024). The policy package is an elaborate guideline for product designing, manufacturing procedures, waste management, and consumer rights, which leads to a well-rounded legal framework that nurtures and supports the proper implementation of circular practices. Among the key aspects of EU policy is the focus on eco-design, where products are designed in a manner that they can be repaired, recycled, and reused. This group of regulatory drivers points to a major shift from voluntary sustainability to enforceable mandates that transform entire markets.

Circularity has been incorporated into environmental and local industrial policies by several governments. China has already adopted the circular economy promotion law, which emphasizes the conservation of resources in various sectors and formulates policies concerning cleaner production and recycling systems. In Latin America, Colombia and Chile are countries where circular strategies are promoted in urban planning, waste disposal, and farming. These policies do not only address environmental protection. However, they are also shown as economic opportunities, sources of employment, and technological advancements. Importantly, regulations are aligning under regional alliances and trade agreements, strengthening cross-border trade and investment circular principles (Straub et al., 2023). However, the effectiveness of these policies depends on the possibility of formulating and carrying them out, enforcing the policies, and incorporating the policies into overall development and industrial strategies. Extended producer responsibility (EPR) is regarded as one of the most effective regulatory mechanisms for the circular economy. EPR policies create a heavy financial and physical burden on manufacturers regarding the proper disposal or treatment of products once individuals have used them. EPR promotes the participation of manufacturers in the full life cycle of their products, which encourages a shift towards circular design, resource-efficient material usage, and the implementation of robust reverse logistics. An example is the EU waste electrical and electronic equipment directive, which requires producers to fund the collection, treatment, and recovery of electronic waste to a certain extent, and therefore encourages design decisions that minimize the costs of disassembly and support material recovery.

Enterprises covered by the ERP are also becoming more creative in reducing compliance costs and increasing operating efficiency. Packaging firms are resorting to the use of more lightweight, recyclable products, electronics makers are developing modular products that are easier to repair, and apparel companies are looking at take-back programs to get their textiles reused. Besides its direct environmental benefits, EPR helps develop secondary markets for refurbished goods, recycled materials, and remanufactured elements. Despite these challenges, inconsistent implementation, consumer ignorance, and diversity in ERP implementation across different regions still exist. In countries with poor governance or informal waste management systems, ERP programs often fail to meet their desired objectives, highlighting the need for strong institutional frameworks and international cooperation. With these regulatory reforms, the concept of public-private partnerships (PPPs) has also become an important mechanism for fostering circular innovation (Tariq, 2025). These collaborative strategies incorporate governments, businesses, civil society, and research institutions to find solutions that address policy requirements but are sensitive to market dynamics. Contrary to the top-down governance structures in traditional models, PPPs in the context of the circular economy are more about collective responsibility, information exchange, and cooperative investment in infrastructural and innovation development. An excellent case is the ce100 network of the Ellen MacArthur Foundation, which gathers companies, cities, and academic institutions and promotes circular solutions by collaborating on pilot projects and innovation laboratories. These platforms can help participants explore innovative business models, share effective practices, and combine resources to solve technical and other market problems.

Governments collaborate with industry associations to create voluntary standards, certification programs, and awareness campaigns that encourage circularity. Public agencies are collaborating with developers and material suppliers in the construction industry to encourage the adoption of circular building practices, including modular construction, material reuse, and digital building passports. In agriculture, PPPs support the growth of regenerative farming methods and bio-based value chains that aim to close nutrient cycles and decrease the reliance on synthetic inputs. These initiatives frequently involve financial tools, such as innovation grants, tax incentives, and co-investment funds, to reduce the risk associated with circular experimentation. Crucially, PPPs also provide a platform for addressing systemic issues, such as fragmented regulations, capacity gaps, and data-sharing challenges, thus enhancing the overall resilience and scalability of circular economy initiatives (Tariq, 2025). One of the growing areas of focus in circular economy governance is the creation of reliable metrics and reporting frameworks to assess circular performance. Precise, comparable, and valuable data are crucial for monitoring progress, pinpointing areas that need improvement, and guiding policy and investment choices. Contrary to traditional financial measures, the indicators within the circular economy are both multidimensional and multifaceted, including material flow analysis, product lifespan, social impact, and resource productivity being some of them. To address this, global organizations and standards bodies are pooling efforts to develop common approaches to circular accounting. The circularity gap report, as an example, provides a global overview of circular material use. However, tools such as the material circularity indicator allow companies to quantify the extent to which their products and processes are restorative.

Circular economy metrics will be increasingly factored into sustainability reporting in the form of broader environmental, social, and governance (ESG) disclosures. Firms must now demonstrate their plans to manage material risks and opportunities related to resource use, waste generation, and climate impact. Important reporting frameworks, such as the Global Reporting Initiative, climate-related financial disclosures (CFD) task force, and EU corporate sustainability reporting directive (CSRD), integrate circularity indicators to improve corporate transparency and accountability. Consequently, companies will include lifecycle assessments, product stewardship metrics, and closed-loop performance indicators in their strategic planning and communication to their stakeholders. Nevertheless, measuring circular performance is challenging. Data availability and quality remain major challenges, especially within sectors with intricate or complex supply chains. Third-party verification systems are still in their infancy; therefore, businesses may lack the resources, knowledge, or incentive to collect and share details on their circular practices. In addition, the broad scope of circular strategies, which include design, materials, logistics, and customer engagement, complicates the opportunity to develop a universal measurement framework. Consequently, there has been a clamor to have indicators that are specific to a particular setting, taking into account the nature of a particular sector, geography, and organizational capacities. Policymakers and standard-setters should be both comprehensive and flexible, as reporting requirements should result in meaningful improvements without placing excessive burdens on businesses. Circular governance should be inclusive and encourage active participation from all stakeholders. This is especially crucial in preventing circular transitions from worsening existing inequalities or creating new environmental injustices. For instance, informal waste workers in numerous developing nations are essential for recovering resources. However, they are frequently excluded from formal programs or initiatives (Perotti et al., 2025). We can enhance equity and efficiency by incorporating these actors into formal circular systems through social protection, skills training, and cooperative models. Similarly, engaging consumers in the policy-making process by conducting public consultations, behavioral studies, and citizen assemblies can cultivate a sense of shared responsibility for achieving circular economy objectives.

## OVERCOMING BARRIERS AND FUTURE TRENDS

The circular economy will result in radical changes in corporate production, consumption and management (Tuboalabo et al., 2024). The potential theoretical advantages of waste reduction, resource efficiency, and environmental regeneration are well known; however, the practical realization of circular principles in all sectors, which encounter numerous challenges, is yet to be achieved. The major issues presented here include financial, technical, and regulatory constraints to widespread adoption. For example, these capital costs appear to be out of control when using short-term measures of profitability, which are typically the focus of traditional business reviews. This has made it difficult for businesses to defend their circular strategies with clear financial incentives, long-term support structures, or quantifiable investments. A large amount of equipment is required to facilitate a circular economy. They are not always accessible, scalable, or interoperable. The time lag and fragmentation of legacy systems and supply chains also complicate the implementation of innovations such as closed-loop production and real-time tracking of material flows.

The technologies needed in some cases in linear systems can be better suited to throughput and efficiency instead of durability and reuse, and there is a central process to circulate the technology. With promising technologies at hand, integration frequently necessitates a dramatic reorganization of the corporate operations model, employee training, and a reconsideration of approaches to customer loyalty. Such a shift to product-as-a-service would entail technological refinements and the need to reengineer valuable promises, support systems, and performance-based revenue models. Current rules and regulations are mainly influenced by the linear nature of production and consumption processes. For example, waste classification schemes may inadvertently penalize the reuse or modification of materials by charging them with waste fees or bureaucracy. Global circulation strategies can be implemented through inconsistent regulations across regions. Despite these challenges, emerging technologies have begun to transform the circular economy into practical and scalable solutions for waste management. One of the most promising areas of innovation is the integration of artificial intelligence (AI) into waste management systems. AI algorithms can optimize the sorting process of recycling systems, distinguish material types with high accuracy, and predict waste generation patterns to improve detection efficiency. AI can also design permanent, repairable, and recyclable products by analyzing lifecycle data and simulating performance results. With these applications, AI promotes higher resource efficiency, reduces operational costs, and accelerates innovation cycles in the circular economy. One of the central obstacles to cyclicity is the difficulty in tracking the origin, composition, and journey of products and materials. Consumer involvement is the missing link in this solution (Al Halbusi et al., 2025). For circular business models to succeed, consumers must accept and actively engage with new consumption methods. This necessitates reliance, ease, and explicit expression of worth. Businesses must create user-friendly take-back programs, repair services, and product customization choices that align with the preferences of today's consumers. When developing branding and marketing strategies, it is essential to emphasize the positive environmental and social impacts of circular offerings while ensuring that they meet or surpass customer expectations regarding quality and performance. Social media, gamification, and community platforms can be used to raise awareness and encourage collective action. Ultimately, empowered consumers can influence the demand for circular products and encourage industries to adopt more sustainable practices.

The journey towards a flourishing circular economy is intricate but attainable. Addressing financial, technological, and regulatory obstacles necessitates collaborative investment, favorable policies, and a shift in traditional business practices (Hojnik et al., 2024). While technology is essential, cultural change, strategic collaboration, and consumer empowerment are equally crucial for success in the field. Companies that actively adopt circular principles and integrate them into their fundamental business strategies will safeguard their future operations and make significant contributions to a sustainable and equitable global economy. Decisive action, innovative thinking, and collaboration are urgently needed to create a sustainable future.

## CONCLUSION

Strategic innovation is significant for transitioning to an environmental economy, reducing environmental degradation, and promoting economic resilience and competitive advantages. This study highlights the need for companies to pursue a systems memorial approach and reinterpret product design, value chains, and customer loyalty. By fostering the integration of digital tools and sector partnerships and prioritizing regeneration practices, companies can transform from the linear "take make despond" paradigm into a circular model focused on reuse, restoration, and recycling. However, achieving large-scale effectiveness requires overcoming political, cultural and infrastructure barriers. Governments, industries, and civil societies must jointly organize incentives, invest in education, and redesign regulatory framework conditions to support circular initiatives. The success of a circular strategy depends on long-term commitment, an innovation ecosystem, and mobility to adapt to changing contexts. Ultimately, embedding circularity at the strategic level provides a way for sustainable development that harmonizes economic prosperity and ecological management.

## REFERENCES


Abu-Bakar, H., Charnley, F., Hopkinson, P., & Morasae, E. K. (2024). Towards a typological framework for circular economy roadmaps: A comprehensive analysis of global adoption strategies. *Journal of Cleaner Production*, *434*, 140066.

Al Halbusi, H., Popa, S., Alshibani, S. M., & Soto-Acosta, P. (2025). Greening the future: Analyzing green entrepreneurial orientation, green knowledge management, and digital transformation for sustainable innovation and circular economy. *European Journal of Innovation Management*, *28*(5), 1916–1942.

Alka, T. A., Raman, R., & Suresh, M. (2024). Research trends in innovation ecosystem and circular economy. *Discover Sustainability*, *5*(1), 323.

Arranz, C. F., Arroyabe, M. F., & de Arroyabe, J. C. F. (2024). Organisational transformation toward circular economy in SMEs. The effect of internal barriers. *Journal of Cleaner Production*, *456*, 142307.

Brogi, S., & Menichini, T. (2024). The pathway towards circular economy: Measuring circular advantage of eco-innovations. *Business Strategy and the Environment*, *33*(4), 3005–3038.

Chabowski, B. R., Gabrielsson, P., Hult, G. T. M., & Morgeson, F. V.III. (2025). Sustainable international business model innovations for a globalizing circular economy: A review and synthesis, integrative framework, and opportunities for future research. *Journal of International Business Studies*, *56*(3), 383–402.

Chau, K. Y., Huang, T., Moslehpour, M., Khan, W., Nisar, Q. A., & Haris, M. (2024). Opening a new horizon in green HRM practices with big data analytics and its analogy to circular economy performance: An empirical evidence. *Environment, Development and Sustainability*, *26*(5), 12133–12162.

Ding, S., Tukker, A., & Ward, H. (2023). Opportunities and risks of internet of things (IoT) technologies for circular business models: A literature review. *Journal of Environmental Management*, *336*, 117662. PMID: 36913854

Fischer, D., Reinermann, J. L., Mandujano, G. G., DesRoches, C. T., Diddi, S., & Vergragt, P. J. (2021). Sustainable consumption communication: A review of an emerging field of research. *Journal of Cleaner Production*, *300*, 126880.

Gu, Y., Wu, P., & Du, R. (2024). Corporate strategic positioning and environmental information disclosure under circular economy: Evidence from China. *Management Decision*, *62*(9), 2791–2813.

Herrador, M., & Van, M. L. (2024). Circular economy strategies in the ASEAN region: A comparative study. *The Science of the Total Environment, 908*, 168280. PMID: 37931812

Herrero, M., Hugas, M., Lele, U., Wirakartakusumah, A., & Torero, M. (2023). A shift to healthy and sustainable consumption patterns. *Science and innovations for food systems transformation, 59.*

Hojnik, J., Ruzzier, M., Ruzzier, M. K., Sučić, B., Soltwisch, B., & Rus, M. (2024). Review of EU projects with a focus on environmental quality: Innovation, eco-innovation, and circular-economy elements. *International Journal of Innovation Studies*, *8*(1), 1–12.

Karaca, F., Tleuken, A., Pineda-Martos, R., Ros Cardoso, S., Orel, D., Askar, R., & Braganca, L. (2024). Cultivating sustainable construction: Stakeholder insights driving circular economy innovation for inclusive resource equity. *Buildings (Basel, Switzerland)*, *14*(4), 935.

Karim, R., Roshid, M., & Waaje, A. (2024). Circular economy and supply chain sustainability. In *Strategic Innovations for Dynamic Supply Chains* (pp. 1–30). IGI Global.

Karim, R., Waaje, A., Roshid, M. M., & Yeamin, M. B. (2024). Turning the waste into wealth: Progressing toward global sustainability through the circular economy in waste management. In Sucheran, R. (Ed.), *Sustainable waste management in the tourism and hospitality sectors* (pp. 507–552). IGI Global Scientific Publishing.

Khan, K., & Khurshid, A. (2024). Are technology innovation and circular economy remedy for emissions? Evidence from the Netherlands. *Environment, Development and Sustainability*, *26*(1), 1435–1449.

Khan, W., Nisar, Q. A., Roomi, M. A., Nasir, S., Awan, U., & Rafiq, M. (2024). Green human resources management, green innovation and circular economy performance: The role of big data analytics and data-driven culture. *Journal of Environmental Planning and Management*, *67*(10), 2356–2381.

Le, T. T. (2023). The association of corporate social responsibility and sustainable consumption and production patterns: The mediating role of green supply chain management. *Journal of Cleaner Production*, *414*, 137435.

Mishra, R., Singh, R. K., & Rana, N. P. (2024). Digital orientation, digital eco-innovation and circular economy in the context of sustainable development goals. *Business Strategy and the Environment*, *33*(4), 2752–2770.

Monciardini, D., Rocca, L., & Veneziani, M. (2024). Virtuous circles: Transformative impact and challenges of the social and solidarity circular economy. *Business Strategy and the Environment*, *33*(2), 642–660.

Nie, P., Dahanayake, K. C., & Sumanarathna, N. (2024). Exploring UAE's transition towards circular economy through construction and demolition waste management in the pre-construction stage–A case study approach. *Smart and Sustainable Built Environment*, *13*(2), 246–266.

Oghazi, P., Mostaghel, R., & Hultman, M. (2024). International industrial manufacturers: Mastering the era of digital innovation and circular economy. *Technological Forecasting and Social Change*, *201*, 123160.

Opoku, A., Deng, J., Elmualim, A., Ekung, S., Hussien, A. A., & Abdalla, S. B. (2022). Sustainable procurement in construction and the realisation of the sustainable development goal (SDG) 12. *Journal of Cleaner Production*, *376*, 134294.

Perotti, F. A., Bargoni, A., De Bernardi, P., & Rozsa, Z. (2025). Fostering circular economy through open innovation: Insights from multiple case study. *Business Ethics, the Environment & Responsibility*, *34*(2), 390–408.

Phonthanukitithaworn, C., Srisathan, W. A., & Naruetharadhol, P. (2024). Revolutionizing waste management: Harnessing citizen-driven innovators through open innovation to enhance the 5Rs of circular economy. *Journal of Open Innovation*, *10*(3), 100342.

Rabby, F., Chimhundu, R., & Hassan, R. (2021). Artificial intelligence in digital marketing influences consumer behaviour: A review and theoretical foundation for future research. *Academy of Marketing Studies Journal*, *25*(5), 1–7.

Sehnem, S., Lara, A. C., Benetti, K., Schneider, K., Marcon, M. L., & da Silva, T. H. H. (2024). Improving startups through excellence initiatives: Addressing circular economy and innovation. *Environment, Development and Sustainability*, *26*(6), 15237–15283. PMID: 37362985

Seyi-Lande, O. B., Layode, O., Naiho, H. N. N., Adeleke, G. S., Udeh, E. O., Labake, T. T., & Johnson, E. (2024). Circular economy and cybersecurity: Safeguarding information and resources in sustainable business models. *Finance & Accounting Research Journal*, *6*(6), 953–977.

Singh, R. K., Mathiyazhagan, K., Scuotto, V., & Pironti, M. (2024). Green open innovation and circular economy: Investigating the role of big data management and sustainable supply chain. *IEEE Transactions on Engineering Management*.

Speranza, M. G. (2018). Trends in transportation and logistics. *European Journal of Operational Research, 264*(3), 830–836.

Straub, L., Hartley, K., Dyakonov, I., Gupta, H., van Vuuren, D., & Kirchherr, J. (2023). Employee skills for circular business model implementation: A taxonomy. *Journal of Cleaner Production, 410*, 137027.

Strippoli, R., Gallucci, T., & Ingrao, C. (2024). Circular economy and sustainable development in the tourism sector-An overview of the truly-effective strategies and related benefits. *Heliyon*. PMID: 39281586

Suphasomboon, T., & Vassanadumrongdee, S. (2022). Toward sustainable consumption of green cosmetics and personal care products: The role of perceived value and ethical concern. *Sustainable Production and Consumption, 33*, 230–243.

Tariq, M. U. (2025). Revolutionizing Urban Congestion: Smart Traffic Management Systems for a Seamless Commute. In Bellam, K., Nagappan, K., Sankaran, N., Chowdhury, S., & Perakovic, D. (Eds.), *Urban Mobility and Challenges of Intelligent Transportation Systems* (pp. 1–16). IGI Global Scientific Publishing., DOI: 10.4018/979-8-3693-7984-4.ch001

Tariq, M. U. (2025). Harnessing Technology to Combat Bias: Innovative Tools and Techniques in HR. In Kanwal, S. (Ed.), *Navigating Bias in Human Resource Management* (pp. 155–184). IGI Global Scientific Publishing., DOI: 10.4018/979-8-3693-7132-9.ch006

Tariq, M. U. (2025). Smart Forecasting With AI: Revolutionizing Energy Consumption and Storage. In Derbali, A. (Ed.), *Integrating Artificial Intelligence Into the Energy Sector* (pp. 165–184). IGI Global Scientific Publishing., DOI: 10.4018/979-8-3693-7112-1.ch009

Tariq, M. U. (2025). Co-Living With Augmented Digital Beings-Navigating the Intersection of AI and Everyday Life. In Moutinho, L., & Martins, L. (Eds.), *Impacts of Sensetech on Society* (pp. 187–206). IGI Global Scientific Publishing., DOI: 10.4018/979-8-3693-7147-3.ch009

Tariq, M. U. (2025). AI-Driven Personalized Learning-Enhancing Student Engagement and Outcomes. In H. Mamede & A. Santos (Eds.), *AI and Learning Analytics in Distance Learning* (pp. 151-178). IGI Global Scientific Publishing. https://doi.org/DOI: 10.4018/979-8-3693-7195-4.ch007

Trivedi, R., Upadhyay, T. K., Khan, F., Pandey, P., Kaushal, R. S., Sonkar, M., & Siddique, M. A. B. (2024). Innovative strategies to manage polluted aquatic ecosystem and agri-food waste for circular economy. *Environmental Nanotechnology, Monitoring & Management*, *21*, 100928.

Tuboalabo, A., Buinwi, U., Okatta, C. G., Johnson, E., & Buinwi, J. A. (2024). Circular economy integration in traditional business models: Strategies and outcomes. *Finance & Accounting Research Journal*, *6*(6), 1105–1123.

Tutore, I., Parmentola, A., di Fiore, M. C., & Calza, F. (2024). A conceptual model of artificial intelligence effects on circular economy actions. *Corporate Social Responsibility and Environmental Management*, *31*(5), 4772–4782.

Williams, J., Prawiyogi, A. G., Rodriguez, M., & Kovac, I. (2024). Enhancing circular economy with digital technologies: A pls-sem approach. [ITEE]. *International Transactions on Education Technology*, *2*(2), 140–151.

Wilson, A., Kask, R., & Ming, L. W. (2024). Exploring circular digital economy strategies for sustainable environmental, economic, and educational technology. [ITEE]. *International Transactions on Education Technology*, *2*(2), 129–139.

Yosep, M. A., Mohamed, M., Yusliza, M. Y., Saputra, J., Muhammad, Z., & Bon, A. T. (2021). Does digital marketing platforms affect business performance? A mini-review approach. *Proceedings of the International Conference on Industrial Engineering and Operations Management*, *12*(3), 4372–4386.

Zorpas, A. A. (2024). The hidden concept and the beauty of multiple "R" in the framework of waste strategies development reflecting to circular economy principles. *The Science of the Total Environment*, *952*, 175508. PMID: 39153637

## KEY TERMS AND DEFINITIONS

**Circular Economy:** Economic model reducing waste, boosting resource efficiency.

**Product-as-a-Service (PaaS):** A business model that provides products via service contracts.

**Reverse Logistics:** Procedure for moving goods backward for reuse.

# About the Contributors

**Rejaul Karim** is an Associate Professor of Finance and Banking in the Department of Business Administration at Varendra University, Rajshahi, Bangladesh. He received his PhD in finance from the Institute of Bangladesh Studies, University of Rajshahi, Bangladesh. He completed the Bachelor of Business Administration and Master of Business Administration in Finance and Banking from the Pabna University of Science and Technology, Pabna, Bangladesh, with outstanding academic performance. Dr. Karim has teaching and research experience for over 10 years in undergraduate and master-level tertiary education at two different universities. His research focused on financial performance, sustainable development, SDGs, financial structure, sustainable finance, the circular economy, and other contemporary issues in finance and economics. Dr. Karim continuously contributes through his field research and publishing many papers in reputed national and international peer-reviewed journals, including book chapters and conference proceedings. Dr. Karim founded and headed BusinessXcel Lab, a center for business research— a voluntary organization that aims to facilitate collaboration among domestic and international undergraduate, graduate, and PhD students and professional researchers. Dr. Karim can be contacted at rkarimreja@gmail.com

**Bablu Kumar Dhar** is working as a faculty member in the Department of Business Administration at Mahidol University International College (ASCSB accredited), Mahidol University (QS World Ranking #255), Thailand. Before joining here, he was a faculty member at the Daffodil International University (Bangladesh), Mediterranea University of Reggio Calabria (Italy), Universitas Airlangga (Indonesia), Universiti Sains Islam Malaysia (Malaysia), Visvesvaraya Technological University (India), and Yantai University (China). Currently, he is pursuing a Post-Doctoral at Mediterranea University of Reggio Calabria (Italy) and accomplished a Post-Doctoral from Universitas Airlangga (Indonesia). Before that, he earned a Ph.D. in Human Resource Management from Universiti Sains Islam

Malaysia (USIM). Before enrolling in Ph.D., he obtained a Master of Philosophy (M.Phil.) in Marketing Audit from the University of Chittagong in Bangladesh, a Bachelor of Commerce (Honors), and a Master of Commerce in Accounting from the National University of Bangladesh. He served as a full-time and part-time faculty member for several universities in Bangladesh, Malaysia, India, Indonesia, Italy, China, and Thailand for over 15 years, He has published several articles in SCIE, SSCI, ESCI, ABS, Scopus, ERA, ABDC, University indexed journals, international daily newspapers, and conferences in Bangladesh, India, Indonesia, Malaysia, Maldives, and Oman.

**Anna A.Kredina** PHD Candidate, Senior Lecturer in UIB. For the last 5 years she has published more than 20 works and have become a participant in several large scientific projects funded by the Committee Science of the Ministry of Science and Higher Education of the Republic of Kazakhstan, recipient of the Soros Foundation Gender Open Economy Fellowship (2022).

**Zerihun Alemu** is a Doctor of Philosophy, Professor (Associate) at Ethiopian Civil Service University Addis Ababa, Ethiopia

**Sofi Badi** is a Lecturer, College of Business and Economics, Department of Management, Bule Hora University, Ethiopia.

**Patcha Bhujanga Rao** is a distinguished professional with a stellar reputation in the realms of human resources (HR) and soft skills development. With over two decades of experience and an extensive academic background encompassing degrees such as M.Com., DCFA., M.Phil., Ph.D., MBA (HR), M.Sc (Psychology), and LL.B, he possesses a comprehensive understanding of multiple disciplines. Currently serving as a professor at Jain Deemed-to-Be University in Bengaluru, Dr. Rao is highly regarded for his expertise in HR management, which has been instrumental in shaping the careers of numerous professionals. His commitment to nurturing essential skills such as effective communication, leadership, and interpersonal abilities reflects his dedication to facilitating personal and professional growth. Dr. Rao's contributions have earned him widespread respect and admiration within the HR and soft skills domains, positioning him as a respected authority and mentor for aspiring professionals seeking to excel in their respective fields.

**Theresia Busagara** is a lecturer in the Department of Finance of the University of Dar Es Salaam Business School. She possesses a Ph.D. in Business Administration with a specialisation in Innovation and Sustainability. She has over Fifteen years of experience teaching, researching, and consulting in Entrepreneurship, leadership, governance, and inclusive finance for women's and youth's growth, she has attended several competency-based courses and international conferences on leadership transformation, communication, entrepreneurship, consultancy skills, adult training, and Public and advocacy policy. Her most extraordinary competence lies in capacity building through training in Leadership, Governance, Communication skills at the workplace, Public Speaking, and Team Building. She attended one of the most prestigious transformational leadership programs for high-potential, mid-career women leaders in East Africa, led by the Sasamani Foundation in the Women Leading Change program. The program unlocked her leadership skills and increased her confidence to lead positive change within family structures, organisations, and communities while increasing her communication skills. Dr Busagara has contributed enormously to women's and youth's participation in decision-making through these programs by improving governance and political processes.

**Md. Faisal-E-Alam** holds two roles as Assistant Professor at the Department of Management Studies at Begum Rokeya University, Rangpur and as Ph.D. Fellow at the Institute of Bangladesh Studies at University of Rajshahi, Bangladesh. His academic career includes publishing 34 journal articles and two book chapters and presenting at four research conferences. His research works find publication in prestigious indexing platforms which include Web of Science (WoS), Scopus, ABDC, and ABS. His research encompasses four areas including human resource management, marketing, tourism and entrepreneurship.

**Meenakshi Gupta** is working as an AssiociateProfessor in the School of Economics, at Shri Mata Vaishno Devi University, Katra. She has teaching and research experience of almost 16 years. She has been a part of many academic and non- academic committees of the university since 2007. Her main areas of interest are Agricultural Economics, Environmental Economics and Development Economics. Her research work has been published in various national and international journals. She has contributed to various books also. She has also presented various articles at National and International Conferences. She is also awarded with many prestigious national and international awards. She is a life Member of Indian Economic Association, Indian Society of Agri- business and Management, Indian Society of Agricultural Economics, TIES, AERA, etc.

**MdAnamul Haque** is a PhD candidate in the Department of Business Administration at Wonkwang University, Republic of Korea. He has obtained a Masters in Marketing and Innovation at Anglia Ruskin University (ARU), Cambridge, United Kingdom (UK). He has extensive work experience in multinational companies including Citi Group and Bloomberg in London, UK. He has also experience of a research associate supporting senior management in people management and customer attraction in the multinational companies operating in London. He was also active in social work during his bachelors majoring in Business and Management. He has also a number of publications in international refereed journals. He also attended in some seminars and conferences in London, Malaysia, Japan, South Korea, and Bangladesh.

**Mehedi Hasan** is an Associate Professor in the Human Resource Management Discipline at Khulna University, Bangladesh. He holds an MBA and an MSc in Management and Leadership, with a strong academic and research background in management. His primary research interests revolve around in the domains of circular economy and sustainability. Mr. Hasan has made substantial contributions to academia through his numerous research publications in prestigious national and international journals. His scholarly work reflects his commitment to advancing the field and addressing critical issues in sustainability and the circular economy.

**Forhad Hossain** has over 10 years of experience in both industry and academia. Their background spans Management and Information Science, with a strong focus on Social Entrepreneurship, SDGs, Digital Health, and Health Innovation. They have worked across sectors to design impactful programs, support startups, and promote sustainable solutions that create real value for communities. Whether through teaching, research, or hands-on projects, they strive to bridge ideas with action to build a more equitable and resilient future.

**Preethi Inampudi** is an Academician with 15years of experience

**S. M. Shafeeul Islam** is affiliated with the Discipline of Human Resource Management at Khulna University, Bangladesh. They hold a Bachelor's degree in Business Administration specialising in Human Resource Management. Their research interests include entrepreneurial ambidexterity, organizational behaviour, research competence, workplace diversity, human cognition, and sustainable business practices. They have conducted field research in light engineering SMEs and among university faculties, experiences that have enriched my understanding of the real-world challenges organizations face. In 2025, They were honoured to receive the Best Paper award at the 6th Nordic Conference on International Business, Export Marketing, International Entrepreneurship, Brand Management, Consumer Behavior, and Tourism—an achievement that reaffirmed their commitment to rigorous, impactful scholarship.

**Ivana Janjić** works as Research Assistant at Innovation Center, University of Niš. She was a scholarship holder at basic academic studies, master academic studies and doctoral academic studies, supported by Ministry Education, Science and Technological Development of the Republic of Serbia. On June 1, 2024, she successfully defended her doctoral dissertation entitled "The impact of R&D investment on enterprise profitability and efficiency of intellectual capital". Her fields of interest are based on the following: R&D management, R&D investment, intellectual capital management, intangible asset management, business performance management, sustainable development, eco-innovation. She has authored and published numerous scientific and research articles in various kinds of publication and she has participated at conferences in the country and abroad. She is a member of Society of Economists Ekonomika from Niš, Serbia and a member of the Publishing council of the journal "Economics of Sustainable Development".

**Shashi Kant** is an Assistant Professor, College of Business and Economics, Department of Management, Bule Hora University, Ethiopia.

**Shivangi Kashyap**, a skillful academician, holds a Bachelor's degree in Commerce with Honours in Accounts and an MBA in Financial Management. She is currently pursuing a PhD at the University of Hyderabad. Earlier, she worked as an Assistant Professor at Brainware University, Barasat, Kolkata, where she passionately imparted her expertise in finance and management to aspiring minds. With a commitment to academic excellence, she also embraces online education, leveraging technology to ensure effective and engaging learning experiences.

**Kenesa Kebede** is a Lecturer, Department of Marketing Management, College of Business and Economics, Bule Hora University Ethiopia

**Pravin Kumar** is an Assistant Professor at Brainware University in Kolkata. With a robust academic background and a dedication to teaching, Mr. Kumar specializes in Marketing Management. He is known for his engaging teaching style and commitment to student success. Mr. Kumar's research interests include marketing, Visual Merchandising, and Consumer Behaviour, and he has contributed to several notable publications in his field. At Brainware University, he actively participates in curriculum development and academic advising, striving to foster a dynamic and inclusive learning environment for all students.

**Md. Tota Miah** is an academic and doctoral researcher whose work focuses on corporate social responsibility (CSR), sustainable management, digital entrepreneurship, and human resource management. He is currently serving as an Assistant Professor in the Department of Business Administration at Varendra University, Rajshahi, Bangladesh. His teaching interests include CSR and sustainability, organizational behavior, business ethics, and management practices. He is presently pursuing Doctor of Philosophy (PhD) degree at the Doctoral

School of Economics and Regional Sciences, Hungarian University of Agriculture and Life Sciences (MATE), Gödöllő, Hungary. Md. Tota Miah has authored and co-authored several peer-reviewed journal articles indexed in Scopus and Web of Science.

**Md.Soleman Mollik** is a lecturer in the Department of Business Administration, Marketing, Faculty of Business Studies, Bangladesh University of Professionals, Mirpur Cantonment, Dhaka-1216, Bangladesh. He completed his MBA and BBA in Marketing from the Department of Marketing, University of Rajshahi, Bangladesh. His research interests include contemporary issues in marketing, service marketing, digital tourism and hospitality marketing, Islamic marketing, circular economy, and sustainable marketing. He has published his research works in several journals, including International Journal of Religious Tourism and Pilgrimage, Technoloy University, Dublin (Scopus indexed, Q1), International Journal of Services and Operations Management, Inderscience (Scopus indexed, Q3) Geojournal of Tourism and Geosites (Scopus indexed, Q3), Global Journal of Management and Business Research, Journal of Business Studies (Pabna University of Science and Technology), and the Journal of integrated marketing communications and digital marketing. He received the "Dean's Award" from the University of Rajshahi in the year of 2018.

**Amir Mohammad Nasrullah** is a Professor of Public Administration at the University of Chittagong, Bangladesh. His research areas include the broader field of governance and administration, specifically Public Policy, Environmental Management, Local Government, and Rural Development. Professor Nasrullah has contributed a number of research articles, and published 6 books, including Mapping Governance Innovations: Perspectives from South Asia, (Co-edited), Routledge Lodnon, and Rethinking Development in South Asia: Issues, Perspective and Practices, (Co-edited), Cambridge Scholars Publishing, London. His recently finished mentionable research projects are: "Fighting with the New Normal Post COVID 19 World: A Study on the Governance Challenges in Bangladesh"; "Assessing Risks in the City of Chittagong, Bangladesh using the Climate and Ocean Risk Vulnerability Index (CORVI)" in collaboration with the OPRI, Japan, and the Stimson Center, USA; and "Evaluating Awareness and Impacts of Heavy Metal Contamination in the Coastal City of Chattogram, Bangladesh: A Cross-Sectional Study" in collaboration with the OPRI, and The University of Tokyo. Dr. Amir is associated with many international organizations such as, Centre Head, GFSRD-Global Centre for Local Government; Research Fellow, Centre for Environment and Sustainability (CES); and Member, South Asian Network of Public Administration (SANPA) etc. He is also a reputed trainer at the Centre on Integrated

Rural Development for Asia and the Pacific (CIRDAP) and African Asian Rural Development Organizations (AARDO).

**Tafese Niguse** is a PhD scholar, Department of Marketing Management, Bule Hora University, Bule Hora, Ethiopia

**Johanna Renny Octavia** is an Associate Professor at the Department of Industrial Engineering, Parahyangan Catholic University, Indonesia. She holds a doctoral degree in Human-Computer Interaction from Hasselt University, Belgium. Her research interests and expertise are ergonomics, human-computer interaction, smart product design, design thinking, social innovation, and circular economy.

**Paulina Permatasari** is a lecturer at Parahyangan Catholic University with qualifications in Sustainability Accounting and Management Accounting.

**Md.Mustaqim Roshid** is a Bangladeshi citizen who has worked as a research associate at BusinessXcel Lab. Mustaqim has a Bachelor of Business Administration in Marketing from Varendra University and is currently pursuing an MBA in Management Studies at Rajshahi University. His research interests include various aspects of business operations, supply chain management, financial performance, agribusiness, neuromarketing, industry revolution, environmental economics, circular economy, and sustainable development goals.

**Alessandro Ruggieri** is Full Professor of Commodity Science at the Department of Economics and Management (DEIM) of University 'Tuscia' in Viterbo (Italy) where he holds the courses of Technology, Innovation and Quality, Quality Management and Certification and Quality Management Techniques. From 21st July 2011 to date is the Director of the Department of Economics and Management (DEIM) at the University of Tuscia. His research has developed mainly under the following guidelines: (i) Quality of goods, products and services. Studies and research on the quality according to the approaches of "Total Quality Management" and "Quality Certification" by "ISO 9000" standards; (i) The relationship between quality and consumer, with particular reference to customer satisfaction; (iii) Quality and innovation in the agro-food industry; (iv) Environmental issues, with particular reference to management and environmental certification; (v) Innovation and technology transfer, with particular attention to the role of innovation for regional development.

**Timcy Sachdeva** holds a doctorate degree in management, specializing in finance from Guru Gobind Singh Indraprastha University. With a teaching and research experience of 14 years, she has various publications in international journals viz., listed in Cabell's Directory, and Ulrich’s Periodical, USA; Scopus indexed and ABDC listed journals. Dr. Sachdeva has participated in many national and international conferences including IIMs, IIT and MDI and has chaired sessions in them as well. She is a sought-after trainer for financial econometrics and modeling and has conducted many FDPs for faculty and research scholars across India. She has also authored book on Fundamentals of Financial Modeling using Excel. Dr. Timcy Sachdeva is currently working as an Assistant Professor in Vivekananda Institute of Professional Studies, affiliated to Guru Gobind Singh Indraprastha University.

**Tanvi Sanotra** is a PG scholar in Economics at school the Shri Mata Vaishno Devi University with a keen interest in exploring contemporary and impactful economic issues. Their academic focus lies in the intersection of digitalization, circular economy and environmental sustainability. They are particularly driven by the potential of innovative, sustainable solutions to address modern economic and environmental challenges through technology and policy integration.

**Ridhima Sharma** is an ex M phil student, thapar University, Patiala

**Cecilia Silvestri** is Researcher of Commodity Science at the Department of Economics and Management (DEIM) at the University of Tuscia, Viterbo (Italy) where she held the course Quality and Customer Relationships. She had her PhD in 2011 in “Economics and local development” at the University of Tuscia. Consistent with carried out studies, her research has developed along the following guidelines: (i) Quality of goods, products and services. Studies and research on the quality according to the approaches of “Total Quality Management” and standards “ISO 9000”, (ii) The relationship between quality and consumers, with particular reference to satisfaction and loyalty, (iii) Quality and innovation in the agro -food, (iv) Innovation and technology transfer, with particular attention to the role of technological innovation for territorial development.

**Amrik Singh** is working as Professor in the School of Hotel Management and Tourism at Lovely Professional University, Punjab, India. He obtained his Ph.D. degree in Hotel Management from Kurukshetra University, Kurukshetra. He started his academic career at Lovely Professional University, Punjab, India in the year 2007. He has published more than 40 research papers in UGC and peer-reviewed and Scopus/Web of Science) journals. He has published 12 patents and 01 patent has been granted in the inter-disciplinary domain. Dr. Amrik Singh participated and acted as a resource person in various national and international conferences, seminars, research workshops, and industry talks. His area of research interest is accommodation management, ergonomics, green practices, human resource management in hospitality, waste management, AR VR in hospitality, etc. He is currently guiding 8Ph.D. scholars and 2 Ph.D. scholars have been awarded Ph.D.

**Miljana Talić** is a research assistant at the Innovation Centre, University of Niš, and a PhD candidate in Business Management at the Faculty of Economics, University of Niš, Serbia. Areas of her scientific interest include entrepreneurship, SMEs, socially responsible business, business management and sustainability. She was part of numerous forums, conferences, study visits and projects and participated in several international summer schools. She is the author and co-author of numerous scientific papers published in national and international journals and is an active participant in scientific conferences in the country and abroad. She was involved in several research projects as a member of project team.

**Muhammad Usman Tariq** has more than 16+ year's experience in industry and academia. He has authored more than 200+ research articles, 110+ case studies, 150+ book chapters and several books other than 4 patents. He is founder and CEO of The Case HQ, a unique repository for courses, narrative and video case studies. Additionally, he is President of Gulf Higher Education (Gulf HE) focused on advancing academic practice. He has been working as a consultant and trainer for industries representing six sigma, quality, health and safety, environmental systems, project management, and information security standards. His work has encompassed sectors in aviation, manufacturing, food, hospitality, education, finance, research, software and transportation. He has diverse and significant experience working with accreditation agencies of ABET, ACBSP, AACSB, WASC, CAA, EFQM and NCEAC. Additionally, Dr. Tariq has operational experience in incubators, research labs, government research projects, private sector startups, program creation and management at various industrial and academic levels. He is Certified Higher Education Teacher from Harvard University, USA, Certified Online Educator from HMBSU, Certified Six Sigma Master Black Belt and has been awarded PFHEA, SFSEDA, SMIEEE, and CMBE.

**Mohammad Shahab Uddin** has over 20 years of teaching experience and is currently working as a professor, at the Department of Management, University of Chittagong, Bangladesh. Prior this position, they have been working as an Assistant Lecturer at the University of Essex, UK since 2021. They are also a Commonwealth Scholar (2020) and pursuing their PhD at the University of Essex, UK. They taught business, management, and entrepreneurship at Department of Management, University of Chittagong, and Principles of Marketing at the University of Essex, UK. They pursued their 2nd master's in the USA and graduated from Prairie View A&M University, Texas, USA, in the year 2017. They completed their BBA and

MBA from the University of Chittagong, Bangladesh. Besides teaching and research, they have also worked different projects conducted by the University of Essex, UK. Their teaching philosophy centers on creating a supportive and inclusive environment that fosters active participation and critical thinking. They integrate up-to-date knowledge from their research into the classroom, encouraging students to bridge theory and practice. They are also founder of a voluntary organization named, 1/24 Social Movement that works for education, health and environment since 2018.

**Abdul Waaje** recently graduated from Michigan Technological University with a Master's in Engineering Management, earning a Quality Engineering Graduate Certification. He also holds a Master's in International Trade and Business from the University of Dhaka and a Bachelor's in Electronics and Telecommunication Engineering from East West University. His research focuses on IoT-based industrial solutions to optimize supply chain operations, ensuring efficient manufacturing with quality assurance in procurement and production part approval. He gained industry experience as a Business Development Officer, specializing in sales planning, market trend analysis, and supplier quality optimization. Abdul interned at a leading commercial laundry manufacturer, managing nonconforming materials and standardizing supply chain processes. He also served as a Teaching Assistant at Michigan Tech. His academic interests include Sustainability, Supply Chain, Operations Management, Circular Economy, Six Sigma, and Industry 4.0. Abdul is a Lifetime Member of Beta Gamma Sigma Honor Society, a Member of ASEM, and a Member of ASCM.

# Index

## A



## B



## C

## D



## E



## F



## G



## I

# M



# P



# R

## S

## T



## U



## W

www.ingramcontent.com/pod-product-compliance
Lightning Source LLC
LaVergne TN
LVHW061528290926
849162LV00004B/706